Ecology and Conservation of the Maned Wolf

Multidisciplinary Perspectives

Ecology and Conservation of the Maned Wolf

Multidisciplinary Perspectives

Edited by
Adriana G. Consorte-McCrea
Eliana Ferraz Santos

CRC Press
Taylor & Francis Group
Boca Raton London New York

CRC Press is an imprint of the
Taylor & Francis Group, an **informa** business

CRC Press
Taylor & Francis Group
6000 Broken Sound Parkway NW, Suite 300
Boca Raton, FL 33487-2742

Printed on acid-free paper
Version Date: 20130624

International Standard Book Number-13: 978-1-4665-1259-7 (Hardback)

Library of Congress Cataloging-in-Publication Data

Ecology and conservation of the maned wolf : multidisciplinary perspectives / edited by Adriana G. Consorte-McCrea and Eliana Ferraz Santos.
pages cm
Includes bibliographical references and index.
ISBN 978-1-4665-1259-7
1. Maned wolf--Ecology. 2. Maned wolf--Conservation. 3. Maned wolf--Habitat. 4. Wolves--Conservation. I. Consorte-McCrea, Adriana G., editor of compilation. II. Santos, Eliana Ferraz, editor of compilation.

QL737.C22E285 2014
599.77--dc23 2013017525

Visit the Taylor & Francis Web site at
http://www.taylorandfrancis.com

and the CRC Press Web site at
http://www.crcpress.com

Contents

Preface ix
About the Editors xi
About the Contributors xiii
List of Contributors xxi
Acknowledgments xxv
Introduction xxvii

PART I The Species and Ecology

Species Status

Chapter 1 The Canidae Family: Setting the Scene for Maned Wolf Conservation 3
Claudio Sillero-Zubiri

Chapter 2 Maned Wolf Population Viability 15
Rogério Cunha de Paula and Arnaud L.J. Desbiez

Chapter 3 Relationships between the Maned Wolf and People 35
Adriana Consorte-McCrea

Chapter 4 The Maned Wolf *Ex Situ* Worldwide 53
Ruben Holland

Chapter 5 The *Ex Situ* Maned Wolf Population in Brazilian Zoos 63
Cecília Pessutti, Cleyde Angélica Ferreira da Silva Chieregatto, and Ralph Vanstreels

Species Biology

Chapter 6 Evolutionary and Conservation Genetics of the Maned Wolf 77
Manoel L. da Fontoura-Rodrigues and Eduardo Eizirik

Chapter 7 Feeding Ecology: A Review 87
José Carlos Motta-Junior, Diego Queirolo, and Adriana de Arruda Bueno

Chapter 8 Reproduction and Factors Affecting Reproductive Success in the Maned Wolf.......99

Nucharin Songsasen, Amy E.M. Johnson, Lauren E. Reiter, and Melissa Rodden

Chapter 9 Conservation Medicine 109

Joares Adenilson May Jr. and Paulo Anselmo Nunes Felippe

PART II Case Studies and Conservation Programs

Chapter 10 Conservation of a Population of Maned Wolves, *Chrysocyon brachyurus*, on a Small Reserve in the Cerrado 131

Flávio H.G. Rodrigues, Arnaud L.J. Desbiez, Raquel L.S.C. Grando, Ana C.R. Lacerda, Adriani Hass, and Wesley R. Silva

Chapter 11 A Case Study: Diet of Maned Wolf and Its Relationship to Seed Dispersal at a Cattle Ranch in Southeastern Brazil 153

Eliana Ferraz Santos, Nivar Gobbi, and Paulo Anselmo Nunes Felippe

Chapter 12 Agricultural Expansion and the Future of the Maned Wolf 165

Carly Vynne

Chapter 13 The Maned Wolf Conservation Project: Serra da Canastra, Minas Gerais, Brazil....... 177

Rogério Cunha de Paula, Flávio H.G. Rodrigues, Marcelo Ximenes Bizerril, Nucharin Songsasen, Eduardo Eizirik, Ronaldo Gonçalves Morato, Jean Pierre Santos, Fernanda Cavalcanti Azevedo, Joares Adenilson May Jr., Carla Cruz Soares, Fabiana Lopes Rocha, and Ricardo Corassa Arrais

Chapter 14 Maned Wolf Conservation in the South of Brazil and Uruguay 193

Diego Queirolo, Cibele Indrusiak, Lorena Coelho, Mariana Cosse, and Graziela Dotta

Chapter 15 Maned Wolf in Argentina 203

Lucía Soler

Chapter 16 Environmental Influences on Maned Wolf Ecology in Bolivia 221

Louise H. Emmons

Chapter 17 The Maned Wolf (*Chrysocyon brachyurus*) in Paraguay 235

José L. Cartes, Anthony J. Giordano, and M. Nathalia Mujica Cameroni

PART III Perspectives for the Future

Chapter 18 Education and Communication for the Conservation of Wild Canids 251

Denise Taylor

Chapter 19 Introduction to the Human Dimension: A Valuable Research Tool to Achieve Wildlife Conservation Objectives and Maned Wolf Conservation 265

Alistair Bath

Chapter 20 The Role of Ethnoscience in the Growth of Ethnoconservation as a New Approach to Nature Conservation in the Tropics: The Case of Brazil 283

Antonio Carlos Diegues

Chapter 21 Policy Intervention in the Cerrado Savannas of Brazil: Changes in Land Use and Effects on Conservation 293

Carlos A. Klink

Chapter 22 Maned Wolf Ecology and Conservation: The Road Ahead 309

Claudio Sillero-Zubiri, Adriana Consorte-McCrea, and Eliana Ferraz Santos

Index 317

Preface

Our maned wolf, one of the most beautiful examples of our masto-fauna, is well deserved of such remembrance. It takes determination to study and make public the facets of its life history, which are at times obscure and difficult, a common feature among generally nocturnal mammals.

Cory Carvalho

DR. CORY CARVALHO: THE INSPIRING JOURNEY OF A PIONEER IN MANED WOLF RESEARCH

Dr. Cory Carvalho is among the few pioneers who meticulously recorded observations of maned wolf behavior and ecology. The association between Dr. Carvalho and the maned wolf began in the mid-1970s and lasted for over 10 years. It was not a conscious choice, but the result of a journey that combined inspiring settings and fortuitous opportunities with a strong will. His career as a mastozoologist was inspired by visits to the National Museum of Rio de Janeiro and studies in taxonomy ("the basis of everything") and taxidermy. He has honored us with a rare insight into his professional trajectory and into how he eventually came to study the maned wolf.

Guided by a need to satisfy his curiosity, he started work experience at the National Museum and Zoo of Rio de Janeiro, under the guidance of Dr. J. Moojen. Later, he was hired as a researcher by the Emilio Goeldi Museum, Pará State (CNPq/INPA), following in the footsteps of great names in the area. Concurrently, he had the chance to work in the mammal section of Belém Zoo, where he spent his time structuring the collection and observing birds in the park owned by the museum. With support from Dr. Paulo Vanzolini, he became curator of the mammal section of the Museum of Zoology of São Paulo. At São Paulo Zoo (FPZSP), he was responsible for improving the husbandry of large animals and creating a database. He also worked at the Department of Hunting and Fishing (animal production) and finally at the Forestry Institute of the Environment Agency of São Paulo State, where his close encounters with the maned wolf took place.

At the Forestry Institute (IF), there was no animal collection at that time, and as he recalls, "there was a lack of understanding about the objectives of a faunal sector. We had to fight for conservation areas and for the rarest animals, threatened with extinction." Meanwhile, as a lecturer in field ecology at the University of Campinas, he supervised graduate research: "We learned much with the '*mateiros*' and field assistants who accompanied the field trips and I hope to have passed some of this knowledge to my students." This is how he recalls his experience at the IF, which "accidentally and with many adversities" directed him toward studying the maned wolf:

> By coincidence, and rare opportunity, the governor of São Paulo state at the time wanted to leave a relevant and impressive reminder of his administration in relation to the natural environment. The IF agreed to fence a vast area with the aim of keeping large native animals. The Ecological Station was established in Águas de Santa Bárbara, next to the Lençois/Sta Bárbara Highway, in an area of former pasture. A rectangular area of 200 ha was fenced off for the "experiment" in 1973/74 with over 1400×1550 linear meters of fences 1.60 m high and about 10 cm off the ground. This area was for the capture and maintenance of deer and rheas, relatively common in the region, as well as partridges, quails, and armadillos, which are in demand for their meat and hunted illegally. An old road leading to other farms and towns crossed the area, and in certain sandy paths after the rain we could see paw prints, indicating a frequent route for animals. We suspended burning and cattle grazing, and made firebreaks around the enclosure. We then set up a small observation tower in the middle and began to

> capture animals. Wolves and jaguars jumped over the walls of the enclosure with ease, and giant anteaters passed under the screen. We often saw tracks of wolves surrounding the enclosure. Feces were left in defined locations, often on top of termite mounds and at the extremities of the enclosure.
>
> Due to difficulties and problems with poachers, hunting dogs, and cougars and issues with monitoring, we left the deer and rhea project in the background and decided to work with the maned wolves, as they frequently walked outside the fenced area and were easier to spot. In the beginning we used traps at strategic points on the farm, capturing wolves and marking their ears with numbered plastic cattle tags. Then we used radiotelemetry collars acquired through friends (J. Dietz) and with our own resources. After many difficulties the state provided a Jeep, bait, and even a "trailer." We were helped by Dr. Faiçal Simon (a veterinarian at São Paulo Zoo) with necropsies and by the director of the Dortmund Zoo in Germany (W. Bartman), who provided us with a blowpipe and tranquilizers to facilitate handling of the animals.
>
> By improvising, we kept four to five pairs within fenced areas in the *Chapada* (São Paulo) and in the backyard of my house in the *Horto*, feeding them and observing their behavior, mating, and so forth. At the same time, we followed the movements of some maned wolves in the field. I spent nights at different spots on the farm writing down notes and observations.

Dr. Carvalho talks about the species in an almost poetic and familiar way, which inspired many of us:

> They are shadow wanderers, solitary and opportunistic, always searching for fruits in their walks, including the *lobeira* (wolf's fruit) amongst others, not turning down nests, eggs, young birds, and even small ground-dwelling animals, when caught out. They are not persecutors but profiteers.

He cautions us that it is not enough to study only their life but that we must also consider their natural environment, without which they cannot exist, and concludes with a poignant message. Retiring after 47 years of service to mastozoology, he leaves

> to the ones who replace us the choice to continue or not the work, hoping that they may find more resources and support, before the extermination of some species.
>
> (Cory T. de Carvalho, D.Sc., Mastozoologist, Manduri)

Ecology and Conservation of the Maned Wolf: Multidisciplinary Perspectives celebrates the work of the many researchers, fieldworkers, and students who have dedicated their time and passion to unveiling aspects of the maned wolf's life. Like Dr. Carvalho, we hope many more will choose to contribute to our understanding and to the persistence of this extraordinary species.

Adriana Consorte-McCrea

About the Editors

Adriana Consorte-McCrea graduated in biological sciences from the Federal University of São Carlos, São Paulo, Brazil, in 1987 and went on to research the captive breeding of maned wolves through an apprenticeship in São Paulo Zoo. After another apprenticeship, with the Golden Lion Tamarin Reintroduction Program (Smithsonian's National Zoo), in Poço das Antas Reserve, Rio de Janeiro, she developed research on maintaining natural behavior in captive mammals for reintroduction purposes toward an MSc in conservation biology (Durrell Institute of Conservation and Ecology, University of Kent, United Kingdom). Adriana has worked in environmental education programs for the education departments of Wildwood Trust, Natural History Museum–Tring, and Zoological Society of London Whipsnade Zoo. While engaging with the Zoological Society of London Carnivore and People Group, she began researching the relationships between the maned wolf and people and their implications for the conservation of the species. Happy to once again focus her attention on this exceptional species, she achieved her PhD in ecology (University of Kent/Canterbury Christ Church University) and began planning the present book.

Adriana lectures at Canterbury Christ Church University, Kent, United Kingdom, where she chairs the interdisciplinary Wildlife and People Group (part of Canterbury Christ Church University's Ecology Research Group). Her main interest areas are attitudes toward wild carnivores and maned wolf conservation. She is a member of the Reintroduction Specialist Group of the International Union for Conservation of Nature's Species Survival Commission.

Eliana Ferraz Santos is a native of Campinas (SP, Brazil), where she has worked for many years in biology and management of wild animals in Campinas Zoo. Eliana graduated in biological sciences, with a master's and doctorate in zoology, from the Biosciences Institute of Universidade Estadual Paulista Júlio de Mesquita Filho, Rio Claro (SP, Brazil). She has coordinated the Jequitibás Woods Zoo of Campinas since 2004, where she also serves as a zoologist. She has been the director of the "Associação Mata Ciliar" at the Canine Department of Jundiaí, and the founder and coordinator of the Project Echoes of the Woods (voluntary environmental education project) in Jequitibás Woods since 1997.

Eliana has experience in zoology, working mainly in the areas of animal behavior, ecology of wild animals, captive animal management, and environmental enrichment. She has been an effective partner at the Paulista Society of Zoos since 1998, where she has directed the biology department since 2002.

About the Contributors

Ricardo Corassa Arrais is a wildlife veterinarian with a master's degree in experimental epidemiology applied to zoonosis, from the College of Veterinary Medicine, University of São Paulo. He has been studying maned wolf diseases since 2010. Today he coordinates the veterinary division of the Cerrado Mammals Conservation Program. He is also an associate researcher of the Maned Wolf Conservation Program in the Serra da Canastra region.

Fernanda Cavalcanti Azevedo is a biologist with a master's degree in ecology, working mainly on mammal conservation with emphasis on wild carnivores. She is co-coordinator of the Cerrado Mammals Conservation Program, which develops field projects on systematic surveys in the Brazilian savanna named the cerrado, gathering information about feeding ecology, behavior, reproduction, parental care, dispersion, spatial distribution, and habitat use by carnivores. The main projects in progress are ecology and conservation of the hoary fox and sympatric canids in the Brazilian cerrado and the Triângulo Mineiro Pumas Project.

Alistair Bath is an associate professor at Memorial University with more than 25 years' experience in human wildlife conflicts and human dimensions in natural resource management issues. He has worked on wolf reintroduction in Yellowstone National Park as well as on wolf control issues in Yukon. Alistair is a member of the Large Carnivore Initiative for Europe of the International Union for Conservation of Nature's Species Survival Commission (http://www.lcie.org/) and has led various research projects throughout Europe on large carnivores. He has taught human dimension courses in conservation biology programs in Italy, Germany, Croatia, Slovenia, Austria, Poland, and Portugal and has been an invited keynote speaker to international wolf and bear conferences. Alistair enjoys working with people and finding solutions to the conservation challenges that we face today.

Marcelo Ximenes Bizerril is a biologist with master's and PhD degrees in ecology. He is currently a professor at University of Brasilia, studying relationships between environmental education, conservation, and community communication. His focus is on the application of education and communication tools to the promotion of citizenship and the conservation of nature. His expertise is on the cerrado's ecology, education, and community communication practices.

Adriana de Arruda Bueno is a biologist with a PhD in ecology. She is currently working with the management plan of reserves and protected areas at Fundação Florestal, in the state of São Paulo, Brazil. Between 1998 and 2003, she was a researcher in two thematic projects of the feeding ecology of the maned wolf in southeast Brazil.

José L. Cartes was born in Asunción, Paraguay. He is a postgraduate from the University of Cordoba, Argentina, under an international program of the U.S. Fish and Wildlife Service. He has been working for the last 15 years in mammals and protected area conservation. His main work on mammals has focused on ungulate ecology and inventories and monitoring related to natural areas. He has worked for the last 12 years as head of the Site Conservation Program in the Guyra Paraguay organization and he is a national researcher in the National Program for Incentives to Researchers of the Consejo Nacional de Ciencias y Tecnologías in Paraguay.

Cleyde Angélica Ferreira da Silva Chieregatto graduated in biology. Since 1986 she has worked in wildlife management in captivity. She is a member of the Canine Working Group (since 1995), a member of the Committee for Conservation and Management of Brazilian Canids (since 2003), coordinator of the management plan for and Brazilian studbook keeper of the bush dog (*Speothos venaticus*), and coordinator of the Department of Conservation Biology of Endangered Animals of the Zoological Society of Brazil (2010/2011).

Lorena Coelho is a biologist and master's student at Universidad de la República, Uruguay. The aim of her thesis is to compare the southern potential distributions of the maned wolf in past, present, and future scenarios, identifying geographic, climatic, and anthropic factors that determine its distributions.

Mariana Cosse is a researcher at the Instituto de Investigaciones Biológicas Clemente Estable, Uruguay. Her work is related to the ecological and genetic knowledge on neotropical mammals, with special attention to conservation aspects of wild populations. Her research has been complemented with projects seeking the involvement of local communities.

Arnaud L.J. Desbiez is a conservation biologist who has been conducting research in the Brazilian Pantanal since 2002. He has a PhD in biodiversity management from the Durrell Institute of Conservation and Ecology, University of Kent, United Kingdom, and has worked on topics ranging from sustainable use of resources and species ecological research to community development programs. He is currently the regional conservation and research coordinator for Latin America for the Royal Zoological Society of Scotland, and convener for the Brazilian network of the Conservation Breeding Specialist Group of the International Union for Conservation of Nature's Species Survival Commission. Through his work with the Conservation Breeding Specialist Group in Brazil, he has conducted conservation planning workshops for over 30 endangered species, including the maned wolf.

Antonio Carlos Diegues is an anthropologist with a diploma from the Institute of Social Studies in the Hague and the Ecole Pratique des Hautes Études in Paris and a PhD in social sciences from the University of Sao Paulo. He is the director of the Institute for Conservation of Wetlands and professor in the graduate course on environmental science at the University of São Paulo. He has written several articles and books on nature conservation involving traditional communities, notably "The modern myth of untamed nature" and "Ethnoconservation: new alternatives for nature conservation in the tropics."

Graziela Dotta is a biologist and has a master's degree in the ecology of agroecosystems. Currently, she is a final-year PhD candidate in the Conservation Science Group at the department of zoology, University of Cambridge, United Kingdom. Her main research interests are the trade-offs between agricultural production and the conservation of biodiversity. Since 2008, she has been a researcher in the group Ecology, Management and Conservation of Carnivorous Mammals group at the Chico Mendes Institute (Brazilian Ministry of Environment, Brazil).

Eduardo Eizirik obtained his undergraduate degree in biology in 1994 from the Federal University of Rio Grande do Sul, Brazil. He received an MSc degree in genetics and molecular biology from the Federal University of Rio Grande do Sul in 1997 and a PhD degree in biology from the University of Maryland (U.S.) in 2002. He is currently an associate professor at the Pontifical Catholic University of Rio Grande do Sul, Brazil, where he leads a research group whose main foci are the genetics, evolution, ecology, and conservation of the Carnivora. He is a member of the Pró-Carnívoros Institute, Brazil, and also a member of the Cat Specialist Group and Small Carnivore Specialist Group of the International Union for Conservation of Nature's Species Survival Commission.

Louise H. Emmons is an adjunct scientist in the division of mammals, National Museum of Natural History (Smithsonian Institution). Her research has included ecology and systematics of tropical mammals in Africa, Borneo, and South America. For the past 15 years she has worked on the interface of cerrado and humid forest biomes in Noel Kempff Mercado National Park in Bolivia, where she has studied a small population of maned wolves.

Paulo Anselmo Nunes Felippe is a titular professor at the University Paulista, Campinas, São Paulo, Brazil, veterinarian and researcher at Campinas Zoological Garden, director of the department of epidemiology at Mata Ciliar Association, member of the committee of wild animals at the Regional Council of Veterinary Medicine, member of the department of veterinary medicine of São Paulo Society of Zoos, member of the Brazilian Wild Animal Veterinary Association, and member of the Ethics Committee on Animal Experiments of the State University of Campinas. He graduated in Veterinary Medicine Sciences from the Federal University of Uberlândia, Uberlândia, Minas Gerais, Brazil, and earned a master's degree and doctorate in genetics and molecular biology from the State University of Campinas.

Manoel L. da Fontoura-Rodrigues obtained his undergraduate degree in biology in 2006, from the Federal University of Rio Grande do Sul, Brazil. He received a master's degree in zoology from the Pontifical Catholic University of Rio Grande do Sul, Brazil, with a thesis investigating the population structure and evolutionary history of the maned wolf. He also conducted parallel projects on the molecular ecology and social structure of the same species. He is currently a PhD candidate at the Pontifical Catholic University of Rio Grande do Sul, working on a taxonomic revision of the genus *Conepatus* (Mammalia, Carnivora). His main areas of interest are mammalian evolution and biodiversity.

Anthony J. Giordano is the executive director of SPECIES, an organization dedicated to the conservation of mammalian carnivore communities around the world. His research interests broadly include the ecology, evolution, and biogeography of carnivores, with a particular emphasis on trophic interactions in tropical ecosystems. For the past 6 years, Anthony has conducted research on jaguars and other carnivores in different parts of Paraguay, including efforts to map the distribution of maned wolves in the humid Chaco. Anthony currently leads the regional Chaco Jaguar Conservation Project, a multinational effort integrating elements of research, monitoring, education and outreach, conflict mitigation, and incentive-based conservation.

Nivar Gobbi is a professor of ecology and environmental law (State University Júlio de Mesquita Filho). Graduated in biological sciences (São Paulo University) and law (Ribeirão Preto University). He has Master degree and PhD (São Paulo State University).

Raquel L.S.C. Grando has an MSc in sustainable development (environmental policy and management) from the Centre for Sustainable Development, University of Brasilia (2007). She has a BSc in biological sciences, specializing in ecology and natural resources, from the Federal University of Bahia (2003). Her areas of expertise are ethnobiology, traditional communities, and social network analysis.

Adriani Hass is a biologist with a PhD in ecology from the University of Campinas (2002). He is currently assistant professor at the University of Brasilia. He has experience in the areas of ecology, bird conservation of the cerrado, and fragmentation.

Ruben Holland was born on 11 May 1979 in Frankfurt am Main, Germany. He studied biology at Goethe University Frankfurt am Main, Germany, and wrote his diploma thesis about the maned wolves in Frankfurt Zoo, with the title: Chronoethologie des Mähnenwolfes (*Chrysocyon brachyurus*) (Chronoethology of the maned wolf [*Chrysocyon brachyurus*]). Afterwards he worked in Frankfurt Zoo, Opelzoo Kronberg, Heidelberg Zoo, and London Zoo before he started to work as curator in Leipzig Zoo. Since 2009, he has been the European Endangered Species Program coordinator for the maned wolf and since 2010 he has also been the International Studbook keeper for the species.

Cibele Indrusiak is a biologist with a master's degree in wildlife management (Universidade Nacional de Córdoba, Argentina). She works at Instituto Brasileiro do Meio Ambiente e dos Recursos Naturais Renováveis, Brazil, in public policies for the conservation and sustainable use of the Pampas biome.

Amy E.M. Johnson received her bachelor's degree in animal sciences from the University of Guelph, Canada. In 2012, she earned her master's degree through the Smithsonian Conservation Biology Institute and George Mason University on the "Development of assisted reproductive techniques for managing maned wolves *ex situ*." Under Dr. Nucharin Songsasen, Amy has been optimizing protocols for the cryopreservation of maned wolf spermatozoa and induction of ovarian activity for use in artificial insemination.

Carlos A. Klink is a native Brazilian. He received his BSc in biology and MSc in ecology from University of Campinas, Brazil, and his MA and PhD in ecology from Harvard University. He was a visiting scholar at University of Maryland, College Park. He is currently Brazil's secretary for climate change, Ministry of Environment. He was a senior program officer of the Sustainable Business Advisory at the International Finance Corporation, where he worked as the team leader for the Amazon-Cerrado initiative. He has 18 years of experience working with land-use and climate change issues in the Amazon and cerrado ecosystems of Brazil. Before joining the International Finance Corporation, Carlos was the agriculture team leader for the Nature Conservancy in Brazil, where he coordinated its strategies on agribusiness and conservation across ecosystems. Carlos is also an adjunct professor of ecology at University of Brasilia (on leave). He also consulted with the World Bank and the Global Environment Facility and has served on the board of Brazilian nongovernmental organizations, governmental institutions, and international scientific committees. Carlos has had 2 books, 40+ scientific articles, and 100+ paper presentations published.

Ana C.R. Lacerda is a biologist with a PhD in animal biology from the University of Brasilia (2008). She is currently a consultant to the Forest Service. She has experience in the area of ecology, with an emphasis in applied ecology.

Joares Adenilson May Jr. is a professor at Universidade do Sul de Santa Catarina and a member of the Pró-Carnívoros Institute. He is an associate veterinarian of the National Center for Research of Carnivores and Global Insular Conservation Society and a member of the International Union for Conservation of Nature's Wildlife Health Specialist Group. A wildlife veterinarian, he received a master's degree from the University of São Paulo in wildlife epidemiology. He worked as a field veterinarian for the Maned Wolf Project in the Serra da Canastra National Park for 5 years. Today he works in the conservation of wild carnivores (foxes, wolves, ocelots, jaguars, cougars, etc.).

Ronaldo Gonçalves Morato is a veterinarian with a master's degree and PhD in veterinary medicine. He has been working with carnivores since 1994 in research projects throughout Brazil. Since 2005, he has been the director of the National Research Center for Carnivores Conservation (Cenap) at the Instituto Chico Mendes for Biodiversity Conservation within the environmental ministry (ICMBio/MMA).

José Carlos Motta-Junior has a PhD in ecology and has been a professor in the department of ecology, Universidade de São Paulo, Brazil, since 1996. For the past 25 years, his studies have included the ecology and behavior of birds and mammals. Currently he is studying bird ecology in the cerrado region, Brazil. Between 1997 and 2003, he was involved in two thematic projects about the feeding ecology of the maned wolf in eight areas of southeast Brazil.

M. Nathalia Mujica Cameroni recently graduated from a master's program at Southern Illinois University, Carbondale. She has worked on mammals in Paraguay for the last 5 years. Her master's thesis, *Habitat Factors Affecting Occupancy and Detection of Mammals in the Paraguayan Chaco*, will soon be published. In the past, she has worked as coordinator of a jaguar conservation project in the Chaco and as a biology teacher at different high school levels.

Rogério Cunha de Paula is a Brazilian biologist, working as a government environmental agent for the National Research Center for Carnivores Conservation (CENAP) at the Chico Mendes Institute for Biodiversity Conservation within the environmental ministry. He coordinates the initiative for the Carnivores' National Action Plans, designing and supervising the implementation of activities for the conservation of Brazilian wild carnivores. He supervises the Human–Carnivore Conflict Control National Program and is directly involved in research projects with maned wolves and jaguars. He is also an associate researcher within the nongovernmental Pró-Carnívoros Institute. He is part of both national and international committees for carnivore conservation, such as the Canid and Cat Specialist Groups within the International Union for Conservation of Nature, coordinating the Maned Wolf Working Group in South America and being a world consultant for conflicts between large felids and humans. He has been a maned wolf specialist since 1999.

Cecília Pessutti graduated in biological sciences (Pontifícia Universidade Católica de Campinas) in 1986. In 1993, she specialized in endangered animals management at the Durrell Wildlife Conservation Trust. Since 1994 she has been the Brazilian Studbook keeper and captivity coordinator of the Brazilian Maned Wolf Survival Plan. She completed a MSc in zoology (Universidade Estadual Paulista Júlio de Mesquita Filho–Botucatu, 1998) and since 1989 she has been the biologist in charge of the fauna and feeding sections at Parque Zoológico Municipal Quinzinho de Barros, Sorocaba, São Paulo, Brazil.

Diego Queirolo is a lecturer at the Universidad de la República, Uruguay. He has worked on maned wolf trophic ecology in the southeastern region of Brazil as well as on maned wolf biogeography. His current research includes the ecology and conservation of mammal species in the Campos region of Uruguay and southern Brazil. He also participates as a researcher in the Assessor Group of the National Plan for Maned Wolf Conservation in Brazil, a Brazilian governmental program established in 2005.

Lauren E. Reiter graduated with her bachelor's degree in zoology from Miami University, Oxford, Ohio. Her passion for animals motivated her to continue working as a veterinary assistant, a cheetah-keeper intern in Africa, and then in a research capacity at the National Zoo's Smithsonian Conservation Biology Institute, Front Royal, Virginia, studying reproductive endocrinology in zoo-housed maned wolves. She earned her master's degree at the George Mason University, Fairfax, Virginia, where her thesis aimed to determine the relationship between reproductive success and adrenal activity in female maned wolves and develop a biomarker for assessing nutritional and health status in this species.

Fabiana Lopes Rocha is a wildlife veterinarian with a master's degree in ecology and conservation and a PhD in parasite biology. She is currently a researcher at the Oswaldo Cruz Foundation and a member of the Brazilian Institute of Conservation Medicine (Tríade), a nongovernmental organization that works at the interface of human, wildlife, and ecosystem health. In 2012, she became a staff member of the Conservation Breeding Specialist Group (Brazil) of the International Union for Conservation of Nature. The main projects in progress are ecology and conservation of hoary foxes and pumas within the Cerrado Mammals Conservation Program and disease ecology of mesocarnivores in the Pantanal wetland (Embrapa Pantanal).

Melissa Rodden joined the staff at the Smithsonian Conservation Biology Institute in 1976, shortly after graduating with a BSc from McGill University in Montreal, Quebec, Canada. She began her long career at the Smithsonian Conservation Biology Institute as a biotechnician on the canid project, assisting several graduate students on studies of social behavior in maned wolves, crab-eating foxes, and bush dogs. Melissa became the coordinator of the Association of Zoos and Aquariums' Maned Wolf Species Survival Plan program when it began in 1985 and retained that role even as her primary job duties shifted to providing information technology support to the staff at the Smithsonian Conservation Biology Institute. She is also a member of the Association of Zoos and Aquariums' Canid Taxon Advisory Group Steering Committee and the International Union for Conservation of Nature's Canid Specialist Group.

Flávio H.G. Rodrigues is a biologist and associate professor at Universidade Federal de Minas Gerais, Brazil, and is associated to the Pró-Carnívoros Institute. His research has included the ecology and conservation of tropical mammals in the Brazilian cerrado, and he has studied maned wolf ecology since 1996. He also participates as a researcher in the Assessor Group of the National Plan for Maned Wolf Conservation in Brazil.

Jean Pierre Santos is a biologist with particular interest in conflicts between carnivores and humans. He is currently working at sites in the Amazon basin but dedicates the majority of his time to education and communication strategies to raise public awareness toward wildlife in the Serra da Canastra region.

Claudio Sillero-Zubiri was born in the Argentine Pampas and studied zoology at the Universidad Nacional de La Plata. He obtained a DPhil at the University of Oxford, where he is currently the deputy director of the Wildlife Conservation Research Unit, and he is a senior research fellow at Oxford's Department of Zoology and Lady Margaret Hall. Claudio's main interest is the biology and conservation of carnivores, and he is the chair of the Canid Specialist Group of the International Union of Conservation of Nature. His work includes the conservation of endangered species and protected area management, with research and conservation projects spanning four continents. He is increasingly involved in the relationships between wildlife and rural communities, biodiversity conservation policy and practices, and the mitigation of conflict between wildlife and human interests. Claudio established the Ethiopian Wolf Conservation Programme in the 1980s, which he still directs. In 1998, he received the Whitley Award for Animal Conservation from the Royal Geographical Society for his work in Ethiopia.

Wesley R. Silva is a professor in zoology and ecology at the State University of Campinas (Brazil), where he leads the Laboratory of Vertebrate–Plant Interactions. This has been his main area of interest since the 1990s and currently he is turning to the application of frugivory and seed dispersal processes to the design of restoration projects in southeastern Brazil.

Carla Cruz Soares is a biologist with expertise in environmental education. She is currently working on implementing environmental education in remote areas of the Atlantic Forest, Brazil.

Lucía Soler is a biologist and teaching assistant at the Universidad Nacional del Sur (Bahía Blanca, Buenos Aires, Argentina). Since 1991, she has been involved in the research of wild carnivores of Argentina, being part of research and conservation projects on *Leopardus guigna*, *L. oreailurus*, and *L. geoffroyi*. She obtained a master's degree in biology (Universidad Nacional del Sur) studying ecological aspects and analyzing conservation perspectives of the maned wolf in Argentina. She is developing a PhD at Universidad Nacional del Sur researching the ecology of *Chrysocyon brachyurus* on a regional scale and developing a model that includes socioeconomic variables to ensure long-term conservation.

Nucharin Songsasen received her doctoral veterinary medicine degree from the Kasetsart University, Thailand, and MSc and PhD from the University of Guelph, Canada. Nucharin has broad experience in reproductive biology and assisted reproductive technologies in a wide variety of mammalian species. This has led her to the Smithsonian Conservation Biology Institute, where she has established and led the Conservation of Rare Canids Program since 2002. Her primary responsibility as the program leader is to develop and direct research projects focusing on improving the understanding about biology, ecology, health, and conservation threats in canids. Nucharin is a reproductive advisor of the Association of Zoos and Aquariums' Maned Wolf Species Survival Plan and Canid Taxon Advisory Group, as well as a member of the International Union for Conservation of Nature's Canid Specialist Group. She also holds adjunct positions at the University of Maryland, George Mason University, and Cornell University.

Denise Taylor is an environmental educationalist, entrepreneur, and wildlife conservationist with many years' experience working with nongovernmental organizations and academic institutions in the United Kingdom and Europe on wildlife conservation projects and issues. In 2003, Denise founded E4C (http://www.education4conservation.org), which provides resources for conservation educators throughout the world. She is also a specialist advisor at the United Kingdom Wolf Conservation Trust and completed her PhD on the Efficacy of Wolf Conservation Education Programmes at the University of Nottingham, England. Denise is the author of numerous books, academic papers, and articles on environmental education and education for the conservation of wild canids. Denise is a panel member of the International Union for Conservation of Nature's Canid Specialist Group (http://www.iucn.org).

Ralph Vanstreels is a wildlife veterinarian. He is currently a PhD candidate in comparative pathology at the Laboratory of Wildlife Comparative Pathology of the School of Veterinary Medicine and Animal Science of the University of São Paulo, and has been a collaborator of the Brazilian Regional Maned Wolf Studbook since 2006.

Carly Vynne is the director of wildlife and habitat conservation for the U.S. National Fish and Wildlife Foundation. Her principal research interest is in understanding the functional connectivity of landscapes from the perspective of wide-ranging mammals. Her PhD research combined fieldwork, DNA and hormone analysis, and spatial modeling to understand the influence of a changing agricultural landscape on the plight of maned wolves and other unique and endangered species of the Brazilian cerrado. Previously, Carly worked at Conservation International as the senior manager for biodiversity analysis and planning, as a research scientist on a lion reintroduction project in South Africa, as the staff scientist for a local Sierra Nevada–based nongovernmental organization, and as a ranger for the National Park Service. Carly completed her bachelor's degree in environmental science and public lands policy at Middlebury College and her PhD in biology at the University of Washington.

ARTISTS

Edu Fortes is a journalist and works as a photographic reporter for the Anhanguera Comunicaiton Network, metropolitan region of Campinas. He also carries out environmental work in the Jalapão region, Tocantins state, Brazil. Edu has 13 years of professional experience, having worked for several media companies.

Marcelo Pereira Pinto has been creating cartoons and illustrations for many magazines and papers in São Paulo since he was 16. For many years he was an illustrator for the staff of the newspaper *O Estado de São Paulo*. In 1996 he went to live in Barcelona, where he collaborated with *El Periódico de Catalunya* and with Biopunt, a studio dedicated to the production of scientific images. Now Marcelo lives in Mallorca and is a caricaturist on the newspaper *Última Hora*. He is also an illustrator of the Museu de Ciències Naturals de les Illes Balears i del Jardí Botànic de Sóller. He is a colorist on the comic editorial staff of *Dynamic Forces*.

List of Contributors

Ricardo Corassa Arrais
Programa de Conservação Mamíferos do Cerrado
Araguari, Brazil

Fernanda Cavalcanti Azevedo
Programa de Conservação Mamíferos do Cerrado
Araguari, Brazil

Alistair Bath
Department of Geography
Memorial University of Newfoundland
St. John's, Newfoundland, Canada

Marcelo Ximenes Bizerril
Faculdade UnB-Planaltina
Universidade de Brasília
Brasília, Brasil

Adriana de Arruda Bueno
Fundação Florestal
Planos de Manejo
São Paulo, Brazil

José L. Cartes
Site Conservation Program
Asociacion Guyra Paraguay
Asunción, Paraguay

Cleyde Angélica Ferreira da Silva Chieregatto
Department of Conservation Biology of Endangered Animals
Zoological Society of Brazil
São Bernardo do Campo, Brazil

Lorena Coelho
Laboratorio de Etología
Instituto de Investigaciones Biológicas Clemente Estable

and

Facultad de Ciencias
Montevideo, Uruguay

Adriana Consorte-McCrea
Department of Geographical and Life Sciences
Canterbury Christ Church University
Canterbury, United Kingdom

Mariana Cosse
Departamento de Genética
Instituto de Investigaciones Biológicas Clemente Estable
Montevideo, Uruguay

Arnaud L.J. Desbiez
Royal Zoological Society of Scotland
Campo Grande, Brazil

Antonio Carlos Diegues
Universidade de São Paulo

and

Cidade Universitária
Sao Paulo, Brazil

Graziela Dotta
Department of Zoology
University of Cambridge
Cambridge, United Kingdom

Eduardo Eizirik
Faculdade de Biociências
Pontifical Catholic University of Rio Grande do Sul
Porto Alegre, Brazil

and

Instituto Pró-Carnívoros
Atibaia, Brazil

Louise H. Emmons
Department of Vertebrate Zoology
Smithsonian Institution
Washington, DC

Paulo Anselmo Nunes Felippe
Mata Ciliar Association
University Paulista
São Paulo, Brazil

Manoel L. da Fontoura-Rodrigues
Laboratório de Biologia Genômica e Molecular
Faculdade de Biociências
Porto Alegre, Brazil

Anthony J. Giordano
Department of Natural Resources Management
Texas Tech University
Lubbock, Texas

Nivar Gobbi
Mata Ciliar Association
University Paulista
São Paulo, Brazil

Raquel L.S.C. Grando
Centre for Sustainable Development
Universidade de Brasília
Brasília, Brazil

Adriani Hass
Department of Ecology
Universidade de Brasília
Brasília, Brazil

Ruben Holland
Zoo Leipzig
Leipzig, Germany

Cibele Indrusiak
Instituto Brasileiro do Meio Ambiente e dos Recursos Naturais Renováveis
Porto Alegre, Brazil

Amy E.M. Johnson
Smithsonian Conservation Biology Institute
Front Royal, Virginia

Carlos A. Klink
Departamento de Ecologia
Universidade de Brasília
Brasília, Brazil

Ana C.R. Lacerda
Instituto Chico Mendes de Biodiversidade
Brasília, Brazil

Joares Adenilson May Jr.
Instituto Pró-Carnívoros
Atibaia, Brazil

Ronaldo Gonçalves Morato
Centro Nacional de Pesquisa e Conservação de Mamíferos Carnívoros
Atibaia, Brazil

and

Instituto Pró-Carnívoros
Atibaia, Brazil

José Carlos Motta-Junior
Departamento de Ecologia
Universidade de São Paulo
São Paulo, Brazil

M. Nathalia Mujica Cameroni
Department of Forestry
Southern Illinois University
Carbondale, Illinois

Rogério Cunha de Paula
Centro Nacional de Pesquisa e Conservação de Mamíferos Carnívoros
São Paulo, Brazil

and

Instituto Pró-Carnívoros
Atibaia, Brazil

Cecília Pessutti
Parque Zoologico Municipal Quinzinho de Barros
São Paulo, Brazil

Diego Queirolo
Centro Universitario de Rivera
Universidad de la República
Rivera, Uruguay

Lauren E. Reiter
Smithsonian Conservation Biology Institute
Front Royal, Virginia

Fabiana Lopes Rocha
Laboratório de Biologia de Tripanosomatídeos
Fundação Oswaldo Cruz
Rio de Janeiro, Brazil

Melissa Rodden
Smithsonian Conservation Biology Institute
Front Royal, Virginia

Flávio H.G. Rodrigues
Department of Biology
Universidade Federal de Minas Gerais
Belo Horizonte, Brazil

and

Instituto Pró-Carnívoros
Atibaia, Brazil

Eliana Ferraz Santos
Zoológico Bosque dos Jequitibás de Campinas
São Paulo, Brazil

Jean Pierre Santos
Instituto Pró-Carnívoros
Atibaia, Brazil

Claudio Sillero-Zubiri
Zoology Department
University of Oxford
Tubney, United Kingdom

Wesley R. Silva
Department of Zoology
University of Campinas
Campinas, Brazil

Carla Cruz Soares
São Caetano do Sul, Brazil

Lucía Soler
Departamento de Biología, Bioquímica y Farmacia
Universidad Nacional del Sur
Buenos Aires, Argentina

Nucharin Songsasen
Smithsonian Conservation Biology Institute
Front Royal, Virginia

Denise Taylor
Education 4 Conservation Limited
Rugby, United Kingdom

Ralph Vanstreels
Pathology Department
University of Sao Paulo
Sao Paulo, Brazil

Carly Vynne
Department of Biology
University of Washington
Seattle, Washington

Acknowledgments

We would like to express our utmost gratitude to all those who contributed to our book. We are grateful to Bob Carling, John Sulzycki, and Jill Jurgensen at CRC Press/Taylor & Francis for their encouragement and advice all along. Special thanks to Melissa Rodden for her positive and thoughtful input to the project. Thanks to the other referees who commented on our proposal. We are grateful to Dr. Cory Carvalho for taking time to write an evocative preface to our book. We are indebted to Claudio Sillero-Zubiri for embarking on this project and making invaluable contributions. We sincerely thank all the contributing authors and artists for their vital, inspiring, often pioneering work. Thank you for your enthusiasm for this project; it has been an absolute pleasure working with you and we look forward to working with all of you in the future. Thanks to Edu Fortes and Marcelo Pinto also for their beautiful and meaningful images of the maned wolf.

Adriana Consorte-McCrea
Many thanks to Stephen Scoffham for his thoughtful advice and valued comments on the manuscript. Thanks for the support of my family, always positive, always present. I am immensely grateful to Robert, Francis, and Felix for inspiring me every day.

Eliana Ferraz Santos
I would like to thank my friend and photographer Edu Fortes for the photos of maned wolves. I also thank Cristina Harume Adania, friend and coordinator of the Mata Ciliar Association of Jundiaí, for kindly letting the maned wolves be photographed in the nursery of the association. I also wish to thank Sabrina Iamamoto for the translations and revisions in English, in all my chapters in the book.

Introduction

It is a great honor and a timely endeavor to gather leading researchers and practitioners from around the globe, and from every country within the maned wolf's distribution, in one book dedicated to the ecology and conservation of the maned wolf. The multidisciplinary nature of our book combines different aspects of the maned wolf *in situ* and *ex situ* together with innovative approaches for the future of conservation strategies. In doing so, it makes a substantial contribution toward meeting the recommendations and targets set out by the Action Plan for Canid Conservation into the 21st Century (Sillero-Zubiri et al. 2004) and by the Maned Wolf Action Plan (Paula et al. 2008).

Wolves are considered controversial figures worldwide: mythical creatures as well as troublemakers, their conservation can polarize extreme views and is often riddled with conflicts. In recent years, much effort has focused on the discussion of how to conserve large carnivores, such as wolves, while addressing public concerns. The maned wolf, however, has mostly eluded this discussion. With its solitary habits and unusual fruit-eating diet, the endangered maned wolf roams the grasslands of South America, playing a vital part in maintaining the integrity of this biodiversity hotspot. Compared with other large canid species, little is known about this unique species, the peculiarities of its relationship with local people and the environment, and the reasons for its decline, making its research an urgent concern.

Considering the maned wolf's uniqueness, issues discussed in this book are also relevant to the conservation of other large carnivores. The elements that contribute to the construction of the relationships and also to the potential conflict between people and wild animals, and particularly carnivores, are very similar worldwide. The ecological role of the maned wolf as a large predator is also similar to that of other predator species: like many large carnivores, maned wolves are keystone species and also potential umbrella and flagship species for the conservation of their habitat. Like other carnivores, they have progressively strived to survive in human-occupied environments, as their natural habitat is encroached by farms. Endangered species of carnivores face similar problems, linking their decline with the value that people place on resources and human activities; protected areas are insufficient to guarantee their survival, and the importance of including local landowners in conservation becomes more evident. These common issues are discussed in the following chapters as researchers and practitioners from different fields disclose the maned wolf's compelling story.

Endangered species have been a major focus of conservation research and of the allocation of financial resources to carry out this research. The establishment of conservation priorities focusing on species such as the maned wolf, however, does not eliminate the problem of endangerment since most of it is caused by habitat loss and anthropic actions. Besides being rapidly destroyed, habitats that occupied large areas are now often divided into small fragments by roads, fields, cities, and many other human activities. Habitat loss and fragmentation can reduce wild populations, causing local extinctions and genetic problems caused by inbreeding (Gaines et al. 1997). An ecological approach is often recommended to conserve ecosystems and habitats of a given region, including all associated biotic communities.

Anthropic connections to the endangerment and to the conservation of wildlife are complex and multifaceted. The decline of carnivore populations in Central and South America seems to be directly related to the value placed on resources and human activities as a result of globalization (Clark et al. 2001). Research indicates that conflicts between people and wildlife are often better understood as people–people or people–state conflicts, as they are affected by tensions and divisions in human society (Knight 2000; Hill 2004). A lack of understanding of the social–cultural dimensions of such conflicts may cause conservation initiatives to see local people as the origin of conflict and their removal from protected areas as the solution, in an attempt to protect wild populations. Most recently, however, the way traditional populations manage wild areas has been linked with increases in biodiversity and wildlife conservation (Arruda 2000). Considering that a large proportion of conservation areas in Latin America are occupied by human populations (Colchester 2000), the exclusion of traditional people with a sustainable way of living can make conservation areas more vulnerable to poaching and logging (Arruda 2000; Guha 2000). Since 1975, the contribution of traditional communities to environmental protection has been recognized by the International Union for Conservation of Nature, which recommends their involvement in conservation strategies without "losing their rights of property and use" (Colchester 2000, 243). Partnerships between traditional communities and scientific experts may offer a way forward in the search for sustainable solutions to the management of natural resources and wildlife conservation.

The inclusion of local people is also advocated by some of the most successful and long-running wildlife conservation programs in Brazil, associated with environmental education programs, where local people have helped to determine local needs and to define directives (Padua and Tabanez 1997). Emerging from their objectives is the need to promote self-esteem associated with valuing the locality, encouraging integration between its cultural and natural resources (including protected areas). Within this framework, appealing flagship species, such as the maned wolf, may be used to attract the attention of the population to the environmental cause, by eliciting support for the preservation of that species' habitat (Dietz and Nagagata 1997).

Although endangerment and extinctions have been part of the dynamics of evolution, we are witnessing an unprecedented decline in biodiversity as a result of the demands of ever-increasing human populations. Within ecological communities, there are certain species that exert a particularly important influence on the persistence of all others. Ecological communities evolve over millions of years, providing unique opportunities for the development of intricate relationships between their biotic and abiotic components. All over the planet, species living in centers of biodiversity and endemism, such as the cerrado in Brazil, have been affected and reduced due to habitat loss and fragmentation, poaching, and the action of new predators and competitors. Keystone species such as the maned wolf can influence the structure and dynamics of the community at a higher level than would be expected based on the quantity of individuals or on their biomass (Terborgh 1976; Janzen 1986). Even the removal of a small number of predators, although constituting a small portion of the biomass of the community, can potentially result in dramatic changes in vegetation and a large loss of biodiversity (Pimm 1991; McLaren and Peterson 1994).

We know that drastic changes in the dynamic relationships between plants and herbivores and between predator and prey can also result in losses, to a greater or lesser extent, in biodiversity and in the quality of life of human populations, particularly those who depend on natural resources for their living. The services provided by biodiversity and interspecific interactions should be seen as important tools in maintaining the biological diversity of fragments. Therefore, taking action toward maintaining viable populations of terrestrial mammals in the medium and long term, including some species of canids such as the maned wolf, is essential in the recovery of degraded or fragmented areas.

The population model created for Brazil by the Maned Wolf Action Plan (Paula et al. 2008) suggests that the major threats to the species are an increase in road mortality and reduction of carrying capacity due to habitat loss. The model indicates that such threats may cause a significant drop in maned wolf numbers resulting in the local extinction of populations, with cascading effects that could contribute to the decline of local ecosystems. Other threats to the maned wolf are the presence of domestic and feral dogs in the vicinity of protected areas, retribution killings and hunting, and killing for the harvesting of their body parts for traditional medicine or magical purposes. Threats to maned wolf populations are a recurring theme in this book, as are reflections and recommendations on how to target and address them to empower conservation strategies for the species.

The book is divided into three parts, preceded by a rare insight into the experiences of one of the pioneers in the study of the maned wolf, Dr. Cory Carvalho, which should inspire newcomers to get to know the species in the same way his work has inspired many of us. The following chapters explore a wide variety of aspects but maintain a common structure based on what is known, or an overview of historical and recent research; and on next steps, look toward future research and areas for further development.

Part I, "The Species and Ecology," starts with "Species Status" to lay out some of the most general issues concerning the present situation of the maned wolf. In Chapter 1, Claudio Sillero-Zubiri offers a general introduction to the species and talks about the maned wolf within the context of the conservation of canids worldwide. Current findings on the population viability of the maned wolf are discussed in Chapter 2 by Rogério Cunha de Paula and Arnaud L.J. Desbiez. In Chapter 3, Adriana Consorte-McCrea reviews and discusses key issues related to the relationships between maned wolves and people. In Chapter 4, the current keeper of the Maned Wolf Studbook, Ruben Holland, talks about the management of the worldwide population of captive maned wolves, its problems, trends, and perspectives, and its role in the knowledge and conservation of the species. In Chapter 5, related aspects linked to the maned wolf *ex situ* in Brazilian zoos are discussed by Cecília Pessutti, Cleyde Angélica Ferreira da Silva Chieregatto, and Ralph Vanstreels.

Part I continues with "Species Biology," where diverse aspects of current knowledge are reviewed. In Chapter 6, Manoel L. da Fontoura-Rodrigues and Eduardo Eizirik review the development, current research, and perspectives of the field of conservation genetics in relation to the maned wolf. José Carlos Motta-Junior, Diego Queirolo, and Adriana de Arruda Bueno provide a comprehensive revision of the maned wolf's feeding ecology, based on the literature on diet and a quantitative analysis of the trophic ecology of the species in central and eastern South America in Chapter 7. In Chapter 8, Nucharin Songsasen et al. look at reproduction and stress and provide an overview of the current understanding of social structure and reproductive behaviors and biology, as well as factors influencing the reproductive success of the maned wolf. In Chapter 9, Joares Adenilson May Jr. and Paulo Anselmo Nunes Felippe examine the current knowledge about conservation medicine in relation to the maned wolf in captivity and in the wild.

Part II, "Case Studies and Conservation Programs," offers an in-depth view of some of the knowledge gathered from past and ongoing programs and field research carried out on maned wolf populations in all countries where the species is currently found. In Chapter 10, Flávio H. G. Rodrigues et al. evaluate the role of small reserves in maned wolf conservation, based on a case study of population viability analysis. In Chapter 11, Eliana Ferraz Santos, Nivar Gobbi, and Paulo Anselmo Nunes Felippe present a case study in feeding ecology, disclosing the field work and analysis involved in determining the diet of the maned wolf, its seasonal variation, and its relationship to seed dispersal on a dairy cattle ranch in the state of Minas Gerais, Brazil. In Chapter 12, Carly Vynne shows how measuring fecal hormone metabolites in scat samples from free-ranging maned wolves in and around a federal nature reserve in Brazil provides a window into the cost of making a living in areas where there is anthropomorphic disturbance or land use change. In Chapter 13, Rogério Cunha de Paula et al. give an insight into the main challenges, findings, and prospects of the remarkable Maned Wolf Conservation Project in Serra da Canastra.

The maned wolf is endemic to South America and its distribution has changed over time, reflecting changes in landscape and habitat use. Although it is considered Near Threatened at global level (IUCN Red List), its status and perspectives change throughout its range. The next few chapters provide a comprehensive overview of the current knowledge and the challenges faced by maned wolf populations in each country, starting with Diego Queirolo et al.'s review of the species in the south of Brazil and Uruguay, in Chapter 14. In Chapter 15, Lucía Soler discusses aspects of the maned wolf in Argentina. In Chapter 16, Louise H. Emmons discusses environmental factors that may influence resource levels, territory, and population size of breeding maned wolves in a study area in Bolivia, which may affect their geographic range. Maned wolf conservation in Paraguay is reviewed in Chapter 17 by José L. Cartes, Anthony J. Giordano, and M. Nathalia Mujica Cameroni.

In Part III, "Perspectives for the Future," we invited contributions from professionals from diverse fields of research that may be presently underexplored in relation to the maned wolf but have increasing importance in its conservation. The next chapters explore challenges and opportunities connecting maned wolf conservation and local communities. Denise Taylor discusses education and communication as tools for the conservation of wild canids, with applications to the maned wolf, in Chapter 18. In Chapter 19, Alistair Bath offers an overview of how to use human dimensions research to explore people's relationships with large carnivores and conflict and mitigation with local groups, and suggests how these could be applied to the conservation of the maned wolf. In Chapter 20, Antonio Carlos Diegues talks about the ethnoconservation perspective and reflects on how this approach may integrate local traditional communities into wildlife conservation plans. In Chapter 21, Carlos Klink talks about the ecology and socioeconomics of the cerrado biome and the challenges to its conservation, which are closely interconnected to the present status and future of the maned wolf.

Claudio Sillero-Zubiri, Adriana Consorte-McCrea, and Eliana Ferraz Santos close the book with Chapter 22, which is an overview of the main lessons from each section that attempts a synthesis of what is in stock for maned wolf conservation, in "Maned Wolf Ecology and Conservation: The Road Ahead."

We hope this book will contribute to a much more precise and global understanding of issues surrounding maned wolf ecology and conservation and its role in the conservation of biodiversity and to the planning of strategies to help guarantee a future for the species. The striking charisma and the essence of the "fox on stilts" have been brilliantly captured by the photos of Edu Fortes and by the drawings of Marcelo Pereira Pinto, perfectly embodying the spirit of the book. Thank you for joining us on this journey!

Adriana Consorte-McCrea and Eliana Ferraz Santos

REFERENCES

Arruda, R. S. (2000). Populações tradicionais e a proteção de recursos naturais. In A. Diegues (ed.), *Etnoconcervação. Novos rumos para a proteção da natureza nos trópicos* (pp. 273–290). Sao Paulo: Ed Hucitec, NUPAUB-USP.

Clark, T. W., Mattson, D. J., Reading, R. P., & Miller, B. J. (2001). Interdisciplinary problem solving in carnivore conservation: An introduction. In J. E. Gittleman (ed.), *Carnivore Conservation* (pp. 223–240). Cambridge: Cambridge University Press.

Colchester, M. (2000). Resgatando a natureza: Comunidades tradicionais e áreas protegidas. In A. Diegues (ed.), *Etnoconservação. Novos rumos para a proteção da natureza nos trópicos* (pp. 225–256). São Paulo: Ed Hucitec, NUPAUB-USP.

Dietz, L. H. & Nagagata, E. Y. (1997). Programa de Conservação do Mico-Leão Dourado: Atividades de educação comunitária para a conservação da Mata Atlântica no Estado do Rio de Janeiro. In S. Padua & M. F. Tabanez (eds), *Educação ambiental: Caminhos trilhados no Brasil* (pp. 133–146). Sao Paulo: Instituto de Pesquisas Ecológicas.

Gaines, M. S., Diffendorfer, J. E., Tamarin, R. H., & Whittam, T. S. (1997). The effects of habitat fragmentation on the genetic structure of small mammal populations. *Journal of Heredity*, 88: 294–304.

Guha, R. (1997). The authoritarian biologist and the arrogance of anti-humanism: Wildlife conservation in the Third World. *The Ecologist*, 27(1): 14–19.

Hill, C. (2004). Farmers' perspectives of conflict at the wildlife-agriculture boundary: Some lessons learned from African subsistence farmers. *Human Dimensions of Wildlife*, 9(4): 279–286.

Janzen, D. H. (1986). The eternal external threat. In M. Soulé (ed.), *Conservation Biology: The Science of Scarcity and Diversity* (pp. 286–303). Sunderland, MA: Sinauer Associates.

Knight, J. (2000). *Natural Enemies: People-Wildlife Conflicts in Anthropological Perspective.* London: Routlege.

McLaren, B. E. & Peterson, R. O. (1994). Wolves, moose, and tree rings on Isle Royale. *Science*, 266: 1555–1558.

Padua, S. M. & Tabanez, M. F. (1997). *Educação ambiental: Caminhos trilhados no Brasil.* Sao Paulo: IPE-Instituto de Pesquisas Ecológicas.

Paula, R. C., Medici, P., & Morato, R. G. (2008). *Plano de ação para a conservação do lobo-guará—análise de viabilidade populacional e de habitat.* Brasília: CENAP/ICMBio.
Pimm, S. L. (1991). *The Balance of Nature.* Chicago: University of Chicago Press.
Sillero-Zubiri, C., Hoffmann, M., & Macdonald, D. W. (2004). *Canids: Foxes, Wolves, Jackals and Dogs.* Status Survey and Conservation Action Plan. Gland: IUCN/SSC Canid Specialist Group.
Terborgh, J. (1976). Island biogeography and conservation: Strategy and limitations. *Science*, 193: 1029–1030.

Part I

The Species and Ecology

Species Status

Drawing of maned wolf by Marcelo Pinto.

1 The Canidae Family
Setting the Scene for Maned Wolf Conservation

Claudio Sillero-Zubiri

CONTENTS

Origins of South American Canids 3
Canid Diversity 5
Canid Adaptability 6
Canid Spatial and Social Organization 6
Relationship with Humans 7
Canid–Human Conflict 8
Canid Status and Conservation 9
References 12

In order to fully understand the biology of maned wolves (*Chrysoscyon brachyurus*) and to appreciate their role in the ecosystem and the threats they face, one must know several things concerning the modern Canidae (Order Carnivora). Their anatomy is adapted for the cursorial pursuit of prey in relatively open environments and often characterized by tall, lithe bodies, a bushy tail, long limbs, and digitigrade, four-toed feet (Macdonald and Sillero-Zubiri 2004a; Sillero-Zubiri 2009). The Canidae family is characterized by a great flexibility of diet, opportunistic and adaptable behavior, and complex social organization with much variation within and between species.

The contemporary canids are the most widely distributed family of the Carnivora, with members on every continent besides Antarctica (Macdonald and Sillero-Zubiri 2004a; Sillero-Zubiri et al. 2004a). Those familiar with taxonomy will not be surprised to know that there is not a simple answer to the question of how many species of canid there are, but we will say there are about 35 (Clutton-Brock et al. 1976; Wozencraft 1989; Sillero-Zubiri 2009 argue for various answers between 34 and 37). We will soon be adding to this list, with the recognition of *Canis lycaon* (Wilson et al. 2000), *C. himalayensis* (*C. lupis chanco*) (Sharma et al. 2004), and even *C. lupaster* (*C. aureus lupaster*) (Rueness et al. 2011).

Africa, Asia, and South America support the greatest canid diversity, with more than 10 canid species each (Johnson et al. 1996). Within any one location, however, canid diversity is usually limited to one to five species. Of 10 genera in the Canidae, 5 are restricted to South America, namely *Atelocynus*, *Cerdocyon*, *Chrysocyon*, *Pseudalopex* (*Lycalopex*), and *Speothos*. They are represented by 11 species, of which a staggering nine are endemic to the continent.

ORIGINS OF SOUTH AMERICAN CANIDS

The South American canids are not a monophyletic group; more likely, they owe their origin to separate invasions. The bush dog (*Speothos*) and maned wolf (*Chrysocyon*) clade is grouped with the wolf-like canids rather than with the South American foxes (*Pseudalopex*) (Wang et al. 2004;

Figure 1.1). The large sequence divergence between the bush dog (*Speothos venaticus*) and maned wolf, and between these taxa and the South American foxes, suggests that they diverged from each other 6–7 million years ago, well before the Panamanian land bridge formed about 2–3 million years ago (Wang et al. 2004). Thus, the phylogenetic distribution of the extant South American species is best explained by three separate canid invasions. Furthermore, within the South American foxes the crab-eating fox (*Cerdocyon thous*) and the short-eared dog (*Atelocynus microtis*) are ancestral

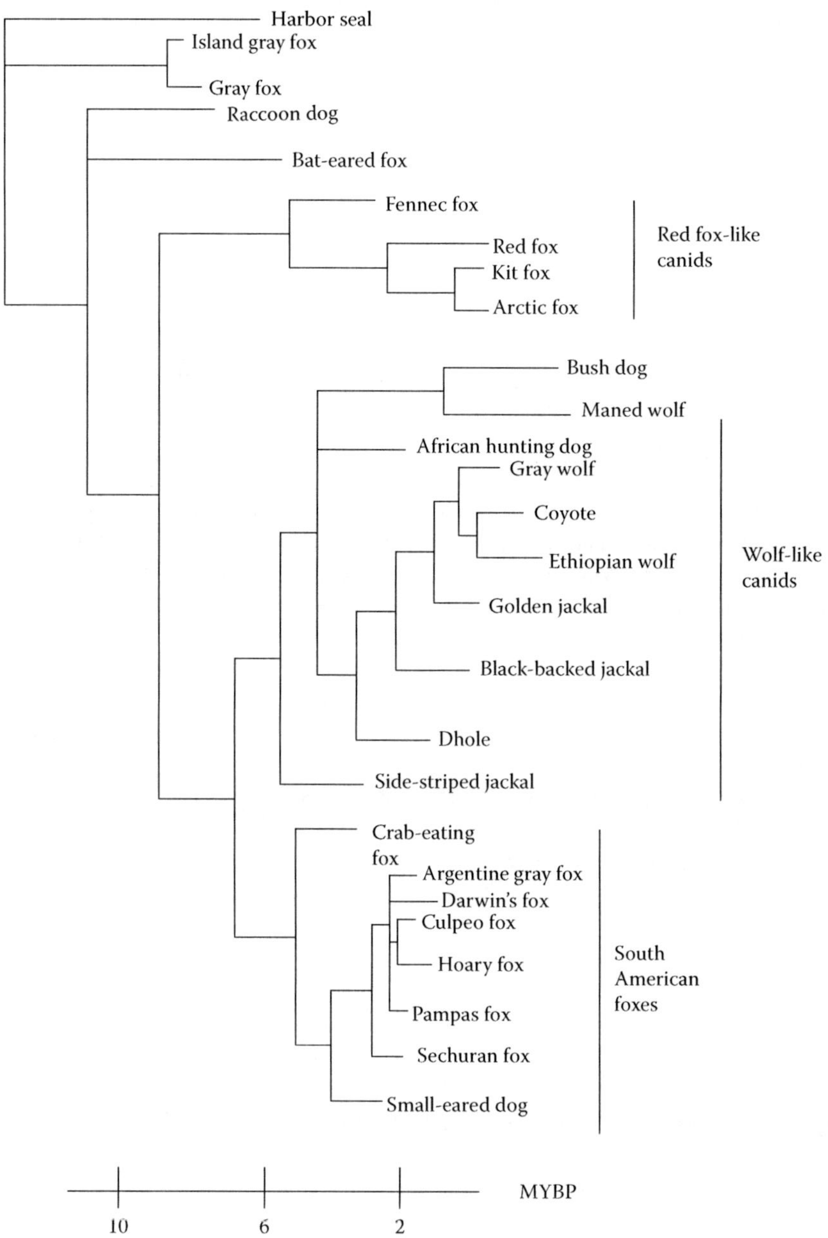

FIGURE 1.1 Consensus tree of 26 canid species (based on the analysis of 2001 bp of DNA sequence from mitochondrial protein coding genes) (From Wayne, R.K. et al., *Systematic Biology*, 46, 622–653, 1997). Time of scale in millions of years before present (MYBP) is based on comparisons of DNA sequence divergence to first appearance times in the fossil record. (Reprinted with permission from the IUCN/SSC Canid Specialist Group.)

(3–6 million years ago). Divergence values between them and other South American foxes suggest they may have diverged before the opening of the Panamanian land bridge as well. The fossil record supports the hypothesis that the crab-eating fox has its origin outside of South America, as the genus has been described from late Miocene deposits of North America (3–6 million years ago). *Pseudalopex* (*Lycalopex* for some authorities) foxes evolved fairly recently (1.0–2.5 million years) and, perhaps together with *Atelocynus*, might have a South American origin.

The now extinct Malvinas zorro or Falklands wolf (*Dusicyon australis*) was first described by Charles Darwin during his visit to the southern archipelago. Based on skull and teeth measurements, these wolf-like foxes were more similar to *Canis* than *Pseudalopex*, but their extermination by pelt-hunters by 1880 sadly relegated them to the status of a biological mystery (Nowak 1999). Some maintain that *D. australis* descended from a domestic canid, pointing out that it had a white-tipped tail, rather than the black one of all surviving *Pseudalopex* species (Clutton-Brock et al. 1976). Slater et al.'s (2009) mitochondrial DNA analysis from museum specimens indicated that their most recent common ancestor lived some 330,000 years ago, making it likely that *D. australis* arrived in the islands long before humans. Intriguingly they found that its closest living relative is the maned wolf (*Chrysocyon brachyurus*).

The directional trend toward the evolution of large, hypercarnivorous forms apparent in groups of dog-like carnivores may be a fundamental feature of carnivore evolution (Wang et al. 2004). The likely cause is the prevalence of interspecific competition among large, sympatric predators. Interspecific competition tends to be more intense among large carnivores because prey are often difficult to capture and can represent a sizable quantity of food that is worth stealing and defending. Hypercarnivory appears late in the history of the Canidae and represents at least several independent radiations in South America, North America, and the Old World.

The South American radiation of large hypercarnivorous canids occurred at a time when cat-like predators were rare or absent. It followed the elevation of the Panamanian land bridge around 2–3 million years ago, which allowed movement between the previously separated continents. The canids that first entered South America found a depauperate predator community. Between 2.5 million years ago and 10,000 years ago, 16 new species of canids appeared in South America, at least 7 of which had trenchant heeled carnassials and clearly were adapted for hypercarnivory. All of these South American hypercarnivorous canids, except the bush dog, went extinct at the end of the Pleistocene. In the Old World, the evolution of hypercarnivorous canines occurred within the last 4 million years and did not coincide with an absence of cats. Wolf-like hyaenids were the dominant dog-like predators of the Old World Miocene, reaching a diversity of 22 species between 9 and 5 million years ago, but then declining dramatically to just 5 species by about 4 million years ago. Their decline may have opened up ecospace for the large canids and favored the evolution of hypercarnivory (Holliday and Steppan 2004).

CANID DIVERSITY

The most obvious thing about canids is their diversity—although their capacity to evoke strong positive or negative feelings in humans becomes quite apparent when dealing with people's attitudes toward wild canids. Canids range in size from Blanford's, fennec, and pale foxes (*Vulpes cana*, *V. zerda*, *V. pallida*), which can weigh less than 1 kg as adults, to the gray wolf (*Canis lupus*), exceeding 60 kg. Their distributions may be highly restricted. Island foxes (*Urocyon littoralis*) are restricted to the small Channel Islands of California, with each of six subspecies occurring only on one island (Wayne et al. 1991); whereas other species span several continents, with the red fox (*V. vulpes*) (Lloyd 1980) having the most extensive natural range of any land mammal, occupying about 70 million km^2. Their diets range from omnivory to strict carnivory—at times almost exclusively frugivorous or insectivorous—and they live in habitats ranging from deserts to ice fields, from mountains to swamps or grasslands, and from rainforests to urban "jungles" (reviews in Johnson et al. 1996; Sillero-Zubiri and Macdonald 2004; Sillero-Zubiri 2009). They may occupy home ranges as small as 0.5 km^2 (island fox) (Roemer et al. 2001a) or as large as 2000 km^2 (African wild dogs, *Lycaon pictus*) (Frame et al. 1979).

The body sizes of wild canids may be explained to some degree by differences in availability of food. Small canids (e.g., pale, fennec, and kit foxes, *V. macrotis*) are usually associated with arid and poor habitats in which only a small body mass can be supported year round, whereas large canids (e.g., Ethiopian wolf, *C. simensis*, and African wild dog) are often associated with habitats in which prey is more abundant. At about 25 kg, the elegant maned wolf is the largest canid in South America. However, for a large canid, it is unusual in its social organization. In contrast to what would be predicted by body size alone, maned wolves are primarily solitary, typically living in pairs, and rely on an omnivorous diet consisting principally of fruits and small to medium vertebrates (Dietz 1985).

Body shape does not vary dramatically in canids, with most species fitting into either a vulpine or a wolf-like morphotype, although some variation exists. Again, the maned wolf is a bit of an outlier, being immediately distinguishable from its fellow canids by its extremely long, thin legs, large ears, and long, reddish-orange fur (Dietz 1985). With the exception of the aptly named maned wolf, canid pelage is relatively short, with dense underfur mixed with longer guard hairs. Coat color is generally tawny brown or gray, but black, white, and shades of ochre also occur. The underparts are usually paler than the rest of the body. The tail is generally bushy, often with a white or black tip and a darker, bristly patch covering the dorsal supracaudal scent gland near the base.

All canids have excellent senses of smell and hearing. Sight is less acute in some species, although always good. The keen sense of smell probably coevolved with scent marking as the dominant means of communication. Most canids have a well-developed vomeronasal organ that allows them to "smell" chemical cues present in scent marks. Sensitive hearing, supported by large ears, helps many species detect food.

CANID ADAPTABILITY

Although canids evolved in relatively open spaces, their great adaptability has enabled them to flourish in most conditions. Wild canids are among the most cosmopolitan groups of mammals and have adapted to living in a variety of landscapes from deserts to rugged mountain environments. Today they occur in all major habitats, spanning tropical forests, woodlands, savannas, deserts, mountains, alpine heathlands, the Arctic, and even cities. Wild canids are found at all altitudes from sea level to over 5000 m above sea level and are present in all continents, except Antarctica, and many oceanic islands (Sillero-Zubiri 2009).

Canid diets are among the most versatile in carnivores, varying from strictly carnivorous to some that contain <5% protein. Canids can be divided into small (<6 kg), medium, and large (>13 kg) species, and there are some general differences in diets in each size group. Most of the smaller species are opportunistic omnivores, eating anything from mammals to birds, reptiles, insects, fruits, and carrion. Larger, group-living species such as gray wolves, dholes (*Cuon alpinus*), and African wild dogs may be more strictly carnivorous, preying on medium to large mammals that often exceed their own body size. The maned wolf, which thrives in the South American grasslands, is unusual for a large canid in feeding mostly on rodents and fruits. As an aside, Geffen et al. (1996) suggested that low food availability probably constrained both the maned wolf's group and litter size (which is low at 2.2).

CANID SPATIAL AND SOCIAL ORGANIZATION

Throughout the world, canids exhibit a wide range of behavioral adaptations and as such have been the focus of intensive study. Variation in ranging behavior and social organization, in particular, have been key to understanding the biology of the family and, in many regions, to solving complex conservation issues.

Wild canids tend to be territorial, relying on scent-marking with urine and feces to demarcate their home ranges and communicate with conspecifics. Marking is involved in establishing and maintaining pair bonds and dominance hierarchies and may provide an indication of a female's reproductive state. Black-backed jackal (*Canis mesomelas*) pairs defend their territory from other jackal pairs by leaving urine and feces on conspicuous objects. Unrelated jackals within this area are chased away vigorously, especially during the breeding season.

Among the medium and large canids, most species have evolved to live in groups or packs. Group living benefits foraging efficiency, breeding success, and survivorship, but it also entails a suite of energetic constraints and behavioral disadvantages. Again, maned wolves seem to be an outlier; maned wolf pairs appear to be facultatively monogamous, like many of the smaller fox species. Pairs form, mate, and raise young together during the reproductive period, but during other times of the year, pairs are rarely seen (Dietz 1984; Rodden et al. 2004).

RELATIONSHIP WITH HUMANS

Wild canids have had a long relationship with people, and canids can be traced back to the earliest recorded history as objects of worship. The human–canine relationship has varied from fear, persecution, and exploitation to the domestication of man's best friend. Today many wild canids are thriving and often find themselves at the wrong end of the stick due to their predisposition for attacking livestock. Some of the more charismatic species are now sought out by growing numbers of wildlife tourists.

The jackal-headed god Anubis was worshipped by ancient Egyptians as keeper of the dead, perhaps in association with the common presence of jackals near abattoirs and cemeteries. In northern cultures there is no animal more enshrined in the myths than the gray wolf. Aesop's fables celebrated its cunning, while the story of Rome's legendary founders Romulus and Remus was only one of many telling of children raised by wolves. Folklore undoubtedly demonizes the wolf, but in the last half century many biological studies have given a more sympathetic view of these mighty predators, which are slowly making their way back from near-extinction a few decades ago. The Indian fox (*V. bengalensis*) is featured in ancient Jataka texts and the Panchatantra, where it is depicted as a clever and sometimes cunning creature. The raccoon dog (*Nyctereutes procyonoides*), which owes its name to an uncanny resemblance to the raccoon, has often appeared in Japanese folklore. In Brazil, the night cry of the maned wolf is believed to portend changes in the weather. Red foxes, which are adaptable and opportunistic foragers, have a long association with humans in Europe and Asia. They have been hunted at least since the fourth century BC. They are wily characters in the popular tales of many cultures, reflecting both their wide distribution and their resourceful behavior.

It is not just the image of wild canids as predators and competitors that influenced our ancestors. Humans have used their skins for clothing, bedding, and shelter, and other body parts for medicinal and shamanistic purposes. Canids have been exploited for subsistence, commercial profit, and hunting for sport (Sillero-Zubiri et al. 2004a). Much of the pioneering history and economies of northern latitudes was built on trade in fur, and much of it was originally worn by foxes and wolves. The extent of commercial exploitation has fluctuated, depending on supply of and demand for furs, which are still used today, although not with the same assiduity. Until the advent of fur farming in the twentieth century, the fur trade concentrated on wild populations, notably of coyote (*Canis latrans*), red fox, and Arctic fox. Very large shipments also originated from three of South America's southern cone foxes (*Pseudalopex* spp.) (Sillero-Zubiri et al. 2004a). Such intensive hunting often affected the density and demography of these populations. In some areas today, trapping is largely done by farmers trying to reduce predation on livestock or poultry, and this affects maned wolves among others. However, even for these people, the sale of fur can be economically important, for example, representing up to a quarter of the annual income of some ranch hands in Patagonia. Although arguably hunting can increase the value of many canid species, their exploitation is becoming increasingly controversial. Wherever the relationship lies on the spectrum from exploitation to conflict, a human dimension underlies almost all canid conservation problems.

The versatility of many canids enables them to flourish in human landscapes such as farmland and even in cities, where they may prey on game species and domestic livestock. This brings them into conflict with humans far beyond the borders of protected areas. Throughout history, wild canids have been seen as competitors for humans' wild prey. In Britain, the practice of predator control to protect game species resulted in the development of large, privately owned sporting estates in the nineteenth century. An estimated 70,000 red foxes are killed by some 3,500 professional gamekeepers in Britain every year to preserve game birds for sport shooting. In the royal hunting preserves of medieval Europe, gray wolves were killed to protect deer. Many deer hunters in Europe and North America believe that competition with canids reduces their hunting opportunities, and today deer hunters are among the most vocal opponents to gray wolf reintroduction. In Alaska, the public harvest of gray wolves to increase deer numbers for hunting has received a very positive response from sport hunters but has been vehemently opposed by conservationists. There are times when canid predation upsets conservationists, such as when red foxes kill capercaillie (*Tetrao urogallus*) in Britain (Baines et al. 2004), California's island foxes kill threatened San Clemente loggerhead shrike (*Lanius ludovicianus mearnsi*) (Roemer and Wayne 2003), and Javan dholes kill banteng (*Bos javanicus*), Asia's threatened wild cattle.

Because of their broad range of sizes, canid species can fill dramatically different ecological roles, and killing them can have varying impacts on their communities. The removal of gray wolves and bears from the Yellowstone ecosystem 150 years ago, for example, triggered a cascade effect by allowing an increased density of moose (*Alces americanus*) (Ripple and Beschta 2012). The resulting increased foraging on vegetation negatively affected the diversity of migrant birds. Coyotes play a keystone role—their removal can lead to an abundance of mesopredators, including gray foxes, and, in turn, to a reduction in the abundance and diversity of prey such as small rodents and jackrabbits (*Lepus californicus*) (Henke and Bryant 1999).

CANID–HUMAN CONFLICT

Predation on domesticated animals, from chickens to cattle, is the root of a deeply ingrained antipathy toward wild canids throughout the world. Hoary foxes (*Pseudalopex vetulus*) in Brazil are killed indiscriminately as predators of domestic fowl, though they may not deserve this reputation since the culprits are likely to be the sympatric crab-eating foxes, which are formidable poultry raiders. The ability of large canids to kill large prey means they can cause substantial economic damage through depredation on livestock. Livestock depredation by canids is highly variable in space and time. For example, 50% of sheep producers in the United States reported annual losses to coyotes that were <5% of their stocks, but nearly a quarter reported losses >15% (in Sillero-Zubiri et al. 2004b).

Although larger canids attract particular wrath, the collective damage of smaller species such as jackals and foxes may be greater, exacerbated by changes in husbandry over the last century that preclude once-traditional livestock-guarding practices. Where stock-guarding traditions have vanished, recolonization by gray wolves now provokes furious public complaint. Region-wide culling of common canids such as coyotes, culpeos, and red foxes over large areas has been widespread in the past, but largely ineffective, with most of these species thriving despite tremendous hunting pressure. However, some canids can be outgunned by high levels of culling. The Falklands wolf disappeared in 1876 due to the activity of fur traders and poisoning by settlers to control sheep predation, worsened by the wolves' unwary behavior.

Some of the larger canids may occasionally maul or even kill people, although these attacks are rare. Healthy gray wolves have caused no human deaths in North America for a century. In contrast, there are many documented cases of wolves killing people in eastern Europe and Russia, although few fatal attacks have occurred during the twentieth century. In India there is good evidence that where gray wolves come into contact with children working as shepherds, they actively select them as prey, and 273 children have been reportedly killed by wolves in the last 30 years (Jhala and Sharma 1997). African wild dogs are feared across Africa as ruthless killers, but their attacks on humans are rare or nonexistent. In the United Kingdom, a few cases of red foxes in urban areas biting or scratching infants have been reported in the press in recent years.

A further reason for canid–human conflict is that canids can be vectors of diseases that are harmful to people and their domestic animals. Paradoxically, their susceptibility to infection derives from the fact that domestic dogs and wild canids share receptivity to rabies, canine distemper, leishmaniasis, and sarcoptic mange, among other pathogens. Rabies is a serious infection that has a major impact on human lives, particularly in developing countries, with 30,000 people estimated to die annually from rabies in India alone. Most recently recorded attacks by wild canids on humans involve rabid animals. With the eradication or reduction of rabies, the incidence of gray wolf attacks has dropped disproportionately in eastern Europe, but cases are still occasionally reported from Asia and the Middle East. In Argentina, rabies in wild canids has been reported only recently, with cases of crab-eating foxes and culpeos that attacked people. The tolerance of some foxes to human encroachment may bring them into close contact with domestic dogs.

There is no doubt that public perception of many carnivore species is beginning to change. Some large carnivores are major attractions for tourists. High-profile and visible canid species such as African wild dogs, gray wolves, Ethiopian wolves, dholes, and maned wolves may be capable of supporting, at least partially, a sustainable tourist trade. However, many other canids are frustratingly secretive. Visitors to Yellowstone in the United States take pleasure in the expectation of seeing or hearing wolves, and more than 20,000 have observed wolves there since 1995. Ethiopian wolves are the chief attraction at the Bale Mountains National Park in southern Ethiopia, and maned wolves are a highlight of tourism to the protected grasslands of Argentina and Brazil. Nonetheless, expectations of revenue should not be exaggerated. For example, in Ethiopia's Bale Mountains, where income from tourism is often given as a justification to the local community for the presence of a park, the number of tourists visiting each year is only in the hundreds, many use their vehicles rather than local guides or horses, and the sums reaching the local community are not great. Furthermore, the susceptibility of tourism to fluctuations in the global economy and to political instability makes it unwise to base conservation entirely on economic values.

CANID STATUS AND CONSERVATION

The current status of wild canids is mixed. Although some threatened species have declined in numbers and distribution, most canids have remained widespread, and a few species have even expanded in distribution. Coyotes, for example, are now more common and more widely distributed than ever before, inhabiting much of North and Central America, reaching as far south as Panama. Golden jackals have spread into western Europe and are becoming a regular sight in countries where they were previously unknown, such as Bulgaria, Italy, and Austria. Gray wolves are another success story, slowly but steadily recolonizing parts of western Europe and some parts of the United States where they had once been eradicated.

The IUCN/SSC Canid Specialist Group (www.canids.org) customarily evaluates the conservation status of all 35 canid taxa for the Red List of Threatened Species (e.g., Sillero-Zubiri et al. 2004a). Six species are currently listed as threatened under the categories Critically Endangered and Endangered, while four species are considered as Near Threatened, including maned wolves (Table 1.1). In general, threatened canids are either naturally rare, with limited distributions, or have become threatened by human factors such as habitat loss, persecution, and disease transmission.

The Critically Endangered Darwin's fox is one of the rarest canids, with a total population of approximately 600 (Jiménez et al. 2008). Darwin's fox was known until recently only from the Island of Chiloé, off the coast of south Chile, where it was collected by Charles Darwin during his HMS *Beagle* voyage. A small population has also been recorded in Nahuelbuta National Park in

TABLE 1.1
Red List Assessments of Threatened Canid Species

Species	Red List 2012.2
Falklands wolf	**EX**
Darwin's fox	**CR** C2a(ii)
Island fox	**CR** A2be + 3e
Red wolf	**CR** D
African wild dog	**EN** C2a(i)
Dhole	**EN** C2a(i)
Ethiopian wolf	**EN** B1ab(iii,v); C1 + 2a(i); D
Bush dog	**NT**
Maned wolf	**NT**
Sechura fox	**NT**
Short-eared dog	**NT**

Source: IUCN. *IUCN Red List Categories and Criteria: Version 3.1.* IUCN, Gland, Switzerland, 2001.

mainland Chile. Darwin's fox, a forest dweller, is known for killing poultry and raiding garbage dumps, apparently with little fear of people, and has even been observed entering houses at night in search of food. Domestic dogs may pose the greatest threat to the species by acting as vectors of lethal diseases or by direct attacks.

Island foxes, also listed as Critically Endangered, are rare and highly range-restricted (Roemer et al. 2008). Island foxes are endemic to the six largest of the eight Channel Islands off southern California. Each island population differs in genetic structure and is considered a separate subspecies. Island fox numbers fell within a decade from approximately 6000 individuals to fewer than 1500 in 2002. They are especially vulnerable to the introduction of canine diseases, including distemper, necessitating more stringent control of domestic dogs brought in by visitors. On the northern Channel Islands, the primary threat is an unusually high level of predation by Golden eagles (*Aquila chrysaetos*). These birds are recent arrivals to the islands, preying on introduced pigs (Roemer et al. 2001b). The conservation prognosis of island foxes has improved recently though, and a downlisting is likely.

The other Critically Endangered canid is the red wolf (*Canis rufus*), which once ranged across most of the southeastern United States (Kelly et al. 2008). There is some debate as to whether red wolves should be considered a distinctive species or a fertile hybrid between coyotes and wolves. One recent line of genetic evidence suggests that it is a unique taxon, while another proposes that red wolves and wolves in southern Ontario (*C. lupus lycaon*) are so genetically similar that they represent a separate canid species, *C. lycaon*. The red wolf was declared Extinct in the Wild by 1980 following years of direct persecution by humans. It was reintroduced into eastern North Carolina and is now common within the reintroduction area of roughly 4000 km^2. Road kills account for 25% of known red wolf deaths in the reintroduced population, but hybridization remains the primary threat.

Of the three wild canids listed as Endangered, African wild dogs are arguably the best known but least understood by people in their range (Woodroffe and Sillero-Zubiri 2012). Formerly distributed throughout much of sub-Saharan Africa, with the exception of rainforests, wild dogs have disappeared from most of their former range—25 of the former 39 range countries no longer support populations—and current population estimates suggest that as few as 5000 free-ranging wild dogs remain. Wild dogs are rarely seen, even where they are common, and it appears that populations have always existed at very low densities. Competition with larger predators has a major impact on their behavior and biology, with lions (*Panthera leo*) being a major cause of natural mortality in many areas. However, more than half of the mortality recorded among adults is a direct result of human activities, even in some of the largest and best-protected areas. Although the status of African wild dogs in southern and eastern Africa is well documented, virtually nothing is known of the remnant dwindling populations in western Africa and central Africa.

The dhole, also known as the Asian wild dog, is another Endangered canid (Durbin et al. 2008). In recent decades, dholes have dwindled in numbers with the shrinking of their native forests. With the exception of a few isolated studies in India, we know very little about them. Dholes provide a harrowing example of another feature of community ecology affecting canid conservation, namely the dilemma of endangered predators killing endangered prey. The rare dholes of Java, a distinct subspecies locally known as ajag, prey upon the calves and cows of the Endangered banteng, a rare wild cow, threatening its population in East Java.

Ethiopian wolves, the rarest living dogs, with a total adult population of less than 500 adults, are restricted to six isolated mountain pockets. The low levels of genetic variation in Ethiopian wolves reflect a long history of population decline due to a gradual post-Pleistocene warming of the continent, compounded by more recent habitat fragmentation and losses to subsistence agriculture. However, a greater and more immediate concern is the vulnerability of the remaining populations to diseases such as rabies. The wolves' last remaining stronghold in the Bale Mountains has been threatened by recurrent rabies epidemics. In addition, Ethiopian wolves have been observed to hybridize with domestic dogs, but this threat seems to be restricted to an area where the wolves coexist with shepherds and their dogs.

Ethiopian wolves were previously listed as Critically Endangered, but recent recovery of their numbers in Bale and better information on other remote, isolated populations in north Ethiopia have permitted their reclassification as Endangered (Marino and Sillero-Zubiri 2011).

The group-living, pack-hunting bush dog has one of the largest distributional ranges of the South American canids, occurring in vast areas of cerrado and rainforest habitats. Intriguingly, this species is never abundant and seems to be naturally rare, independent of human disturbance. Rapid habitat conversion and human encroachment in those habitats, however, are causes for concern for this Near Threatened species, considered by some as the ultimate forest hunter (DeMatteo et al. 2011). Another Amazonian dweller, the short-eared dog, is similarly listed as Near Threatened due to the rapid advance of the deforestation arc on the Amazon (Leite-Pitman and Williams 2011). Sechuran foxes in northern Peru and maned wolves complete the list of species that are likely to become formally threatened in the near future.

Maned wolves are threatened by the drastic reduction of the grasslands and scrub forests they inhabit due to conversion to agriculture and pastureland (Rodden et al. 2008). Their low reproductive output makes them particularly vulnerable to anthropogenic mortality. Maned wolves are targeted as killers of small livestock and poultry and are often killed by domestic dogs and, in some areas, by vehicles on highways. Native folklore contributes to local people's attitudes, which range from tolerance to fear and dislike. Although it is one of the largest carnivores in the grasslands, the species is apparently not well known to a large segment of the population.

Ignorance of distribution and abundance, and even taxonomic status, makes it hard to be confident about the conservation status of many of the lesser-known canid species. In several cases—the bush dog and the short-eared dog paramount among them—repeated but failed attempts to find them and survey their populations give grounds for pessimism. A similar situation applies to the pale, Ruppell's, and fennec foxes of the Sahara, whose distribution and status can only be guessed from scant reports.

That is why, for example, surveys of the attitudes of local people to maned wolves and other threatened canids are considered a priority. Some problems, such as road kills, cannot simply be talked out of existence—they require action. Hence the positioning of road signs and traffic-calming measures to protect Ethiopian wolves and African wild dogs and efforts to study how the impact of urban development and highways on San Joaquin kit foxes can be mitigated.

Another expression of the human dimension is the emergence of disease as an important threat to wild canids and the role wild canids play as vectors (Woodroffe et al. 2004). Rabies, canine distemper, anthrax, and other pathogens have caused dramatic die-offs and even local extinctions of several canid populations, notably Island foxes, Ethiopian wolves, and African wild dogs. Infectious pathogens cannot persist in small, isolated populations, but may be transmitted by closely related species such as domestic dogs. The threat of rabies to Ethiopian wolves was demonstrated all too dramatically in 1991, 2003, and 2008, when up to three quarters of an infected subpopulation perished. Conservation priority has since been given to protecting the wolves from disease by vaccinating domestic dogs within the wolf's range against rabies and canine distemper.

One hope of many canid conservationists is that people formerly hostile to canids will come to value them. This aspiration is behind developments in tourism around African wild dogs on game farms in South Africa. This tourism ties in nicely with the conservation strategy for wild dogs in South Africa. Because no sufficiently large piece of land remains in the country that can support a self-sustaining, unmanaged wild dog population, a constellation of smaller, satellite (mostly private) reserves has been managed for small dog populations to function as a natural metapopulation. This involves intervention to manage movements of individuals between these subpopulations to obviate the adverse effects of crowding and inbreeding. Steps are taken to minimize the damage done by the dogs to commercially valuable game and to maximize the revenue generated by the dogs themselves.

Overall, the Canidae, at least several canid species, have been studied in detail in the last few decades. The captivating processes, patterns, and details revealed in several reviews of the Canidae family (see Macdonald and Sillero-Zubiri 2004b) fire in our imagination the desire to know even more. We are able to answer many questions relevant to maned wolves and other relatives and to ask many more. But, do we know enough to be able to successfully conserve maned wolves and the habitats they live in and, as a related matter, to resolve—or at least manage—the conflicts that they face with people? If not, what must be discovered? And do the answers at hand include those necessary to underpin conservation action? Many of these questions will be revealed in this informative volume. Read on!

REFERENCES

Baines, D., Moss, R. and Dugan, D. 2004. Capercaillie breeding success in relation to forest habitat and predator abundance. *Journal of Applied Ecology* 41:59–71.

Clutton-Brock, J., Corbet, G.B. and Hill, M. 1976. A review of the family Canidae with a classification by numerical methods. *Bulletin of the British Museum of Natural History and Zoology* 29:119–199.

DeMatteo, K., Michalski, F. and Leite-Pitman, M.R.P. 2011. *Speothos venaticus*. In: IUCN 2012. *IUCN Red List of Threatened Species. Version 2012.2*. www.iucnredlist.org. Downloaded on 5 December 2012.

Dietz, J.M. 1984. Ecology and social organization of the maned wolf (*Chrysocyon brachyurus*). *Smithsonian Contribution to Zoology* 392:1–51.

Dietz, J.M. 1985. *Chrysocyon brachyurus*. *Mammalian Species* 234:1–4.

Durbin, L.S., Hedges, S., Duckworth, J.W., Tyson, M., Lyenga, A. and Venkataraman, A. (IUCN SSC Canid Specialist Group—Dhole Working Group) 2008. *Cuon alpinus*. In: IUCN 2012. *IUCN Red List of Threatened Species. Version 2012.2*. www.iucnredlist.org. Downloaded on 5 December 2012.

Frame, L.H., Malcolm, J.R., Frame, G.W. and van Lawick, H. 1979. Social organization of African wild dogs (*Lycaon pictus*) on the Serengeti Plains. *Zeitschrift für Tierpsychologie* 50:225–249.

Geffen, E., Gompper, M.E., Gittleman, J.L., Luh, H.K., Macdonald, D.W. and Wayne, R.K. 1996. Size, life-history traits, and social organization in the Canidae: A reevaluation. *American Naturalist* 147:140–160.

Henke, S.E. and Bryant, F.C. 1999. Effects of coyote removal on the faunal community in western Texas. *Journal of Wildlife Management* 63:1066–1081.

Holliday, J.A. and Steppan, S.J. 2004. Evolution of hypercarnivory: The effect of specialization on morphological and taxonomic diversity. *Paleobiology* 30:108–128.

IUCN. 2001. *IUCN Red List Categories and Criteria: Version 3.1*. IUCN Species Survival Commission. IUCN, Gland, Switzerland, and Cambridge, UK.

Jhala, Y.V. and Sharma, D.K. 1997. Childlifting by wolves in eastern Uttar Pradesh, India. *Journal of Wildlife Research* 2:94–101.

Jiménez, J.E., Lucherini, M. and Novaro, A.J. 2008. *Pseudalopex fulvipes*. In: IUCN 2012. *IUCN Red List of Threatened Species. Version 2012.2*. www.iucnredlist.org. Downloaded on 5 December 2012.

Johnson, W.E., Fuller, T.K. and Franklin, W.L. 1996. Sympatry in canids: A review and assessment. In: J.L. Gittleman, ed., *Carnivore Behavior, Ecology and Evolution*, pp. 189–218. Cornell University Press, Ithaca, NY.

Kelly, B.T., Beyer, A. and Phillips, M.K. 2008. *Canis rufus*. In: IUCN 2012. *IUCN Red List of Threatened Species. Version 2012.2*. www.iucnredlist.org. Downloaded on 5 December 2012.

Leite-Pitman, M.R.P. and Williams, R.S.R. 2011. *Atelocynus microtis*. In: IUCN 2012. *IUCN Red List of Threatened Species. Version 2012.2*. www.iucnredlist.org. Downloaded on 5 December 2012.

Lloyd, H.G. 1980. *The Red Fox*. Batsford, London.

Macdonald, D.W. and Sillero-Zubiri, C. 2004a. Wild canids—An introduction and dramatis personae. In: D.W. Macdonald and C. Sillero-Zubiri, eds, *The Biology and Conservation of Wild Canids*, pp. 3–36. Oxford University Press, Oxford.

Macdonald, D.W. and Sillero-Zubiri, C. 2004b. Conservation: From theory to practice, without bluster. In: D.W. Macdonald and C. Sillero-Zubiri, eds, *The Biology and Conservation of Wild Canids*, pp. 353–372. Oxford University Press, Oxford.

Marino, J. and Sillero-Zubiri, C. 2011. *Canis simensis*. In: IUCN 2012. *IUCN Red List of Threatened Species. Version 2012.2*. www.iucnredlist.org. Downloaded on 5 December 2012.

Nowak, R.M. 1999. *Walker's Mammals of the World*, 6th edn. Johns Hopkins University Press, Baltimore, MD.

Nowak, R.M. and Paradiso, J.L. 1983. *Walker's Mammals of the World*, 4th edn. Johns Hopkins University Press, Baltimore, MD.

Ripple, W.J. and Beschta, R.L. 2012. Trophic cascades in Yellowstone: The first 15 years after wolf reintroduction. *Biological Conservation* 145:205–213.

Rodden, M., Rodrigues, F. and Bestelmeyer, S. 2004. Maned wolf (*Chrysocyon brachyurus*). In: C. Sillero-Zubiri, M. Hoffmann, and D.W. Macdonald, eds, *Canids: Foxes, Wolves, Jackals and Dogs*, pp. 38–44. IUCN/SSC Canid Specialist Group, Gland, Switzerland, and Cambridge, UK.

Rodden, M., Rodrigues, F. and Bestelmeyer, S. 2008. *Chrysocyon brachyurus*. In: IUCN 2012. *IUCN Red List of Threatened Species. Version 2012.2*. www.iucnredlist.org. Downloaded on 5 December 2012.

Roemer, G.W. and Wayne, R.K. 2003. Conservation in conflict: The tale of two endangered species. *Conservation Biology* 17:1251–1260.

Roemer, G.W., Fuller, T.K., List, R. and Wayne, R.K. (IUCN SSC Canid Specialist Group—Island Fox Working Group) 2008. *Urocyon littoralis*. In: IUCN 2012. *IUCN Red List of Threatened Species. Version 2012.2*. www.iucnredlist.org. Downloaded on 5 December 2012.

Roemer, G.W., Smith, D.A., Garcelon, D.K. and Wayne, R.K. 2001a. The behavioural ecology of the island fox (*Urocyon littoralis*). *Journal of Zoology London* 255:1–14.

Roemer, G.W., Coonan, T.J., Garcelon, D.K., Bascompte, J. and Laughrin, L. 2001b. Feral pigs facilitate hyperpredation by golden eagles and indirectly cause the decline of the island fox. *Animal Conservation* 4: 307–318.

Rueness, E.K., Asmyhr, M.G., Sillero-Zubiri, C., Macdonald, D.W., Bekele, A., Atickem, A. and Stenseth, N.C. 2011. The cryptic African wolf: *Canis aureus lupaster* is not a golden jackal. *PLoS ONE* 6(1):e16385.

Sharma, D.K., Maldonado, R.E., Jhala, Y.V. and Fleischer, R.C. 2004. Ancient wolf lineages in India. *Proceedings of the Royal Society London Biological Sciences* 271(Suppl. 3):S1–S4.

Sillero-Zubiri, C. 2009. The Canidae. In: D.E. Wilson and R.A. Mittermeier, eds, *The Handbook of the Mammals of the World*, Vol. 1, pp. 352–446. Carnivora, Lynx Edicions, Barcelona.

Sillero-Zubiri, C., Hoffmann, M. and Macdonald, D.W. (eds) 2004a. *Canids: Foxes, Wolves, Jackals and Dogs. Status Survey and Conservation Action Plan*, 2nd edn. IUCN Canid Specialist Group, Gland, Switzerland and Cambridge, UK.

Sillero-Zubiri, C., Reynolds, J. and Novaro, A. 2004b. Management and control of canids near people. In: D.W. Macdonald and C. Sillero-Zubiri, eds, *The Biology and Conservation of Wild Canids*, pp. 107–122. Oxford University Press, Oxford.

Slater, G.J., Thalmann, O., Leonard, J.A., Schweizer, R.M., Koepfli, K.-P., Pollinger, J.P., Rawlence, N.J., Austin, J.J., Cooper, A. and Wayne, R.K. 2009. Evolutionary history of the Falklands wolf. *Current Biology* 19:R937–R938.

Wang, X., Tedford, R.H., Van Valkenburgh, B. and Wayne, R.K. 2004. Ancestry: Evolutionary history, molecular systematics, and evolutionary ecology of Canidae. In: D.W. Macdonald and C. Sillero-Zubiri, eds, *The Biology and Conservation of Wild Canids*. Oxford University Press, Oxford.

Wayne, R.K., George, S.B., Gilbert, D., Collins, P.W., Kovach, S.D., Girman, D. and Lehman, N. 1991. A morphologic and genetic study of the island fox, *Urocyon littoralis. Evolution* 45:1849–1868.

Wayne, R.K., Geffen, E., Girman, D.J., Koepfli, K.P., Lau, L.M. and Marshall, C.R. 1997. Molecular systematics of the Canidae. *Systematic Biology* 46:622–653.

Wilson, P.J., Grewal, S., Lawford, I.D., Heal, J.N.W., Granacki, A.G., Pennock, D., et al. 2000. DNA profiles of the eastern Canadian wolf and the red wolf provide evidence for a common evolutionary history independent of the gray wolf. *Canadian Journal of Zoology* 78:2156–2166.

Woodroffe, R. and Sillero-Zubiri, C. 2012. *Lycaon pictus*. In: *IUCN 2012. IUCN Red List of Threatened Species. Version 2012.2*. www.iucnredlist.org. Downloaded on 5 December 2012.

Woodroffe, R., Cleaveland, S., Courtenay, O., Laurenson, M.K. and Artois, M. 2004. Infectious disease in the management and conservation of wild canids. In: D.W. Macdonald and C. Sillero-Zubiri, eds, *The Biology and Conservation of Wild Canids*, pp. 123–142. Oxford University Press, Oxford.

Wozencraft, W.C. 1989. Classification of the recent Carnivora. In J.L. Gittleman, ed., *Carnivore Behavior, Ecology and Evolution*, pp. 569–593. Cornell University Press, Ithaca, NY.

2 Maned Wolf Population Viability

Rogério Cunha de Paula and Arnaud L.J. Desbiez

CONTENTS

Introduction 15
Vortex Simulation Parameters 16
Baseline Model Parameters 16
Baseline Model Results 19
Deterministic Output 19
Stochastic Baseline Results 19
Sensitivity Testing 19
Demographic Rates 19
Primary Threats 20
Conclusions 25
Specific Models: Brazilian Case Study 25
Baseline Model 25
Initial Population Size 25
Dispersal Rates 26
Mortality Rates 26
Harvest 27
Brazil Baseline Model Results 27
Testing Different Scenarios 27
Final Remarks 32
References 33

INTRODUCTION

The maned wolf (*Chrysocyon brachyurus*) is the largest of the South American canids. The species inhabits the grasslands and scrub forests of central South America. Its present geographic range covers about 5 million km^2 and six countries (Argentina, Bolivia, Brazil, Paraguay, Uruguay, Peru) (Queirolo et al. 2011). The main threat to maned wolf conservation is the drastic reduction of suitable habitats. The seriousness of this situation can be seen when it is considered that much of the species' distribution range has already been converted to farmland. This situation creates other problems. Human development increases mortality of individuals through road kills, killings as a result of conflicts in rural areas, and health problems related to diseases carried from domestic animals (Deem & Emmons 2005). These forces reduce population size, with resultant demographic instability that can lead to local population extinction.

Although there is information available on different aspects of the species, there is still a lack of knowledge on the species' ecology and population dynamics, among other scientific knowledge important to improve their conservation in altered landscapes and even in protected areas (de Paula et al. 2008).

Computer modeling is a valuable and versatile tool for assessing the risk of decline and extinction of wildlife populations (Lacy 1993). Complex and interacting factors that influence population persistence and health can be explored, including natural and anthropogenic causes. Models can also be used to evaluate the effects of alternative management strategies to identify the most effective conservation actions for a population or species and to identify research needs. Such an evaluation of population persistence under current and varying conditions is commonly referred to as a population viability analysis (PVA) (Dietz et al. 2000; Ballou 2006).

PVA has been an extremely useful tool for investigating current and future risk for maned wolf population decline and predictions on altered and protected landscapes. In addition, the need for and consequences of alternative management strategies can be modeled to suggest which practices may be the most effective in managing populations (de Paula et al. 2008).

PVA methodologies such as the Vortex simulation model are not intended to give absolute and precise "answers," since they are projecting the interactions of many randomly fluctuating parameters used as model input and because there is often considerable measurement uncertainty in wildlife population demography data sets and in the estimates of demographic rates. Because of these limitations, some authors have cautioned against the sole use of PVA results to promote specific management actions for threatened populations (Beissinger & McCullough 2002; Ellner et al. 2002; Ludwig 1999; Reed et al. 2002).

Instead, the true value of an analysis of this type lies in the assembly and critical analysis of the available information on the species and its ecology and in the ability to compare the quantitative metrics of population performance that emerge from a suite of simulations, with each simulation representing a specific scenario and its inherent assumptions about the available data and a proposed method of population and/or landscape management. Interpretation of the output depends on our knowledge of the biology and distribution of a species, the environmental conditions and anthropogenic impacts affecting it, and possible future changes in these conditions (Beissinger & McCullough 2002).

Central to this process is the use of Vortex, a computer software simulation model of wildlife population dynamics that performs a risk assessment and provides a tangible focus for quantitative evaluation of conservation options for a species; it is also a vehicle for integrating diverse biological and human sociological data (Lacy 2000). The tool was used here as a mechanism to study the effect of a number of maned wolf life history and population parameters on population viability, to explore which demographic parameters were the most sensitive to data uncertainty or alternative management practices, and to test the effects of selected management scenarios. Threats and management options were then explored for the countries the species inhabits according to its specificities.

VORTEX SIMULATION PARAMETERS

Generally, many of the parameters included in the models are best guesses due to the lack of current data. Although the input data cannot be regarded as accurate, the exercise was valuable in highlighting gaps in knowledge, critical problems, and the main threats and, most importantly, providing a general overview of the species situation and persistence.

One of the main parameters, crucial for the modeling process, is the minimum viable population. A common criterion for what constitutes a minimum viable maned wolf population has proven to be quite elusive. Although the subject has been extensively discussed, for the simulations presented in this chapter we use a consensual definition among maned wolf specialists: a maned wolf population is considered viable if it has a probability of extinction over 100 years equal to, or smaller than, 1% *and* will lose 10% or less of the original genetic variation over the same 100 years (de Paula et al. 2008).

Baseline Model Parameters

Due to the variation of various parameters between countries, regions, and populations, we used a general baseline model that could then be tailored to countries or specific regional populations. The baseline population model was designed to investigate the viability of a nonexistent but biologically accurate maned wolf population. Alternative values for demographic parameters were then explored through sensitivity testing.

The baseline model was developed in the First International Workshop for Maned Wolf Conservation (Serra da Canastra, Brazil, 2005) based on published information and discussions among maned wolf researchers (Rogério Cunha de Paula and Flávio H.G. Rodrigues, Brazil; Lucia Soler, Argentina; Luiz Cartes, Paraguay; Louise Emmons, Bolivia). The following population characteristics and model inputs were accepted as realistic for a free-ranging maned wolf population.

The parameters used are listed in Table 2.1. However, the particularities that define some of them are presented below:

- The population was modeled for 100 years (approximately 15 generations) so that long-term population trends can be observed but not so far into the future that results become too uncertain and selection starts to play a more important role.
- Extinction is defined in the model as no animals of either sex.
- In the baseline model only one population is considered (i.e., no metapopulation dynamics are explored).
- In the baseline model, only one population is considered, with no immigration or emigration.

TABLE 2.1
Summary of Parameter Input Values Used in the Baseline Model

Parameter	Baseline Value
Number of populations	1
Initial population size	100
Carrying capacity	100
Inbreeding depression	3.14 LE
Percentage of the effect of inbreeding due to recessive lethal alleles	50
Breeding system	Long-term monogamy
Age of first reproduction (♀/♂)	2 years
Maximum age of reproduction	10 years
Annual percentage of adult females reproducing (SD)	60 (10)
Density-dependent reproduction?	No
Maximum litter size	5
Distribution of Offspring per Breeding Female per Year (%)	
1 offspring	21
2 offspring	24
3 offspring	28
4 offspring	19
5 offspring	8
Overall offspring sex ratio	50:50
Percentage of adult males in breeding pool	90
Percentage of mortality, age 0–1 (SD)	60 (10)
Percentage of mortality, age 1–2 (SD)	20 (5)
Percentage of mortality, age 2+ (SD)	10 (2)
Catastrophe	None
Harvest	None
Supplementation	None

- The carrying capacity was considered as the same as the initial population ($N_0 = 100$). No environmental variation (EV) was added to the carrying capacity, as variations in population size are accounted for by EV in reproduction and survival.
- Inbreeding is thought to have major effects on reproduction and survival, especially in small populations, and so was included in the model. The impact of inbreeding was modeled as 3.14 lethal equivalents, the median value estimated from an analysis of studbook data for 40 captive mammal populations (Ralls et al. 1988), with 50% of the effect of inbreeding due to recessive lethal alleles. Inbreeding was implemented in the model as reduced first-year survival of inbred individuals.
- EV is the annual variation in reproduction and survival due to random variation in environmental conditions. EV affects not only maned wolves directly but also prey populations and fruit production, which in turn affect maned wolf survival and reproduction. EV for survival and reproduction was linked in the model (i.e., good years for reproduction are also good years for survival).
- Maned wolves form stable, long-lasting, pair bonds, in which an individual mate seems to be only replaced at death (Dietz 1985).
- The age of first offspring is 2 years for both sexes. (Vortex defines reproduction onset as the time the first litter is born rather than sexual maturity. This parameter represents the average age of first reproduction, not the earliest age at which reproduction can occur.)
- The maximum age of reproduction is 10 years (according to the maned wolf studbook). Vortex assumes that animals can reproduce throughout their adult life, and it does not model reproductive senescence.
- The maximum number of progeny per year was estimated at five for the baseline model to produce a more conservative estimate, although in captivity maned wolves give birth to one to seven pups per litter. Field observations show, on average, a much lower number of pups per litter. In the Serra da Canastra region, Brazil, five litters were observed with four pups each and one litter with five pups (R.C. de Paula, personal observations). These were all observed 1–3 weeks after delivery.
- The percentage of females breeding each year was modeled at 60% (mean interbirth interval less than 2 years), with a standard deviation (SD) due to EV of 10%.
- The distribution of offspring produced per breeding female in 1 year was specified using data from the studbook, discussion among specialists, and model refinement.
- The percentage of adult males in the breeding pool was chosen to only include an average of 90% of males reproducing each year. It was thought that this would be a more conservative and therefore appropriate value, in light of the lack of additional field information. Young males might be sexually mature, but because they are still dispersing or have not established their own territory they might not be an effective part of the breeding pool.
- Mortality rates for natural populations are practically unknown. Therefore, different values based on personal observations of specialists were investigated in the model, and final values were chosen based on those that yielded conservative and reasonable deterministic growth rates for a carnivore of the size of a maned wolf. Additionally, we accounted that first-year mortality in captivity is generally reported to be around 50% (Maia & Gouveia 2002; Veado 1997). The values do not take into account mortality due to road kills, which may have a significant impact in wolves ≥1 year old. Road kill mortality was not incorporated into the baseline model, because road kill frequency and impact vary between countries, regions, and populations. Road kill was included under harvesting in sensitivity testing of the general model (as the annual removal of a fixed number of wolves) and in the country-specific models (as the annual removal of a percentage of wolves). The mortality rates do take into account subadult mortality due to natal dispersal. The relative effect of mortality values on overall population demographic rates was explored through sensitivity analysis.

- No harvest was included in the baseline model. Causes and intensity of harvest varied between countries, regions, and populations. Values for this parameter were explored through sensitivity testing and were explored for each country and for populations within each country.

Baseline Model Results

Results reported for each Vortex modeling scenario include (Miller & Lacy 2003):

r_{stoch} (SD): The mean rate of stochastic population growth or decline and standard deviation, demonstrated by the simulated populations, averaged across years and iterations, for all simulated populations. This population growth rate is calculated each year of the simulation, prior to any truncation of the population size due to the population exceeding the carrying capacity.

$P(E)_{100}$: Probability that the population will go extinct. Extinction is defined in the model as no animals of either sex remaining. $P(E)_{100}$ is determined by the proportion of 500 iterations within a given scenario that go extinct within 100 years.

MTE: Mean time to population extinction, in years, over a 100-year period.

N_{100} (SD): Mean (standard deviation) population size at the end of the simulation, averaged across all simulated populations, including those that go extinct.

GD_{100}: The gene diversity or expected heterozygosity of the extant populations, expressed as a percent of the initial gene diversity of the population. Fitness of individuals usually declines proportionately with gene diversity.

Deterministic Output

The demographic rates (reproduction and mortality) included in the baseline model can be used to calculate deterministic characteristics of the model population. These values reflect the biology of the population in the absence of stochastic fluctuations (both demographic and environmental variation), inbreeding depression, limitation of mates, and immigration/emigration. The baseline model results in a deterministic growth rate (r_{det}) of 0.091 ($\lambda = 1.096$). This represents an annual potential growth rate of about 10%. Generation time (the average age of reproduction) is 5.3 years for both males and females. Adult sex ratio of adult males to adult females is 1. It is suggested that maned wolf populations have the potential to grow quickly in the absence of additional threats or stochastic events.

Stochastic Baseline Results

It is important to use caution in interpreting the results described below, which represent a general maned wolf population based on the parameters previously described. No harvest rates, no increase in mortality due to road kill, and no catastrophes are included in this general model.

Results of the baseline model project that a population of 100 maned wolves is likely to persist over the next 100 years. The stochastic growth rate (r_{stoch}) is 0.062, enabling the population to grow when below carrying capacity. There is zero probability of extinction in 100 years, and the mean population size at 100 years is 95 wolves with 81.8% gene diversity remaining. The loss of gene diversity is in part due to the relatively small population size and because we modeled a closed population in which no immigration of unrelated animals can occur.

Sensitivity Testing

Demographic Rates

Sensitivity analysis is a tool used to evaluate the robustness of a model to variations in parameter values. The more robust the model is to variations in a particular parameter, the less sensitive the model's results are to the input values of that parameter. This tool is used, in the current context, to uncover particularly sensitive parameters that could significantly alter the results and conclusions derived from the model. Here, sensitivity analyses were performed to evaluate the effect of model parameters on the stochastic growth rate (r_{stoch}) of maned wolf populations.

The analyses consisted of varying one demographic parameter at a time to a value lower or greater than what was assigned as the baseline value. Parameters and values used are outlined in Table 2.2.

Analyses on the sensitiveness of parameters in the models were performed. The three most sensitive parameters in the model are female mortality (in the order FM3, FM1, and FM2), with a 50% increase in mortality driving the stochastic growth rate close to or below 0. Changes in the percentage of females breeding each year and in the maximum age of reproduction had the next largest effect on stochastic growth rate, although population growth remained positive across the values tested. However, male mortality in general seems to be very robust, showing a much smaller variation in r_{stoch} over the same range of input values used for female mortality. In particular, decreases in male mortality did not have the same significant increases in r_{stoch} as were observed for female mortality.

At first glance, the sensitivity results for age of first reproduction in females might seem surprising, with earlier reproduction leading to a slightly lower growth rate relative to the baseline. However, this alteration might be a reflex of Vortex itself and not necessarily biologically relevant. Lowering the age of first reproduction alters the stable age class distribution calculated by Vortex, increasing the number of individuals in the first age class relative to the baseline. Since this age class also suffers the highest mortality rate, a slight (nonsignificant) decrease in r_{stoch} was observed.

These results indicate the parameters to which the model is most sensitive across plausible values. In some cases, the country models explored even higher levels of mortality. Research efforts to better estimate the values of the most sensitive parameters would strengthen the ability to project the future viability of maned wolf populations. Also, management actions that can improve female mortality and the percentage of females breeding (e.g., adequate food resources for good health, adequate denning sites free from disturbance, underpasses for secure road crossings) would benefit the population.

Primary Threats

Maned wolf populations are spread out across an extensive area of South America, encompassing six countries. Because of this, a more general approach was chosen to model threats, rather than

TABLE 2.2
Parameter Values Used for the Sensitivity Analyses

Parameter	Low	Baseline	High
Juvenile male mortality (MM1) (%)	30	60	70
Subadult male mortality (MM2) (%)	10	20	40
Adult male mortality (MM3) (%)	5	10	20
Juvenile female mortality (FM1) (%)	30	60	70
Subadult female mortality (FM2) (%)	10	20	40
Adult female mortality (FM3) (%)	5	10	20
Percentage of adult females breeding (FBR) (%)	50	60	70
Age of first reproduction for males (A1OM)	1	2	4
Age of first reproduction for females (A1OF)	1	2	4
Percentage of males in the breeding pool (MBP) (%)	75	90	100
Maximum age of reproduction (MAR)	8	10	12
Inbreeding (lethal equivalents) (INB)	0	3.14	6.28

modeling each known population. As such, each threat was first modeled separately across different levels of threats and different initial population sizes to evaluate the effect of each threat on its own across a possible range of maned wolf population sizes. Subsequently, we modeled both threats simultaneously by varying both threats in a factorial design, but keeping a constant initial population size of 100, to gauge the effect of both threats acting together on a midsize maned wolf population. This produced results for general maned wolf populations that may guide maned wolf managers in estimating the level of threat for a specific population.

Five different measures of population status were examined: stochastic growth rate (r_{stoch}); probability of extinction over 100 years ($P(E)_{100}$); mean population size after 100 years (N_{100}); average retention of gene diversity (GD); and average time to extinction (MTE). The results are summarized in Figure 2.1.

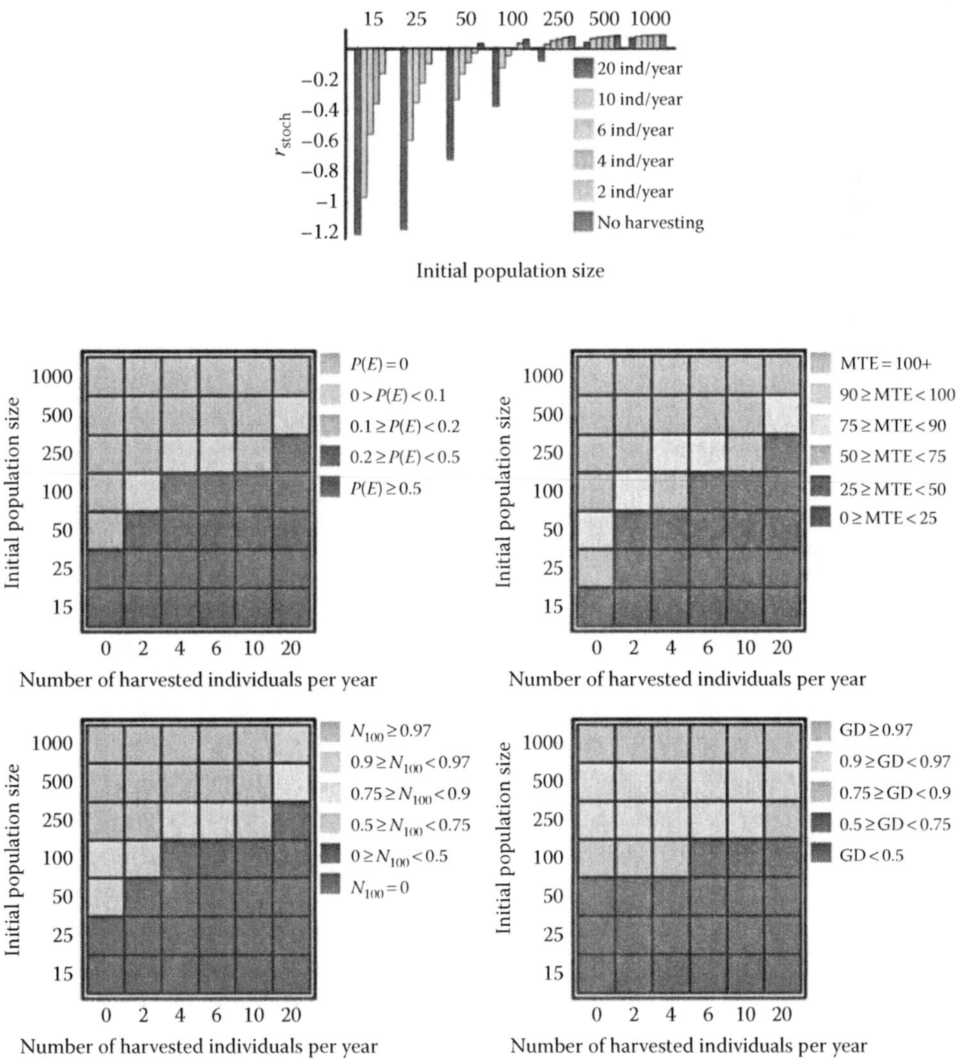

FIGURE 2.1 Summary of results obtained from modeling different harvesting levels against populations with different sizes. First row (bar graph): N_{init} versus r_{stoch}; second row (top two matrices): N_{init} versus $P(E)_{100}$; N_{init} versus MTE; third row (bottom two matrices): N_{init} versus N_{100} shown as the proportion N_{100}/N_{init}; N_{init} versus GD.

Considering important threats to maned wolf populations, we modeled two of them for the general analysis. Threats and modeling conditions are as follows:

1. *Deaths of wolves due to road kill, poaching, and shooting from conflicts.* All of these categories fit under the harvest module within Vortex, which allows for a specific number of males and/or females to be removed per year over a determined number of years. Here, we modeled different numbers of males and females being removed annually over the entire period modeled of 100 years.
2. *Habitat loss due to land conversion and human encroachment.* The proximate result of habitat loss is a decrease in the carrying capacity of the environment. As such, we chose to model this threat as a linear decrease in carrying capacity over 100 years.

Harvesting versus Initial Population Size

The effect of harvesting on maned wolf populations of different sizes was evaluated (N_{init} = 15–1,000 individuals). The objective was to gain insight into the relative effect of different harvesting levels on different populations. In total, six different levels of harvesting were tested: 0 (i.e., no harvesting), 2, 4, 6, 10, and 20 adult wolves removed from the population per year (equal number of females and males removed).

As expected, small populations are more vulnerable to harvesting. Because harvest is modeled as a fixed number of wolves removed, the same harvest level represents a greater proportion of smaller populations. For example, removing 20 individuals represents 80% of a population of 25, but only 2% of a population of 1000. The greater the proportion of the population that is removed each year, the greater the impact of harvesting on the population. With a deterministic annual growth rate of only 9%, populations cannot biologically sustain harvest at this level or greater, even in the absence of stochastic effects. Therefore, even large populations can suffer from harvesting if too many animals are harvested. Furthermore, as seen in the sensitivity analysis, maned wolf growth rates seem to be more sensitive to female mortality than to male mortality, implying that harvesting can be more or less pervasive depending on which animals are removed.

The smallest population modeled (N_{init} = 25) showed negative stochastic growth rates at any level of harvesting; even with no harvest, a population of 25 has a high probability of extinction, with a small population size and low level of gene diversity for those populations that do persist. Only populations with 500 or more individuals display positive growth rates at all harvest levels tested, which is mirrored with positive results in all other population measures examined. Populations of 100 appear to be most sensitive to the different harvest rates tested (removal of 2%–20% of N_{init}); the loss of only a couple of additional wolves each year can have a large effect on population viability.

Loss of Habitat versus Initial Population Size

In the previous analysis, carrying capacity remained constant ($K = N_{init}$), meaning that populations were not limited by habitat resources but by the removal of breeders from the population, affecting the potential of the population to grow to *K*. In this analysis, the effect of habitat loss (permanent reduction in *K*) was investigated for populations of different sizes (15–1000 wolves). Three different levels of habitat loss were modeled: 5%, 25%, and 50% loss over 100 years. This was modeled as a linear decrease in carrying capacity over time (i.e., a total of 5% of the carrying capacity was lost over a 100-year period). This effectively reduced the maximum size that a population could attain, meaning that in the most extreme example of 50% habitat loss, maximum final population sizes were 7, 12, 25, 50, 125, 250, and 500 wolves. All scenarios were run for 100 years, with 500 iterations of each. The same population measures were evaluated, and the results are summarized in Figure 2.2.

Again, small populations are affected more than large populations are. Populations of N_{init} of 25 or fewer wolves show negative growth rates due to their vulnerability to stochastic processes. Although populations of N_{init} = 50–100 wolves have a relatively low risk of extinction, they lose a

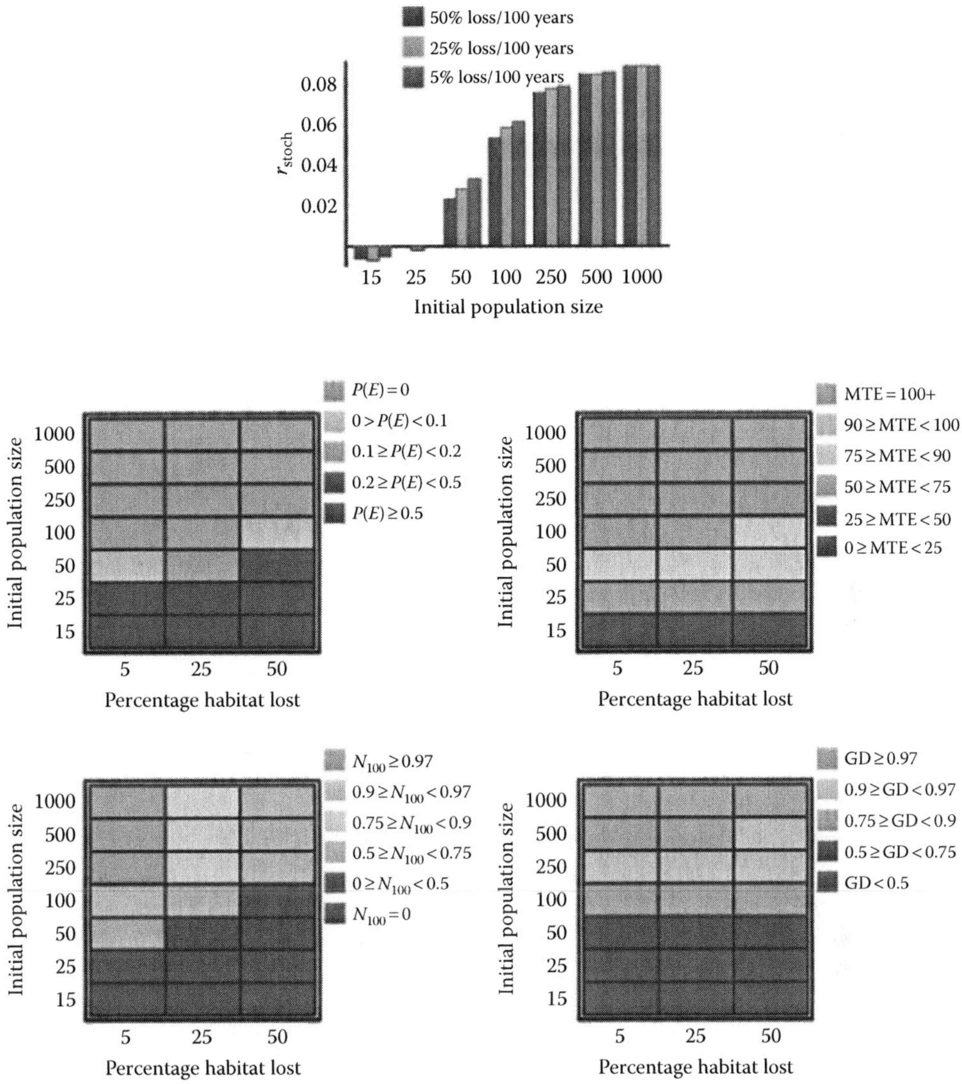

FIGURE 2.2 Summary of results obtained from modeling different levels of habitat loss against populations with different sizes. First row (bar graph): N_{init} versus r_{stoch}; second row (top two matrices): N_{init} versus $P(E)_{100}$; N_{init} versus MTE; third row (bottom two matrices): N_{init} versus N_{100} shown as the proportion N_{100}/N_{init}; N_{init} versus GD.

significant amount of genetic variation. These scenarios represent cases in which final K is 25–95 wolves, suggesting that the long-term genetic health of populations below 100 wolves may be poor. Initial populations of 250–1000 wolves that decline to a K of 125–950 wolves show better viability over 100 years in the absence of additional losses (harvest) due to road kill, poaching, and other sources of removal. As discussed earlier, these additional threats reduce the viability of maned wolf populations, particularly populations with $K < 500$ wolves.

Harvesting versus Habitat Loss

Throughout the species' range, populations are suffering some form of harvest and habitat loss at the same time. Earlier analyses suggest that populations of around 100 wolves may be the most sensitive to such threats. To assess the synergistic effect of both threats, a factorial design was used to model the interaction of different levels of harvest and habitat loss, while keeping initial population size (N_{init}) and K constant at 100 wolves. Five different levels of harvesting (2, 4, 6, 10, and 20 individuals/year) and three different levels of habitat loss (5%, 25%, and 50%) were analyzed (for a total of 15 scenarios). All models were run for 100 years, with 500 iterations each. The results are summarized in Figure 2.3.

The loss of habitat modeled results in final carrying capacities (maximum population sizes) of 95, 75, and 50 wolves, respectively. It is not surprising, therefore, that the results of this analysis are similar to those summarized in Figure 2.2 for population sizes of 50–100 and harvest >0. Harvest has an overriding impact over habitat loss; this means that varying maximum population size between 50 and 95 wolves has less impact on population viability than varying annual harvest tenfold.

Populations of about 50–100 wolves cannot withstand the harvest of more than two adults each year, as this results in a negative stochastic growth rate and relatively high probability of extinction within 100 years. Even with the lowest level of harvest, substantial genetic diversity is lost. The

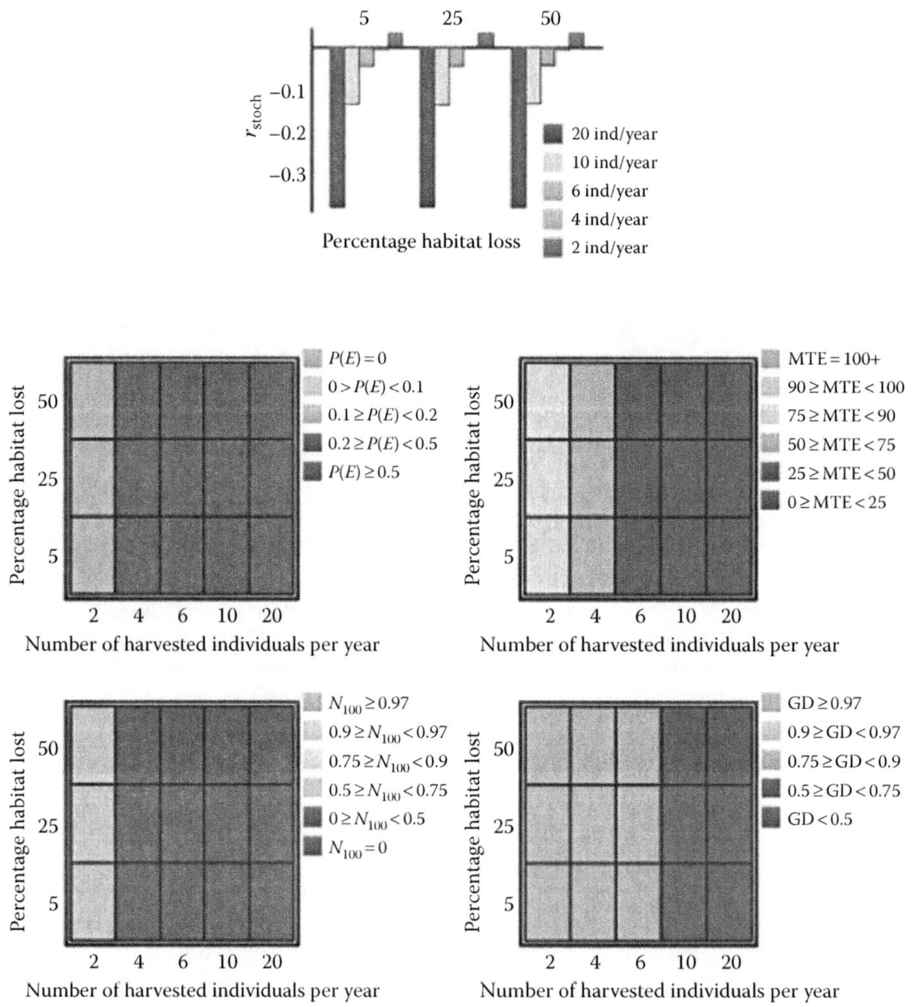

FIGURE 2.3 Summary of results obtained from modeling different levels of harvesting against different levels of habitat loss. First row (bar graph): N_{init} versus r_{stoch}; second row (top two matrices): N_{init} versus $P(E)_{100}$; N_{init} versus MTE; third row (bottom two matrices): N_{init} versus N_{100} shown as the proportion N_{100}/N_{init}; N_{init} versus GD.

viability of populations of 100 wolves or fewer is poor over all harvest and habitat loss levels modeled here. Larger populations are more likely to be able to withstand such threats and may yield different results.

Conclusions

As the effects of harvest and habitat loss on maned wolf populations have been explored, it is observed that populations of 25 or fewer decline and have a high probability of extinction within 100 years, and populations of 50–100 may persist but will retain low levels of gene diversity. Several hundred wolves may be needed to maintain a long-term, viable population, depending upon the severity of threats and the management goals and acceptable level of risk for wildlife managers.

These results may serve as a guide to the relative effects of each threat to different-sized populations. Furthermore, the results have been displayed in a way that might facilitate a maned wolf conservationist to identify the relative viability of a specific wolf population, based on specific viability criteria and management goals.

SPECIFIC MODELS: BRAZILIAN CASE STUDY

Considering the baseline model and data from known populations, models that are more specific can be generated. This modeling approach might help stimulate an in-depth regional analysis of current maned wolf knowledge. Unfortunately, this approach is more general and provides only a limited assessment of the population viability of specific populations. For specific populations, a combination of a country model (general information on metapopulation) and the sensitivity testing analysis may shed some light on probable future viability.

Baseline Model

A baseline model for all Brazilian maned wolf populations was developed during the First International Workshop for Maned Wolf Conservation (Serra da Canastra, Brazil, 2005). Parameter input values entered into the Brazilian baseline model are presented in Table 2.3, and details for other parameters are given next.

For modeling purposes, populations were separated into nine main populations, even though, as shown in Figure 2.4, populations of maned wolves can occur outside these areas.

Initial Population Size

Maned wolf populations in the cerrado biome were defined following the main natural areas, based on the remaining fragments. Nine main populations of maned wolves were identified. The populations from São Paulo State or from the south of Brazil were considered isolated and very small; therefore we did not include them in this modeling exercise. Also, we are not including animals dispersed throughout the distribution range. This choice was made for practical reasons and does not mean those populations or individuals are not important, but including them would result in unfeasible models. The viability of these smaller populations can be assessed based on results from the initial baseline model and the sensitivity analysis.

To estimate initial population size, we first calculated the areas of polygons drawn on available cerrado habitats to maned wolves based on the cerrado's remaining vegetation map (Machado et al. 2004).

Density estimates for each area were based on the literature and the experience of several researchers, and were used to calculate a "best-guess" population estimate for maned wolves in Brazil. Alternative estimates of population size of maned wolves were made by using a general high-density estimate of 0.1 individuals/km^2 and a general low population density of 0.02 individuals/km^2 to calculate high and low population estimates, respectively.

TABLE 2.3
Parameters Entered in the Model

Parameter	Baseline	Brazil
Number of populations	1	Changed[a]
Initial population size	100	21,745[a]
Carrying capacity	100	21,745[a]
Dispersal among populations	None	Yes[a]
Inbreeding depression? (%)	3.14 LE	3.14 LE
Percentage of the inbreeding effects due to recessive lethal alleles	50	100
Breeding system	Long-term monogamy	Long-term monogamy
Age of first reproduction (♀/♂)	2 years	2 years
Maximum age of reproduction	10 years	10 years
Annual percentage of adult females reproducing (SD)	60 (10)	60 (10)
Density-dependent reproduction?	No	No
Maximum litter size	5	5
Distribution of Offspring per Female per Year (%)		
1 offspring	21	21
2 offspring	24	24
3 offspring	28	28
4 offspring	19	19
5 offspring	8	8
Overall offspring sex ratio	50:50	50:50
Percentage of adult males in breeding pool	90	90
Percentage of mortality, age 0–1 (SD)	60 (10)	60 (10)
Percentage of mortality, age 1–2 (SD)	20 (5)	Changed[a]
Percentage of mortality, age 2+ (SD)	10 (2)	10 (2)
Catastrophe	None	None
Harvest	None	Yes[a]
Supplementation	None	None

[a] Parameters have been modeled for different values from baseline model.

In the following models, the best-guess population estimates were used unless specified otherwise. Populations 6 and 6a are considered to be one population, but were separated into two areas to obtain initial population figures since the density of maned wolves in these areas were different.

The area of occupation for each population, the densities, and the results of the population estimates are presented in Table 2.4. The population map is presented in Figure 2.4. Because the initial population size relies on habitat availability and density estimates, the carrying capacity was assumed to be equal to the initial population.

Dispersal Rates

It was estimated that there is connectivity between the different populations shown in Figure 2.4. It was considered that some wolves could disperse between populations, as shown in Table 2.5. In the model, both subadult males and subadult females could disperse and had a 50% survival rate during dispersal.

Mortality Rates

Subadult mortality due to road kill increased the estimated mortality rates from those used in the baseline model. The impact of road kill varied between populations and is presented in Table 2.6.

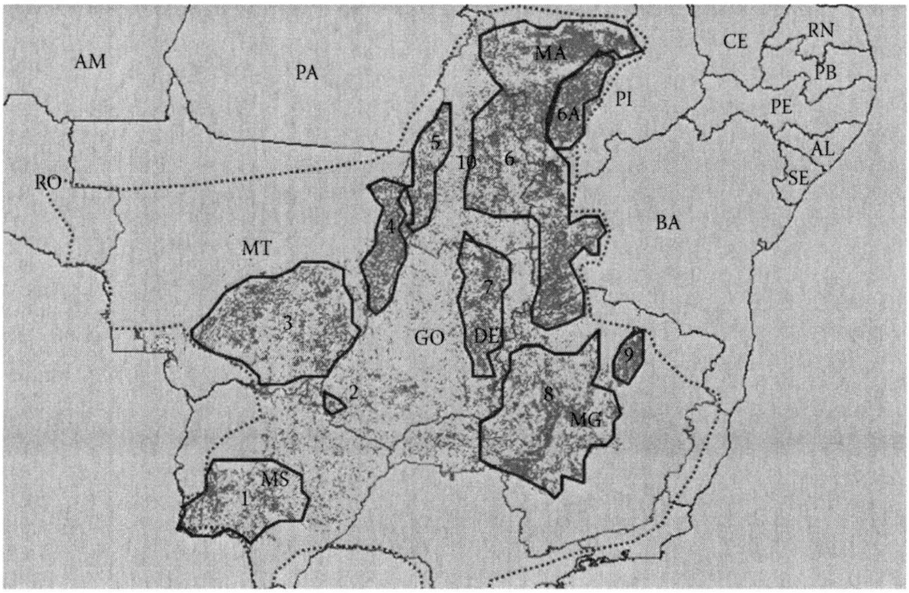

FIGURE 2.4 Map of the distribution of the main populations of maned wolves in Brazil.

Harvest

Although there are no incipient data on direct killing as retaliation for livestock depredation or cultural beliefs, this might be an important threat in some areas, and thus, we model based on field information from Serra da Canastra region (Minas Gerais state) and the Paraiba Valley (eastern Sao Paulo state). Therefore, harvest was modeled as 2% of the subadults and adults (male and female) removed each year for 100 years from each population and 5% for Populations 8 and 9. Persecution is assessed only through anecdotal reports. This may be an important variable for population reduction, but will vary within regions and populations, being insignificant in many areas.

Brazil Baseline Model Results

The mean stochastic rate of population growth (or decline), the probability of population extinction, the mean population size after 100 years, and the remaining gene diversity are presented for the three different scenarios in Table 2.7.

The estimated initial population size of maned wolves was 21,746. After 100 years, model projections indicate that, on average, 18,168 maned wolves still survive. However, this does not include factors such as habitat loss that were modeled separately or any removals (road kills or hunting), other catastrophes, or diseases.

Testing Different Scenarios

Defining Initial Population Estimates

Estimating the initial population of maned wolves in Brazil was difficult. Scenarios were developed based on the best-guess population estimates, as well as the lowest and highest population estimates (Table 2.4). The probability of population extinction, the mean population size at 100 years, and the population reduction are given in Table 2.8.

These results show how different estimates in initial population size can affect the projected number of maned wolves remaining in 100 years, since there are six times more maned wolves in the high population estimate after 100 years than in the low population estimate (Table 2.8).

TABLE 2.4
Population Estimates for Maned Wolves in Brazil

	Population										
	1	2	3	4	5	6	6a	7	8	9	Metapopulation
Area available (in km^2)	26,134	1,268	44,112	35,289	27,088	133,180	35,364	27,167	12,880	51,519	394,000
Density estimate (best guess)	0.03	0.10	0.04	0.04	0.06	0.07	0.03	0.08	0.03	0.06	
Population estimate (best guess)	784	127	1,764	1,412	1,625	9,323	1,061	2,173	386	3,091	21,746
Population estimate (high estimate)	2,613	127	4,411	3,529	2,709	13,318	3,536	2,717	1,288	5,152	39,400
Population estimate (low estimate)	523	25	882	706	542	2,664	707	543	258	1,030	7,880

TABLE 2.5
Dispersal Estimates among Populations of Maned Wolves in Brazil in a Given Year

	Population 1	Population 2	Population 3	Population 4	Population 5	Population 6	Population 7	Population 8	Population 9
Population 1	99	1	0	0	0	0	0	0	0
Population 2	1	98	1	0	0	0	0	0	0
Population 3	0	1	98	1	0	0	0	0	0
Population 4	0	0	1	99	0	0	0	0	0
Population 5	0	0	0	0	100	0	0	0	0
Population 6	0	0	0	0	0	98	1	0	1
Population 7	0	0	0	0	0	1	98	0	1
Population 8	0	0	0	0	0	0	0	99	1
Population 9	0	0	0	0	0	1	1	1	97

TABLE 2.6
Mortality Rates of Maned Wolves from Different Populations in Brazil

Mortality (%)	Baseline Single Population	Population 1	Population 2	Population 3	Population 4	Population 5	Population 6	Population 7	Population 8	Population 9
Age 0–1 (SD)	60 (10)	60 (10)	60 (10)	60 (10)	60 (10)	60 (10)	60 (10)	60 (10)	60 (10)	60 (10)
Age 1–2 (SD)	20 (5)	40 (5)	40 (5)	40 (5)	24 (5)	40 (5)	36 (5)	40 (5)	24 (5)	40 (5)
Age 2+ (SD)	10 (2)	10 (2)	10 (2)	10 (2)	10 (2)	10 (2)	10 (2)	10 (2)	10 (2)	10 (2)

TABLE 2.7
Results of the Baseline Model after 100 Years for Maned Wolf Populations in Brazil

Brazil Baseline	N_{init}	r_{stoch}	SD(r_{stoch})	$P(E)_{100}$	MTE	N_{100}	SD(N_{100})	GD_{100}
Population 1	784	0.011	0.131	0	—	534	203.1	0.963
Population 2	127	0.017	0.140	0	—	98	26.0	0.947
Population 3	1,764	0.013	0.130	0	—	1,328	400.0	0.986
Population 4	1,412	0.065	0.123	0	—	1,349	106.5	0.987
Population 5	1,625	0.016	0.131	0	—	1,224	353.7	0.984
Population 6	10,383	0.027	0.127	0	—	8,985	1,646.3	0.998
Population 7	2,173	0.020	0.130	0	—	1,801	381.7	0.994
Population 8	386	0.066	0.124	0	—	370	28.9	0.971
Population 9	3,091	0.017	0.129	0	—	2,481	568.6	0.995
Metapopulation	21,745	0.030	0.088	0	—	18,168	2,329.0	0.999

Although in all scenarios no maned wolf population extirpates, it is very important to obtain more information on the species demography to evaluate how the decline affects smaller or isolated populations. We noticed that the tendency of highest population decline (35% in 100 years) is exactly on the minimum population estimate (Table 2.8).

Elimination of Road Kill and Harvest Variables

Additionally, we simulate the population viability with some threat variables removed, that is, when

1. There is harvest and road kill as estimated in the Brazil baseline model
2. There is no harvest, but there is road kill
3. There is no road kill, but there is harvest

The model confirmed what was previously predicted: road kill is generally of higher conservation concern than harvest in Brazil. Tables 2.9 and 2.10 summarizes some model results. Mean population size is highest after elimination of road kill. However, eliminating this source of mortality may be much more difficult than campaigning to decrease harvest, which is a factor specific to

TABLE 2.8
Impact of Initial Population Size Estimates on Mean Population Size and Probability of Survival over 100 Years

	$P(E)_{100}$	N_{init}	N_{100}	Population Reduction (%)
Minimum population estimate	0	7,880	5,152	35
Best-guess population estimate	0	21,746	18,168	16
Maximum population estimate	0	39,400	30,983	21

TABLE 2.9
Impact of Harvest and Road Kill on Mean Population Size and Probability of Survival

	$P(E)_{100}$	N_{init}	N_{100}	Population Reduction (%)
Brazilian baseline model	0	21,746	18,168	16
No harvest	0	21,746	19,918	8
No road kill	0	21,746	21,039	3

TABLE 2.10
Impact of Road Kill on Mean Population Size and Probability of Extinction

	$P(E)_{100}$	N_{init}	N_{100}	Population Reduction (%)
High road kill estimates	0.002	21,746	893	96
Current road kill estimates	0	21,746	18,168	16
No road kill	0	21,746	21,039	3

certain areas. It must also be stressed that the harvest estimates were considered low and in some regions may be much higher. However, these results encouraged us to analyze both impacts—of hunting and road kill—more closely.

Impact of Road Kill

As habitats become more fragmented and density of roads increases, the mortality of maned wolves due to road kill will likely also increase. We therefore tested what would happen if we increased the mortality of subadult maned wolves by 20% in the different populations. This 20% increase is due to road kill of dispersing subadults.

Considering the model outputs, it can be concluded that as road density increases and more maned wolves are killed, the higher the risk of extinction of local maned wolf populations. The mean population size of maned wolves greatly decreased when subadult mortality due to road kill increased. The Brazilian metapopulation has a very low probability of extinction in 100 years, even with increased road kill, but this level of increased mortality will cause individual populations throughout Brazil to go extinct and dramatically reduce the remaining number of wolves by 96% (Table 2.11). When looking at the specific populations, the model indicates that five of nine populations have a 100% reduction in number of individuals after 100 years of road kills at that rate (Table 2.11). It is indeed critical, but we need to consider that the high road kill estimates assume a highly complex road network, which is most extensive in central and southeastern Brazil. This is exactly where there are large overlaps between the road system and viable maned wolf habitats.

Habitat Loss

The Brazilian cerrado, the main area of maned wolf distribution, is one of the most endangered biomes on Earth (Myers et al. 2000), and habitat loss continues every year. The rates of habitat loss in the cerrado vary year-by-year and according to the human ecomomic activities in defined areas. We calculated three possibilities of habitat reduction based on reports and predictions (Table 2.12). The impact of habitat loss rates (Table 2.13) over the next 30 years was modeled to evaluate the potential effect on maned wolf populations.

Predictably, habitat loss had a high impact on maned wolf populations, with the mean number of maned wolves decreasing proportionally to the amount of habitat loss, as shown in Table 2.13.

Considering the information given by the model and the generation period (7 years), we found that the species will be subjected to a population decrease of at least 29% in the next 21 years (three generations). Considering that we use a general deforestation rate (for the best guess, with the

TABLE 2.11
Impact of Road Kill on Mean Population Size and Probability of Extinction of the Different Maned Wolf Populations

	$P(E)_{100}$	N_{init}	N_{100}	Population Reduction (%)
Population 1	0.984	784	0	100
Population 2	0.994	127	0	100
Population 3	0.358	1,764	9	99
Population 4	0.010	1,412	597	58
Population 5	0.886	1,625	2	100
Population 6	0.100	10,383	158	98
Population 7	0.574	2,173	8	100
Population 8	0.106	386	107	72
Population 9	0.472	3,091	10	100
Metapopulation	0.020	21,746	891	96

TABLE 2.12
Habitat Loss Estimates for Maned Wolves in Brazil

Population	Low Habitat Loss (%)	Best Guess (%)	High Habitat Loss (%)
Population 1	0.7	1	1.4
Population 2	None	None	None
Population 3	1	1.5	2
Population 4	0.5	1	0.5
Population 5	1	1	2
Population 6	0.8	1	1.6
Population 7	0.7	1	1.4
Population 8	1	1.5	2
Population 9	0.7	1	1.4

TABLE 2.13
Impact of Habitat Loss on Mean Population Size and Probability of Survival

	$P(E)_{100}$	N_{init}	N_{100}	Population Reduction (%)
No habitat loss	0	21,746	18,168	16
Low habitat loss	0	21,746	13,803	37
Best guess	0	21,746	12,299	43
High habitat loss	0	21,746	9,573	56

exception of Populations 3 and 8, which were higher) and that the official deforestation information for the period 2002–2008 is 1.34% per year, the population decrease would be even higher, which is critical. When observing the results of a wolf population reduction of 56% after 100 years of high habitat loss (considering that agricultural activities and their expansion in Brazil are located and predicted to be exactly in the areas of pristine wolf habitats), we suggest that the model indication is an underestimate and the scenario might be even worse for the maned wolf in the cerrado biome.

FINAL REMARKS

The purpose of this exercise was to examine the viability of maned wolves in Brazil as a whole and therefore build a baseline model that best reflects this. We did not intend to model specific populations but to observe the tendencies of a country metapopulation. Therefore parameters had to be an average best guess for the population. Models for individual, well-studied populations can and should be adapted from the baseline model.

It appeared from this exercise that the main current threats to maned wolves in Brazil are increased mortality due to road kill and reduction in carrying capacity due to habitat loss. Facing important factors when looking specifically at population level, hunting due to mainly livestock losses and the unassessed impact of diseases from domestic dogs are not being considered here since they are problems in specific areas. Although the model results indicate that these threats are not likely to lead to the extinction of maned wolves from Brazil, given the estimated mortality rates modeled, these four threat variables independently may lead to significantly fewer wolves in certain areas. Acting together, they can cause more harm to the populations, perhaps even leading to local extinction in some regions of the country.

REFERENCES

Ballou, J.D. (2006). Population viability analysis and conservation planning. In: Morato, R.G., Rodrigues, F.H.G., Eizirik, E., Mangini, P.R., Azevedo, F.C.C. and Marinho-Filho, J. (eds), *Manejo e conservação de carnívoros neotropicais*. Edições Ibama, Brasília.

Beissinger, S.R. and McCullough, D.R. (2002) *Population Viability Analysis*. Chicago University Press, Chicago.

Deem, S.L. and Emmons, L.H. (2005) Exposure of free-ranging maned wolves (*Chrysocyon brachyurus*) to infectious and parasitic disease agents in the Noel Kempff Mercado National Park, Bolivia. *Journal of Zoo and Wildlife Medicine*, **36**, 198–197.

Dietz, J.M. (1985) *Chrysocyon brachyurus*. *Mammalian Species*, **234**, 1–4.

Dietz, J.M., Baker, A.J. and Ballou, J.D. (2000). Demographic evidence of inbreeding depression in wild golden lion tamarins. In: Young, A.G. and Clarke, G.M. (eds), *Genetics, Demography and the Viability of Fragmented Populations*. Cambridge University Press, Cambridge.

Ellner, S.P., Fieberg, J., Ludwig, D. and Wilcox, C. (2002) Precision of population viability analysis. *Conservation Biology*, **16**, 258–261.

Lacy, R.C. (1993) Vortex: A computer simulation model for population viability analysis. *Wildlife Research*, **20**, 45–65.

Lacy, R.C. (2000) Structure of the Vortex simulation model for population viability analysis. *Ecological Bulletins*, **48**, 191–203.

Ludwig, D. (1999) Is it meaningful to estimate a probability of extinction? *Ecology*, **80**, 298–310.

Machado, R.B., Ramos Neto, M.B., Pereira, P.G.P., Caldas, E.F., Gonçalves, D.A., Santos, N.S., Tabor, K. and Steininger, M. (2004). Estimativas de perda da área do Cerrado brasileiro. Relatório técnico. Conservação Internacional, Brasília.

Maia, O.B. and Gouveia, A.M.G. (2002) Birth and mortality of maned wolves *Chrysocyon brachyurus* (Illiger, 1811) in captivity. *Brazilian Journal of Biology*, **62**, 25–32.

Miller, P.S. and Lacy, R.C. (2003). *Vortex: A Stochastic Simulation of the Extinction Process*. Version 9 User's Manual. Conservation Breeding Specialist Group (SSC/IUCN), Apple Valley, MN.

Myers, N., Mittermeier, R.A., Mittermeier, C.G., Fonseca, G.A.B. and Kent, J. (2000). Biodiversity hotspots for conservation priorities. *Nature*, **403**, 853–858.

de Paula, R.C., Médici, P. and Morato, R.G. (org.) (2008). *Plano de ação para a conservação do lobo-guará: Análise de viabilidade populacional e de habitat*. Edições IBAMA, Brasília.

Queirolo, D., Moreira, J.R., Soler, L., Emmons, L.H., Rodrigues, F.H.G., Pautasso, A.A., Cartes, J.L. and Salvatori, V. (2011). Historical and current range of the Near Threatened maned wolf (*Chrysocyon brachyurus*) in South America. *Oryx*, **45**(2), 296–303.

Ralls, K., Ballou, J.D. and Templeton, A.R. (1988) Estimates of lethal equivalents and the cost of inbreeding in mammals. *Conservation Biology*, **10**, 769–775.

Reed, J.M., Mills, L.S., Dunning, J.B., Menges, E.S., McKelvey, K.S., Frye, R., Beissinger, S.R., Anstett, M. and Miller, P.S. (2002) Emerging issues in population viability analysis. *Conservation Biology*, **16**, 7–19.

Veado, B.V. (1997) Parental behaviour in maned wolf *Chrysocyon brachyurus* at BeloHorizonte Zoo. *International Zoo Year Book*, **35**, 279–286.

3 Relationships between the Maned Wolf and People

Adriana Consorte-McCrea

CONTENTS

Introduction....35
Methodology and Study Areas....36
Results and Discussion....37
Human Encroachment and Habitat Protection....37
People in the Research Areas....38
The Value of the Maned Wolf....38
People and Carnivores' Multifaceted Relationship: The Maned Wolf within the Universe of Local People....39
Relationship between Human-Occupied Landscapes and the Maned Wolf....42
People's Livelihoods....42
Conflict and Discourse: Changes in Relationship between Local People and Nature over Time....43
Anthropogenic Causes of Maned Wolf Mortality....44
Environmental Awareness and People in the Vicinity of Conservation Areas....44
Conflicts between Rural and Urban Interests....45
Human Dimensions: Local People and Support for the Conservation of the Maned Wolf....46
Next Steps in Research and Conservation....47
Toward Coexistence and Mitigation of Conflict....47
Clarifying Misconceptions and Addressing Negative Attitudes....47
Attitudes toward Other Interest Groups....48
Acknowledgments....49
References....49

INTRODUCTION

The maned wolf is a large carnivore, the largest one in South America. Although it was described as a wolf by early agents of the Portuguese crown, it is monophyletic and therefore difficult to mistake as any other species. The fact that the maned wolf is unique and distinctive is important if its image is to represent the cerrado biome and all of its dwindling biodiversity. Do people's relationships with the maned wolf make it charismatic and likable enough to earn the badge of flagship species for cerrado conservation?

The validity of the one-species approach to conservation has been questioned in favor of a focus on ecosystems. However, the single species has its merits when the species in question "plays the role of keystone or umbrella species" (Boitani et al. 2004; McNeely 2000; Gittleman et al. 2001; Sergio et al. 2006). The maned wolf is, arguably, a charismatic carnivore with a wide home range, qualifying it to fit both roles. Between 2005 and 2008, I conducted a field study on local people's attitudes toward the maned wolf in the state of São Paulo, which helped to answer this question. The objectives of my study were to shed some light on the associations between the way local people perceive the maned wolf and its decline, and to enable a comparison between such associations in urban and rural areas. The study aimed to improve the understanding of people's relationships with the maned wolf to help the planning of conservation strategies for the species.

In this chapter, I review and discuss key issues that are instrumental in the wide range of local people's attitudes toward the maned wolf in the southeast of Brazil, referring to data from my research as evidence.

METHODOLOGY AND STUDY AREAS

The data compiled in this research were gathered from a literature survey; local research programs and official records; and fieldwork based on questionnaires (total = 725) and interviews, combining quantitative and qualitative elements. Target groups were identified according to their involvement in the conservation of the maned wolf either as opinion makers or as supporters, or because of their potential exposure to contact and possible conflict with the maned wolf, in both rural and urban settings (Table 3.1).

São Paulo state is the most populous state in Brazil and is of high political and economic importance in the national context. Furthermore, it houses some of the most significant remnants of maned wolf habitat within conservation areas. The research was carried out in three areas: Greater São Paulo, São Carlos, and Low Mogiana regions, complemented by the input of conservation units (CUs) and zoo professionals as well as students from a number of other locations within the state of São Paulo (see map, Figure 3.1). The questionnaires targeted four different groups (Table 3.2). The questions were designed to identify selected attitudes, beliefs, and knowledge of target groups in relation to the maned wolf, wildlife, conservation, and other target groups and to compare these factors in urban and rural areas among sociodemographic groups and between different locations in São Paulo State (Consorte-McCrea and Rubin 2009; Consorte-McCrea 2011).

TABLE 3.1
Interest Groups Sampled in the Research

Target Groups	
Students year eight (12–13 years old)	176
Students sixth form (16–17 year olds)	174
Zoo visitors	110
CU visitors	39
School professionals	36
Zoo professionals	66
CU professionals	31
Rural population	54
International paper	35
Other	04
Total	**725**

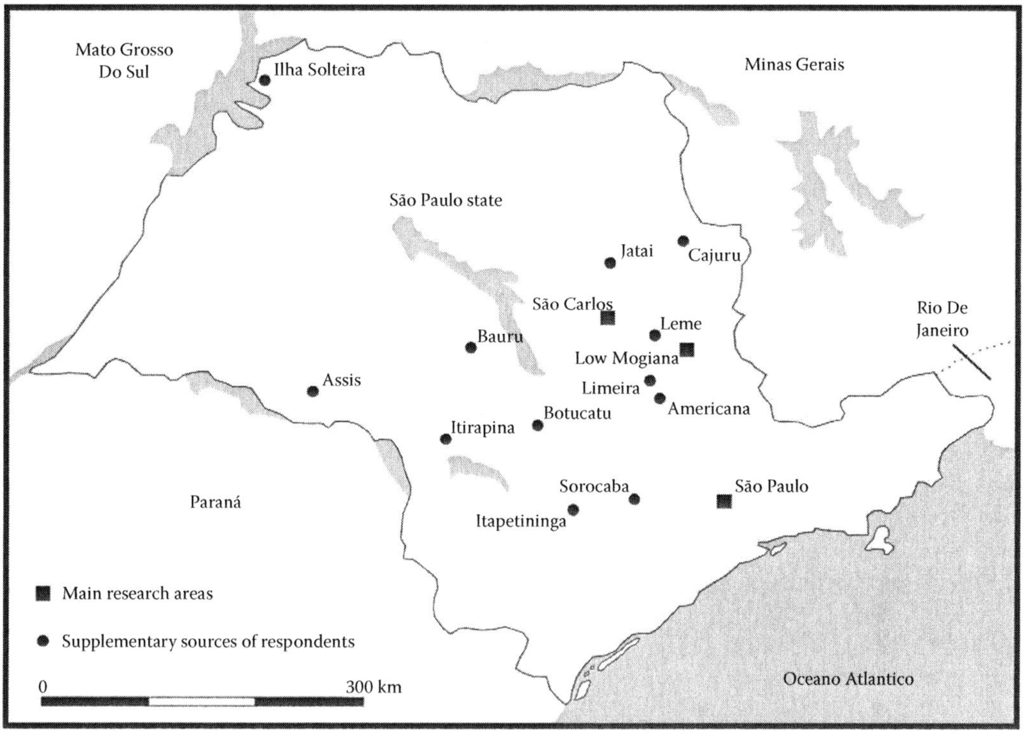

FIGURE 3.1 Research sites (main and supplementary information sources). (Map by John Hills, 2012.)

TABLE 3.2
Respondents for Each Separate Questionnaire

Type of Questionnaire	Frequency	Percentage
Q1: General public and students who recognized the maned wolf	451	62.2
Q2: General public and students who did not recognize the maned wolf	77	10.6
Q3: Biology, conservation, and education professionals	143	19.7
Q4: Rural population	54	7.4
Total	**725**	**100.0**

RESULTS AND DISCUSSION

The study of the relationships between carnivores and people is of interest worldwide. The ability of people to coexist peacefully with wild predators is a frontier in conservation, therefore understanding the multifaceted elements influencing people's attitudes toward carnivores is paramount. The following key issues that emerged from my study are transferable to the relationship between people and wild carnivores worldwide.

Human Encroachment and Habitat Protection

The relationships between maned wolves and people may have been altered with changes in land use and population growth. Fossil records indicate that the maned wolf evolved in central Brazil, during the Pleistocene Age or even earlier, and it was already established in South America when the first human settlers arrived during the last Ice Age (Langguth 1975, quoted in Dietz 1984; for historical and current distribution see Queirolo et al. 2011).

Indigenous populations were familiar with the feeding ecology of the maned wolf* and named it after its large size and reddish coat. Although Portuguese settlers soon identified the species as some sort of wolf (when wolves were being intensely persecuted in Europe), it apparently experienced no significant impact from the time of the arrival of the Portuguese settlers in 1500 until the mid-twentieth century (de Miranda 2003). This time was marked by human population growth and increasing deforestation involving large areas of the southeast of Brazil.

People in the Research Areas

The population of the state of São Paulo has its roots in both the Portuguese and the indigenous people thrown together during the *bandeiras* movement to explore inland Brazil (between the sixteenth and eighteenth centuries, anticipating the "march towards the west" movement of the twentieth century) (Arruda 2000; Candido 2001). This group originated from the small-scale farmers in the areas of mineral extraction and coffee expansion (Ribeiro 1995, quoted in Arruda 2000) that gave birth to most of the cities in inland São Paulo. The Portuguese culture lost ground in favor of the indigenous culture, which was better adapted to the environment and to mobility. After the abolishment of slavery in 1888, the black population who previously worked the plantations migrated to urban areas (Consorte 2007). The early twentieth century saw an influx of immigrants, mostly from Europe, tied up with the coffee boom and the later development of industry and commerce, adding new roots to the local population.

After the Proclamation of the Republic (1889), a "march towards the west" to explore the country's inland disseminated ideas of urbanization as *progress*; in contrast, wild habitats were seen as backward (Villas-Bôas and Villas-Bôas 2004; de Miranda 2003)—a concept that may remain in the research areas. As a consequence of the capitalist expansion, the *caipira*† had to work harder: this pressure resulted in individualization and detachment from the social and economic harmony with the local environment toward a regional and national context (Candido 2001). The relationship between this group and the natural environment was thus altered and with it the knowledge and use of natural resources, giving way to an unprecedented level of environmental impact. The expansion and modernization of agriculture in the cerrado contributed to the decline of the populations of maned wolves.

Conservation awareness is not new to Brazil. Wildlife protection is supported by a hunting ban, in place for over 45 years. Nevertheless, only 6.3% of Brazil is protected by federal CUs (only 2.8% have integral protection), and according to Costa et al. (2005) many protected areas may suffer from inadequate management, making them vulnerable to illegal hunting, human settlement, and logging. Today many endemic terrestrial carnivores are threatened with extinction because of the destruction, fragmentation, or reduction of habitat, and hunting (IBAMA 2003, quoted in de Miranda 2003). As an added pressure, new technology creates shortcuts in labor and pressure to increase production and, in turn, it may lower farmers' tolerance to damage as it also distances them from nature (Knight 2000; Candido 2001; Hill 2004).

The Value of the Maned Wolf

In Brazil, the maned wolf is on the National List of Threatened Species of Brazilian Fauna (MMA 2008). In recent years, its status has changed from Vulnerable to Near Threatened by the

* Kamaiurá indigenous people, inhabiting the high Xingú River and grasslands in central Brazil, refer to the maned wolf in one of their traditional tales as an animal that eats only fruits, or "auaratsim" (Carvalho 1976; Miller 1930). See also Felix de Azara's description of the species in Paraguay from the 1700s (Cartes et al., Chapter 17, this volume).

† Traditional population of inland São Paulo.

International Union for Conservation of Nature (IUCN) Red Data Book (IUCN 2009), and it is included in the Convention on International Trade in Endangered Species of Wild Fauna and Flora (CITES; Appendix II). Maned wolves, like many other predators, are valued for a variety of reasons. Carnivores are cited as potentially good for tourism, being an important part of cultural heritage; as a "keystone" species, the maned wolf plays an important role in ecosystem dynamics. The species aids in the dispersal of various species of fruits in the cerrado, including "wolf's fruit" (*Solanum lycocarpum*), which shows improved germination rates after passing through the maned wolf's digestive system, gabiroba (*Campomanesia pubescens*), which are enjoyed by cattle, people (Lombardi and Motta-Junior 1993; Courtenay 1994; Ferraz 2000; Santos et al. 2003), and many others (see Motta-Junior et al., Chapter 7, this volume). The maned wolf provides a service by feeding on insects and rats that are carriers of diseases such as hantavirus and leptospirosis (Dietz 1984; Motta-Junior 2000; Anic 2002).* Arguably, as a "flagship" and "umbrella" species, it has the potential to champion efforts to protect the whole ecosystem where it is found.

What is the value of the maned wolf to local people? Most local people consulted in this research (64.5%) considered the maned wolf valuable (as opposed to *worthless*). Most did not value the maned wolf for its body parts or as game. The great majority of respondents who considered the maned wolf as valuable also believed that it needs to be protected (88.3%, $N = 451$, see Table 3.3 for statistical analysis) and that it has an important role in nature (86.9%, $N = 451$). This suggests they awarded the species an intrinsic value that deserves protection. The value local people award to the maned wolf places it in a strong position to be a champion for the cerrado biome. But is it also considered charismatic and likable?

People and Carnivores' Multifaceted Relationship: The Maned Wolf within the Universe of Local People

Although the maned wolf was first described using its indigenous (Guarani) name, *yaguaraçú* (meaning big red canid) (Padre José de Anchieta [1534–1597] 1988, quoted in de Miranda 2003; Gabriel Soares de Souza 1587, quoted in Ribeiro 2003), it was later categorized as a "common species," already known, by the crown naturalists and named *lobo guará* (*guará* wolf or red wolf). The name "wolf" might have sealed a future of unjustified antipathy on the part of the colonizers toward the maned wolf. Kruuk (2002) suggests that we have evolved innate antipredator responses, which may manifest themselves toward "look-alikes"—carnivores similar to the ones known by previous experiences to have caused harm or material loss. In addition, negative values related to experiences of attacks on poultry (by foxes and wolves in Europe) might have traveled with the European settlers and been transferred to their relationship with the maned wolf. This might also be true in the attribution of magical and pharmaceutical value to some of its body parts (similar to foxes and wolves in Europe; Freefy 1983; Bush 1995; Kruuk 2002). In fact, an alarming 33% of respondents believed that maned wolves' body parts were used for either fashion, medicinal, or magical purposes, and 13% ($n = 7$) of rural respondents believed that maned wolves were killed because of the sale or use of their body parts. This potential area of conflict has been noted previously by Puglia (1978), Silva (2000), Silva and Nicola (1999), Anic (2002), and Emmons (Chapter 16, this volume). Beliefs about the use of maned wolf body parts have been collected by Dietz (1984), Figueira (1995), Anic (2002), Rodrigues (2002), Sillero-Zubiri et al. (2004), and Villas-Bôas and Villas-Bôas (2004), including the use of canine teeth, pieces of the hide, hearts, eyes, and dried feces to cure diseases and to bring good luck and protection (see review in Consorte-McCrea 2011).

Traditionally in the literature, the maned wolf has been described as timid, shy, scarce, cowardly, weak, frightened, and not nearly as strong and brave as its European counterpart, except

* The loss of food grains and crops to rodents has been estimated somewhere between 2 and 200 million tons worldwide per year; in South American countries the loss of sugarcane production to rodents has been estimated between 1 and 12,500 tons (in case of rodent outbreak) per year (Conover 2001).

TABLE 3.3
Summary of Strong Statistical Associations per Question Category

Aspect	Result	Chi Squared Statistics
Knowledge		
Positive identification of maned wolf from photo	All respondents from São Paulo State	20.6, df = 5, $p = .001$
	Rural respondents > zoo visitors	
Wrongly identified chicken as favorite food item	Educated up to sixth form < other education groups	40.4; df = 6; $p < .001$
	Rural respondents > other groups	95.8; df = 15; $p < .001$
Knowledge Score		
High	(Q1) Saw maned wolf live in nature > no use of this source	13.5; df = 1; $p < .001$
Low	Saw maned wolf live in nature < no use of this source	23.9; df = 1; $p < .001$
	Information from school or environmental education course < no use of this source	11.8; df = 1; $p = .001$
Attitudes Toward Maned Wolf		
Do not care about maned wolf	Rural residents > urban residents	12.1; df = 1; $p = .001$
	Educated up to year 10 > more educated	23.2; df = 3; $p < .001$
Value associated with biodiversity	Belief that maned wolf needs protection	39.3; df = 3; $p < .001$
	It plays a role in conserving the ecosystem	33.0; df = 3; $p < .001$
Negative attitudes		
Maned wolf scares and attacks people	Rural respondents > other target groups	11.6; df = 5; $p = .041$*
Maned wolf attacks chicken pens and livestock	Rural respondents > sixth-form students, CU visitors	101.2; df = 5; $p < .001$
	Rural residents > urban residents	41.9; df = 1; $p < .001$
Sources of Information about Maned Wolf		
Saw the maned wolf live in nature	Belief that maned wolf attacks chicken pens and livestock	15.5; df = 1; $p < .001$
	Belief that maned wolf must be protected	17.8; df = 1; $p < .001$
	Belief that maned wolf attacks and scares people	11.9; df = 1; $p = .001$
Zoos, museums and CU	Belief that maned wolf must be protected	19.9; df = 1; $p < .001$

when defending pups (see Magalhães 1939; Ihering 1968; Carvalho 1976; Dietz 1984, 1987; Ribeiro 2003). Traditional people from central Brazil (*sertanejos*) associate the eerie-sounding maned wolf's call with feelings of melancholy and sometimes of bad luck (Magalhães 1939; Villas-Bôas and Villas-Bôas 2004). Consistently research has indicated that local ranchers and rural people consider them harmless, show no negative feelings or fear toward them, and have sound knowledge of several aspects of their ecology and habitat (Carvalho 1976; Dietz 1984, 1987; Bestelmeyer 2000; Anic 2002). However, beliefs about the maned wolf's diet seem to be inaccurate. The diet of the maned wolf is composed of a similar amount of items of plant and animal origin, and the occurrence of poultry remains is very rare in scat samples (0%–1.5%) (Motta-Junior 2000; Aragona and Setz 2001; Anic 2002; Bueno et al. 2002; Juarez and Marinho-Filho 2002; Rodrigues 2002; Santos et al. 2003; Bueno and Motta-Junior 2004). Many respondents (39.8%) in this research correctly identified the maned wolf's favorite food items as wolf's fruit, fruits, and rats; however many rural people (24%) erroneously identified chicken as its favorite item, indicating a considerable level of misconception relating to a carnivorous diet. In field data, poultry remains were found in only 0.3%–1.5% of analyzed scat samples. In rural areas, this misconception may be associated both with inherited European beliefs about "wolves" and with the infrequent but striking occurrence of predation events.

In spite of the maned wolf's crepuscular habits and shy nature, most people (86.7%, $n = 504$) who answered the questionnaires, and particularly rural people, were able to identify it by looking at a picture. However, significantly fewer visitors of zoos where the maned wolf is found recognized it, indicating that the potential of zoos to disseminate awareness about the maned wolf is yet to be fulfilled (see also Bizerril and Andrade 1999). At the other end of the spectrum, the data analysis shows that rural respondents knew more about the maned wolf's habits and ecology (except for feeding habits) than most other target groups did, suggesting that a familiarity with the natural environment might have persisted in the rural population.

Respondents were not only familiar with the maned wolf, but most of them also had a positive view of its behavior and appearance (67.8%, $n = 1701$, out of a total of 2513 evaluative responses) (see Table 3.4). According to Kellert et al. (1996), this perception could contribute to positive attitudes toward the maned wolf and its conservation. Perceptions of urban and rural residents were similar, challenging some of the most negative traditional portrayals of the species in the literature and from earlier accounts. The maned wolf, though considered brave, powerful, and strong, is not perceived as a threat to people. Also, its large size was not associated with fear or dislike for the species, as suggested in relation to other carnivores (Macdonald and Sillero-Zubiri 2004), but quite the opposite: respondents who considered the maned wolf big also considered it good, beautiful, brave, strong, valuable, and powerful, similar to the findings of Kellert (1985) about other charismatic carnivores. Positive perceptions toward the maned wolf and its successful coexistence with local people have been reported from fieldwork carried out by other researchers (Carvalho 1976; Dietz 1984; Figueira 1995; Silva 2000; Bestelmeyer 2000; Anic 2002) and seem related in part to its overall shy behavior in avoiding people.

TABLE 3.4
(Q1) Respondents' Beliefs about the Maned Wolf According to Evaluative and Potency Factors

	Ex: Bad 1; 2; 3; 4; 5; 6; 7 Good			
What Do You Think about the Maned Wolf?	**Negative < 4**	**Neither**	**Positive > 4**	**No Answer**
Evaluative Factors				
Bad/good	22	52	300	77
Ugly/beautiful	38	51	297	65
Cowardly/brave	21	66	254	110
Dangerous/harmless	77	78	193	103
Aggressive/defensive	59	77	204	111
Worthless/valuable	22	50	291	88
Ferocious/tame	109	90	162	99
Evaluative factors total (respondents = 451; responses = 2513)	348	464	1701	653
Potency Factors				
Weak/strong	27	88	253	83
Powerless/powerful	105	85	142	119
Small/big	70	119	163	99
Potency factors total (respondents = 451; responses = 1015)	202	292	521	301
Total (respondents = 451; responses = 3528)				

Source: Ajzen, I. and Fishbein, M., *Understanding Attitudes and Predicting Social Behaviour*, Prentice-Hall, Englewood Cliffs, NJ, 1980.

Notes: Respondents were asked to rate objects in terms of bipolar adjectives, based on Osgood's Semantic Differential Method.

Only a small minority of respondents did not like (1.7%, $n = 8$) or did not care (6.5%, $n = 33$) for the maned wolf. Within the latter group there were many rural residents ($n = 14$, $N = 470$) and the least educated ($n = 24$, $N = 483$). Overall, the results show that the more people knew about the maned wolf, the better their beliefs and attitudes toward them were. This was particularly true among rural respondents. The lack of an association between knowledge about the maned wolf and negative attitudes toward it (as found to be the case for other carnivores, in Kaczensky et al. 2004) may be due to a general absence of perceived threat in relation to the species. Research suggests a correlation between positive attitudes toward large carnivores, low fear level, and higher support for their protection (Bath and Farmer 2000), suggesting positive implications for maned wolf conservation.

Overall, a majority of both urban and rural residents in this research demonstrated a high degree of positive attitudes toward the maned wolf. In a minority of cases, negative feelings were associated with potency variables, as suggested by Kellert (1985), and maned wolves were seen as dangerous, strong, ferocious, and powerful. Results suggest that a lack of positive attitudes relate to a lack of knowledge and to misconceptions (cognitive in nature), rather than to negative experiences of the species or values about wildlife conservation. Therefore, the circulation of accurate information about the characteristics of the maned wolf, such as size, weight, ecology, feeding habits, and population numbers, should improve public attitudes toward it by building more complete and informed images in people's minds, as suggested by Bath and Majic (2001).

Relationship between Human-Occupied Landscapes and the Maned Wolf

The pressures of habitat impoverishment and fragmentation caused by human expansion may encourage some less specialized carnivores with opportunistic habits to explore human-occupied landscapes, even expanding their populations. This is likely to be the case of the maned wolf, whose range seems to be shifting toward the southeast (Queirolo et al. 2011). Research in Serra da Canastra National Park indicates that maned wolf populations may tolerate low-level impact from human colonization; however long-term and high-impact activities, such as farming and tourism, may be potentially significant and might cause changes in the behavior and ecology of their populations (Dietz 1984; Rodrigues and de Paula 2007; see also Silva and Talamoni 2004).

The maned wolf makes use of pastures, clearings, and abandoned human habitations. This behavior certainly helps the capacity of the species to disperse and not depend exclusively on natural areas to move between habitat patches. *Lobeira* plants are abundant in altered areas, especially pastures, allowing maned wolves to use such areas for dispersal and even expanding their distribution in certain regions, making use of areas where native forests have been cleared (Dietz 1984; Motta-Junior 2000; Rodrigues 2002; Santos et al. 2003).

People's Livelihoods

Despite the risks for maned wolves, in times of need human-occupied environments can provide an opportunity not to be missed for readily available food, in concentration, in the minimum possible time. However, this might not be achieved without conflict. Although apparently rare due to their shy nature (Magalhães 1939; Ihering 1968; Carvalho 1976; Dietz 1984), field research has recorded occurrences of maned wolves approaching chicken pens and causing great losses during the winter months and from September to November when pups are more likely to be at den sites (Puglia 1978; Dietz 1984). On such occasions, wolves were reported to be trapped, poisoned, shot at, imprisoned, or killed (Puglia 1978).

The way people perceive risk may be strongly influenced by rare and extraordinary or extreme events, such as "worst-case scenarios" (Hill 2004), and the widespread belief that maned wolves attack chicken pens with blind rage may fall into this category, even among people who have never witnessed such an event. In fact, 28.1% ($n = 142$) of respondents believed that maned wolves attack chicken pens and livestock, among them a majority of rural respondents ($n = 45$; $N = 502$).

A significant number of rural respondents also believed that maned wolves are a threat to people (Table 3.3); this may be related to isolated and rare incidents of females protecting pups (Magalhães 1939; Ihering 1968; Carvalho 1976; Dietz 1984, 1987). Could these few negative events have damaged people's attitudes toward the species?

My results suggest that, despite the negative feelings of a small minority, attitudes among the rural respondents in the research areas indicated tolerance toward the likelihood of a maned wolf being found attacking livestock. In fact, as suspected, the great majority of the respondents and their families had not suffered damage caused by the maned wolf (91.9%, $n = 464$), including rural respondents (77.8%, $n = 42$). Among rural respondents, 18.5% ($n = 10$) alleged suffering damage caused by the maned wolf to themselves or to their families. Nevertheless, only one of them believed a maned wolf found attacking livestock must be killed or trapped (see Table 3.5). Retaliation killing was not popular among people who had lost livestock to the maned wolf. However, killing maned wolves is outlawed and some people may have refrained from expressing opinions in support of killing. The most frequent responses among this group were handing the culprit over to the authorities or scaring it away (33.3%, $n = 4$).

The tolerance observed among rural respondents toward maned wolf consumption of domestic stock may reflect an overall sympathy for it and may be an indication of the remnants of traditional integration between local people and the natural environment, as suggested by Knight (2000). The high level of tolerance toward maned wolves and their raids (albeit possibly rare) may also indicate that the conservation of the species is compatible with the local economy of the research sites. The fact that the increased value of cerrado habitat has apparently not lowered the threshold of tolerance toward raids is an indication that maned wolf raids do not affect production.

Conflict and Discourse: Changes in Relationship between Local People and Nature over Time

Human activities in the vicinity of protected areas may affect the populations of maned wolves in direct and indirect ways. The population model created for Brazil by the Maned Wolf Action Plan (Paula et al. 2008) suggests that the major threats to the species lie in an increase in road mortality and reduction of the carrying capacity due to habitat loss. Other human-related threats are the presence of domestic and feral dogs in the vicinity of protected areas (Pontes-Filho et al. 1997; Anic 2002; Rodrigues 2002), deaths by retribution for attacks on poultry (Dietz 1984; Rodrigues 2002) and hunting, killing for the harvesting of body parts for magical and medicinal purposes

TABLE 3.5
Rural Respondents' (Q4) Attitudes Toward Maned Wolves Found Attacking Livestock, by Damages

If a Maned Wolf is Found Attacking Livestock	Experience of Damage: No Damage	Experience of Damage: Damage Caused by Maned Wolf or Other Animal	Total Responses
It must be trapped	2	1	3
It must be killed	3	1	4
It must be handed over to the authorities	18	4	22
Body parts must be harvested	0	0	0
One must set the dogs after it	2	3	5
One must scare it away	9	4	13
One must reinforce the chicken pen	2	0	2
Total responses	36	13	49
Total respondents	42	12	54

(Puglia 1978; Silva 2000; Silva and Nicola 1999; Anic 2002), and loss of genetic variability. Louise Emmons (Chapter 16, this volume) also points out the threat of climate change and extended droughts, as they may affect food resources and disease susceptibility.

Anthropogenic Causes of Maned Wolf Mortality

Some believe that road killings of maned wolves and other wild animals may be caused intentionally (Carvalho 1976; Rodrigues 2002). When asked, "Why do people kill the maned wolf?", rural respondents in this research also suggested that maned wolves are killed intentionally, out of people's *badness*, because people *feel pleasure in killing*, because they *dislike it*, or due to *ignorance*. Ignorance about the species may result in a dislike for it. This may be due to fear, cultural and historical antipathies, association with more carnivorous species—such as *Canis lupus*, and blame for attacks on poultry, which in the long term may result in negative perceptions and attitudes toward the maned wolf, even as far as harming or killing. It is also possible that the dissemination of a rural development agenda may resonate in people's feelings of dislike toward wilderness and wild predators. Such negative feelings have been linked to endemic species of the cerrado in Brazil (Carvalho 1976; Klink and Machado 2005) and may be connected with feelings of dislike toward the maned wolf. However, killing for pleasure does not derive from the same causation.

The most popular belief about human-related mortality among rural people in this sample was that maned wolves are killed for sport (40.7%, $n = 22$). Despite a lack of evidence from data collection in the research areas to establish the true number of maned wolves killed by sport hunters, it is possible to make some inferences about the nature of the hunting issue.

Hunting, fishing, and gathering have been described as indigenous practices adopted by the *caipira* culture as a means of subsistence and of reinforcing bonds with the natural environment (Candido 2001; de Miranda 2003). Hunting has also been described in a similar way by modern Native Americans, who also associate hunting with stress relief, excitement, and the strengthening of social bonds (Daigle et al. 2002). Antônio Candido (2001) suggests that with the modern hunter however, for whom hunting is not part of a subsistence economy, hunting has taken on different functions. With hunting bans in place and an increasing dependence on urban centers for manufactured goods and food, rural people almost totally limit hunting to the defense of plantations and livestock. In other instances, recreational hunters pursue wild carnivores and birds whose meat is not edible. As an added dimension, claims of damage caused by certain species have been used as justification for hunting them down, even when landowners admit having no bad feelings toward the species and admiring their presence, as is the case with the red fox in Britain (Marvin 2000).

The belief of 20.4% ($n = 11$) of rural respondents that maned wolves are killed in retaliation is consistent with the alleged experiences of 18.5% of maned wolves causing damages to themselves or family members. Bearing in mind the endangered status of the species, those numbers are still worthy of consideration and suggest the need to address conflict between the maned wolves and local farmers. However, rather than being a primary reason for maned wolf mortality, retaliation and prevention of attacks on livestock may have been used as a form of justification by people who actually kill the maned wolf for sport.

Environmental Awareness and People in the Vicinity of Conservation Areas

Over a decade ago, research conducted by the Forestry Institute showed that 73% of all CUs of over 10,000 hectares in the state of São Paulo were inhabited by traditional populations, most also housing nontraditional populations (Arruda 2000). As conflicts between people and carnivores tend to center on human dimensions, canid education worldwide has focused on fighting misinformation and improving people's understanding of the biology and management of species, making use of the formal school system, thus targeting younger generations to foster changes in negative views (Taylor 2004). Brazilian educators such as Ligia Moreira da Rocha (1997) have emphasized that both the success of conservation initiatives and the effective long-term conservation of natural resources depend on public support, and movement toward the inclusion of local people in the planning of conservation strategies has taken shape (Strauss 2004).

Some of the problems faced by environmental education programs relating to protected areas spring from the fact that local people do not seem to be aware of the existence of neighboring protected areas, or know very little about their aims or the benefits that they can generate, and are consequently uninvolved in their protection (Padua and Tabanez 1997; de Fiori 2006). Indeed, only 37.6% (n = 190) of the respondents were familiar with the local CU, and only 6.3% (n = 12) who knew it attended an environmental education course there. Environmental educators, however, advocate their practice as a way to increase knowledge, change values, and improve skills; these are conditions considered as essential for the development of environmentally aware attitudes as part of the quest for an improved quality of living (Padua and Tabanez 1997; Sorrentino 1997). Some of the most successful and long-running conservation programs in Brazil have aimed to promote self-esteem associated with the valuing of the locality, encouraging integration between its cultural and natural resources (including the protected area) (see Almeida 1997; Castilhos et al. 1997; Dietz and Nagagata 1997; Padua and Padua 1997; Bizerril et al. 2011). Variables such as "ownership" (personal involvement with issues) and "empowerment" (a sense that one has the power to make important changes) have been considered essential tools as information alone is not enough to promote long-lasting changes in behavior (Hungerford and Volk 1990). Within this framework, appealing endemic flagship species have been used to attract the attention of the population to the environmental cause, by eliciting support for the preservation of that species' habitat (Dietz and Nagagata 1997).

Results indicate the existence of the potential of environmental education courses to effectively address misconceptions in important areas of the maned wolf ecology. Most respondents who had heard about it in *school or an environmental education course* (n = 20, N = 505) displayed good levels of knowledge about the maned wolf, indicating that these were effective, though limited in reach, as sources of accurate information about its feeding ecology.

Zoos may also contribute to the long-term construction of positive attitudes and opinions toward wild carnivores by bringing the plight of species to public attention and by providing education programs and accurate information about individual species and their ecology (Consorte-McCrea 1994; see Projeto Cerrado Nossa Casa, Almeida 1997; Bowkett 2009; Songsasen and Rodden 2010; Vanstreels and Pessutti 2010). In fact most of the respondents (69.1%, n = 349) were familiar with the local zoo, although only 4.3% (n = 15) of them where engaged in an environmental education course offered by it. A surprising 35.8% (n = 181) had seen the maned wolf live in nature and 29% of rural respondents (16 out of 51) had seen it live on their own properties. However, while urban residents who had seen a maned wolf live in nature displayed more knowledge and positive attitudes and beliefs toward the species (as suggested by Kellert 1984; Kvaalen 1998), rural residents who had seen a maned wolf live on their property did not (as suggested by Bangs et al. 2005). Many within this group tended to believe that maned wolf numbers were increasing and that they were a threat to livestock and to people, suggesting that the experience triggered feelings of insecurity and fear in landowners. Similar feelings have been observed when wild animals invade people's personal space (Bangs et al. 2005; Michelle et al. 2005).

Conflicts between Rural and Urban Interests

Conflicts between rural people and urban administrations are a common issue that affects the relationships between people and protected wildlife. In São Paulo State, Arruda (2000) has suggested that the use of natural resources by local people within CUs and neighboring areas leads to conflict with the administration.

In the sample, many rural people were *sitiantes** (or part of their family units), inhabiting small holdings and selling their produce to large agricultural companies (Figure 3.2). Such lifestyles

* Small rural producers.

FIGURE 3.2 *Left*, Orange plantations in the Low Mogiana region are the main crop in rural areas surrounding CUs, followed by sugarcane plantations; *right*, sugarcane plantations in rural São Carlos are responsible for most of the agricultural revenue.

might provide small rural producers with enough economic support to free them from reliance on supplementing their income with resources from the wild areas, giving them some leeway to withstand eventual losses to wildlife, thus reducing potential conflict with CU management. There may be associations between income, reliance on natural resources, and vulnerability to wildlife-related damage and attitudes toward conservation, which could be investigated further. My results, however, provided no indication of a strong antagonism between local rural and urban interests in areas concerning the conservation of the maned wolf.

Human Dimensions: Local People and Support for the Conservation of the Maned Wolf

It is understood that predation affects landowners differentially according to their sociodemographic conditions (Knight 2000; Breck 2004; Hill 2004). Hence, the importance of understanding landowners' perceptions of conflict goes hand in hand with the importance of conservation on private land: support for the latter cannot be enlisted without consideration of the former. Large carnivores cannot survive only in protected zones as they require a vast home range, so their survival depends on linked areas of healthy habitat and on the tolerance and support of neighboring landowners and residents of rural areas (Clark et al. 2001; Macdonald and Sillero-Zubiri 2004; Michelle et al. 2005; Woodroffe et al. 2005; Sollmann et al. 2010). Within this scenario, rural people are not the problem, but are part of the solution, and their beliefs, values, and culture must be taken into account by conservation programs (Taylor 2004, 2009).

Results from this research suggest that the great majority of local people in the sample across all target groups, locations, and demographic characteristics believed that the maned wolf must be protected. The fact that coexistence between local people and the maned wolf has been consistent and uninterrupted may facilitate acceptance and tolerance toward the species, as suggested by Bath (2009). Kaczensky et al.'s (2004) findings suggest that in view of their positive attitudes toward the maned wolf, local people may support actions favorable to them, tolerate damage caused by them (as was indicated by rural respondents), and maintain their position in cases of conflict. Research on other carnivores indicates that positive choices may occur as long as local people can make informed decisions in support of maned wolf conservation (Kellert et al. 1996; Kaczensky et al. 2004; Andersone 2005; Randveer and Mäe 2005; Bath 2009). In addition to information, related peer group expectations and perceived control over their behaviors must be considered when ascertaining the strength of people's intentions (as suggested by the theory of planned behavior, Daigle et al. 2002) toward maned wolf conservation.

NEXT STEPS IN RESEARCH AND CONSERVATION

The success of maned wolf conservation depends on large areas of healthy habitat,* public support, coexistence, and the reduction of risks to them and to people to further a successful coexistence, within and outside protected areas. Understanding people's attitudes toward maned wolves involves investigating sociocultural elements pervading this relationship, and this understanding is essential for any attempt to change or influence behavior. In this chapter, I suggest areas for further investigation and concern that emerged from my and other people's research and are relevant to the maned wolf conservation scenario.

Toward Coexistence and Mitigation of Conflict

Public perceptions of the damaging impacts of a species cannot be shifted by research and knowledge alone. It is necessary to identify measures that can be perceived as effective by the local people who are involved in conflict (Woodroffe et al. 2005; Bath 2009; Majic and Bath 2010). Methods to resolve people–carnivore conflict around the world have been addressed by many authors (Musiani et al. 2002; Treves and Karanth 2003; Breck 2004; Macdonald and Sillero-Zubiri 2004; Mattson 2004; Sillero-Zubiri et al. 2004) and can be grouped as follows (Boitani et al. 2004):

1. Prevention: Means of avoiding conflict, by preventing carnivore attacks on livestock, including changes in husbandry; by avoiding harmful contact between people and carnivore species
2. Mitigation: Compensation for losses, making them more acceptable to people; improving tolerance of species
3. Control: Means of controlling carnivore populations, including translocations and removal of individuals.

Research into how to apply these methods to the maned wolf scenario may offer some of the answers to increase support for conservation among affected groups. A review of measures to address and mitigate conflict in relation to the maned wolf scenario can be found in Consorte-McCrea (2011).

Clarifying Misconceptions and Addressing Negative Attitudes

Although most respondents displayed positive attitudes toward the maned wolf, the impact of the negative attitudes of minority groups cannot be ignored, considering the status of the species. According to Bath (2009), for some species the conflict with people may have a cognitive nature (based on a lack of knowledge or misconceptions), while for others it may be rooted in a hierarchy of values (pitching conservation against livelihood), be related to costs and benefits, or spring from a lack of trust and conflicts between interest groups themselves. In the case of the maned wolf, results suggest that negative attitudes are associated with misconceptions concerning feeding habits and learned associations with *Canis lupus*; with a perception of threat related to the proximity of the maned wolf to people's properties; and possibly with values undermining the importance of local wildlife. Therefore, conservation education programs for the maned wolf must target knowledge as well as values.

* "Heterogeneous matrix of woody fruiting plants and grassland rodent habitat" (Castro and Emmons 2012, 55).

Negative attitudes that have a cognitive nature (such as misconceptions), on the other hand, may be targeted by disseminating clear and accurate information about specific beliefs regarding the species (size, weight, population numbers, density and distribution patterns, threatened status) among some groups. Dissociating its image from that of the European wolf and clarifying misapprehensions about its feeding habits may contribute to the elimination of fears toward the species (Dietz 1987; Motta-Junior 2000; Anic 2002; Bath 2009).

The spread of specific information about the potential benefits that maned wolves may bring to the lives of local people (the dissemination of seeds of cerrado fruits and pest control) may contribute to increased support for the conservation of the species (Motta-Junior 2000; Anic 2002). Behavioral change may be brought about using information associated with feelings of connection with the maned wolf. This may be achieved by enlisting nationalistic feelings and developing local pride in order to help preserve a species that belongs to the national heritage (Dietz 1984; Hungerford and Volk 1990; Sillero-Zubiri et al. 2004). The effectiveness of any educational strategies will also depend on the credibility of the information source among local people and on the appropriate choice of medium.

Cooperation between accurate information sources (zoos, museums, CUs, and environmental education courses) and the most far-reaching media (TV and radio) would help to improve the quality of information and could increase support for maned wolf conservation.

Maned wolf mortality was associated with lack of appreciation and ignorance toward the species, which may be linked to an agenda for rural development for the cerrado. Hunting, nevertheless, may be ingrained in indigenous roots and cultural traditions of local people, or sport. Further research into local people's attitudes toward hunting may help to identify which groups are the potential foci of conflict with maned wolf conservation.

Neutral attitudes were more common than negative attitudes in my research (see Table 3.4) and were particularly strong among sixth-form students. They provide a good opportunity for advances in coexistence and are the most likely to be influenced positively, making them a good target for conservation programs.

Based on most people's positive attitudes toward the species, its attractive characteristics, and its potential as an umbrella species, the maned wolf could be an effective flagship for the conservation of the cerrado habitat. Local people's self-esteem may be promoted by fostering integration between natural resources and cultural values; existing bonds between species such as the maned wolf and the environment that contribute to feelings of ownership, themselves important components in behavioral change, could be strengthened in favor of wildlife conservation.

Attitudes toward Other Interest Groups

The ability of conservation programs to foster change to the present scenario of decline relies on the relationships between bio/education professional and local people. Bonding may be achieved by the inclusion of diverse local groups in the development of plan directives and in the identification of local needs, which may also be addressed by conservation programs. Assistance from rural people who display the highest levels of positive attitudes toward maned wolf conservation should be enlisted to help foster positive changes in attitudes among their peer group. By enlisting local expertise, conservation professionals can promote a rich exchange of information with people who hold traditional knowledge about the local natural environment.

Professionals from schools, zoos, and CUs, in this sample, consistently displayed negative expectations of the attitudes of people living in the neighborhood of reserves and such feelings must also be addressed. This is to prevent the creation of a negative bias in their contact with local people, with consequent alienation (Taylor 2004). Thus, educational initiatives will need to target bio/education professionals as well as local communities. Information, reflection, and critical thinking are favored as a way forward together with emphasis on reciprocity and cooperation on the exchange of expertise about the local environment.

ACKNOWLEDGMENTS

I would like to thank the CCCU for funding my research and Dr Mike Nicholls and Rona Rubin for supervising. Thanks to Stephen Scoffham for his insights and for revising this chapter and to John Hills for making the map. Thanks to São Paulo Zoo, the IF-SP, Juquery State Park, UFSCar, and all the organizations and individuals who took part in the research for their support and infrastructure and for responding to the questionnaires.

REFERENCES

Ajzen, I. and Fishbein, M., 1980. *Understanding Attitudes and Predicting Social Behaviour*. Prentice-Hall, Englewood Cliffs, NJ.

Almeida, C.X., 1997. Importância da auto-estima no Projeto Cerrado, Casa Nossa, do Jardim Botânico de Brasília. In: Padua, S.M. and Tabanez, M.F. (eds), *Educação ambiental: Caminhos trilhados no Brasil*. Instituto de Pesquisas Ecológicas (IPE), Brasil, pp. 201–210.

Anchieta, J. de, 1988. Cartas. Informações, fragmentos históricos e sermões. EDUSP, São Paulo. In: de Miranda, E.E. 2003. *Natureza, Conservação e Cultura*. Metalivros, São Paulo.

Andersone, Z., 2005. People and carnivores in Latvia: Attitude survey. *Wolf Print* 23:17.

Anic, C.C., 2002. Conhecimentos e crenças de algumas populacões rurais no sudeste do Brasil frente ao lobo-guará *Chrysocyon brachyurus*, Illiger, 1815 (Mammalia; Canidae). Dissertacão de Mestrado, Instituto de Biociências da USP, Departamento de Ecologia.

Aragona, M. and Setz, E.Z., 2001. Diet of the maned wolf, *Chrysocyon brachyurus* (Mammalia; Canidae), during wet and dry seasons at Ibitipoca State Park, Brazil. *Journal of Zoology*, 254(1):131–136.

Arruda, R.S., 2000. "Populações tradicionais" e a proteção dos recursos naturais em unidades de conservação. In Diegues, A.C.S. (ed.), *Etnoconservação. Novos rumos para a protecão da natureza nos trópicos*. São Paulo, SP. Editora Hucitec, NUPAUB-USP, pp. 273–290.

Bangs, E.E., Fontaine, J.A., Jimenez, M.D., Meier, T.J., Bradley, E.H., Niemeyer, C.C., Smith, D.W., Mack, C.M., Asher, V., and Oakleaf, J.K., 2005. Managing wolf-human conflict in the northeastern United States. In: Woodroffe, R., Thirgood, S., and Rabinowitz, A. (eds), *People and Wildlife: Conflict or Co-Existence?* Conservation Biology 9. Cambridge University Press, pp. 340–356.

Bath, A.J., 2009. Working with people to achieve wolf conservation in Europe and North America. In Musiani, M., Boitani, L., and Paquet, P.C. (eds), *A New Era for Wolves and People: Wolf Recovery, Human Attitudes and Policy*. Energy, Ecology and the Environment Series. University of Calgary Press, Calgary, AL, pp. 173–199.

Bath, A. and Farmer, L., 2000. Europe's carnivores—A survey of children's attitudes towards wolves, bears and otters. WWF UK Campaign for Europe's Carnivores 7. WWF-UK Report, March 2000.

Bath, A. and Majic, A., 2001. Human dimensions in wolf management in Croatia: Understanding attitudes and beliefs of residents in Gorski Kotar, Lika and Dalmatia toward wolves and wolf management. Report, Large Carnivore Initiative for Europe.

Bestelmeyer, S.V., 2000. Solitary, reproductive, and parental behaviour of maned wolves (*Chrysocyon brachyurus*). PhD Dissertation, Department of Biology, Colorado State University.

Bizerril, M.X.A. and Andrade, T.C.S., 1999. Knowledge of the urban population about fauna: Comparison between Brazilian and exotic animals. *Ciência e Cultura*, 51(1):38–41.

Bizerril, M.X., Soares, C.C., and Santos, J.P., 2011. Linking community communication to conservation of the maned wolf in central Brazil. *Environmental Education Research*, 17(6):815–827.

Boitani, L., Asa, C., and Moehrenschlager, A., 2004. Conservation tolls for Canids. In: Macdonald, D. and Sillero-Zubiri, C. (eds), *The Biology and Conservation of Wild Canids*. Oxford University Press, pp. 143–162.

Bowkett, A.E., 2009. Recent captive-breeding proposals and the return of the ark concept to global species conservation. *Conservation Biology*, 23(3):773–776.

Breck, S.W., 2004. Minimizing carnivore-livestock conflict: The importance and process of research in the search for coexistence. In Fascione, N., Delach, A., and Smith, M., 2004. *People and Predators: From Conflict to Coexistence*. Island Press, Washington DC, pp. 13–27.

Bueno, A.A. and Motta-Junior, J.C., 2004. Food habits of two syntopic canids, the maned wolf (*Chrysocyon brachyurus*) and crab-eating fox (*Cerdocyon thous*), in southeastern Brazil. *Revista Chilena de historia Natural* 77:5–14.

Bueno, A.A., Belentani, S.C.S., and Motta-Junior, J.C., 2002. Feeding ecology of the maned wolf, *Chrysocyon brachyurus* (Illiger, 1815) (Mammalia: Canidae), in the ecological station of Itirapina, São Paulo state, Brazil. *Biota Neotropica* 2(2).

Bush, R.H., 1995. *The Wolf Almanac.* The Globe Pequot Press, Guilford, CT.

Candido, A., 2001. *Os parceiros do Rio Bonito.* 9th edn, Coleção Espi'rito Cri'tico, Duas Cidades, Editora 34, Brasil.

Carvalho, C.T., 1976. Aspectos faunísticos do cerrado- o lobo guará (*Chrysocyon brachyurus*). IFSP, Bol. Tec., São Paulo, 21:1–18.

Castilhos, J.C., Alves, D.A.R., and Silva, A.C.C.D., 1997. Resgate cultural e conservação de tartarugas marinhas. In: Padua, S.M. and Tabanez, M.F. (eds), *Educação ambiental: Caminhos trilhados no Brasil.* Instituto de Pesquisas Ecológicas (IPE), Brasil, pp. 147–156.

Castro, J.M. and Emmons, L.H., 2012. Variation in diet and resources. The maned wolves of Noel Kempff Mercado National Park. *Smithsonian Contributions to Zoology* 639:37–55.

Clark, T.W., Mattson, D.J., Reading, R.P., and Miller, B.J., 2001. Interdisciplinary problem solving in carnivore conservation: An introduction. In: Gittleman, J. et al. (eds), *Carnivore Conservation.* Cambridge University Press, Cambridge, UK, pp. 223–240.

Conover, M., 2001. *Resolving Human Wildlife Conflicts: The Science of Wildlife Damage Management.* CRC, Boca Raton, FL.

Consorte, J.G., 2007. The black question in Brazil: An issue denied. In: Hamilton, R.S. (ed.), *Routes of Passage: Rethinking the African Diaspora*, vol. 1, part 2. African Diaspora Series Endowment at Michigan State University.

Consorte-McCrea, A.G., 1994. The maned wolf in captivity. *Canid News*, 2:8–12. The IUCN/SSC Canid Specialist Group.

Consorte-McCrea, A.G., 2011. Conservation of the maned wolf (*Chrysocyon brachyurus*): Carnivore and people relationships in the southeast of Brazil. Canterbury: Thesis submitted to the University of Kent for the degree of Doctor of Philosophy, CCCU.

Consorte-McCrea, A.G. and Rubin, R., 2009. Attitudes of zoo visitors and professionals towards the conservation of the maned wolf in southeast Brazil. *Canid News* 12.1 (online). The IUCN/SSC Canid Specialist Group.

Costa, L.P., Leiter, Y.L.R., Mendes, S.L., and Ditchfield, A.D., 2005. Conservação de mamíferos no Brasil. *Megadiversidade* 1(1):103–112.

Courtenay, O., 1994. Fruitful relations in a changing environment. *Canid News*, 2:41–43.

Daigle, J.J., Hrubes, D., and Ajzen, I., 2002. A comparative study of beliefs, attitudes, and values among hunters, wildlife viewers, and other outdoor recreationists. *Human Dimensions of Wildlife*, 7(1):1–19.

Dietz, J.M., 1984. Ecology and social organization of the maned wolf (*Chrysocyon brachyurus*). *Smithsonian Contributions to Zoology*, 392.

Dietz, J.M., 1987. Grass roots of the maned wolf. *Natural History*, 96(3):52–59.

Dietz, L.A.H. and Nagagata, E.Y., 1997. Programa de Conservação do Mico-Leão Dourado: Atividades de educação comunitária para a conservação da Mata Atlântica no Estado do Rio de Janeiro. In: Padua, S.M. and Tabanez, M.F. (eds), *Educação ambiental: Caminhos trilhados no Brasil.* Instituto de Pesquisas Ecológicas (IPE), Brasil, pp. 133–146.

Ferraz, E., 2000. Ecologia alimentar e dirpesão de sementes realizada por *Chrysocyon brachyurus* em uma região rural do sudeste do Brasil (Carnivora – Canidae). Protocolo de manejo, III workshop do lobo-guará *Chrysocyon brachyurus*, Sociedade de Zoológicos do Brasil- SZB, p. 42.

Figueira, C.J.M., 1995. Ocorrência, relações gerais com residentes e comportamento alimentar do lobo-guará (*Chrysocyon brachyurus*) em zona rural do estado de Minas Gerais. UNESP, Rio Claro, SP. Monografia (Bacharelado em Ecologia).

de Fiori, A., 2006. A percepção ambiental como instrumento de apoio de programas de educação ambiental da Estação Ecológica de Jataí (Luiz Antonio, SP). Tese de doutorado, UFSCar, São Carlos, SP.

Freefy, R., 1983. *Man and Beast: The Natural and Unnatural History of British Mammals.* Blandford Books, UK.

Gittleman, J.L., Funk, S.M., Macdonald, D., and Wayne, R.K., 2001. *Carnivore Conservation.* Cambridge University Press, Cambridge, UK.

Hill, C., 2004. Farmers' perspectives of conflict at the wildlife-agriculture boundary: Some lessons learned from African subsistence farmers. *Human Dimensions of Wildlife*, 9(4):279–86.

Hungerford, H.R. and Volk, T.L., 1990. Changing learner behavior through environmental education. *Journal of Environmental Education*, 21(3):8–21.

Ihering, R., 1968. *Dicionário dos animais do Brasil.* Brasília, Ed. Universidade de Brasília.

IUCN, 2009. Red Data Book. www.iucnredlist.org/.

Juarez, M.K. and Marinho-Filho, J., 2002. Diet, habitat use, and home ranges of sympatric canids in central Brazil. *Journal of Mammalogy*, 83(4):925–933.

Kaczensky, P., Blazic, M., and Gossow, H., 2004. Public attitudes towards brown bears (*Ursus arctos*) in Slovenia. *Biological Conservation*, 118:661–674.

Kellert, S.R., 1984. Urban American perceptions of animals and the natural environment. *Urban Ecology*, 8:209–228.

Kellert, S.R., 1985. Public perceptions of predators, particularly the wolf and coyote. *Biological Conservation*, 31:167–189.

Kellert, S.R., Black, M., Rush, C.R., and Bath, A.J., 1996. Human culture and large carnivore conservation in North America. *Conservation Biology*, 10(4):977–990.

Klink, C.A. and Machado, R.B., 2005. Conservation of the Brazilian Cerrado. *Conservation Biology*, 19(3):707–713.

Knight, J., 2000. *Natural Enemies: People-Wildlife Conflicts in Anthropological Perspective*. Routledge, London.

Kruuk, H., 2002. *Hunter and Hunted, Relationships between Carnivores and People*. Cambridge University Press, Cambridge, UK.

Kvaalen, I., 1998. Acceptance of lynx by sheep farmers: A sociological comparison. In: Breitenmoser-Würsten, C., Rohner, C., and Breitenmoser, U. (eds), *The Re-introduction of the Lynx into the Alps*. Council of Europe Publishing, Strasbourg, pp. 59–64.

Lombardi, J.A. and Motta-Junior, J.C., 1993. Seed dispersal of *Solanum lycocarpum* St.Hil. (Solanacea) by the maned wolf, *Chrysocyon brachyurus* Illiger (Mammalia, Canidae). *Ciencia e Cultura*, 45(2):127.

Macdonald, D.W. and Sillero-Zubiri, C. (eds), 2004. *The Biology and Conservation of Wild Canids*. Wildlife Conservation Research Unit, University of Oxford, Oxford University Press.

Magalhães, A.C., 1939. *Ensaio sobre a fauna brasileira*. Secr. Agric. Ind. Com. SP. Brasil.

Majic, A. and Bath, A.J., 2010. Changes in attitudes toward wolves in Croatia. *Biological Conservation*, 143:255–260.

Marvin, G. 2000. The problem of foxes: Legitimate and illegitimate killing in the English countryside. In: Knight, J. (ed.), *Natural Enemies: People-Wildlife Conflicts in Anthropological Perspective*. Routledge, London, pp. 189–211.

Mattson, D.J., 2004. Living with fierce creatures? An overview and models of mammalian carnivore conservation. In Fascione, N., Delach, A., and Smith, M. (eds), *People and Predators: From Conflict to Coexistence*. Island Press, Washington, DC, pp. 151–178.

McNeely, J.A., 2000. Practical approaches for including mammals in biodiversity conservation. In Entwistle, A. and Dunstone, D. (eds), *Priorities for the Conservation of Mammalian Diversity: Has the Panda Had Its Day?* Cambridge University Press, Cambridge, UK, pp. 355–367.

Michelle, D., Nikolaev, I., Goodrich, J., Litvinov, B., Smotnov, E., and Suvorov, E., 2005. Searching for the coexistence recipe: A case study of conflicts between people and tigers in the Russian Far East. In Woodroffe, R., Thirgood, S., and Rabinowitz, A. (eds), *People and Wildlife: Conflict or Co-existence? Conservation Biology 9*. Cambridge University Press, Cambridge, UK, pp. 305–322.

Miller, F.W., 1930. Notes on some mammals of southern Matto Grosso, Brazil. *Journal of Mammalogy*, 11:10–22.

Ministério do Meio Ambiente (MMA), 2008. Accessed 20 October 2008. http://www.mma.gov.br/sitio/.

de Miranda, E.E., 2003. *Natureza, Conservação e Cultura*. Metalivros, São Paulo.

Motta-Junior, J.C., 2000. Variação temporal e seleção de presas na dieta do lobo-guará, *Chrysocyon brachyurus* (Mammalia: Canidae), na Estação Ecológica de Jataí, Luiz Antônio, SP. In: Santos, J.E. and Pires, J.S.R. (eds), *Estudos integrados em ecossistemas. Estação Ecológica de Jataí*. Vol. 1. Rima Editora, São Carlos, Brazil, pp. 331–346.

Musiani, M., Mamo, C., Boitani, L., Callaghan, C., Cormack Gates, C., Mattei, L., Visalberghi, E., Breck, S., and Volpi, G., 2002. Wolf depredation trends and the use of fladry barriers to protect livestock in western North America. *Conservation Biology* 17(6):1538–1547.

Padua, S.M. and Padua, C.V., 1997. Um programa integrado para a conservação do mioc-leão-preto (*Leontopithecus chrysopygus*): Pesquisa, educação e envolvimento comunitário. In Padua, S.M. and Tabanez, M.F. (eds), *Educação ambiental: Caminhos trilhados no Brasil*. Instituto de Pesquisas Ecológicas (IPE), Brasil, pp. 119–132.

Padua, S.M. and Tabanez, M.F. (eds), 1997. *Educação ambiental: Caminhos trilhados no Brasil*. Instituto de Pesquisas Ecológicas (IPE), Brasil.

Paula, R.C., Medici, P., and Morato, R.G. (org.), 2008. *Plano de ação para a conservação do lobo-guará: Análise de viabilidade populacional e de habitat*. Brasília: Centro Nacional de Pesquisas para Conservação dos Predadores Naturais-CENAP/ICMBio.

Pontes-Filho, A., Silva, C.B.X., Lange, R.R., and Cavalcanti, R.K., 1997. *Projeto Lobo-guará: Contribuição à conservacão ambiental dos Campos Gerais do Paraná, Brasil.* Congresso Brasileiro de Unidades de Conservacão, Curitiba, PR., vol II:848–860.

Puglia, L.R.R., 1978. O lobo guará em cativeiro. *Revista de Zoológico*, n.1, Sorocaba, 1978.

Queirolo, D., Moreira, J.R., Soler, L., Emmons, L.H., Rodrigues, F.H.G., Pautasso, A.A., Cartes, J.L., and Salvatori, V., 2011. Historical and current range of the Near Threatened maned wolf *Chrysocyon brachyurus* in South America. *Oryx* 45:296–303.

Randveer, T. and Mäe, K., 2005. Whose afraid of the big bad wolf in Estonia? *Wolf Print*, 23:12–13.

Ribeiro, R.F., 2003. Bestiário Brasílico: A nossa fauna no imaginário colonial. PUC Minas Gerais, Department of Antropologia. www.anppas.org.br/encontro_anual/encontro2/GT/GT16/gt16_ricardo_ferreira.pdf.

Rocha, L.M., 1997. Unidades de conservação e organizações não-governamentais em parceria: Programas de educação ambiental. In Padua, S.M. and Tabanez, M.F. (eds), *Educação ambiental: Caminhos trilhados no Brasil*. Instituto de Pesquisas Ecológicas (IPE), Brasil, pp. 237–246.

Rodrigues, F.H.G., 2002. Biologia e conservação do lobo-guará na Estação Ecológica de Águas Emendadas, DF. Tese de doutorado em Ecologia. Instituto de Biologia da Universidade Estadual de Campinas.

Rodrigues, F. and de Paula, R., 2007. Biologia comportamental e conservação do lobo-guará (*Chrysocyon brachyurus*) no Cerrado de Minas Gerais. Pró-Carnívoros, Intituto para a conservação dos carnívoros neotropicais. http://www.procarnivoros.org.br/acoes.php.

Santos, E.F., Setz, E.Z.F., and Gobi, N., 2003. Diet of the maned wolf (*Chrysocyon brachyurus*) and its role in seed dispersal on a cattle ranch in Brazil. *Journal of Zoology*, 260:203–208.

Sergio, F., Newton, I., Marchesi, L., and Pedrini, P., 2006. Ecologically justified charisma: Preservation of top predators delivers biodiversity conservation. *Journal of Applied Ecology* (43):1049–1055.

Sillero-Zubiri, C., Hoffmann, M., and Macdonald, D.W. (eds), 2004. *Canids: Foxes, Wolves, Jackals and Dogs. Status Survey and Conservation Action Plan*. IUCN/SSC Canid Specialist Group. Gland, Switzerland and Cambridge, UK.

Silva, C.B.X., 2000. Projeto lobo guará—Uma contribuição à conservação dos Campos Gerais no Paraná, Brasil. Em: *Workshop do lobo guará*, 3, 1998, São Bernardo do Campo. Protocolo de manejo do lobo guará mantidos em cativeiro. Sorocaba: Sociedade de Zoológicos do Brasil, pp. 43–49.

Silva, C.B.X. and Nicola, P.A., 1999. Observações sobre o lobo guará no Parque Estadual de Vila Velha, Ponta Grossa, Paraná, Brasil. In *Congresso Brasileiro de Conservação e Manejo da Biodiversidade*, 23, Ribeirão Preto, anais.

Silva, A.J. and Talamoni, S.A., 2004. Core area and centre of activity of maned wolves, *Chrysocyon brachyurus* (Illiger) (Mammalia, Canidae), submitted to supplemental feeding. *Revista Brasileira de Zoologia* 21(2):391–395.

Sollmann, R., Furtado, M.M., Jácomo, A.T.A., Tôrres, N.M., and Silveira, L., 2010. Maned wolf survival rate in central Brazil. *Journal of Zoology* 282:207–213.

Songsasen, N. and Rodden, M.D., 2010. The role of the species survival plan in maned wolf *Chrysocyon brachyurus* conservation. *International Zoo Yearbook* 44:1–13.

Sorrentino, M., 1997. Educação Ambiental e Universidade: Um estudo de caso. In Padua, S.M. and Tabanez, M.F. (eds), *Educação ambiental: Caminhos trilhados no Brasil*. Instituto de Pesquisas Ecológicas (IPE), Brasil, pp. 43–54.

Strauss, A., 2004. Carnivore education: Beyond biology. Carnivores 2004: Expanding partnerships in carnivore conservation. *5th Defenders of Wildlife Conference*. November 14–17, Santa Fe, NM, p. 38.

Taylor, D., 2004. Conservation education and its relevance to wild canids. In Sillero-Zubiri, C., Hoffmann, M., and Macdonald, D.W. (eds), *Canids: Foxes, Wolves, Jackals and Dogs. Status Survey and Conservation Action Plan*. IUCN/SSC Canid Specialist Group. Gland, Switzerland and Cambridge, UK, pp. 298–305.

Taylor, D., 2009. Education as a conservation strategy: Exploring perspective transformation. In Musiani, M., Boitani, L., and Paquet, P.C. (eds), *A New Era for Wolves and People: Wolf Recovery, Human Attitudes and Policy*. Energy, Ecology and the Environment Series. University of Calgary Press, Calgary, AL, pp. 161–172.

Treves, A. and Karanth, K., 2003. Human–carnivore conflict and perspectives on carnivore management worldwide. *Conservation Biology* 17(6):1491–1499.

Vanstreels, R.E.T. and Pessutti, C., 2010. Analysis and discussion of maned wolf *Chrysocyon brachyurus* population trends in Brazilian institutions: Lessons from the Brazilian studbook, 1969–2006. *International Zoo Yearbook* 44:121–135.

Villas-Bôas, O. and Villas-Bôas, C., 2004. *A marcha para o Oeste: Epopeia da expedição Roncador-Xingu*. 6ª edição, Editora Globo, Brasil.

Woodroffe, R., Thirgood, S., and Rabinowitz, A. (eds), 2005. *People and Wildlife: Conflict or Co-existence?* Conservation Biology 9. Cambridge University Press, Cambridge, UK.

4 The Maned Wolf *Ex Situ* Worldwide

Ruben Holland

CONTENTS

History of the ISB and Some Regional Programs 53
Function of Studbooks 54
EEP and SSP 54
The *Ex Situ* Maned Wolf Population 55
Europe 57
North America 57
Australia 58
Asia 58
Africa 59
Maned Wolf Observations in Zoos 59
Next Steps in Research and Conservation 61
Product Mentioned in the Text 61
References 61

Maned wolves have a long tradition *ex situ*. In many zoos all over the world, people can view this species. Over the last 80 years, more than 3250 individuals have been listed in the International Studbook (ISB) for the maned wolf. These animals have been shown in more than 320 different institutions globally.

HISTORY OF THE ISB AND SOME REGIONAL PROGRAMS

The ISB for the maned wolf was established in 1972 at Heidelberg University and the studbook keeper was Peter Löben. In 1980, the ISB went to Frankfurt Zoo, Germany, and Dr. Bernd Matern (1981–2002) was the new coordinator. After his retirement, Dr. Sabine Hilsberg (2003–2005) was the studbook keeper for a short while before Dr. Rüdiger Dmoch (2006–2008) took over the role. His successor was Dr. Thomas Wilms. He did the studbook keeping for a short while until 2010, when the ISB was taken over by Ruben Holland (2010) from Leipzig Zoo, Germany. In addition to this studbook, regional studbooks were established in some associations.

In 1985, the American Zoo and Aquarium Association (AZA) arranged the Species Survival Program for the Maned Wolf (SSPMW), which has been maintained by Melissa Rodden from the National Zoo Conservation and Research Center in Front Royal (Fletchall et al. 1995).

The European Association of Zoos and Aquaria (EAZA) founded the European Endangered Species Program (EEP) for the Maned Wolf in 1989. This program was coordinated at Frankfurt Zoo by Dr. Bernd Matern and his colleagues until 2010, when it was transferred, together with the ISB, to Ruben Holland and Leipzig Zoo.

Also, in Brazil, the Brazilian Studbook for the Maned Wolf (BSMW) was started in 1989 and was maintained by the Parque Zoologico Municipal Quinzinho de Barros (Sorocaba, Brazil) and in person by Cecilia Pessutti (Vanstreels & Pessutti 2010).

FUNCTION OF STUDBOOKS

Studbooks in general are databases for different species where diverse information is stored. These data are important for the life history of any species controlled in such a studbook. In addition to genetic data and demographic data, individual markings and transport history are shown. The Single Population Analysis and Records Keeping System (SPARKS) program is used by the ISB of the maned wolf to capture every individual of the species in captivity worldwide and every maned wolf that is entered receives an individual studbook number. Data that are filled in include birth date, birth place, parents, rearing type, identifications (transponders, tags, tattoos, etc.), transfers, death date, post mortem reports, and any other special notes that are important. All this information helps to observe population trends and allows management of the population in a scientific way. Demographic and genetic data are especially important to assess if the population is in good condition or if action is needed to increase breeding or to build new breeding pairs.

To keep the studbook up to date every year, it is very important that all zoos that keep maned wolves participate in collecting all relevant data for every individual in captivity. In former times, the studbook keeper sent questionnaire letters to every institution to ask for all necessary changes in the maned wolf livestock at the beginning of every year. In present times, these requests are sent via e-mail. Every institution can fill in an attached Word document or send an up-to-date Animal Records Keeping System (ARKS) taxon report. In North America, Europe, and Africa, almost every participant answers the requests of the studbook keeper, but in Asia and South America, this is not the case. For example, the reply rates in Brazil range from 40% to 70%, depending on the year (Vanstreels & Pessutti 2010). On the one hand, this is a language problem because the requests are written in English with only a few translations into other languages. On the other hand, in many countries it is not a requirement to answer those requests and participation depends on the willingness and commitment of the institutions. Zoos that use International Species Information System (ISIS) regularly send their data via ARKS to the ISIS data bank. If the participating institutions do not answer the requests but update their ARKS data frequently, the studbook keeper can get these data directly from ISIS.

ISIS was developed in 1973; in 1974, 51 North American and 4 European zoos participated. Now, 835 zoos in 80 countries on 6 continents are using this system (Bingaman Lacky 2010). With ARKS, all participating institutions can record, maintain, and analyze data for every animal of every species in the zoos and send those data to ISIS.

EEP AND SSP

The EEP in Europe and the SSP in the United States are powerful programs to manage maned wolves in these regions so that both populations are genetically healthy and demographically stable. The most important tool in these scientifically managed *ex situ* programs is the studbook (Althaus et al. 2010). The coordinator of the program obtains all data needed to manage the population successfully from the studbook. After entering all the data into SPARKS, many different data analysis possibilities are available. You can export the studbook files to create a studbook for the region, or you can export genetic and demographic files to bring them into PMx (Ballou et al. 2011), which is a population management program that can analyze the data to create good breeding pairs and prevent inbreeding. After finding the right pairings, the coordinator discusses the recommendations with the members of the species committee, which consists of elected representatives of institutions that keep maned wolves in the appropriate region. If they agree to the suggestions, the coordinator contacts the participants who will swap their animals with other institutions or authorize breeding. A big problem of *ex situ* management is the space that is available to a population. For example, in Europe there are about 70 institutions keeping maned wolves at the moment. If every institution kept one breeding pair, this would be 140 animals altogether and the chance of higher breeding than death numbers is most likely. The fact that maned wolves do not live in groups means that for most surplus animals the coordinator has to find a new home. In this case, many new institutions would need to start keeping the species. Because this is not possible in the majority of the cases, the coordinator has to recommend breeding stops in pairs that are genetically less important than others.

THE *EX SITU* MANED WOLF POPULATION

As mentioned above, 3288 maned wolves were listed in the ISB at the beginning of 2012: 1426 males, 1291 females, and 571 unsexed animals. Over the last 80 years, these animals have been kept in more than 320 institutions all over the world. Maned wolves have been and are still kept in Africa, Asia, Europe, North America, and South America. The first maned wolf in the studbook was mentioned in the United States in 1933. In the following years, the studbook never mentioned more than five maned wolves altogether in the zoos of the world. In 1965, there were over ten individuals in captivity and from this point the numbers increased up to a maximum of 436 specimens in 2011. In the first years of keeping maned wolves in zoos, most of the animals were wild caught and the first captive birth was in 1963 in Frankfurt Zoo, but all the puppies died the day they were born. The first animal that reached adulthood was a female born in Frankfurt in 1967. From this time on, more and more successful captive births have taken place and in 1978, more captive-born animals lived in zoos than wild-caught animals. The numbers of captive-born animals increased in the following years to more than 350, while the number of wild-caught animals only increased slowly and has remained stable at about 50 animals for the last few decades (Figure 4.1). Most of these wild-caught animals either were ill or were puppies that had lost their parents, and were found in their wild range and brought to zoos in South America, mainly in Brazil. In the years 1991–2011, between 5 and 17 animals were captured each year. Few animals came to zoos with unknown origin. Most of these animals were transferred between 1933 and 1970, long before the studbook was started. These animals were entered in the studbook belatedly and their places of origin could not be found in the documents, if they had any documents at all.

In 1988, 100 or more puppies were born in 1 year for the first time since the studbook was established. In the following years, an average of 100 puppies were born per year until 1998. In 1990, the highest number was reached, with 116 births in 1 year. In recent years, the number of births has decreased to 40–50 because in Europe and North America strict breeding controls have been realized because of a lack of enclosures. Reproduction in maned wolves starts when the animals reach the age of 1 and the oldest reproductive animals are about 14 years old.

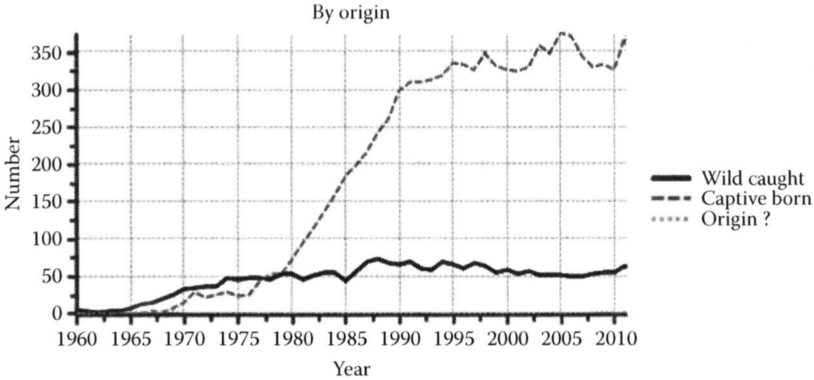

FIGURE 4.1 Maned wolves' origins. Wild-caught animals (black), captive-born animals (dark gray dashes), and animals of unknown origin (light gray dots). Before 1960, only single individuals were held in zoos, which are not shown in the graph. Until 1978, more wild-caught animals lived in zoos than did captive-born animals. But since 1978, the number of captive-born animals has increased, while the number of wild-born animals has remained at about 50 individuals.

The sex ratio over the years has been balanced. From 1975 to 1981, both males and females have been kept in the same numbers. From 1981 until 1994, a few more males were shown in the zoos. From 1995 until 1999, the sex ratio was balanced again and from 2000 on up to 30 more females were kept than males (Figure 4.2). Unsexed animals usually only appear in the studbook when puppies are born and disappear again before they can be sexed, because their parents have eaten them.

Maned wolves' births take place at different times of the year, depending on where they are kept. In the Northern Hemisphere (North America, Europe, Asia), births take place between November and March. In the Southern Hemisphere (South America and Australia), births take place between May and August.

In times when many animals lived in the zoos and were born there, many animals of course died there too. Between 1989 and 1997, about 100 animals died per year. But it has to be said that many of those deaths were puppies that did not survive for a longer time. Within the first 30 days of life, 41% of the puppies died. In recent years, birth rates and death rates have been approximately the same and the population is stable. Of all animals, 25% reach an age of 9.4 years and about 1% survive up to 16.6 years. The oldest animal that ever survived in a zoo was a female maned wolf that died at the age of 17 (6535 days) in Asson, France.

When an animal dies in one of the institutions, the responsible zoo normally sends a postmortem or autopsy report to the studbook keeper. This helps to gain a good overview of all relevant illnesses in the population and trends are visible. Due to this fact, veterinarians have the chance to do research on different frequent diseases to prevent their recurrence in the future.

A big advantage in the maned wolf programs is that every living individual in the studbook has 100% known origin. Through precise data collection and the help of every zoo, giving as much information as possible, the pedigree of every maned wolf in the studbook is known back to the ancestors who were captured in the wild. Thus it is possible to make exact specifications about the genetics of every individual and inbreeding can be avoided. This also helps to see which bloodline or individual is genetically more represented than others; the coordinator can then manage the population well. Based on the global population of maned wolves, the genetic diversity was 0.969 at the beginning of 2012, which is a very high value. The value of the wild population is 1. For example, the gene diversity at the beginning of 2012 of the European population was 0.9442. The EEP's aim is to make sure the value does not go below 0.9 in the next 100 years. Reasons for decrease can be the number of individuals in the population, the current gene diversity, the maximum capacity for the species in the participant zoos, the generation length, the pairings in the

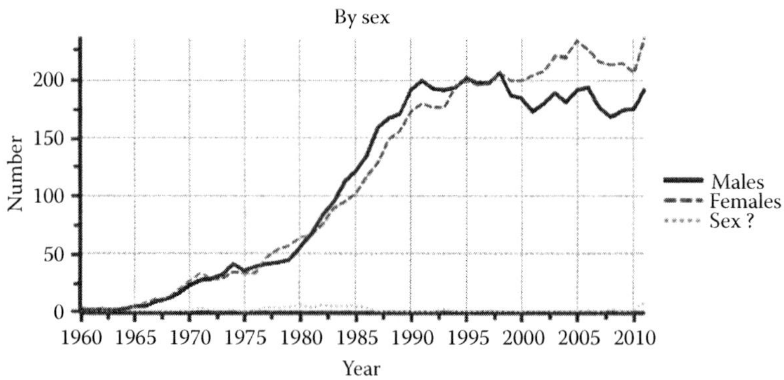

FIGURE 4.2 Maned wolves' sex. Males (black), females (dark gray dashes), and unsexed animals (light gray dots). Before 1960, only single individuals were held in zoos, which are not shown in the graph. Numbers of both sexes are nearly the same over the years, but since 1997, more females than males have lived in zoological institutions.

population, and of course the input of new individuals from other regions or from the wild. One way of increasing the gene diversity is to bring more wild-caught animals into breeding situations. From the 64 wild-caught animals that lived in zoos at the end of 2011, only four ever successfully bred. This is only 6%. Altogether, 633 wild animals were captured, from whom only 77 (12%) successfully bred. From these animals, 60 were still potential breeders in 2012.

Of the 3288 maned wolves that have ever lived in zoos, 685 (21%) have bred. The four animals with the most offspring all lived in Prague Zoo, Czech Republic. One male produced 37 offspring, a male and a female had 35 puppies, and one female had 33 puppies. The average number of puppies in all breeding animals is 7.7. The average size per litter ($n = 975$) is 2.7 puppies. Litters vary from one to seven puppies (1 puppy, 20%; 2, 26%; 3, 28%; 4, 18%; 5, 6%; 6, 1%; 7, <1%) and the bigger the litter, the better the survival of the puppies. If only one puppy is born, its chances of reaching the age of 1 year or older are only 23%. If five or six puppies are born in one litter, 51% of them reach an age of one or older.

The average age of first reproduction is 4 years in females and 3 years and 11 months in males and the average age of final reproduction is 6 years and 7 months in females and 6 years and 6 months in males. The average period of reproduction is 5 years and 9 months in females and 5 years and 8 months in males. A total of 2397 puppies were reared by their parents and only 256 (10%) were hand reared.

In the beginning of 2012, a total of 150 institutions held 394 (182 males, 211 females, 1 unknown) maned wolves.

Given that global management of the population has not yet taken place, an overview of the different regions is given here.

EUROPE

A total of 1582 (680, 592, 310) manes wolves were specified in the ISB for the European region. All countries of the European continent and the countries of Russia, Kazakhstan, Estonia, Latvia, Lithuania, and Ukraine are included in this region.

The first four animals were wild-caught animals that were transferred to Frankfurt in 1960. Over the years, 69 wild-caught animals were imported to European zoos, 38 of which were reproductive. A further 160 animals, which were captive bred, were imported from North and South America. This is 10% of the overall population in Europe. A total of 1353 animals were born in Europe, many of which were transferred to South and North America, Africa, Asia, and Australia.

Frankfurt Zoo was the first institution to breed maned wolves in captivity. On 4 February 1963, the first four puppies were born, but they did not survive more than 1 day. A total of 638 individuals did not survive the first month, but 521 of all puppies born survived longer than 1 year. Of all animals born in Europe, 100 were hand reared, which amounts to 7%. In the European population, 379 animals were reproductive. One of these animals, in Prague Zoo, had 37 offspring in its lifespan of 13 years.

The oldest animals were 17 years at death. One of these animals was born in 1987 in Halle and lived until death in 2004 in Asson. The other one was born in Moscow in 1989 and died in Katowice in 2006.

At the beginning of 2012, 156 maned wolves were listed in the ISB in 72 European institutions.

NORTH AMERICA

In the ISB for the maned wolf, 547 (222, 217, 108) individuals were listed in North America. The first animal noted in the ISB was transferred to Smithsonian National Zoological Park, Washington, in 1933. The origin of this animal was unknown. In 1944, the last animal died and no maned wolf was shown in the United States until 1957, when one animal was imported to the Philadelphia Zoo, but it died 1 year later. Permanent animals lived in the US zoos from 1964. Of the animals that were imported to North American zoos, 18 were wild-born animals. All captive animals that were brought to the United States, a total of 59, came from Europe and South America.

In 1968, the first two puppies were born in a US zoo, in Los Angeles Zoo, but both died the same day. Since then, 470 maned wolves have been born in zoos in North America. Of all maned wolves in North America, 86% were captive born in one of the national zoos and only 14% were imported from overseas. Of all the maned wolves born in North America, 247 survived beyond 1 year, but 187 (40%) did not survive beyond 30 days. Ninety-three (20%) of the puppies were hand reared, eleven (2%) were not reared at all, and the rest of the population were parent reared.

Reproduction took place in 152 animals of the whole North American population and 33 of them gave birth to at least 10 puppies. Four animals even gave birth to 26 puppies.

The oldest animals in North American zoos reached the age of 16 in the zoos of San Antonio, Chicago (Lincoln Park), and Denver. One of these animals was wild born.

At the beginning of 2012, there were 99 individuals in 30 North American institutions.

AUSTRALIA

In Australia, 87 (35, 25, 27) maned wolves are noted in the ISB. Six different zoos hold or held the species. The first three animals were transferred from the German zoos of Berlin and Dortmund to Perth Zoological Gardens in 1981. Since this time, 15 animals have been imported to Australia, from South America, North America, and Europe. All these animals were captive born. These 15 animals were the breeding stock for 72 puppies that have been born in Australia since 1983. The first four puppies were born in Perth from two of the first imported animals from Berlin and Dortmund, but they did not survive more than 2 days. Of the 72 puppies that were born in Australia, 28 survived beyond 1 year, but there were also 34 puppies that did not survive the first 30 days.

Birth seasonality is visible in Australia. Most of the births take place in June and July (53 births = 74%). Only one birth (1%) took place in May, three (4%) in August, ten (14%) in September, two (3%) in October, and three (4%) in December.

The oldest animal that lived in Australia was 15 years old. It was born in Port Lympne in the United Kingdom and was transferred to Melbourne at the age of eight.

In 2012, only the Altina Wildlife Experience held this species, with eight (3, 5) individuals.

ASIA

In Asia, 95 (45, 42, 8) maned wolves are noted in the ISB. This species is recorded in China, India, Indonesia, Japan, Malaysia, Singapore, South Korea, Sri Lanka, Thailand, Tajikistan, Uzbekistan, and Vietnam. In total, 15 different institutions in Asia presented maned wolves to the public. Some of these zoos only showed the species once and for a short time, but others even had breeding success. Animals that were imported came from all parts of the world. Wild-caught animals were brought from Brazil and Paraguay to Asia and captive-born animals were brought from South American, European and North American zoos.

The first maned wolf arrived in Ueno Zoological Gardens, Tokyo, Japan, in 1969 and it was a wild-caught animal. From this time on, maned wolves were regularly transported to Asia. The first birth took place in Zoo Negara Malaysia, but all the puppies died within the first 3 months of life. Up to 2011, 29 puppies were born in Asia. Most of them died shortly after birth. Twelve (41%) Asian captive-born animals never reached the age of 30 days. Only 7 (24%) of the 29 reached the age of 1 year. Breeding success took place in Ueno Zoological Gardens Tokyo, Japan, Shanghai Zoological Gardens, China, Singapore Zoological Gardens, Singapore, and Zoo Negara Malaysia. Four (14%) of the puppies that were born in Asia were hand reared at Shanghai Zoological Gardens with full success.

Births do not take place the whole year round. From December to February, 22 (76%) births took place and between March and May there were seven (24%).

The oldest animal in Asia was transferred at a young age from Rotterdam Zoo, the Netherlands, to Singapore Zoo and lived there till it died at the age of 15 years.

In 2011, there were still 7 institutions in Asia which showed 20 (8, 12) animals altogether.

AFRICA

In 1984, the first two maned wolves came to Africa. More precisely, a young pair was transferred from Zoo Dortmund in Germany to Pretoria National Zoological Gardens in South Africa. These two animals produced 23 puppies in eight broods, but more than half of the offspring died shortly after birth (14 did not survive). In 1991, two female maned wolves from Munich Zoo, Germany, were imported by Pretoria. One of these females was paired with a male from Sao Paulo Zoo, Brazil, which was imported in 1993. The pair produced twelve puppies in three broods. The other female produced six puppies in two broods with a male that was born in Pretoria from the Dortmund pair. In 2005, Pretoria imported another male maned wolf from Frankfurt Zoo, Germany. This male produced seven puppies in three broods with a female born in Pretoria.

There was another South African zoo called African Game Service Skeerport, which imported a female maned wolf from Zoo Osnabrück, Germany, in 2002, but this animal never produced any offspring.

In 2000, Zoo Plzen in the Czech Republic exported the only two maned wolves, a pair, to a zoo in Africa that was not located in South Africa. These two animals were transferred to Parc Zoologique National de Rabat in Morocco. They never produced any offspring and died 1 year after arrival.

The only animal from Africa to leave the continent was a female born in Pretoria. It was transferred to Cologne Zoo in Germany in 1997.

In total, 65 (32, 26, 7) maned wolves lived in Africa, 63 (31, 25, 7) of whom were located in South Africa's zoos. In addition to Pretoria and Skeerport, five other institutions showed maned wolves in their collections.

Of the nine animals that were imported to Africa, only six ever produced offspring. These animals were directly involved in 50 of the 56 offspring born in South Africa since 1984. Only two pairs where both animals were born in Africa ever produced offspring, with a total of six.

Most births (43 = 77%) in Africa take place in June. A few puppies were born at the end of May (8) or at the beginning of July (5). Twenty (36%) animals that were born in Africa were hand reared. Of these animals, only one male produced offspring, having six puppies with one of the Munich females.

Only five animals (8%) reached the age of 10 years or older and the oldest animal in Africa was 13 years when it died. Only 19 (34%) animals born in Africa reached their first birthday; 30 (54%) animals never reached the age of 1 month and 21 (38%) of them did not even survive beyond 4 days.

In 2011, only one pair with one female offspring lived in African zoos.

MANED WOLF OBSERVATIONS IN ZOOS

As everybody knows, maned wolves are very shy animals and are hard to observe in their natural range. For this reason, zoos can play a key role in behavioral observations. In 1981, Bartmann and Nordhoff observed that not only the female but also the male is involved in rearing the puppies (Bartmann and Nordhoff 1984). Both animals protect the young and play with them, and as soon as they are weaned, both parents bring them food. This observation had never been made in nature. Hämmerling and Lippert (1975) gave detailed descriptions of mating behaviors and mother–child bonds (Figures 4.3 and 4.4).

In zoos, the growth of puppies can be accurately charted, because they can be weighed regularly (Figure 4.5).

FIGURE 4.3 Mating maned wolves. Many behaviors can be observed in zoos much easier than in the field.

FIGURE 4.4 Maned wolf with puppy. If maned wolves with puppies are disturbed, they search for another place to hide them.

FIGURE 4.5 Maned wolf puppies. In captivity it is easier to control puppies, so you can create weight graphs and other important data.

NEXT STEPS IN RESEARCH AND CONSERVATION

The numbers of maned wolves *ex situ* are high enough to establish a healthy self-sustainable zoo population over the next 100 years. Furthermore, the animals that came into zoos from the wild could help to make the population stronger yet. The problem at the moment is that there are enough animals in captivity but the exchange between the different regions is not working well. All three regional breeding programs work more or less only in their regions. The ISB provides the opportunity to bring all these single programs to a common aim, as in other species. For example, the EEP and the SSP of the okapi (*Okapia johnstoni*) work closely together to obtain a higher number of animals in the managed population (Byers 2008). A problem that occurs when different regions want to exchange animals is that the transfer regulations of every country are different. Obtaining the necessary documents and transport papers involves much bureaucracy. Also, strict veterinary requirements have to be met before an animal can be sent from one continent to another. All these issues are responsible for the difficulty of such transport and often it takes more than a year to organize and complete such a transfer.

If all the institutions and studbook keepers together keep the population of the maned wolf healthy, genetically and demographically, there may be a chance to reintroduce some of the individuals back to the wild if necessary.

Maned wolves in zoos are also representatives for the wild-living individuals of the species and visitors can see what wonderful animals they are. By seeing the beauty of these animals in zoos and by learning more about the species, people could be motivated to save the maned wolves and their habitat.

Stephanie V. Bestelmeyer wrote in 1998 that zoos are increasingly supportive of research in a variety of areas, both within zoos and in the field. For example, she wrote: "Research on captive Maned Wolves provided an ethogram and numerous questions to test in the field." All research *ex situ* can help to better understand the species and protect them from extinction.

PRODUCT MENTIONED IN THE TEXT

SPARKS: Single Population Analysis and Records Keeping System, version 1.6, developed by International Species Information System (ISIS), Egan, MN 55121, USA. Data current through 31.01.2012.

REFERENCES

Althaus, T., Bingaman Lacky, L., Fisken, F.A., & Morgan, D. (2010) The role of international studbooks in conservation breeding programmes. *Building a Future for Wildlife: Zoos and Aquariums Committed to Biodiversity Conservation* (eds. Dick, G. & Gusset, M.), pp. 49–52. Gland: WAZA Executive Office.

Ballou, J.D., Lacy, R.C., & Pollak, J.P. (2011) PMx: Software for demographic and genetic analysis and management of pedigreed populations (Version 1.0.20110726). Chicago Zoological Society, Brookfield, IL. Available from http://www.vortex9.org/PMx.html.

Bartmann, W. & Nordhoff, L. (1984) Paarbindung und Elternfamilie beim Mähnenwolf. Zeitung des Kölner Zoos pp. 63–71.

Bestelmeyer, S.V. (1998) Linking zoo research and field research for the conservation of the endangered maned wolf (*Chrysocyon brachyurus*). *AZA Annual Conference Proceedings*, pp. 48–53.

Bingaman Lacky, L. (2010) Records, studbooks, regional zoo associations, and ISIS. *Wild Mammals in Captivity: Principles and Techniques for Zoo Management*, (eds. Kleiman, D.G., Thompson, K.V., & Kirk Baer, C.), 2nd edn, pp. 504–510. Chicago, IL: University of Chicago Press.

Byers, O. (2008) Global programmes for species management. *EAZA News 64. Collection Planning Special*, p. 26.

Dmoch, R. (2006–2008) *International Studbook for the Maned Wolf* Chrysocyon brachyurus *(Illiger, 1811)*. Frankfurt: Zoologischer Garten Frankfurt.

Fletchall, N.B., Rodden, M., & Taylor, S. (eds) (1995) *Husbandry Manual for the Maned Wolf* (Chrysocyon brachyurus). Bethesda, MD: American Association of Zoos and Aquariums, Maned Wolf Species Survival Plan Committee.

Hämmerling, F. & Lippert, W. (1975) Beobachtung des Geburts-und Mutter-Kind-Verhaltens beim Mähnenwolf über ein Nachtsichtgerät (Infrarot Beobachtungsanlage). *Der Zoologische Garten* 45: 393–415.

Hilsberg, S. (2003–2005) *Internationales Register und Zuchtbuch für den Mähnenwolf* Chrysocyon brachyurus *(Illiger, 1811)*. Frankfurt: Zoologischer Garten Frankfurt.

Holland, R. (2010) *International Studbook 2010 for the Maned Wolf* Chrysocyon brachyurus. Leipzig: Zoo Leipzig.

Matern, B. (1981–2002) *Internationales Register und Zuchtbuch für den Mähnenwolf* Chrysocyon brachyurus (*Illiger, 1811*). Frankfurt: Zoologischer Garten Frankfurt.

Vanstreels, R.E.T. & Pessutti, C. (2010) Analysis and discussion of maned wolf *Chrysocyon brachyurus* population trends in Brazilian institutions: Lessons from the Brazilian studbook, 1969–2006. *International Zoo Yearbook* 44: 121–135.

5 The *Ex Situ* Maned Wolf Population in Brazilian Zoos

Cecília Pessutti, Cleyde Angélica Ferreira da Silva Chieregatto, and Ralph Vanstreels

CONTENTS

History of Brazilian Management Plan for the Maned Wolf 64
Characteristics and Demography of the *Ex Situ* Maned Wolf Population in Brazil 64
Research Conducted on the Brazilian *Ex Situ* Maned Wolf Population 68
Environmental Education through the Brazilian *Ex Situ* Maned Wolf Population 68
Husbandry Protocols for Maned Wolves 69
  Enclosure Characteristics 69
  Environmental Enrichment 70
  Nutritional Management 70
  Management of Reproduction 71
  Hand Rearing of Cubs 71
  Handling and Husbandry of Wild-Caught Cubs and Adults 72
Conclusions and Future Perspectives 72
References 73

Before discussing animals in captivity and their conservation, it is essential to examine the history of *ex situ* animal collections and how far they date back in human history. The oldest known zoos were built in Egypt and date back to 4000 BC (Rose, 2010), and for thousands of years these collections of exotic animals were an exclusivity of pharaohs, emperors, and kings. It was not until the eighteenth century that the first exotic animal exhibitions were organized for the leisure of the general public; these gradually developed into modern zoos as we know them. The first zoos in Europe were those of Vienna, Paris, and London; in North America, those of Boston, New York, and Chicago; and in Brazil, the Belém and Rio de Janeiro zoos were founded in the nineteenth century (Costa, 2004).

The increasing habitat destruction since last century has given zoos worldwide a new role as stewards of genetically and demographically viable wild animal populations. For that purpose, zoos have had to adapt and restructure, seeking to provide knowledge on species biology, to promote the welfare of their captive specimens, and to participate in collaborative networks for wildlife conservation (Mallinson and Feistner, 1995; WAZA, 2012).

Zoological associations and specialist groups have been created throughout the world, such as the World Association of Zoos and Aquariums (WAZA), the Association of Zoos and Aquariums (AZA), the European Association of Zoos and Aquaria (EAZA), the Brazilian Zoological Society (SZB), and the Conservation Breeding Specialist Group (CBSG). These associations and groups share expertise, techniques, and efforts with local institutions and groups for the conservation of threatened species, combining both *in situ* and *ex situ* strategies (WAZA, 2005; Songsasen and Rodden, 2010). Conservation Assessment and Management Plan (CAMP) and Population and Habitat Viability Assessment (PHVA) workshops, along with specific training programs in veterinary medicine, captive husbandry, environmental education, environmental enrichment, and fundraising, among others, are some of the tools used by these groups and associations to contribute to the protection of wildlife species and habitats.

Population recovery and species conservation programs as we know them were developed in the early twentieth century, following the examples of the reproduction programs conducted for the European bison (*Bison bonasus*) (Channell and Lomolino, 2000) and the Californian condor (*Gymnogyps californianus*) (Snyder and Snyder, 1989; Meretsky et al., 2000). Similar programs were also developed for several canids, including the red wolf (*Canis rufus*), the Mexican wolf (*Canis lupus baileyi*), the African wild dog (*Lycaon pictus*), the bush dog (*Speothos venaticus*), and the maned wolf (*Chrysocyon brachyurus*) (Bauman et al., 2004; Songsasen and Rodden, 2010).

Of the six canid species occurring in Brazil, only the bush dog and the maned wolf are considered threatened with extinction; both are listed as vulnerable by the Brazilian Institute of Environment and Renewable Natural Resources (IBAMA, 2003) and are classified as Near Threatened by the International Union for Conservation of Nature (IUCN, 2011).

HISTORY OF BRAZILIAN MANAGEMENT PLAN FOR THE MANED WOLF

Coordinated efforts for the recovery and captive breeding of threatened species in Brazil began in 1989, following the first edition of the course "Curso de Manejo Integrado Cativeiro—Natureza" at the Sorocaba Zoo, São Paulo state.

In 1990, the Brazilian Management Plan for the Maned Wolf (Plano de Manejo Brasileiro para o Lobo Guará) was created with the support of the Brazilian Zoological Society and a few years later, in 1995, it received official governmental recognition from the Brazilian Institute of Environment and Renewable Natural Resources (IBAMA) and was renamed the Canid Workgroup (Grupo de Trabalho de Canídeos). Since the creation of the Brazilian Management Plan for the maned wolf in 1990, several annual or biannual technical meetings have been held at the conferences of the Brazilian Zoological Society and the São Paulo Zoological Society, and three national workshops (1990, 1994, 1998) were organized by the group.

In 2003, IBAMA revoked the Canid Workgroup and created a new group, the Committee for the Conservation and Management of Brazilian Canids (Comitê para Conservação e Manejo dos Canídeos Brasileiros), which was also dismissed shortly after. Since then, no official group or management plan for the conservation of the maned wolf or Brazilian canids has been appointed by the governmental authorities. However, many of the original contributors to the Brazilian Management Plan for the Maned Wolf have continued their efforts as an unofficial network. In 2006, a PHVA workshop was organized for the species with the support of the North American Maned Wolf Species Survival Plan (SSP). The PHVA workshop had representatives from Latin American countries in which the species occurs, as well as from the North American SSP, and aimed to diagnose problems and provide solutions to *in situ* and *ex situ* professionals working for the conservation of the species.

CHARACTERISTICS AND DEMOGRAPHY OF THE *EX SITU* MANED WOLF POPULATION IN BRAZIL

Maned wolves were first admitted in Brazilian zoos in the 1960s and 1970s, and since then have been recorded by the regional and international studbook databases for the species (Pessutti, 1994; Dmock, 2007). In the late 2000s, studbook data were used to revise and evaluate the demographics of the Brazilian *ex situ* population, and these analyses were provided in detail in Vanstreels and Pessutti (2010). We herein summarize and highlight the most critical aspects and characteristics of this captive population during the first two decades of the existence of the regional studbook.

Overall, the current status of this population is similar to that of the first decades of most captive large mammal populations when first established throughout the world: it is highly dependent on the influx of wild-caught specimens and has overall poor breeding success. Almost one thousand maned wolves have lived in captivity in Brazil since 1969, a population that has had a slow but stable growth, reaching a plateau of approximately 120–140 animals in the late 1990s (Figure 5.1). This plateau likely reflects the limits of enclosure availability for the species, as institutions tend to become uninterested in receiving new animals after they have acquired a few animals (in December 2010, each institution housed 3.1 ± 1.83 individuals).

Most maned wolves captive in Brazil spend all their lives in the same institution (78%) and transfers between Brazilian institutions have been rare. Such transfers are even less frequent on an international level: Brazil has exported 5% of its maned wolf captive population, mostly to Europe and North America, and has never received any international transfers. This apparent lack of interest in exchanging specimens results from the relatively high availability of wild recruits. Wild maned wolves were most frequently brought into captivity from 1985 to 1995, with 20–30 wild individuals captured per year. After that, however, the average number of captured maned wolves dropped considerably (Figure 5.2), and in the recent past typically <10 wild individuals are brought into captivity per year. It should be made clear that in the vast majority of cases those wild recruits are not captured intentionally for the purpose of being brought into zoos, but are instead captured due to road accidents, conflicts with human activities, and so forth.

Not all of these captured maned wolves will thrive in captivity, as 42% of the wild-caught individuals will die in their first captive year, yet the influx has been sufficient to make maned wolves relatively common in Brazilian zoos. Unfortunately, capture locations were not recorded or reported in the studbook for most animals (52%); for those animals for which capture location is known, central Brazil (particularly the states of São Paulo, Goiás, and Minas Gerais) is the most frequent origin.

On the other hand, captive breeding has met with relatively poor success, mostly because approximately only 21% of the captive-born cubs survive their first year. This startlingly elevated cub mortality is thought to result largely from infanticide. Greater efforts were made to breed maned wolves in Brazilian institutions around 1988–1998, resulting in a rapid increase in the captive-born population, especially in 1995–1998 (Figures 5.1 and 5.2). Since 1999, efforts to breed the species in captivity have decreased considerably. This probably occurred because by that time the species' captive population started to reach the plateau of 120–140 individuals, and as maned wolves started to become perceived by administrators as common (if not overabundant) animals in the zoo collections, institutions decreased their efforts to breed the species.

The low breeding success is further complicated by the fact that it does not maximize genetic management: 20% of the breeders were vastly overrepresented in the population (i.e., had ten or

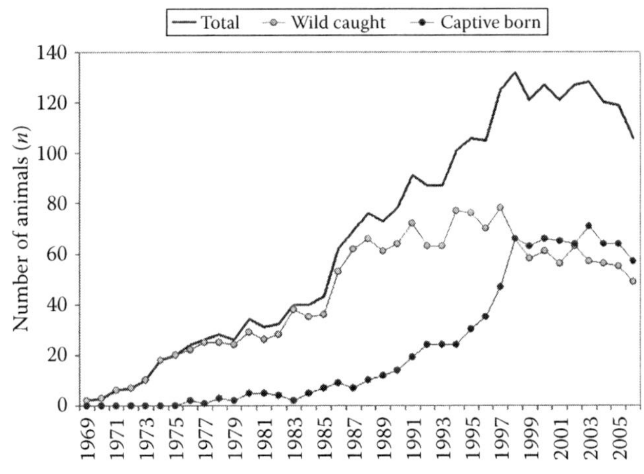

FIGURE 5.1 Historical series of the studbook-recorded captive population of maned wolves in Brazil.

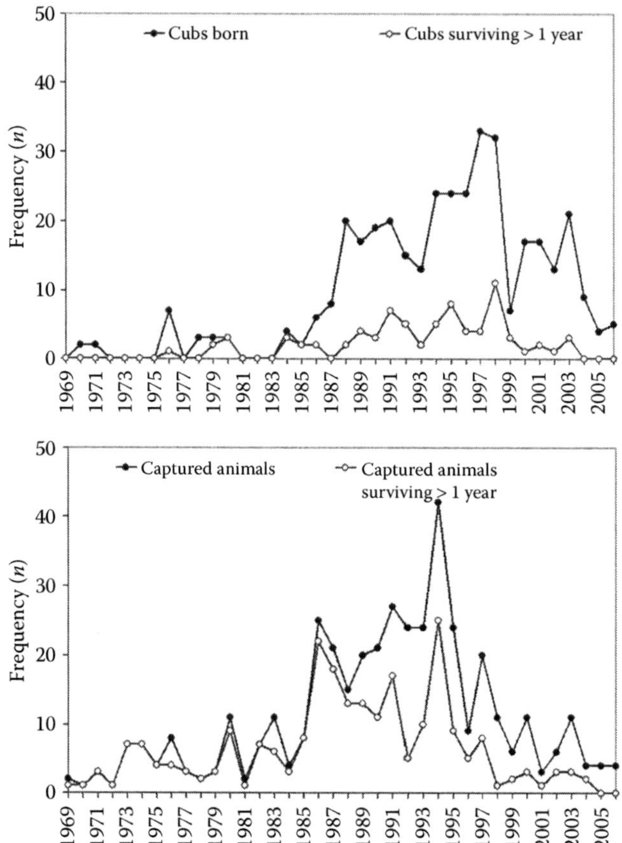

FIGURE 5.2 Historical series of captive births and incidental captures as recorded by the studbook of maned wolves in Brazil.

more cubs in their lifetime). This probably occurred because institutions struggled to find a pair of maned wolves that would not kill their cubs, and when such a good breeding pair was found, they would be encouraged to breed year after year, regardless of their overrepresentation in the population. Favoring cub productivity over genetic management, however, meant that all captive-born cubs are direct or indirect descendants of just over 40 wild-caught founders. Another unfortunate result of this strategy is that when these highly productive breeding pairs retire or die, as has been happening in the past decade, the success of future breeding attempts will be low.

Vortex population viability analysis conducted with the demographic parameters summarized in Vanstreels and Pessutti (2010) suggested that the Brazilian *ex situ* population might decline dramatically without a steady influx of wild recruits, unless considerable improvements were made in terms of decreasing cub mortality and increasing reproductive rates (Figure 5.3; Ralph Vanstreels and Cecília Pessutti, unpublished data). Figure 5.3a presents the simulations if the current rates of births, cub survival, and wild recruitment are maintained, and the simulated population remains viable. On the other hand, Figure 5.3b presents the simulations for the same parameters except without the wild-caught recruitment influx (which was simulated as five wild-caught animals per year), and the difference is dramatic: all simulated populations collapse within 20–50 years. Moreover, the simulations suggest that if no wild-caught animals were to be admitted in the future, the stability of the captive population would require dramatic husbandry changes, such as a 66% reduction in cub mortality (Figure 5.3c) or the combination of that 66% cub mortality reduction with a 33% increase in births (Figure 5.3d).

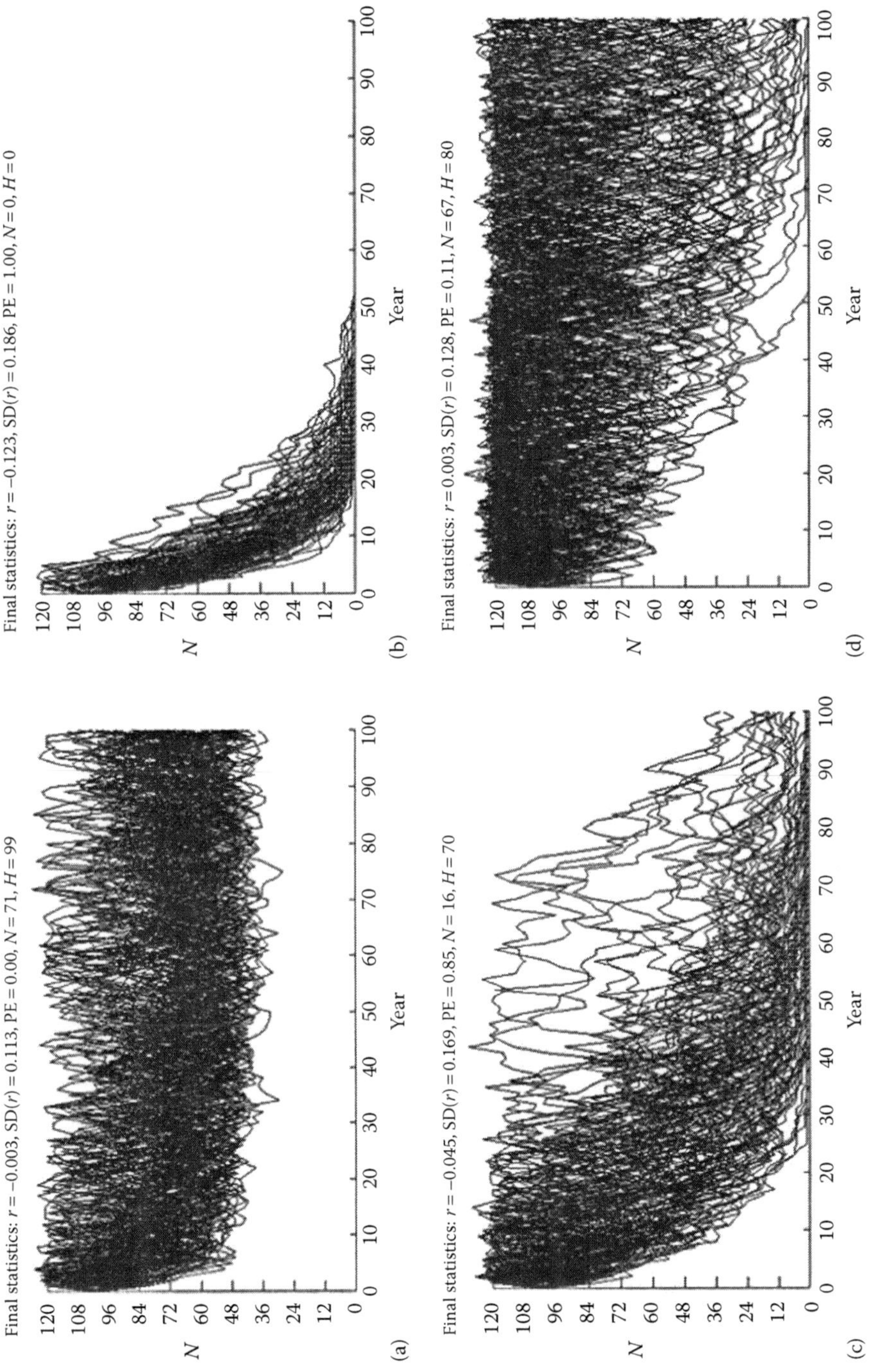

FIGURE 5.3 (a–d) Population viability simulations for different scenarios of wild recruitment and husbandry. (See text, Ralph Vanstreels and Cecília Pessutti, unpublished data.)

In other words, the captive population has been demographically unstable, yet maned wolves have remained generally abundant in Brazilian institutions thanks to the constant recruitment of wild individuals. It is our belief that because most present zoo administrators have seen wild-caught maned wolves very frequently brought into zoos in the recent past (late 1980s and early 1990s), there is a still a generalized perception that the species is abundant and no efforts of captive breeding are required. It may be true that the captive population is likely to be near its carrying capacity (i.e., enclosure availability); however, a lack of efforts to improve captive breeding protocols and husbandry may result in a population crash if the influx of wild recruits comes to a halt (and it has been dropping considerably in recent years). For these reasons, the improvement of husbandry and captive breeding protocols should remain a priority if the *ex situ* population is expected to sustain itself in a future in which wild maned wolves would no longer be incidentally captured and brought into captivity.

RESEARCH CONDUCTED ON THE BRAZILIAN *EX SITU* MANED WOLF POPULATION

The first compilation of scientific studies conducted in captive maned wolves in Brazil was prepared during the Third Maned Wolf Workshop in 1998. Research was conducted by independent groups and by the Maned Wolf Program of the São Paulo Zoological Society on a broad variety of topics such as genetic variability of the captive population, demography, immunology, parasitology, anesthesiology, pathology, and digestive physiology; moreover, research conducted on captive specimens built a bridge between *ex situ* and *in situ* populations through studies on the feeding ecology, seed dispersal, and historical distribution and status in the Paraná state (Santiago et al., 2000).

Endocrinology research was pioneered in the use of steroid hormone dosage to detect and characterize the species' reproductive cycle (Velloso et al., 1998); ethological research elucidated its circadian rhythm (Pessutti et al., 1999); immunological investigations tested the postvaccination antibody responses to canine distemper and parvovirus vaccines, providing safer immunization protocols (Maia et al., 1999); and physiological research provided insights on the nutrient digestibility and absorption, improving the species alimentary husbandry in captivity (Pessutti, 1997). In these early years, it became clear that collaboration with the North American SSP was essential to direct the Brazilian research and captive management programs, and numerous research collaborations were developed at the time.

These studies contributed to improving the species' husbandry and captive management, thus indirectly contributing to its conservation. The Maned Wolf PHVA Workshop, in 2006, provided further opportunities for the compilation and dissemination of technical and scientific knowledge among *in situ* and *ex situ* researchers. Moreover, the PHVA Workshop identified and prioritized problems for the species and designed collaborative long-term conservation strategies, providing researchers with recommendations for research on the *ex situ* population (de Paula et al., 2008). Lastly, the analysis of the regional and international studbook databases has provided demographic and husbandry details on the *ex situ* maned wolf population that may to some extent be extrapolated for the study of the species' ecology in the wild (Maia and Gouveia, 2002; Vanstreels and Pessutti, 2010).

ENVIRONMENTAL EDUCATION THROUGH THE BRAZILIAN *EX SITU* MANED WOLF POPULATION

Among the environmental education strategies proposed at the Third Maned Wolf Workshop was the creation of a "National Maned Wolf Day." October 12 was chosen, and the day aimed to inform, involve, and sensitize zoo visitors on the ecological importance of maned wolves, the problems faced by the species in the wild and in captivity, and the strategies for the conservation of its habitat, the cerrado biome (Figure 5.4).

FIGURE 5.4 National Maned Wolf Day has been celebrated on October 12 in Brazil every year since 1998. On this day, zoos throughout the country promote educational activities to sensitize visitors to the ecological importance of maned wolves and the cerrado.

The management plan proposed and donated didactic and ludic materials to the zoos in the first years, including informative booklets, treasure hunt games for children involving topics of maned wolf biology, behavioral observation sheets, maned wolf ear tiaras and masks, documentaries and films prepared by Brazilian zoos and/or captive breeders, and comic books featuring maned wolves (featuring Chico Bento, a very popular character in children's comic books), among others. Zoos gradually adapted these materials and produced their own, and many also took initiatives to carry these environmental education efforts outside the zoo. A good example was the initiative by the RioZoo Foundation in presenting an adapted version of the theater play *Red Riding Hood and the Maned Wolf* by Angelo Machado at Central do Brasil, the largest train station in the country. This play became very popular and has been performed in zoos across the country on National Maned Wolf Day.

HUSBANDRY PROTOCOLS FOR MANED WOLVES

Among other actions proposed by the management plan, the preparation of husbandry protocols was one of the greatest contributions for the improved management and care of the species' *ex situ* population. These husbandry protocols were compiled by experienced professionals and included an extensive survey of the husbandry practices and facilities of each institution housing maned wolves, including veterinary, biological, and educational aspects. Among others, these protocols provide recommendations on exhibit structure, cub and adult diet, veterinary care, reproduction, cub hand-rearing, transport, initial care of wild-caught individuals, research methods, behavior, environmental enrichment, and so forth.

In particular, the nutrition of captive maned wolves is a great challenge. The traditionally protein-rich diets used for other canids may lead to cystinuria in maned wolves, which may in turn produce various health complications and death (Mussart and Coppo, 1999; Childs-Sanford, 2001); as a result, alternative diets have been attempted and evaluated, and although much progress has been made, the nutrition of the species remains a challenge. Analyses on the digestibility of experimental diets containing food items most frequently provided to maned wolves in Brazilian zoos revealed the high metabolic rates of maned wolves, as is typical in carnivores, and the high digestibility and absorption of vegetal food items. As a result, the diet of captive maned wolves has been recommended to simulate the fruit-rich diet these animals have in the wild, with approximately 60% plant items and 40% animal items (Pessutti, 1997). We herein summarize the husbandry recommendations of the Brazilian Management Plan for the Maned Wolf; further details may be found in Santiago et al. (2000).

Enclosure Characteristics

Exhibits should be on the same level as or above the public; moat exhibits are considered inadequate. Total enclosure area should be at least 400 m^2, as enforced by the Brazilian legislation

(IBAMA IN 169/2008). Such an enclosure may house two adults and up to four cubs, and may be surrounded by fences (height ≥ 2 m), glass, or dry or wet moats. A holding pen (at least 2×3 m^2) should be available to handle the animals; it should be externally protected by nylon netting to prevent other animals from entering and have at least one wall socket. A whelping area (at least 4 m^2) is recommended; it should have access to the holding pen and include at least one nesting box and a source of warmth (heat lamps or pads). Guillotine doors may be used inside the holding facilities, but only sliding doors should control access to the outdoor enclosure. The enclosure should have two or more masonry or wooden shelters; dog houses may be adapted for that purpose (approximately 1 m^2, with an 80×60 cm entrance), and they may be partially burrowed or slightly above ground level (10 cm platform). These shelters should be as numerous as possible; they should be distant from the visiting public and allow the husbandry staff to examine its interior without disturbing the animal (through cameras or peepholes). The outdoor enclosure should have approximately 30% sunshade, with abundant bushes, trees, tree hollows, dirt mounds, burrows, and water ponds to simulate the cerrado biome. Grass, dead leaves, and dirt may be used as substrate. The public should only have access to one side of the exhibit and should remain at least 2 m from the exhibit fence; thorny bushes (such as the Sansão-do-campo *Mimosa caesalpiniifolia*) may be used to discourage visitors from approaching, while maintaining a natural appearance. If two exhibits are contiguous, natural or artificial barriers should be used to prevent visual and physical contact between the animals.

Environmental Enrichment

The following may be used directly or indirectly as environmental enrichment: bones (particularly long bones and the patella), PVC tubes with prey or meat inside them, food hidden in the enclosure, animal blood and/or feces spread in the enclosure, live prey, lobeira fruits (*Solanum lycocarpum*), fragments of decaying wood, and other items that might be stimulating to the animals (Figures 5.5 and 5.6).

Nutritional Management

Zoos should generally offer fruit items as a greater proportion of the diet, taking advantage of seasonal fruits that might be accepted by the animals. Live prey such as ducks, chicken, guinea pigs, rats, and mice should be offered weekly or more. Any change in diet, whether it be adding new food items or removing others, should be evaluated and carefully monitored by a responsible technician. The dietary composition of cubs and adults should generally follow these proportions: 25%–30% dog food, 25%–30% meat (red meat, chicken, and live prey), and 40%–50% fruits (particularly seasonal fruits and lobeira fruits). When the cubs reach the age of weaning, milk with crushed dog

FIGURE 5.5 Herbivore feces may be used as olfactory enrichment for maned wolves.

FIGURE 5.6 A paperboard box filled with food items (dog food, bovine meat, chicken heads, seasonal fruits) and tightly closed with string may be used as cognitive and feeding enrichment for maned wolves.

food or commercial weaning diets may be used for approximately 1 week to encourage them to adapt to an adult diet.

Management of Reproduction

Pairs should be formed a few months before the breeding season, to increase the chances of successful breeding. Potential pairs should be approached gradually and with great care, under the supervision of an experienced observer; if aggressive reactions occur, the pair should be separated and an interval should be given before the next attempt to approach the animal. A pair may sometimes be completely behaviorally incompatible, in which case the animals should not be paired and different individuals should be found to attempt approximation.

Behavioral observations before and during the breeding season are encouraged, as they may provide important information on the species' ethology. The zookeeper should be attentive to the behavioral changes typical of the breeding period, and any unusual observations should be promptly communicated to the responsible staff. The exhibit, husbandry procedures, and zoo-keeping staff should be changed as little as possible during this period. Enclosure cleaning frequency should be reduced during the first weeks after parturition. The dietary composition should not be changed after the birth of the cubs, but the amount of food offered will have to be adjusted; the offering of excessively bloody meat and live prey should be suspended until the cubs are older.

As previously mentioned, at least two shelter options (burrows and/or dog houses) should be placed away from areas with visitors or zoo keepers, in the quieter edges of the enclosure. If public visitation is thought to have caused aggressive behavior of the parents toward their cubs, the zoo should provide visual and acoustic isolation of the exhibit during the early stages of cub rearing, through plastic canvas or by limiting visitation.

The separation of the pair is not mandatory, as in many cases it has been found that in captive maned wolves the male may invest nearly as much as the female in rearing the cubs. However, if there are signs that the male may attack the female or the cubs, he should be isolated in a different pen until the cubs are older to prevent infanticide. Otherwise, the cubs may be housed with their parents until the next breeding season (1 year).

Cubs should only be separated from their parents for hand rearing if the female presents abnormal behavior, such as constantly moving the cubs among nests, excessively cleaning them, or attempting to burrow them. In these cases, half the litter should be removed first and the maternal behavior further monitored; if abnormal behavior continues, then the remaining cubs should also be removed.

Hand Rearing of Cubs

After the removal of the cubs, they should be subjected to an initial physical examination and weighing, then fed and placed in a warm room. Hydration should be performed with saline in the first 12–24 h, then a 1:1 solution of milk and saline in the following 12 h, then milk in the following days. Nursing bottles (or gastric tubes, if necessary) may be used to feed the cubs, and feeding should be done at 2-h intervals during the first week, 3-h intervals during the second week, and 4-h intervals during the fourth week. Milk for maned wolf cubs may use either homemade or commercial formulas; strict hygiene should be used during its preparation and the nursing bottles must be thoroughly sanitized in between feedings. Cubs should be held standing when being fed with nursing bottles; after feeding, abdominal massage and anal stimulation with moist gauze are recommended to encourage defecation.

The cubs should be kept in an incubator or a warm nest box, with a substrate of paper, cloth, or straw. Heating devices such as lamps, heating pads, and hot water bottles should be covered with cloths. Electric heaters should be used with care, and the environment should be kept relatively moist to prevent respiratory complications or hair loss in the cubs.

The box should be cleaned as often as necessary to avoid moisture or the accumulation of feces or urine. Detailed daily records should be kept on each individual's food consumption, and cubs should be measured and weighed at least three times a week. If cubs have to be separated from their mother, they should be hand raised together to promote normal behavioral development. Because the immune system of the cubs is incomplete and may be vulnerable, they should not be kept with other animals (particularly domestic), to prevent risks of disease exposure.

Handling and Husbandry of Wild-Caught Cubs and Adults

It is essential to determine the area from which the animal has been captured and to obtain as much detail as possible on the capture history. If, upon arrival, the animal presents no external traumas or lesions, no physical or chemical restraint should be attempted, and the animal should be allowed to rest. The animal should be housed in an enclosure away from public visitation, with a minimum area of 6 m^2; the enclosure should have a "bed" (dry straw or grass, thick wood shavings) to avoid bruises, and the animal must have access to sunlight. Electrolytes should be offered in the drinking water, and special care should be given to detect ectoparasite infestation.

After approximately 2 days, a more detailed clinical examination is recommended, including the collection of morphometric information, weighing, oral cavity inspection, and blood sampling. At this point the animal should also be identified (microchip or otherwise). Nutritional husbandry for wild-caught animals is similar to that of captive-raised animals, however the rejection of some food items may occur and alternatives might be initially required (e.g., live prey). Further details on vaccination and sanitary and health protocols are available in Santiago et al. (2000).

CONCLUSIONS AND FUTURE PERSPECTIVES

Despite the great efforts and progress made in the 30 years since the creation of the management plan, important difficulties persist and limit the conservation contribution of the Brazilian *ex situ* population of maned wolves. So far, Brazilian zoos and captive breeders have largely been the recipients of incidentally captured maned wolves, as human populations invade and destroy the species' natural habitat; however, the *ex situ* population has yet to fulfill its mission as a genetic stock for the species.

However, we should not forget that Brazilian zoos generally lack financial resources and experience with multi-institutional management plans for endangered species; in that context, the Brazilian Management Plan for the Maned Wolf has been a precursor for the development of similar plans for other species in the country and bears the burden and the advantages of that pioneering. The management plan was instrumental in launching the conservation of maned wolves and the cerrado biome as topics of discussion in the scientific community and general public during the early 1990s. Captivity provides outstanding opportunities to convey the message for species and cerrado conservation, either directly through visitation in zoos and educational programs or indirectly through television programs and magazines. It also provides possibilities to encourage new researchers and conservationists and to promote scientific meetings between *in situ* and *ex situ* professionals.

Courageous steps have been taken in Brazil toward turning these captive maned wolves into ambassadors for environmental education and the conservation of the cerrado biome, and they have also been used as subjects for a variety of scientific studies. However, unification of these efforts under a management plan is still lacking. Our main challenge and mission for the coming years therefore is to develop a more cohesive network that embraces the management plan as a unified and collaborative effort for the contribution of these captive collections to the species' conservation.

REFERENCES

Bauman, K.L., Asa, C.S., Grisham, J., and Verberkmoes, W. 2004. Captive canid conservation. In: *Canids: Foxes, Wolves, Jackals and Dogs. Status Survey and Conservation Action Plan*, eds. C. Sillero-Zubiri, M. Hoffmann, and D.W. Macdonald. IUCN/SSC Canid Specialist Group. Gland, Switzerland and Cambridge, UK, pp. 208–287.

Channell, R. and Lomolino, V.M. 2000. Dynamic biogeography and conservation of endangered species. *Nature* 403:84–85.

Childs-Sanford, S.E. 2005. The captive maned wolf (*Chrysocyon brachyurus*): Nutritional considerations with emphasis on management of cystinuria. Doctoral dissertation, University of Maryland.

Costa, G.O. 2004. Educação ambiental: Experiências dos zoológicos Brasileiros. *Revista Eletrônica do Mestrado de Educação Ambiental.*

Dmock, R. 2007. *International Register and Studbook for the Maned Wolf* (Chrysocyon brachyurus). Frankfurt, pp 62–182.

IBAMA 2003. Lista das espécies da fauna Brasileira ameaçadas de extinção. MMA.

IBAMA 2008. Instrução Normativa n° 169. MMA.

IUCN 2011. IUCN Red List of Threatened Species. Version 2011.2. www.iucnredlist.org. Downloaded on 18 April 2012.

Maia, O.B. and Gouveia, A.M.G. 2002. Birth and mortality of maned wolves *Chrysocyon brachyurus* (Illiger, 1811) in captivity. *Brazilian Journal of Biology* 62:25–32.

Maia, O.B., Gouveia, A.M.G., Souza, A.M., and Barbosa, E.F. 1999. Avaliação pós-vacinal de lobos guarás *Chrysocyon brachyurus* (Illiger, 1811) contra os vírus da cinomose e parvovirose caninas. *Arquivo Brasileiro de Medicina Veterinária e Zootecnia* 51(5):415–420.

Mallinson, J.J.C. and Feistner, A.T.C. 1995. Captive propagation and effective conservation. *Biodiversity and Conservation* 4:535.

Meretsky, V.J., Snyder, N.F.R., Beissinger, S.R., Clendenen, D.A., and Wiley, J.W. 2000. Demography of California condor: Implications for reestablishment. *Conservation Biology* 14(4):957–967.

Mussart, N.B. and Coppo, J.A. 1999. Cystine nephrolithiasis in an endangered canid, *Chrysocyon brachyurus* (Carnivora: canidae). *Revista de Biologia Tropical* 47:623–625.

de Paula, R.C., Médici, P., and Morato, R.G. (eds). 2008. Plano de ação para conservação do lobo guará: Análise de viabilidade populacional e de habitat. Instituto Brasileiro do Meio Ambiente e Recursos Naturais Renováveis, Brasília.

Pessutti, C. 1994. Studbook regional para o lobo guara *Chrysocyon brachyurus*. Sociedade de Zoológicos do Brasil, Sorocaba.

Pessutti, C. 1997. Aspectos da fisiologia digestiva de lobo guará *Chrysocyon brachyurus* e cachorro-do-mato *Cerdocyon thous*. Dissertação (Mestrado em Ciências Biológicas), Instituto de Biociências, Universidade Estadual Paulista, Botucatu.

Pessutti, C., Valdermarin, H.F., Andriolo, A., and Sridhara, S. 1999. Circannual rhythms of the maned wolf (*Chrysocyon brachyurus*) in captivity. XXVI International Ethological Conference, Bangalore, India, August 2–9, Sridhara, Shakunth ala, abstracts only.

Rose, M. 2010. World's first zoo—Hierakonpolis, Egypt. *Arqueology* 63(1). Accessed on 16 April 2012. http://www.archaeology.org/1001/topten/egypt.html.

Santiago, B.E.M., Pessutti, C., Chieregatto, C.A., and Gomes, M.S. 2000. *Protocolo de manejo, III Workshop para o lobo guará* Chrysocyon brachyurus. Sociedade de Zoológicos do Brasil.

Snyder, N.F. and Snyder, H.A. 1989. Biology and conservation of the California condor. *Current Ornithology*, 6:175–267.

Songsasen, N. and Rodden, M.D. 2010. The role of the species survival plan in maned wolf *Chrysocyon brachyurus* conservation. *International Zoo Yearbook* 44:136–148.

Vanstreels, R.E.T. and Pessutti, C., 2010. Analysis and discussion of maned wolf *Chrysocyon brachyurus* population trends in Brazilian institutions: Lessons from the Brazilian studbook, 1969–2006. *International Zoo Yearbook* 44:121–135.

Velloso, A.L., Wasser, S.K., Monfort, S.L., and Dietz, J.M. 1998. Longitudinal fecal steroid excretion in maned wolves (*Chrysocyon brachyurus*). *General and Comparative Endocrinology* 112(1):96–107.

WAZA. 2005. Building a future for wildlife: The world zoo and aquarium conservation strategy. World Association of Zoos and Aquariums, Bern.

WAZA. 2012. World Association of Zoos and Aquarium website. www.waza.org/en/site/about-waza/what-we-do. Accessed on 23 April.

Species Biology

6 Evolutionary and Conservation Genetics of the Maned Wolf

Manoel L. da Fontoura-Rodrigues and Eduardo Eizirik

CONTENTS

Introduction 77
Phylogenetics and Evolution 78
Maned Wolf Genetic Diversity and Population Structure 80
Conservation Issues and Future Research 82
Concluding Remarks 84
References 84

INTRODUCTION

When one thinks about the application of genetics in the context of conservation biology, it may be difficult to imagine all the possible ways in which molecular markers can be employed. People that are not familiar with this field tend to think about the "genetic health" of a given population as the main—and perhaps the only—concern of geneticists working with species conservation. This view is rather narrow, as today molecular approaches are some of the main tools employed by any research project that aims to deal with diverse aspects of evolution, taxonomy, population structure, and species ecology. Likewise, the field of genetics is continuously growing and diversifying, constantly providing new tools for several areas of conservation biology, and going far beyond aspects such as population "health" and long-term viability (e.g., see Frankham et al., 2002; Allendorf and Luikart, 2007).

In the field of evolutionary biology, the use of molecular markers not only allows the generation of much larger data sets relative to traditional approaches, but also opens up the possibility of choosing among fast-evolving, moderate, and slow-evolving genomic segments, rendering the tools particularly suitable to the question to be tackled. Furthermore, modern analyses are capable of estimating divergence times among lineages, integrating fossil and/or biogeographic evidence with molecular data sets, and helping to elucidate more thoroughly the evolutionary history of the group under study. Altogether, these tools are currently contributing considerably to the integrated study of evolution, phylogeny, phylogeography, and taxonomy of several taxa, which are basic requirements for any conservation effort on their behalf.

In addition to their application in phylogenetics, phylogeography, and population genetics, molecular tools are also being used with increasing frequency in intraspecific studies focusing on behavioral or ecological questions. The field known as molecular ecology represents a relatively new and growing area in which novel approaches complement classical ecological techniques, allowing studies to probe further into the biology of any species (Beebee and Rowe, 2008). Molecular tools allow researchers to precisely draw population genealogies and to exactly identify species, individuals, and their sex, which is especially useful when working with noninvasive samples such as feces. With these data in hand, it is possible to study in detail many ecological phenomena, including food habits, parasite load and composition, mating systems, individual home ranges, social structure, and dispersal patterns. Although molecular approaches have their own limitations, they tend to enable precise insights that are usually not possible solely on the basis of direct observations, such as confirmed parenthood and the exact composition of gut microbiota. Jointly with classical ecological surveys, molecular ecology tools hold great potential to generate detailed studies that contribute to the fine-level understanding of complex interactions among individuals within populations, as well as among species and their environments in intricate biological communities.

In this chapter we review the contribution of molecular genetic approaches to our current understanding of the maned wolf (*Chrysocyon brachyurus*), including the evolutionary history of the lineage leading up to the present-day species, as well as studies addressing population-level issues of this canid. We also present a brief overview of ongoing research targeting this species, and point out relevant avenues for investigation in the future.

PHYLOGENETICS AND EVOLUTION

The phylogeny of the family Canidae has been hotly debated. Some monophyletic groups are clearly identifiable and were recovered early in the history of phylogenetic efforts targeting this family, but the high morphological and ecological plasticity of canids has led to several uncertainties that persisted for decades (for reviews of this history see Zrzavy and Ricancova, 2004; Bardeleben et al., 2005). The phylogenetic position of maned wolves was one of such long-standing problems, likely due to its highly derived morphology and behavior. It was only by the end of the 1990s that molecular tools began to resolve this issue, revealing that *Chrysocyon* is the sister-clade of *Speothos*, presently represented by the South American bush dog (*S. venaticus*). Although today some morphological similarities uniting these taxa have been recognized, it is not surprising that earlier phylogenies were unable to recover such phylogenetic proximity: maned wolves are large and slim; have long legs, large ears and tails, orange fur, and solitary habits; and are grassland-dwelling animals (Dietz, 1984). In contrast, bush dogs are much smaller and stocky; have short ears and tails, brown fur, and very strong social habits; and seem to be strongly associated with forested areas (Beisiegel and Zuercher, 2005). They are therefore as distinct as any possible pair of extant wild canids, a remarkable fact that has obscured their close evolutionary relationship for many years.

The monophyletic group containing *Chrysocyon* and *Speothos* was first suggested by a study that relied on 2001 base pairs (bp) of the mitochondrial DNA (mtDNA) (Wayne et al., 1997), but the position of this clade within the Canidae, as well as other major relationships within the whole family, remained unclear for several years. Finally, in 2005, two studies (Bardeleben et al., 2005; Lindblad-Toh et al., 2005) that utilized a large number of molecular characters from both mtDNA and nuclear segments reached a more robust phylogenetic hypothesis for the Canidae, which reveals several interesting patterns and points to a resolution with respect to the evolutionary origin of the maned wolf.

The currently accepted phylogeny of Canidae (Lindblad-Toh et al., 2005) provides good resolution to most portions of the tree connecting present-day species. There is evidence indicating that *Urocyon* is the most basal genus, having diverged from other extant canids prior to the separation among three main internal groups. One of these groups is formed by the fox-like canids, including some well-known genera such as *Vulpes* and *Alopex*, and presents a very broad Northern Hemisphere and African distribution. The second group comprises the wolf-like (or dog-like) canids, and also occurs in most parts of North America, Africa, and Eurasia. Finally, the third clade includes all South American canids, encompassing ten extant species placed in five genera (Figure 6.1). In addition to these living taxa, several extinct species belonging to the genera *Protocyon*, *Theriodictis*, and *Dusicyon* have also been placed within this monophyletic neotropical assemblage on the basis of phylogenetic analyses employing fossil and/or molecular data sets (Prevosti, 2009; Slater et al., 2009; Perini et al., 2010).

Within this South American group, a major subdivision has been observed (Lindblad-Toh et al., 2005): one subgroup comprises the South American foxes (genera *Cerdocyon*, *Atelocynus*, and *Lycalopex*), while the other subgroup is made up of *Chrysocyon* and *Speothos*. A phylogenetic analysis including several extinct taxa (Prevosti, 2009) has indicated that fossils assigned to *Theriodictis*

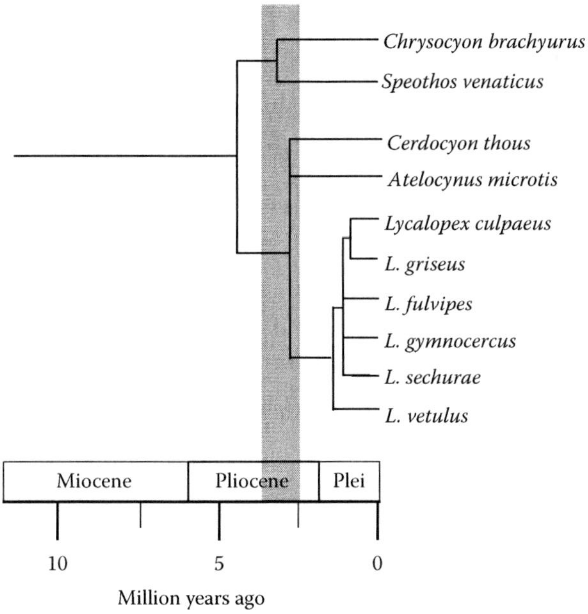

FIGURE 6.1 Phylogenic relationships among South American canids. A timescale is shown at the bottom, indicating the inferred timing of their diversification. The shaded bar indicates the time frame of the Great American Biotic Interchange. Plei, Pleistocene. (Modified from Eizirik, E., *Bones, Clones, and Biomes: An Extended History of Recent Neotropical Mammals*, eds. Patterson, B.D., Costa, L.P., pp. 123–142. University of Chicago Press, Chicago, IL, 2012.)

and *Protocyon* are also contained in the latter subclade, with a closer relationship to *Speothos* than to *Chrysocyon*. Another study, focusing on the extinct Falklands wolf (*Dusicyon australis*) and employing molecular analyses of museum material (Slater et al., 2009), indicated that it also belonged in this subclade, being actually more closely related to *Chrysocyon* than to any other living taxon. Such findings illustrate the complex history of this canid group in South America, and raise interesting questions regarding the patterns and processes that led to the evolution of its morphological and ecological diversity.

This endemic radiation of South American canids is a pattern also observed in other groups of carnivores, such as felids (genus *Leopardus*; Johnson et al., 2006) and mephitids (genus *Conepatus*; Wang and Carranza-Castañeda, 2008). This is consistent with an invasion of South America by several North American groups after the closure of the Panama Isthmus about 3 MYA, the geological event that led to the Great American Biotic Interchange (GABI) (Marshall et al., 1982; see Eizirik 2012 for a recent review). It is important to note, however, that a recent study has dated the coalescence of South American canids at about 4.3 MYA (Perini et al., 2010). This result suggests that the most recent common ancestor of this lineage may have originated in North or Central America before the land bridge was completely formed, giving rise to at least two different species that invaded South America separately during the GABI.

After the invasion of South America by one of these putative ancestral species, a rapid divergence between *Chrysocyon* and *Speothos* seems to have happened, as Perini et al. (2010) dated this split at 3 MYA, and it had probably taken place in this subcontinent, given these species' present distributions. During the time since this separation, the two genera have evolved substantial morphological differences, likely driven by divergent adaptation to grasslands and forested habitats. Analyzing in detail the genetic, ecological, physiological, and behavioral traits that are associated with such morphological divergence will be a very interesting avenue for future research, given the striking differences that have evolved between these lineages in a relatively short time frame. Moreover, genomic analyses using newly developed approaches (e.g., next-generation DNA sequencing and high-performance bioinformatics techniques) should allow the identification of the very genes that underlie such strong morphological differentiation, helping to shed light on the adaptive processes that drive this divergence in canids and mammals in general.

MANED WOLF GENETIC DIVERSITY AND POPULATION STRUCTURE

Genetic structure and diversity of a species is a very important issue to any conservation effort. Structure studies provide taxonomic information, as well as insights regarding the need of some populations for special attention. Two modern concepts broadly employed in these studies are evolutionarily significant units (ESUs) and management units (MUs) (Moritz, 1994). ESUs refer to geographically defined populations that are genetically distinct from other groups, as assessed by molecular genetic approaches (e.g., monophyletic with respect to mtDNA lineages) and that usually also bear distinctive morphological or ecological features. These characteristics suggest a unique and independent evolutionary history, resulting in a unit with taxonomic status. In fact, ESUs are considered by many authors as a modern equivalent of the subspecies concept, and should be considered as an independent unit in terms of conservation. On the other hand, the MU concept is different, referring to populations that have no taxonomic status (i.e., not exhibiting sufficiently deep divergence from other groups to qualify as an ESU) but are still important to the conservation planning of the species. A good example of a MU is the set of jaguar (*Panthera onca*) populations of the inner Atlantic Forest in Brazil, Paraguay, and Argentina. Although not presenting a clearly independent evolutionary history relative to other regions, they are currently isolated in remnant forest fragments that probably could not be recolonized if the species went extinct locally and therefore need to be managed as a separate operational unit.

Molecular diversity indices are also an important intraspecific parameter that can be estimated using molecular tools. Usually, low diversity levels are seen as a sign of a problem for the viability of a given population in the long term. Low variability, however, must be correctly interpreted, as there are many different measures of genetic diversity. As previously mentioned, in the whole genome there are different regions with very different rates of evolution. Slow-evolving segments, which can be represented by nuclear exons of conserved genes, take several million years to accumulate mutations, and thus have no useful information for intraspecific surveys. Even segments that evolve at a moderate pace, such as nuclear introns and mtDNA, must be analyzed carefully, as low diversity in these regions may have no direct connection to present-day population dynamics. Fast-evolving segments, on the other hand, are very good markers for population-level studies. The genetic diversity of regions such as microsatellites is directly affected by recent and current demographic processes, and may indicate population isolation, inbreeding, or unequal sex contributions over several generations. Cases of atypically low microsatellite diversity (relative to large, outbred populations of the same species or closely related species) must be further investigated, as there may be some demographic problem with the studied population. If intense enough, even very recent events are detectable, so human impact can be measured and thus regulated, a vital point in most conservation efforts (for a review regarding molecular diversity indices and the application of various molecular markers, see Avise, 2004; Frankham et al., 2002; Allendorf and Luikart, 2007).

Regarding maned wolves, there are still few published studies addressing the species' population structure and genetic diversity, although some research has already been conducted. Structure surveys are very important, as maned wolves occur in several countries and different biomes. Although there are no described subspecies or any evidence of population discontinuities, morphological traits usually take a long time to become evident. Moderate- to fast-evolving molecular markers, on the other hand, can accumulate differences very quickly and thus could detect an incipient structure pattern more accurately.

Although there are no published works regarding the genetic structure and diversity of the species as a whole, a few studies with local populations have already been conducted and provide insights on this issue. Moreira et al. (1998) and De Mattos et al. (2004) investigated isozyme diversity on different geographic scales. Diversity levels were considerable in both surveys, especially when compared with other species of the family. The variability found was even higher than that of gray wolves (*Canis lupus*) and island foxes (*Urocyon littoralis*). On the other hand, both studies obtained very little signal of population structure (i.e., differentiation among geographic regions). Moreira et al. (1998) obtained very low genetic distance values between populations from different regions of the Brazilian cerrado, including a total of 100 samples collected in São Paulo, Minas Gerais, and Goiás States in Brazil. De Mattos et al. (2004) studied a local population from an area of approximately 20,000 km^2 crossed by two rivers in São Paulo State and found no evidence of isolation between the three areas separated by these watercourses.

More recently, three additional studies that used a DNA-based molecular marker—microsatellites—reached similar results. Salim et al. (2007) and Fontoura-Rodrigues et al. (2008) tested the suitability of microsatellite loci developed for the domestic dog (*Canis familiaris*) for use in the maned wolf. Although the main objective of those studies was not to measure genetic diversity, the results of the locus amplification efficiency did provide information on variation in this species. Both investigations corroborated the previous molecular surveys involving the maned wolf. Salim et al. (2007) found four out of five polymorphic loci, based on testing of 69 individuals coming from the Brazilian states of São Paulo, Minas Gerais, and Goiás. The polymorphic markers showed considerable levels of heterozygosity, very similar to or higher than those observed in other canid species, such as the gray wolf (*Canis lupus*), African wild dog (*Lycaon pictus*), and Ethiopian wolf (*Canis simensis*). Fontoura-Rodrigues et al. (2008) also tested domestic dog loci for cross-amplification in some species of South American canids, including the maned wolf. A total of 10 loci and 25 maned wolves from a single population of the Brazilian cerrado were included in the survey, and the results once again presented considerable levels of heterozygosity. Compared with the other three South American canid species that were also investigated in that study, genetic diversity found at each locus for the maned wolf was very similar to, although slightly lower than, the overall mean heterozygosity (mean among all four species surveyed).

Finally, Lion et al. (2011) used the microsatellite loci tested in the studies mentioned above to investigate the levels of connectivity versus isolation of distinct maned wolf populations inhabiting three nearby protected areas separated by a dense human-dominated matrix near Brasilia, the capital of Brazil. Again, the results showed high diversity levels and no lack of connectivity among the three populations. To some degree, the intervening impacted areas included in the study seem to be permeable to dispersal by the maned wolf, given the inference that it can maintain some level of gene flow across this matrix. Another possibility is that the inferred gene flow does not occur directly through the intervening matrix, but rather around it through less-modified areas, via a longer and more complex route of potential dispersal.

Considering all the studies mentioned above, it is possible to see that a considerable area of the Brazilian cerrado has already been sampled for genetic analyses of the maned wolf. As this biome seems to account for most of the species' distribution, these surveys provide a good insight into the genetic diversity of the maned wolf. All studies are concordant in pointing to the lack of any dramatic isolation of local populations and to a high variability of the species, especially when assessed with fast-evolving molecular markers. This indicates that populations are not suffering yet from severe genetic problems given the intense human activity that accounts for the destruction of more than 80% of the original cerrado biome (Machado et al., 2004). High diversity levels are compatible with good "genetic health," as low values are usually linked with reduced individual fitness and congenital diseases induced by inbreeding or the founder effect.* Also, these results point to

* In population genetics, the founder effect is defined as a low level of genetic variability due to a populational foundation by a small number of individuals (Mayr, 1942, quoted in Provine, 2004). A possible consequence of such an event is the increased frequency of a given character (which can be a congenital disease, a condition, or a neutral phenotypic trait) caused by the particular genetic frequencies of the populations founders.

no dramatic processes of isolation in vegetation patches or population reduction to critical numbers. However, a more general and detailed survey is highly recommended to better evaluate the population as a whole and to allow direct comparisons among different populations. Besides, although the results show no local structuring, it cannot be inferred that there is no species structure on a larger scale, such as the existence of subspecies. Only a survey that uses proper molecular markers and a sampling strategy that includes most parts of the species distribution, allowing for a general comparison, can elucidate this issue. Also, it cannot be stated that the species is not under anthropogenic pressure or that conservation concern can be prescinded. Although dramatic processes such as population isolation and reduction in the number of individuals to inviable levels are not likely, the great habitat loss caused by human activity can lead to a population reduction that may reach critical levels in the near future. Furthermore, the lack of connectivity between habitat patches could be a very recent event, preventing even fast-evolving markers from detecting it yet. Thus, the possible conclusions, based on the available genetic surveys conducted so far, are as follows: (1) Maned wolves seem to present a good genetic diversity level in fast-evolving molecular markers when compared with other canid species, which suggests a considerable present *effective size** and consequently a good number of living individuals. (2) Apparently there is no sign of local population isolation, which indicates that maned wolves can still cross the anthropized matrix and sustain genetic flow on a small geographic scale. However, there is only one study using this approach and more surveys with similar results are needed to make broad conclusions. (3) Although the genetic diversity results are auspicious for conservation purposes, more efforts are needed to better understand the current demographic dynamic of the species and prevent a possible population reduction to critical levels before it happens. (4) Broad-scale studies are needed, especially regarding species structure, as well as a large-scale, comparable diversity survey.

CONSERVATION ISSUES AND FUTURE RESEARCH

Some points that deserve immediate attention have already been mentioned in this volume. In this chapter, we will discuss these points more thoroughly and provide some insights into ongoing and future research.

The most basic issue for any conservation effort is to know what to preserve. This implies the need for good taxonomic knowledge, at all possible levels. As the maned wolf is a very unique and conspicuous species, there is no apparent difficulty in terms of species delimitation. However, below this scale there are no studies that investigate intraspecific structure. This is very important, since the existence of possible ESUs (or subspecies) and MUs can determine specific units that need their own conservation plan. To achieve such results, it is fundamental to employ a good sampling strategy. All the different geographic regions, including all the different biomes, composing the maned wolf's distribution should be sampled. Grassland areas with confirmed occurrence of the maned wolf, such as the cerrado of Brazil, the Chaco of Paraguay, and the pampas of Argentina, Uruguay, and Southern Brazil, would be regions of interest for such a survey. An ongoing study regarding this issue is being conducted by the combined efforts of two groups of researchers, one from Argentina and other from Brazil. Using mtDNA as the main molecular marker, this team is exploring the structure patterns and past demographic events that shaped the current genetic diversity of the species. So far, some preliminary results already published in meeting reports seem to point to very low mtDNA diversity, with no or little signal of population structure concerning the regions already sampled. A more detailed analysis of past events, population structure, and contextualization in an evolutionary perspective is expected to be published by these authors in the near future.

* Effective size is one of the deepest and most abstract concepts in population genetics. It is defined as the size of an idealized population that bears genetic properties (e.g., rate of loss of genetic diversity) that are the same as those of the actual population. The idealized population would have several simplifying and standardized features, such as constant size over time, random mating, 1:1 sex ratio, and predictable reproductive success across individuals, i.e., average of one offspring per adult, with a variance of 1 (Hartl and Clark, 1997).

Studies regarding population connectivity on a finer scale, similar to the one published by Lion et al. (2011), are also needed. Fast-evolving segments, as previously mentioned, are adequate to investigate recent and ongoing events, a point that cannot be reached by moderate-evolving markers such as mtDNA. Thus, processes of total or partial isolation potentially caused by human activity can be investigated with such fast-evolving markers and are of vital interest to conservation plans. A team led by Eduardo Eizirik is working on this scale, with a broad geographic sampling strategy, focusing on the Brazilian cerrado. So far, the results corroborate the trend discussed in the section above: fast-evolving markers seem to present a very high diversity in maned wolves, and no signals of population structure or lack of connectivity between local populations are evident. Such findings reinforce the idea that maned wolves are tolerant at some level to impacted areas, since the cerrado is currently a highly fragmented biome.

Putting both preliminary results together, an evolutionary history starting in the moderate past and reaching the present time can be outlined. Theoretically, a very low diversity suggests a low effective number, while high diversity suggests the contrary. Once maned wolves seem to present low mtDNA and high microsatellite variability, an increase in the total number of individuals from the moderate past to the present is a likely scenario. Furthermore, a very low past diversity is compatible with a dramatic reduction in the number of individuals, and a dating analysis could shed light on which past geological context this possible demographic event could have occurred in. A general analysis, addressing all the mentioned issues, is expected to be published soon, so researchers will be able to rely on more detailed information to design more precise conservation efforts.

Regarding ecology, a considerable contribution of genetics is expected. A pioneer study in this sense is the one of Lion et al. (2011). Investigating connectivity between isolated preserved areas, the authors relied on different sampling methods, including feces. To correctly assign the scats to the target species, the first step of the survey was to sequence a small diagnosal fragment of the mtDNA. After that, microsatellite loci optimized for use in low-quality DNA (which is expected from feces) were amplified. This was the first study with maned wolves that successfully employed these strategies of sample collection and lab techniques, enabling a new set of approaches to investigate this species. Sampling methods that rely on biologic material that are collected with no necessity of direct contact with the animals, such as feces, urine, and hair, are called noninvasive. This kind of approach became practicable not long ago due to recent methodological advances in the area of DNA isolation and handling and allows a remarkable new set of sample collection strategies (for a review of noninvasive sampling see Marucco et al., 2011). Such innovation is very auspicious for the ecological study of many taxa, especially the large ones such as large mammals. These species often have very elusive habits and low demographic density, hampering surveys that depend on direct observation. Finally, when combined with genetics, noninvasive sampling methods are a powerful tool that permits different approaches to classic ecological questions. The successful sampling strategy of Lion et al. (2011) makes the potential employment of this approach clear in studies concerning maned wolves.

Referring to classic maned wolf ecological studies, the major contribution to this area was made by Dietz (1984). This author studied a broad variety of questions, including the species mating system. He was the first to postulate the famous monogamic system of the species. However, despite constructing a very good hypothesis based on radio-tracking analysis, it was impossible for Dietz to test his mating and parenthood assumptions. A similar ecologic approach to that of Dietz in the early 1980s but with the contribution of modern genetics would provide more precise information about this important question. In fact, a very similar study is being conducted with the Serra da Canastra National Park (Minas Gerais State, Brazil) population. A large number of free-living individuals were caught and tagged with radio collars. The animals were tracked for several years, and blood samples were obtained from all of them. A precise pedigree based on microsatellite markers is being reconstructed by our research team and will be published soon. Combined with classic ecological analyses that are being conducted in parallel, this survey can test Dietz's hypothesis more precisely, besides providing important insight into questions such as territorial sharing and dynamics and individual dispersion.

Finally, the combination of the methods mentioned in the last section, which have already been successfully employed in the pioneer study of Lion et al. (2011), can lead to very important studies concerning the maned wolf. As an example, the precise identification of the species by its scats can be a powerful tool to record the presence of maned wolves in areas where their occurrence is not certain, such as some regions of the pampas of southern Brazil and Uruguay. Furthermore, the ecological knowledge concerning maned wolves comes mostly from studies conducted in the Brazilian cerrado. As the distribution area of the species includes different biomes, with different rainfall levels and fauna and flora composition, the basic ecological aspects of maned wolves of such distinct areas must also be investigated and compared with the data gathered from cerrado studies. In this case, noninvasive sampling combined with molecular tools can be a very useful approach, especially in areas where the population density of the species seems to be very low or its existence is uncertain.

CONCLUDING REMARKS

The utilization of genetics in the study of the maned wolf is still incipient and no global results have been published so far. However, local studies that have been done show that the species presents a substantial level of diversity, as assessed by fast-evolving molecular markers, which is indicative of good population size and genetic health. No sign of local decrease in gene flow has been verified so far, but global studies are needed to confirm this hypothesis and to better understand the flow of individuals in an environment where human impacts grow each day. A broad study of population structure is also imperative, as taxonomic units need to be delimited to correctly drive conservation efforts. On a larger, phylogenetic scale, the position of the maned wolf in the Canidae tree seems to have been well resolved by molecular and morphological approaches, although further investigation is recommended, especially to refine the dating of its divergence from *Speothos* and to integrate additional fossil evidence, as it becomes available.

Finally, noninvasive sampling and molecular tools together are a powerful approach that can facilitate the study of several aspects of the species' behavior and ecology. This is especially relevant for regions where direct observation is hampered by low demographic densities and that lack local ecological investigations. As this approach has already been successfully employed in maned wolf studies, there is a good prospect for its utilization in further surveys. Such investigations are critical not only for conservation efforts in areas that are far from the Brazilian cerrado, but also for the species as a whole. Overall, we expect the next decade to see an increase in genetic studies of the maned wolf, revealing details of its evolutionary history and population structure, as well as providing an adequate basis to design conservation strategies on its behalf.

REFERENCES

Allendorf FW, Luikart G (2007). *Conservation and the Genetics of Populations*. Blackwell, Malden, MA.

Avise JC (2004). *Molecular Markers, Natural History, and Evolution*, 2nd edn. Sinauer, Sunderland, MA.

Bardeleben C, Moore RL, Wayne RK (2005). A molecular phylogeny of the Canidae based on six nuclear loci. *Molecular Phylogenetics and Evolution*, 37: 815–831.

Beebee T, Rowe G (2008). *An Introduction to Molecular Ecology*, 2nd edn. Oxford University Press, New York.

Beisiegel BM, Zuercher G (2005). *Speothos venaticus*. *Mammalian Species*, 783: 1–6.

De Mattos PSR, Lama MA, Toppa RH, Schwantes AR (2004). Populational genetic structure of free-living maned wolves (*Chrysocyon brachyurus*) determined by proteic markers. *Brazilian Journal of Biology*, 64: 639–644.

Dietz JM (1984). Ecology and social organization of the maned wolf (*Chrysocyon brachyurus*). *Smithsonian Contributions to Zoology*, 392: 1–51.

Eizirik E (2012). A molecular view on the evolutionary history and biogeography of neotropical carnivores (mammalia, carnivora). In *Bones, Clones, and Biomes: An Extended History of Recent Neotropical Mammals* (eds. Patterson BD, Costa LP), pp. 123–142. University of Chicago Press, Chicago, IL.

Fontoura-Rodrigues ML, Lima-Rosa CAV, Tchaicka L, Valdez FP, Rodrigues FHG, de Paula RC, Gough MP, Johnson WE, Bonatto SL, Eizirik E (2008). Cross-amplification and characterization of 13 tetranucleotide microsatellites in multiple species of neotropical canids. *Molecular Ecology Resources*, 8: 898–900.

Frankham R, Ballou JD, Briscoe D (2002). *Introduction to Conservation Genetics*. Cambridge University Press, Cambridge.

Hartl D, Clark AG (1997). *Principles of Population Genetics*, 3rd edn. Sinauer, Sunderland, MA.

Johnson WE, Eizirik E, Pecon-Slattery J, Murphy WJ, Antunes A, Teeling E, O'Brien SJ (2006). The late miocene radiation of modern felidae: A genetic assessment. *Science*, 311: 73–77.

Lindblad-Toh K, Wade CM, Mikkelsen TS, Karlsson EK, Jaffe DB, Kamal M, Clamp M, et al. (2005). Genome sequence, comparative analysis and haplotype structure of the domestic dog. *Nature*, 438: 803–819.

Lion MB, Eizirik E, Gardan AA, Fontoura-Rodrigues ML, Rodrigues FHG, Marinho-Filho JS (2011). Conservation genetics of maned wolves in a highly impacted area of the Brazilian Cerrado biome. *Genetica*, 139: 369–381.

Machado RB, Ramos Neto MB, Pereira PGP, Caldas EF, Gonçalves DA, Santos NS, Tabor K, Steininger M (2004). *Estimativas de perda da área do Cerrado Brasileiro*. Conservation International do Brasil, Brasília.

Marshall LG, Webb SD, Sepkoski JJ, Raup DM (1982). Mammalian evolution and the great American interchange. *Science*, 215: 1351–1357.

Marucco F, Boitani L, Pletscher DH, Schwartz MK (2011). Bridging the gaps between non-invasive genetic sampling and population parameter estimation. *European Journal of Wildlife Research*, 57: 1–13.

Mayr E (1942). *Systematics and the Origin of Species*. Columbia University Press, New York.

Moreira JR, Klatau-Guimarães N, Pilla EJS, Contel EPB, De Bem AR (1998). Estudo preliminar da variabilidade genética do lobo-guará (*Chrysocyon brachyurus*). *Comunicado Técnico Embrapa*, 30: 1–10.

Moritz C (1994). Defining "evolutionarily significant units" for conservation. *TREE*, 9: 373–375.

Perini FA, Russo CAM, Schrago CG (2010). The evolution of South American endemic canids: A history of rapid diversification and morphological parallelism. *Journal of Evolutionary Biology*, 23: 311–322.

Prevosti FJ (2009). Phylogeny of the large extinct South American canids (Mammalia, Carnivora, Canidae) using a "total evidence" approach. *Cladistics*, 25: 1–26.

Provine WB (2004). Ernst Mayr: Genetics and speciation. *Genetics*, 167: 1041–1046.

Salim DC, Akimoto AA, Carvalho CB, Oliveira SF, Grisolia CK, Moreira JR, Klautau-Guimarães MN (2007). Genetic variability in maned wolf based on heterologous short-tandem repeat markers from domestic dog. *Genetics and Molecular Research*, 6: 248–257.

Slater GJ, Thalmann O, Leonard JA, Schweizer RM, Koepfli KP, Pollinger JP, Rawlence NJ, Austin JJ, Cooper A, Wayne RK (2009). Evolutionary history of the Falklands wolf. *Current Biology*, 19: R937–R938.

Wang X, Carranza-Castañeda O (2008). Earliest hog-nosed skunk, *Conepatus* (Mephitidae, Carnivora), from the early Pliocene of Guanajuato, Mexico and origin of South American skunks. *Zoological Journal of the Linnean Society*, 154: 386–407.

Wayne RK, Geffen E, Girman DJ, Koepfli KP, Lau LM, Marshall CR (1997). Molecular systematics of the Canidae. *Systematic Biology*, 46: 622–653.

Zrzavy J, Ricancova V (2004). Phylogeny of recent Canidae (Mammalia, Carnivora): Relative reliability and utility of morphological and molecular datasets. *Zoologica Scripta*, 33: 311–333.

7 Feeding Ecology
A Review

José Carlos Motta-Junior, Diego Queirolo, and Adriana de Arruda Bueno

CONTENTS

Introduction 87
Study Areas 88
General Diet 89
Prey and Fruit Size Distribution 91
Food Intake and Bioenergetics 92
Seasonality and Functional Response 92
Prey and Fruit Selection 93
Trophic Relationship with Other Carnivores 94
The Maned Wolf as a Seed Disperser 94
Habitat of Prey and Fruits 94
Further Studies on Feeding Ecology 94
Acknowledgments 95
References 96

INTRODUCTION

General biological studies on the maned wolf (*Chrysocyon brachyurus*) (Illiger, 1815), particularly on its feeding ecology, have mostly involved captive individuals (Silveira, 1968; Kleiman, 1972; Brady & Ditton, 1979; Garcia 1983; Dietz 1984; Barboza et al. 1994). Few have been done in the wild; Dietz's (1984) work is the most representative of these and until recently was one of the most complete. Previously, Azara (1801), Goeldi (1893), Krieg (1940), Lund (1950), Cabrera (1958), Langguth (1975), Carvalho (1976), Garcia (1983), and Bartmann and Bartmann (1986), among others, mentioned the feeding habits of the species in short observations or analyses of small amounts of feces or stomach contents, but no systematic study had been done. Dietz (1985), in a review on the biology of the species, made a concise statement on the maned wolf's diet: "diet consisted largely of fruits and small mammals, and changed with seasonal availability of food."

After the end of the 1980s, a set of studies appeared focusing on the species' diet in the wild in several localities of the species' distribution, mainly situated in central and southeast Brazil (Motta-Junior et al. 1996; Motta-Junior 1991, 1997, 2000; Bianchi et al. 2000; Ferraz 1999; Aragona & Setz 2001; Juarez & Marinho-Filho 2002; Jácomo et al. 2004; Bueno et al. 2002; Santos et al. 2003; Silva & Talamoni 2003; Bueno & Motta-Junior 2004, 2006, 2009; Belentani et al. 2005; Cheida 2005; Queirolo & Motta-Junior 2007; Rodrigues et al. 2007; Massara 2009), Argentina (Soler & Casanave 2007), and Bolivia (Castro & Emmons 2012). Most of these areas were natural parks or reserves, with fewer studies being conducted at disturbed sites including ranches and farms with different kinds of intensive use.

The feeding habits were studied in terms of percentage of occurrence of each food item with regard to the total number of feces (Schaller 1983; Juarez & Marinho-Filho 2002) or in terms of the total occurrences of all food items (Dietz 1984; Motta-Junior et al. 1996; Bueno et al. 2002; Silva & Talamoni 2003; Jácomo et al. 2004; Queirolo & Motta-Junior 2007). Although we can find several studies based on occurrences, variation in the number of samples (20–1673 feces) and level of prey identification makes comparisons difficult. The numbers of occurrences and prey items vary in studies not only because of the diet itself but also due to the level of identification of prey items.

Another aspect of the maned wolf's diet frequently considered is the estimated consumed biomass (Motta-Junior et al. 1996; Motta-Junior 1997, 2000; Belentani 2001; Bueno et al. 2002; Juarez & Marinho-Filho 2002; Santos et al. 2003; Silva & Talamoni 2003; Bueno & Motta-Junior 2004, 2009; Amboni 2007; Queirolo & Motta-Junior 2007; Rodrigues et al. 2007; Massara 2009; Castro & Emmons 2012). However, quantification of biomass differs among studies, preventing comparison in some cases (Rodrigues et al. 2007).

Despite the amount of information about the maned wolf's diet, few studies quantify food availability (Dietz 1984; Motta-Junior 2000; Motta-Junior & Martins 2002; Bueno & Motta-Junior 2006; Queirolo & Motta-Junior 2007; Rodrigues et al. 2007), and only two attempted to test if food availability is a powerful factor contributing to the species' movement pattern (Silva & Talamoni 2003; Amboni 2007). Studies including prey availability and carnivore functional responses increase the interpretability of the results obtained (Jaksic et al. 1992; Corley et al. 1995). However, for canid species in the neotropics, limited data are available regarding functional responses and selectivity in the diet (Iriarte et al. 1989; Castro et al. 1994; Corley et al. 1995). Particularly for the maned wolf, studies on prey selectivity and functional responses have seldom been conducted (except for Motta-Junior 2000; Bueno & Motta-Junior 2006; Queirolo & Motta-Junior 2007; Rodrigues et al. 2007).

Thus, in this chapter we provide a review of the literature on diet and a quantitative analysis of the trophic ecology of the maned wolf in central and east South America, considering 18 localities: eight from our own published or original not yet published data and ten from other researchers, obtained from the literature. All diet data were obtained from feces collection and its analysis; more details on techniques of feces analyses are described in each study citation. Our aims in this chapter are (1) to identify quantitative patterns in diet composition, by both occurrence and biomass consumption; (2) to identify animal prey size distribution in the diet; (3) to search for possible seasonal patterns in the diet; (4) to review information from the literature on prey selection and functional response in the diet of the maned wolf; and (5) to suggest future feeding ecology research perspectives.

STUDY AREAS

The localities studied were as follows: Fazenda Fortaleza (FFOR; original data), Fazenda São Luis (FSLU; Santos et al. 2003), Reserva Natural da Serra do Caraça (RSCA; Silva & Talamoni 2003), Reserve of Universidade Federal de São Carlos campus (UFSC; original data), Estação Experimental Itapetininga (EITA; Bueno & Motta-Junior 2004), Rio Pratudão Ranch (RPRA; Juarez & Marinho-Filho 2002), Fazenda Salto e Ponte I (FSPI; Belentani 2001), Parque Estadual Ibitipoca (PEIB; Aragona & Setz 2001), Estação Ecológica Jataí (EJAT; Motta-Junior 2000), Estação Ecológica Águas Emendadas (EEAE; Rodrigues et al. 2007), Fazenda Água Limpa (FALI; Motta-Junior et al. 1996), Parque Nacional Chapada dos Guimarães (PNCG; Aragona 2002), Parque Nacional Emas (PNEM; Jácomo et al. 2004), Estação Ecológica Águas de Santa Bárbara (ESTA; original data), Parque Nacional Serra da Canastra (PNSC; Queirolo & Motta-Junior 2007), Estação Ecológica Itirapina (EITI; Bueno et al. 2002), Fazenda Monte Alegre (FMAL; Cheida 2005), and Parque Nacional Noel Kempff Mercado (PNKM; Castro & Emmons 2012). Except for the last study, in Bolivia, all the others were performed in central to southeast and south Brazil.

The selection of these 18 studies was based on the following criteria: the collection of 50 feces or more covering at least 1 year and with raw data at least at frequency of occurrence level (*sensu*

Dietz 1984). Some studies of the above list that overestimated fruit biomass in the diet (Juarez & Marinho-Filho 2002; Santos et al. 2003) were not considered for biomass analyses. We calculated the food-niche breadth according to Levins' standardized measure (Krebs 1999) for the best taxonomic levels of identification of the food items, avoiding the artificial inflation of the measure generated by major taxonomic levels of order or class (Greene & Jaksic 1983).

GENERAL DIET

Qualitatively the maned wolf is a highly generalist and omnivorous species: for all 18 areas, we identified at least 301 food items in the feces, including 116 plant and 178 animal species/morphospecies, besides 7 organic/inorganic items. Quantitatively, however, the standardized food-niche breadth for best taxonomic resolution among these 18 sites revealed a relatively narrow diet (mean $B_{st} \pm$ SE, 0.217 ± 0.011), ranging from 0.119 to 0.289. Among the 301 items, only a few are consumed in high proportions, including wolf's fruit (*Solanum lycocarpum*, Solanaceae), grass leaves, "unidentified small rodents," *Calomys* spp. (Muridae), and *Necromys lasiurus* (Muridae) for many areas, and *Campomanesia* spp. (Myrtaceae), *Alibertia edulis* (Rubiaceae), *Syagrus romanzoffiana* (Arecaceae), *Parinari obtusifolia* (Chrysobalanaceae), *Ophiodes* spp. (Anguidae), and *Clyomys bishopi* (Echimyidae) for a few areas. As predicted by Greene and Jaksic (1983), the food-niche breadth measure based on Levins' index is biased in the direction of higher values when class or order levels of identification are used for the food items. On the contrary, consistently lower and more real values are reported when we use the best taxonomic level possible (Table 7.1).

By frequency of occurrence relative to total number of occurrences, animals (mean ± SE; 47.9% ± 1.6%) and plants (50.7% ± 1.9%) were similarly distributed in the diets of the 18 study sites. In nine sites, the frequency of occurrence of plant material was higher than that of animals, but never above 60.0% of the total food item occurrences (Motta-Junior 2000; Rodrigues et al. 2007). The most important groups were miscellaneous fruits, wolf's fruit, and rodents, which together yielded 57.2% (Table 7.1 and Figure 7.1). In contrast, armadillos, medium to large mammals, fishes, reptiles, and anurans were of minor relevance according to occurrence.

On the other hand, on average, the estimated consumed biomass of animal prey (63.5% ± 3.8%) was more important than plants (35.8% ± 4.3%). Armadillos, wolf's fruits, and small mammals (mostly rodents, but also opossums and rabbits) are the dominant items (Figure 7.1). When we analyze the diet by estimated biomass consumed, only armadillos, wolf's fruits, and rodents yield the bulk (60.2%) of the diet (Table 7.1 and Figure 7.1). Conversely, arthropods, grasses, leaves, and stems are insignificant.

In general, the relative importance of particular food categories may vary. In many studies miscellaneous fruits were the most commonly consumed item (Motta-Junior 2000; Aragona & Setz 2001; Bueno & Motta-Junior 2004; Jácomo et al. 2004; Queirolo & Motta-Junior 2007), whereas wolf's fruit was more important in others (Dietz 1984; Motta-Junior et al. 1996; Rodrigues 2002; Santos et al. 2003), and small mammals in some sites (Juarez & Marinho-Filho 2002; Silva & Talamoni 2003; Bueno et al. 2002). Among animals, birds were regular and rather important components (Table 7.1). On the other hand, arthropods were not very well represented in the maned wolf's diet in these studies, with the exception of Aragona and Setz (2001), Aragona (2002), Santos et al. (2003), and Castro and Emmons (2012), where insects represented more than 10% of occurrences. Reptiles in general accounted for a maximum of 4%–5% of occurrences, except for the Queirolo and Motta-Junior (2007) study, with almost 10% reptiles. The frequency of occurrence of birds was similar between sites (8%–10%), apart from the UFSC study site (15.5%). Armadillos were much more relevant in estimated consumed biomass than frequency of occurrence, which varied between sites, probably according to species presence/abundance. The maned wolf is known to feed on medium-sized mammals, mostly deer (Motta-Junior et al. 1996; Bestelmeyer & Westbrook 1998; Juarez & Marinho-Filho 2002; Jácomo et al. 2004; Castro & Emmons 2012), although at low frequency.

TABLE 7.1
Diet of Maned Wolf in 18 Localities of Central and East South America by Percentage of Frequency of Occurrence

Major Taxonomic Groups/ Study Areas	FFOR	FSLU	RSCA	UFSC	EITA	RPRA	FSPI	PEIB	EJAT	EEAE	FALI	PNCG	PNEM	ESTA	PNSC	EITI	FMAL	PNKM
Solanum lycocarpum and *S. gomphodes* fruits	7.37	29.22	5.00	23.50	17.20	31.94	26.93	6.10	21.54	27.08	25.66	17.23	17.97	18.27	9.89	16.60	10.90	4.62
Miscellaneous fruits	15.06	7.56	5.00	7.00	27.49	9.42	8.26	20.10	23.08	23.97	9.21	24.76	36.28	22.34	29.72	20.59	35.03	33.46
Grasses and leaves	23.08	17.13	20.56	13.00	12.42	9.42	16.92	19.28	15.08	8.21	11.84	13.11	3.24	19.80	9.83	11.95	10.76	1.35
Grains and other anthropic items	3.53	0.00	9.63	0.00	0.00	0.00	0.00	7.25	0.77	0.22	0.00	0.00	0.00	0.00	0.00	0.09	4.41	0.00
Insects and other arthropods	3.21	12.09	11.11	5.00	6.57	3.66	4.23	14.50	4.92	5.77	1.97	19.90	1.65	2.54	4.32	4.84	12.83	22.31
Fishes	0.32	0.25	0.00	0.50	0.10	0.00	0.00	0.00	0.00	0.00	0.00	0.00	0.00	0.00	0.00	0.38	0.00	0.38
Amphibians	0.32	0.00	0.19	0.50	0.00	0.00	0.00	0.00	0.00	0.00	0.00	0.00	0.00	0.00	0.00	0.28	0.00	0.00
Lizards	1.28	1.01	1.30	1.00	0.34	0.52	0.72	1.81	1.08	0.11	2.63	0.73	0.64	0.00	7.40	0.19	0.55	3.46
Snakes	2.56	0.76	3.33	4.00	1.06	1.05	2.48	1.48	1.69	0.00	0.00	4.13	2.51	1.52	1.78	3.42	2.90	0.96
Birds	8.65	11.08	10.00	15.50	8.89	8.38	16.41	6.75	6.15	10.77	13.82	10.44	11.30	12.18	12.61	11.86	3.17	7.12
Anteaters	0.00	0.00	0.00	0.00	0.05	0.00	0.00	0.00	0.15	0.00	0.00	0.00	0.00	0.00	0.00	0.00	0.00	0.00
Armadillos	3.85	6.30	1.67	1.00	3.53	1.57	2.27	2.47	2.77	6.77	9.21	1.46	2.11	4.06	0.89	5.50	2.48	0.00
Opossums	2.56	0.50	1.30	1.00	0.43	0.52	1.03	1.98	2.31	4.11	2.96	1.21	0.07	1.52	1.42	0.28	2.62	7.69
Rodents	26.60	14.11	25.74	25.50	21.01	32.46	20.12	14.33	18.31	12.43	22.04	6.80	23.99	16.75	22.02	22.68	11.17	18.27
Rabbits	0.64	0.00	1.67	0.50	0.77	0.00	0.00	3.95	2.00	0.00	0.00	0.00	0.00	0.00	0.00	1.04	0.28	0.00
Carnivores	0.00	0.00	0.19	0.00	0.10	0.52	0.00	0.00	0.00	0.33	0.00	0.00	0.15	0.00	0.00	0.00	0.41	0.00
Peccaries	0.00	0.00	0.00	0.00	0.00	0.00	0.00	0.00	0.00	0.11	0.00	0.24	0.00	0.00	0.00	0.00	0.41	0.00
Deer	0.64	0.00	0.00	0.00	0.05	0.52	0.10	0.00	0.15	0.11	0.66	0.00	0.09	1.02	0.00	0.00	0.14	0.38
Other medium to large mammals	0.32	0.00	3.33	2.00	0.00	0.00	0.52	0.00	0.00	0.00	0.00	0.00	0.00	0.00	0.12	0.28	1.93	0.00
Total occurrences	312	397	540	200	2070	191	969	607	650	901	304	412	4540	197	1689	1054	725	520
Number of feces	80	150	230	52	614	70	280	141	191	328	105	129	1673	77	399	325	200	108
Number of items (spp. and morphospecies)	57	24	55	34	71	33	63	50	55	62	43	61	38	40	69	69	76	75
Occurrences/feces	3.9	2.6	2.3	3.8	3.4	2.7	3.5	4.3	3.4	2.7	2.9	3.2	2.7	2.6	4.2	3.2	3.6	4.8
Bst major taxonomic groups	0.337	0.483	0.410	0.373	0.329	0.298	0.418	0.579	0.391	0.352	0.556	0.505	0.293	0.549	0.455	0.385	0.302	0.377
Bst species/morphospecies	0.289	0.259	0.221	0.273	0.209	0.153	0.119	0.240	0.198	0.147	0.210	0.210	0.189	0.229	0.249	0.200	0.238	0.270

Note: Standardized food-niche breadths (Bst) were calculated for the major taxonomic groups and for the best taxonomic level obtained in each study. Locality abbreviations are given in the section on Study Areas.

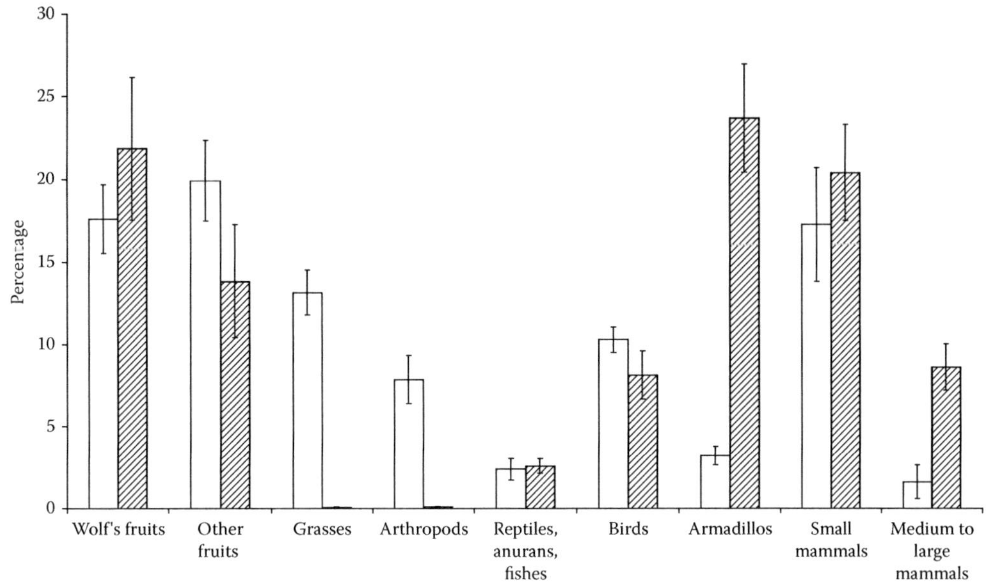

FIGURE 7.1 Percentages of frequency of occurrence (*white bars*) and estimated consumed biomass (*hatched bars*) of major food item categories found in the diet of the maned wolf in central–eastern South America. Whiskers are standard errors. $N = 18$ localities for occurrences and $N = 12$ sites for biomass.

Anthropic items such as organic remains of garbage, bovine meat offered by humans (e.g., Silva & Talamoni 2003), cultivar fruits, and sugarcane were on average of low importance, both in percent occurrence (mean ± SE; 3.8 ± 1.22) and in biomass (4.61 ± 1.48). Not surprisingly, higher levels of anthropic items were found in more disturbed sites such as FFOR (unpublished data) and RSCA (Silva & Talamoni 2003). Poultry (mostly *Gallus gallus* and occasionally *Cairina moschata* and *Numida meleagris*) had a low incidence in the diet, both by occurrence (0.45 ± 0.15) and by biomass (2.15 ± 0.76).

Our assessment revealed and confirmed unequivocally that wild individuals have mixed diets, with high proportions of fruits. Comparatively, captive animals are frequently fed on higher proportions of meat, and even when equal amounts of fruit and meat are offered, they tend to consume more meat and small quantities of fruits. The excessive intake of animal protein may exacerbate the development of cystinuria and gingivitis (Barboza et al. 1994). Thus, we suggest that meat or animal protein intake of captive individuals be reduced (as already proposed by Barboza et al. 1994) and that the proportion of fruits in their diet should be increased to at least 30%–40% of daily food biomass intake.

PREY AND FRUIT SIZE DISTRIBUTION

According to the 14 comparable studies, by frequency of occurrence small animals (10–100 g) comprise most of the animal prey consumed by the wolves (Figure 7.2), basically small mammals and birds. Solitary habits and predation on sparsely distributed food items may explain the consumption of relatively small prey for a canid weighing 20–30 kg. Nevertheless, active predation on large prey by the maned wolf has been reported (e.g., Bestelmeyer & Westbrook 1998). Conversely, by consumed biomass, most of the animal prey was represented by larger individuals above 1 kg (Figure 7.2), which included mainly armadillos and occasionally deer. Fruits weighing between 0.1 and 1 kg were the bulk of the diet from nine localities in central and southeast Brazil, mostly wolf's fruits (see figure 19.1 in Motta-Junior & Martins 2002).

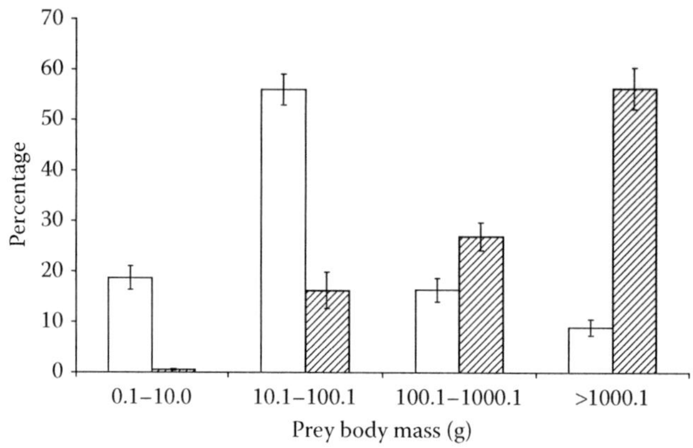

FIGURE 7.2 Prey size distribution in the diet of the maned wolf: mean percentage of frequency of occurrence (*white bars*) and biomass (*hatched bars*) of animal prey as a function of body mass found in 14 sites from central and eastern South America. Whiskers are standard errors.

FOOD INTAKE AND BIOENERGETICS

Few studies have estimated the food intake and bioenergetic constraints of the maned wolf (Barboza et al. 1994; Pessutti 1997; Castro & Emmons 2012). A study of captive maned wolves' ability to digest food concluded that they are able to digest fruits similarly to omnivorous animals, but their glycemic metabolism is characteristic of carnivorous animals (Pessutti 1997).

Castro and Emmons (2012) provided the first estimates of energy values of actual prey and fruit for the maned wolf. Based on the baseline for maintenance without travel of captive individuals of 1297 kcal/day (Barboza et al. 1994), but inputting costs related to travel distances measured by telemetry, Castro and Emmons (2012) estimated an overall average energy cost of about 1,580 kcal/day. These authors estimated an average energy intake of 1436 kcal/day represented by four daily scats from the estimated 1492 g of consumed biomass.

Carbone et al. (1999) hypothesized that carnivores above 21 kg could not be entirely supported by a diet of small prey. The maned wolf appears to ironically support this hypothesis, since this large canid probably survives as a predator because almost half of its diet is fruit (Castro & Emmons 2012).

SEASONALITY AND FUNCTIONAL RESPONSE

In 11 sites where seasonality in the diet was studied on occurrence data (Figure 7.3), we found that maned wolves consumed significantly more wolf's fruits (8 sites) and small mammals (6 sites) in the dry season, whereas other miscellaneous fruits (8 sites) and arthropods (4 sites) were eaten significantly more during the wet season. The availability of wolf's fruit is either higher in the dry months or constant all the year, whereas miscellaneous fruits are more common in the wet season (Motta-Junior 2000; Belentani 2001; Silva & Talamoni 2003; Queirolo & Motta-Junior 2007; Rodrigues et al. 2007). In general, small mammals are more abundant in the dry months in some of the studied areas and in central Brazil (Alho et al. 1986; Motta-Junior 2000; Belentani 2001; Queirolo & Motta-Junior 2007). Thus, the consumption patterns found by season appear to reflect natural cycles of abundance of the respective food resources, showing a temporal trophic opportunism by the maned wolf with regard to these major groups of food.

Functional responses in the diet were the subject of few studies. A positive and significant correlation between availability and consumption of miscellaneous fruits by maned wolves was found in two areas (Motta-Junior 2000; Queirolo & Motta-Junior 2007). Also, a significant and negative

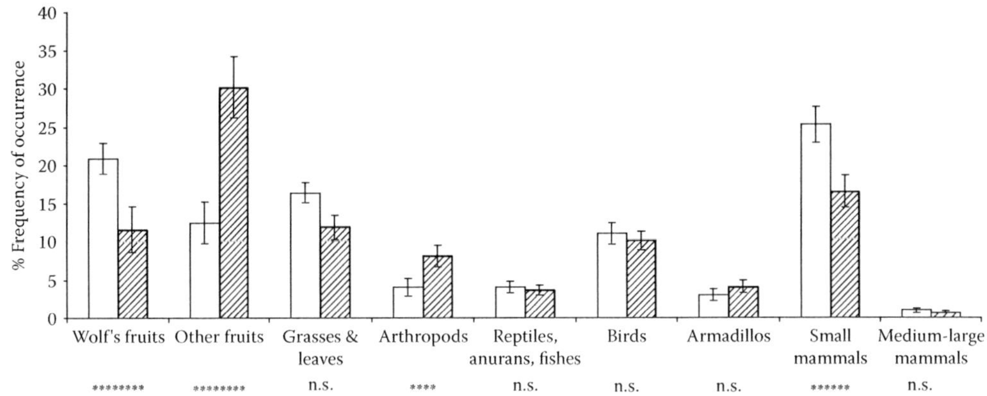

FIGURE 7.3 Seasonality by frequency of occurrence in the diet of the maned wolf in central–eastern South America. White bars are means for dry season and hatched bars are means for wet season; whiskers are standard errors. $N = 11$ study localities. Asterisks represent the number of localities with seasonal significant difference and n.s. indicates the nonsignificant results, both according to the chi-square test.

association between the frequency of wolf's fruit and other fruits was also reported (Motta-Junior 2000; Motta-Junior & Martins 2002; Queirolo & Motta-Junior 2007). The negative correlation between consumption of wolf's fruit and miscellaneous fruits suggests that wolves use the different fruits to maintain an approximately constant rate of fruit intake in the face of seasonal variation in community-wide fruit availability (Motta-Junior & Martins 2002).

Even though the consumption of small mammals by the maned wolf changes with seasons (Figure 7.3), only one study reported a significant correlation between variations in the abundance of small mammals and consumption by the maned wolf (Silva & Talamoni 2003). The only attempt to find a correlation for birds was unsuccessful (Motta-Junior 2000).

PREY AND FRUIT SELECTION

The common perception in the literature that maned wolves consume most food items according to their availability (e.g., Rodrigues et al. 2007; Castro & Emmons 2012) appears to be an oversimplification. Even though wolves are temporal opportunists for some major food groups such as fruits and small mammals, and consume some items according to availability, our review has found examples of a diet selective for small mammals and fruits.

In the five studies that evaluated small mammal selection, a selective diet was always detected, but no consistent pattern was found (Motta-Junior 2000; Belentani 2001; Bueno & Motta-Junior 2006, 2009; Queirolo & Motta-Junior 2007). Of four studies where *Calomys tener* occurred, it was positively selected in three of them, relative to the most common small savanna mammals, and consumed less than expected in one. Opossums (*Didelphis albiventris*, *Gracilinanus* spp., and *Monodelphis* spp.) were consumed according to their abundance in three studies, but were selected in one. In three studies, *Oligoryzomys* spp. (*O. nigripes* and *O. flavescens*) were either less consumed (one study) or consumed according to their abundance (two studies). Finally, *Necromys lasiurus* was either more consumed than expected (two studies), less consumed (one study), or preyed upon according to its abundance (two studies).

This discrepancy may result from different methodologies and sampling efforts among studies, masking the real frequencies of species in the communities: trapping methods may not reflect the real availability of small mammals in the field (Jaksic 1989). Also, different biogeographic areas may have different community structures, imposing differences on wolves' predation patterns.

Fruit selection was evaluated only in the review of frugivory in the maned wolf by Motta-Junior and Martins (2002). Wolf's fruit was preferred in the five study sites in which a selection analysis was possible. Other fruit species appeared to be either avoided or consumed in proportions similar to those that occurred in the field (Motta-Junior & Martins 2002). This result emphasizes the relevant role wolf's fruit plays in the ecology of the maned wolf, supporting Courtenay's (1994) suggestion that this plant is a possible indicator of a suitable habitat for the maned wolf.

TROPHIC RELATIONSHIP WITH OTHER CARNIVORES

Few comparative studies on the feeding ecology between the maned wolf and other carnivores have been published. Apart from general and brief discussions without empirical data, such as those of Dietz (1984) and Castro and Emmons (2012), quantitative comparisons with other syntopic species are scarce (Motta-Junior 2000; Juarez & Marinho-Filho 2002; Bueno & Motta-Junior 2004; Jácomo et al. 2004). The maned wolf has a more omnivorous diet, based on smaller animals, when compared to the exclusively carnivorous cougar (*Puma concolor*) (Motta-Junior 2000). When compared with the crab-eating fox (*Cerdocyon thous*) and hoary fox (*Lycalopex vetulus*), wolves eat more fruits and a larger spectrum of animal prey sizes than the foxes, which capture many arthropods (Juarez & Marinho-Filho 2002; Jácomo et al. 2004).

THE MANED WOLF AS A SEED DISPERSER

A consequence of extensive consumption of fruits by the wolves is seed dispersal. Most of the seeds of all 16 fruit species tested passed through the wolf's gut unharmed, but most of the *Allagoptera campestris* seeds were destroyed in the feces (Lombardi & Motta-Junior 1993; Courtenay 1994; Motta-Junior & Martins 2002; Santos et al. 2003). The germination percentage was either similar to or higher than the control (no gut passage). These studies showed that maned wolves are not only legitimate seed dispersers of wolf's fruit and other plants of open habitats, but are also efficient, as they mostly defecate in open sites suitable for seed germination and seedling establishment.

HABITAT OF PREY AND FRUITS

Some studies have reported that the presumable sites of capture of food items by wolves are mostly grasslands and savannas (cerrado), based on known habitat use of identified prey and fruits (Motta-Junior & Martins 2002; Motta-Junior et al. 1996; Belentani 2001; Bueno et al. 2002; Queirolo & Motta-Junior 2007). By occurrence and biomass, the food items varied, respectively, between 48.0% and 69.1% and between 43.3% and 63.6% among all identified food items to genus and/or species level. An analysis of this kind, specifically for fruit species in eight study sites from southeast Brazil, revealed that, despite the high variation in percentage of land covered by grassland savanna, the maned wolves consistently consumed high proportions of cerrado fruits (see figure 19.3 Motta-Junior & Martins 2002). These findings suggest that wolves search for food resources more intensively in cerrado patches than would be expected by chance and this should be taken into consideration for habitat management of this species.

FURTHER STUDIES ON FEEDING ECOLOGY

As the diet of the maned wolf is well known, new studies on this topic should focus on regions where its distribution has not yet been evaluated, such as northeast and south Brazil, as well as Uruguay and Paraguay. More information is also needed from Argentina and Bolivia. Data gathered should include not only frequency of occurrence, but also minimum number of prey and fruits (based on mean number of seeds per fruit species) and biomass consumption estimates. Prey and fruit identification should attain the best level of identification possible with the help of reference collections and museum specialists. Also, studies in unprotected areas are needed for comparison with wolves' diets in reserves.

Silveira's hypothesis (1968) on the therapeutic properties of wolf's fruit to control the incidence of parasitic worms in the maned wolf remains untested. Unfortunately, Silveira's (1968) study was biased (see Carvalho 1976; Dietz 1984), since not only fruits were offered as food, but also a mixture of stems, roots, flowers, and leaves, which are not consumed by wolves in the wild. The inhibition of wolf mortality could be attributed to any component, not necessarily to the pulp of wolf's fruit. A more controlled study design is needed.

Data on functional response of the maned wolf are relatively recent in the literature and cover only 1 or 2 years of fieldwork. The lack of correlations could just be a result of the short periods of study. We suggest long-term research of at least 5 years, but ideally covering 10 years, which can reveal not only seasonal fluctuations, but also possible interannual ones (e.g., Jaksic et al. 1997; Bonecker et al. 2009). In the same context, we suggest that the maned wolf density or abundance should be measured simultaneously to prey and fruit density or abundance, to permit a numerical response approach which has never been studied before for this canid species.

The selectivity of the maned wolf's diet is also a relatively new topic, focusing almost only on small mammals. As fruits are a large portion of the maned wolf's diet, further studies should focus particularly on this group of food resources, which are perhaps more easily evaluated in the field than animals (Motta-Junior& Martins 2002). Future studies on small mammal selectivity should carefully avoid trap and bait selectivity, using traps less likely to be selective, as pitfalls (Santos-Filho et al. 2006; Umetsu et al. 2006). Future studies on diet selection should adopt the protocol and definitions proposed by Jaksic (1989) regarding opportunism versus selectivity. Additionally, more detailed information on morphological and behavioral traits of prey is necessary to understand their differential predation by maned wolves (e.g., Bueno 2003; Corley et al. 1995).

For a more comprehensive understanding of the maned wolf's digestive physiology and bioenergetics, which is important for a quantitative evaluation of the role this canid plays in trophic chains, we suggest experiments of food digestibility and assimilation include food items consumed by wild populations, not only dog rations or few food categories, such as chicken and cultivar fruits (Barboza et al. 1994; Pessutti 1997). Also, experiments should take into account energy constraints of reproductive versus nonreproductive periods. These data would be useful for the management of animals in captivity.

Further telemetry studies focusing particularly on foraging activity hours are necessary to test our indirect evidence revealing that most identified fruits and prey in the diet of the wolves are from grasslands and cerrado vegetation, irrespective of landscape cover. As already commented, the results can be of relevance for conservation management policies on this species.

ACKNOWLEDGMENTS

We thank the Instituto Brasileiro do Meio Ambiente e Recursos Naturais Renováveis (IBAMA) for permission to study in PNSC and the Instituto Florestal de São Paulo, which provided facilities at the ESI. We also thank A. Percequilo, G. I. Ximenes, M. J. J. Silva, V. Fagundes, and J. M. Pagnozzi for identifying small mammals; M. Martins for identifying reptiles; and J. A. Lombardi for identifying botanical specimens. We also thank Fundação de Amparo a Pesquisa do Estado de São Paulo (FAPESP Process 00/12339-2 and 97/06090-7), Conselho de Aperfeiçoamento de Pessoal do Ensino Superior (CAPES), and WWF Brasil (CSR 123-99) for financial support. Finally, we would like to express our gratitude to all the researchers and colleagues who made our project possible by helping us in fieldwork and laboratory analysis as well as making invaluable comments and suggestions on the project's execution. This chapter is dedicated to the memory of César Ades, a gentleman as a person and a national and international authority as an animal behavior scientist.

REFERENCES

Alho, C. J. R., Pereira, L. A. & Costa, A. P. 1986. Patterns of habitat utilization by small mammal populations in Cerrado biome of central Brazil. *Mammalia* 50: 448–460.

Amboni, M. P. M. 2007. Dieta, disponibilidade alimentar e padrao de movimentacao do lobo-guara, *Chrysocyon brachyurus*, no Parque Nacional da Serra da Canastra, MG. MSc. Dissertation. Universidade Federal de Minas Gerais, Belo Horizonte, Brazil.

Aragona, M. 2002. Ecologia alimentar do lobo-guará (*Chrysocyon brachyurus,* Illiger, 1811) no Parque nacional de Chapada dos Guimarães, MT. MSc. Dissertation. Universidade Federal do Mato Grosso, Cuiabá, Brazil.

Aragona, M. & Setz, E. Z. F. 2001. Diet of the maned wolf, *Chrysocyon brachyurus* (Mammalia: Canidae), during wet and dry seasons at Ibitipoca State Park, Brazil. *Journal of Zoology (London)* 254: 131–136.

Azara, F. 1801. *Essais sur l'histoire naturelle des quadrupeds de la province du Paraguay*. Vol. 1. C Pougens, Paris.

Barboza, P. S., Allen, M. E., Rodden, M. & Pojeta, K. 1994. Feed intake and digestion in the maned wolf (*Chrysocyon brachyurus*): Consequences for dietary management. *Zoo Biology*, 13: 375–781.

Bartmann, W. & Bartmann, C. 1986. Mähnenwölfe (*Chrysocyon brachyurus*) in Brasilien—Ein Freiland-Bericht. *Z. Kolner Zoo* 29: 165–176.

Belentani, S. C. S. 2001. Ecologia alimentar do lobo guará, *Chrysocyon brachyurus* (Mammalia: Canidae) no Parque Florestal Salto e Ponte, município de Prata, MG. MSc. Dissertation. Universidade de São Paulo, São Paulo.

Belentani, S. C. S., Motta-Junior, J. C. & Talamoni, S. A. 2005. Notes on the food habits and prey selection of the maned wolf (*Chrysocyon brachyurus*) (Mammalia, Canidae) in southeastern Brazil. *Biociências* 13(1): 95–98.

Bestelmeyer, S. V. & Westbrook, C. 1998. Maned wolf (*Chrysocyon brachyurus*) predation on pampas deer (*Ozotoceros bezoarticus*) in central Brazil. *Mammalia* 62: 591–595.

Bianchi, C. A., Bagno, M. A. & Juarez, K. M. 2000. Predation of *Ara ararauna* and *Amazona aestiva* (Psittaciformes, Psittacidae) by *Chrysocyon brachyurus* (Carnivora, Canidae) in the Cerrado, Brazil. *Ararajuba* 8(1):49–50.

Bonecker, S. T., Portugal, L. G., Costa-Neto, S. F. & Gentile, R. 2009. A long term study of small mammal populations in a Brazilian agriculture landscape. *Mammalian Biology*, 74: 467–477.

Brady, C. A. & Ditton, M. K. 1979. Management and breeding of maned wolves at the National Zoological Park. *International Zoo Yearbook* 19: 171–176.

Bueno, A. A. 2003. Vulnerabilidade de pequenos mamíferos de áreas abertas à vertebrados predadores na Estação Ecológica de Itirapina, SP. MSc. Dissertation. Universidade de São Paulo, São Paulo, Brazil.

Bueno, A. A. & Motta-Junior, J. C. 2004. Food habits of two syntopic canids, the maned wolf (*Chrysocyon brachyurus*) and the crab-eating fox (*Cerdocyon thous*), in southeastern Brazil. *Revista Chilena de Historia Natural* 77: 5–14.

Bueno, A. A. & Motta-Junior, J. C. 2006. Small mammals selection and functional response in the diet of the maned wolf, *Chrysocyon brachyurus* (Mammalia: Canidae), in southeast Brazil. *Mastozoología Neotropical* 13: 11–19.

Bueno, A. A. & Motta Junior, J. C. 2009. Food habits of the maned wolf, *Chrysocyon brachyurus* (Canidae) in southeastern Brazil. *Studies on Neotropical Fauna and Environment* 44(2): 67–75.

Bueno, A. A., Belentani, S. C. S., & Motta-Junior, J. C. 2002. Ecologia alimentar do lobo-guará, *Chrysocyon brachyurus* (Illiger, 1815) (Mammalia: Canidae), na Estação Ecológica de Itirapina, Estado de São Paulo. *Biota Neotropica* 2(2): 1-–9.

Cabrera, A. 1958. Catálogo de los mamíferos de América del Sur. v. 1. Revista del Museo Argentino de Ciencias Natutales "Bernardino Rivadavia". *Ciencias Zoológicas* 4(1): 1–307.

Carbone, C., Mace, G. M., Roberts, S. C. & Macdonald, D. W. 1999. Energetic constraints on the diet of terrestrial carnivores. *Nature* 402: 286–288.

Carvalho, C. T. 1976. Aspectos faunísticos do Cerrado—O lobo guará (Mammalia: Canidae). *Boletim Técnico do Instituto Florestal* 21: 1–18.

Castro, J. M. & Emmons, L. H. 2012. Variation in diet and resources. *Smithsonian Contributions of Zoology* 639: 37–55.

Castro, S. A., Silva, S. I., Meserve, P. L., Gutiérrez, J. R., Contreras, L. C. & Jaksic, F. M. 1994. Frugivoría y dispersión de semillas de Pimiento (*Schinus molle*) por el zorro Culpeo (*Psedalopex culpaeus*) en el Parque Nacional Fray Jorge (IV Región, Chile). *Revista Chilena de Historia Natural* 67: 169–176.

Cheida, C. C. 2005. Dieta e dispersão de sementes pelo lobo-guará *Chrysocyon brachyurus* (Illiger 1815) em uma área com campo natural, Floresta Ombrófila mista e silvicultura, Paraná, Brasil. MSc. Dissertation. Universidade Federal do Paraná, Curitiba, Brazil.

Corley, J. C., Fernandez, G. J., Capurro, A. F., Novaro, A. J., Funes, M. C. & Travaini, A. 1995. Selection of cricetine prey by the Culpeo fox in Patagonia: A differential prey vulnerability hypothesis. *Mammalia*, 59(3): 315–325.

Courtenay, O. 1994. Conservation of the maned wolf: Fruitful relations in a changing environment. *Canid News* 2: 41–43.

Dietz, J. M. 1984. Ecology and social organization of the maned wolf (*Chrysocyon brachyurus*). *Smithsonian Contributions to Zoology* 392: 1–51.

Dietz, J. M. 1985. *Chrysocyon brachyurus*. *Mammalian Species* 234: 1–4.

Ferraz, E. 1999. Ecologia alimentar e dispersão de sementes pelo lobo-guará (*Chrysocyon brachyurus*, Illiger, 1811) em uma área rural no sudeste do Brasil (Carnívora: Canidae). MSc. Dissertation. Universidade Estadual de São Paulo, Rio Claro, Brazil.

Garcia, A. 1983. On the social behaviour of maned wolves (*Chrysocyon brachyurus*). Universidade de São Paulo, *Boletim de Zoologia*, 6: 63–77.

Goeldi, E. 1893. *Os mamíferos do Brasil*. Livraria Alves, Rio de Janeiro.

Greene, H. W. & Jaksic, F. M. 1983. Food-niche relationships among sympatric predators: Effects of level of prey identification. *Oikos* 40(1): 151–154.

Iriarte, J. A., Jimenez, J. E., Contreras, L. C. & Jaksic, F. M. 1989. Small-mammal availability and consumption by the fox, *Dusicyon culpaeus*, in central Chilean scrublands. *Journal of Mammalogy* 70(3): 641–645.

Jácomo, A. T. A., Silveira, L. & Diniz-Filho, J. A. F. 2004. Niche separation between the maned wolf (*Chrysocyon brachyurus*), the crab-eating fox (*Dusicyon thous*) and the hoary fox (*Dusicyon vetulus*) in central Brazil. *Journal of Zoology (London)* 262: 99–106.

Jaksic, F. M. 1989. Opportunism vs selectivity among carnivorous predators that eat mammalian prey: A statistical test of hypotheses. *Oikos* 56: 427–430.

Jaksic, F. M., Jiménez, J. E., Castro, S. A. & Feinsinger, P. 1992. Numerical and functional responses of predators to a long-term decline in mammalian prey at a semi-arid neotropical site. *Oecologia* 89: 90–101.

Jaksic, F. M., Silva, S. I., Meserve, P. L. & Gutiérrez, J. R. 1997. A long term study of vertebrate predator responses to an El Niño (ENSO) disturbance in western South America. *Oikos* 78: 341–354.

Juarez, K. M. & Marinho-Filho, J. 2002. Diet, habitat use, and home ranges of sympatric canids in central Brazil. *Journal of Mammalogy* 83: 925–933.

Kleiman, D. G. 1972. Social behavior of the maned wolf (*Chrysocyon brachyurus*) and bush dog (*Speothos venaticus*): A study in contrast. *Journal of Mammalogy* 53: 791–806.

Krebs, C. J. 1999. *Ecological Methodology*, 2nd edn. Benjamin/Cummings, Menlo Park, CA.

Krieg, H. 1940. Im lande des Mähnenwölfe. *Der Zoologische Garten* 12: 257–269.

Langguth, A. 1975. Ecology and evolution in the South American canids. In: Fox, M. W. (ed.), *The Wild Canids: Their Systematics, Behavioral Ecology and Evolution*. Van Nostrand Reinhold, New York, pp. 192–206.

Lombardi, J. A. & Motta-Junior, J. C. 1993. Seed dispersal of *Solanum lycocarpum* St. Hil. (Solanaceae) by the maned wolf, *Chrysocyon brachyurus* Illiger (Mammalia, Canidae). *Ciência e Cultura* 45: 126–127.

Lund, P. W. 1950. *Memórias sobre a paleontologia brasileira revistas e comentadas por Carlos de Paula Couto*. Instituto Nacional do Livro, Rio de Janeiro.

Massara, R. L. 2009. Dieta, uso do habitat e endoparasitas fecais do lobo-guará na Serra da Calçada, região metropolitana de Belo Horizonte, Minas Gerais. MSc. Dissertation. Pontificia Universidade Católica de Minas Gerais, Belo Horizonte, Brazil.

Motta-Junior, J. C. 1991. Predação de *Micropygia schomburgkii* (Aves: Rallidae) por *Chrysocyon brachyurus* (Mammalia: Canidae) no Distrito Federal, Brasil. *Ararajuba* 2: 87–88.

Motta-Junior, J. C. 1997. Ecologia alimentar do lobo guará, *Chrysocyon brachyurus* (Mammalia: Canidae). *Anais de Etologia* 15: 197–209.

Motta-Junior, J. C. 2000. Variação temporal e seleção de presas na dieta do lobo-guará, *Chrysocyon brachyurus* (Mammalia: Canidae), na Estação Ecológica de Jataí, Luiz Antônio, SP. In: Santos, J. E. & Pires, J. S. (eds), *Estudos integrados em ecossistemas. Estação ecológica de Jataí*. Rima Editora, São Carlos, pp. 331–346.

Motta-Junior, J. C. & Martins, K. 2002. The frugivorous diet of the maned wolf, I, in Brazil: Ecology and conservation. In: Levey, D. J., Silva, W. R. & Galetti, M. (eds), *Seed Dispersal and Frugivory: Ecology, Evolution and Conservation*. CABI Publishing, Wallingford, pp. 291–303.

Motta-Junior, J. C., Talamoni, S. A., Lombardi, J. A. & Simokomaki, K. 1996. Diet of the maned wolf, *Chrysocyon brachyurus*, in central Brazil. *Journal of Zoology (London)* 240: 277–284.

Pessutti, C. 1997. Aspectos da fisiologia digestiva de lobo guará *Chrysocyon brachyurus* e cachorro do mato *Cerdocyon thous*. MSc. Dissertation, Instituto de Biociências, Universidade Estadual Paulista, Botucatu, Brazil.

Queirolo, D. & Motta-Junior, J. C. 2007. Prey availability and diet of *Chrysocyon brachyurus* in Serra da Canastra National Park, Brazil. *Acta Theriologica* 52: 391–402.

Rodrigues, F. H. G. 2002. Biology and conservation of maned wolf at Águas Emendadas Ecological Station, DF. PhD Thesis, Universidade de Campinas, Campinas.

Rodrigues, F. H. G., Hass, A., Lacerda, A., Grando, A. C. R., et al. 2007. Feeding habits of the maned wolf (*Chrysocyon brachyurus*) in the Brazilian Cerrado. *Mastozoologia Neotropical* 14: 37–51.

Santos, E. F., Setz, E. Z. F. & Gobbi, N. 2003. Diet of the maned wolf (*Chrysocyon brachyurus*) and its role in seed dispersal on a cattle ranch in Brazil. *Journal of Zoology (London)* 260: 203–208.

Santos-Filho, M., Silva, D. J. & Sanaiotti, T. M. 2006. Efficiency of four trap types in sampling small mammals in forest fragments, Mato Grosso, Brazil. *Mastozoologia Neotropical* 13: 217–225.

Schaller, G. B. 1983. Mammals and their biomass on a Brazilian ranch. *Arquivos de Zoología* 31(1): 1–36.

Silva, J. A. & Talamoni, S. A. 2003. Diet adjustments of maned wolves, *Chrysocyon brachyurus* (Illiger) (Mammalia Canidae), subjected to supplemental feeding in a private natural reserve, southeastern Brazil. *Revista Brasileira de Zoologia* 20: 339–345.

Silveira, E. 1968. Notes on the care and breeding of the maned wolf *Chrysocyon brachyurus* at Brasília Zoo. *International Zoo Yearbook* 8: 21–23.

Soler, L. & Casanave, E. B. 2007. Hábitos alimenticios del aguará guazú (*Chrysocyon brachyurus*) en Corrientes, Argentina. Libro de resúmenes y programa—XXI Jornadas Argentinas de Mastozoología/6 al 9 de noviembre de 2007, Tafí del Valle—Tucumán, Argentina.

Umetsu, F., Naxara, L. & Pardini, R. 2006. Evaluating the efficiency of pitfall traps for sampling small mammals in the neotropics. *Journal of Mammalogy* 87: 757–765.

8 Reproduction and Factors Affecting Reproductive Success in the Maned Wolf

Nucharin Songsasen, Amy E.M. Johnson, Lauren E. Reiter, and Melissa Rodden

CONTENTS

Introduction 99
Social Structure and Reproductive Behaviors 99
Reproductive Endocrinology 101
Seminal Traits 102
Factors Affecting Reproductive Success 103
In Situ Populations 103
Ex Situ Populations 104
Assisted Reproductive Technologies 105
Future Priorities 105
In Situ Populations 105
Ex Situ Populations 105
Acknowledgments 106
References 106

INTRODUCTION

The maned wolf is unique among canid species regarding its solitary nature and distinct reproductive biology. Although research on the reproductive biology of this species was initiated in the late 1990s, recent research has provided more clues into social behaviors and reproductive mechanisms of this charismatic canid. This chapter provides an overview of current understanding of its social structure, reproductive behaviors, and biology, as well as factors influencing reproductive success in the maned wolf. Future priorities both for *ex situ* and *in situ* studies are also discussed.

SOCIAL STRUCTURE AND REPRODUCTIVE BEHAVIORS

The maned wolf, a solitary canid, is facultatively monogamous (Dietz, 1984; Jácomo et al., 2009; Emmons et al., 2012). Although breeding pairs share a large home range, ranging on average from ~20 km^2 (Dietz, 1984) to >80 km^2, they mostly spend time together during breeding seasons (Jácomo et al., 2009; Emmons et al., 2012b). Males and females form a long-term pair, although this is not always lifelong (Emmons et al., 2012b). Recent research conducted in Bolivia has shown that adult males sometimes switch to pair with other females, even after several years of pairing (Emmons et al., 2012b).

To date, there are no data on the precise onset of sexual maturity. However, both sexes can reproduce before 2 years of age, although this rarely occurs (Songsasen and Rodden, 2010). Pups can be produced from 10 to 12-year-old parents, but this is also uncommon. For both sexes, breeding predominantly occurs from 3 through 8 years of age, an interval generally considered prime for reproduction (Songsasen and Rodden, 2010).

Maned wolves breed seasonally. During the reproductive season, females enter estrus once (Rodden et al., 2004) and males produce sperm (Comizzoli et al., 2009). It has been suggested that the reproductive season in the maned wolf is regulated by photoperiod because there is a 6-month displacement in reproduction between individuals living in southern versus northern hemispheres (Velloso et al., 1998). Specifically, the peak breeding season generally extends from March through June, but pairs may breed as early as November and as late as July in South America (Dietz, 1984; Emmons et al., 2012). In North America and Europe, the breeding season for captive maned wolves begins in October and continues till February (Maia and Gouveia, 2002); however, females can enter estrus as late as April if pairs are separated during the breeding season (M. Rodden, personal communication). The gestation length of this species is ~65 days (Dietz, 1985). For wild maned wolves, births begin in February and peak in June and July, but can occur as late as September or even October (Dietz, 1984; Vynne, 2010; Emmons et al., 2012). In the northern hemisphere, parturition occurs as early as November and peaks in December and January (Maia and Gouveia, 2002). Litter size ranges from one to seven pups with an average of 2.6 (Maia and Gouveia, 2002). Females play a major role in pup rearing, especially before the pups emerge from the den. Cumulative information obtained from both *in situ* and *ex situ* studies indicates that males play a major role in pup rearing by providing food to females and young as well as defending the territory (Bestelmeyer, 2000; Melo et al., 2007; Emmons et al., 2012). Males have been observed to stay near pup dens (Emmons et al., 2012). A male maned wolf was also observed regurgitating to three pups after the mother died (Jácomo et al., 2009). Recent studies have indicated that juveniles may stay within the parents' home range beyond the birth of the next litter and help with the care of their younger siblings (Melo et al., 2007; Emmons et al., 2012).

Scent marking has been identified as a communicative signal within and between breeding pairs (Brady, 1981; Dietz, 1984; Emmons et al., 2012). It has been shown that the rate of scent marking does not differ between males and females (Rodden et al., 1996, 2004). Breeding and nonbreeding males scent mark at the same rate throughout the reproductive season, suggesting that male maned wolves primarily utilize scent marking as a territorial signal (Rodden et al., 1996). This speculation was confirmed by a study conducted in wild wolves, which showed that floater individuals do not scent mark (Bestelmeyer, 2000). Female wolves may also use scent marking as a communication cue to mates about reproductive stage and as a territorial signal. It has been shown that female wolves that successfully produce pups have higher levels of scent marking both before and after the estrus period than nonbreeding individuals (Rodden et al., 1996).

Maned wolves also use "roar barks," which are loud, deep-throated vocalizations to communicate with conspecifics. Roar barks are heard during both the day and night, but more frequently at night when maned wolves are most active (Kleiman, 1972; Brady, 1981; Bestelmeyer, 2000; Emmons et al., 2012). An individual wolf exhibits a bout of barking ranging from 20 to 30 barks spaced 5–8 or more seconds apart (Kleiman, 1972). The caller often waits for replies before calling again (Emmons et al., 2012). Earlier research has suggested that maned wolves use roar barks for individual spacing (i.e., territorial signal; Kleiman, 1972; Brady, 1981). However, a recent study has shown that roar barks occur more frequently when young are present, indicating that this signal also serves as intrafamily communication across the home range (Emmons et al., 2012).

During the estrus period, the amount of time a pair spends in close proximity increases significantly (Rodden et al., 1996). Females exhibit swelling of the vulva, and courtship behavior is characterized by frequent approaches, mutual anogenital investigation, and playful interactions (Rodden et al., 1996). Mounting may occur frequently during estrus. Successful breeding includes a copulatory tie that may last several minutes (Rodden et al., 1996, 2004).

REPRODUCTIVE ENDOCRINOLOGY

Current knowledge of reproductive endocrinology in the maned wolf is mostly gleaned from studies conducted *ex situ* utilizing noninvasive fecal endocrine monitoring (Wasser et al., 1995; Velloso et al., 1998; Songsasen et al., 2006), although some data on wild individuals are available (Maia et al., 2008). Both free-ranging and captive maned wolves exhibit seasonal variations in reproductive steroid hormones that reflect their reproductive seasonality (Maia et al., 2008). Specifically, serum testosterone concentrations in both captive and free-ranging males living in Brazil are high in autumn compared to other seasons (Maia et al., 2008). Longitudinal monitoring of fecal gonadal metabolites in breeding males housed in North American zoos also demonstrates seasonal variations in testosterone metabolites. In breeding males, testosterone metabolite concentrations peak during December and January, sharply decline in February, reach the lowest levels in April and May, and then gradually increase as the breeding season approaches (Figure 8.1). Evaluations of serum progesterone and fecal progestagen metabolites in female maned wolves living *in situ* and *ex situ* in Brazil also show seasonal variations in hormone concentration. Specifically, concentration of the steroid hormone is elevated from April through June and sharply decreases in August (Maia et al., 2008; Vynne, 2010).

Longitudinal monitoring of patterns of female gonadal hormones reveals that profiles of steroid hormones are similar to those of domestic dogs and other wild canids, especially during the preovulatory period (Wasser et al., 1995; Velloso et al., 1998; Songsasen et al., 2006). Specifically, estrogen concentration gradually increases beginning about 2 weeks before breeding and declines to baseline on the day of lordosis and copulation. Progestagen steadily rises during late proestrus and throughout estrus and remains elevated throughout the pregnant or nonpregnant luteal phase (Velloso et al., 1998; Songsasen et al., 2006). When sexual behaviors are correlated with gonadal hormone profiles, copulatory ties are observed as early as 2 days before or 9 days after the estrogen peak (Velloso et al., 1998; Songsasen et al., 2006). Nonpregnant females excrete overall lower concentrations of progestagen than their pregnant counterparts (Velloso et al., 1998; Songsasen et al., 2006). The latter finding is consistent with a study conducted on the domestic dog (Gudermuth et al., 1998), but differs from that one conducted on the red wolf (*Canis rufus*; Walker et al., 2002). It has been demonstrated that there are no differences in gonadal steroid concentrations between pregnant and nonpregnant red wolves (Walker et al., 2002).

With the exception of the Island fox (*Urocyon littoralis*; Asa et al., 2007b), all canid species studied to date are spontaneous ovulators, that is, females ovulate regardless of the presence or absence

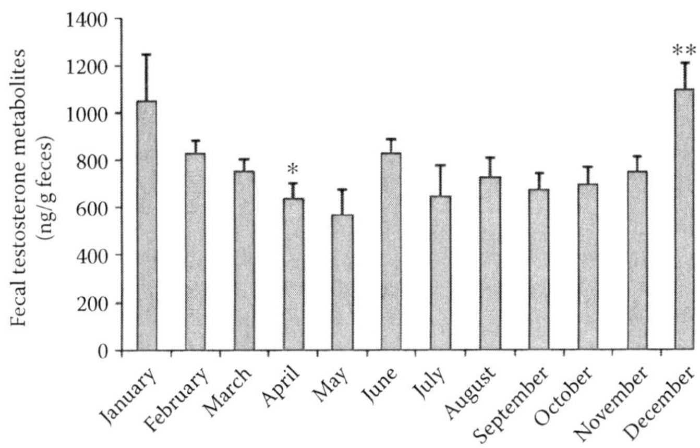

FIGURE 8.1 Means ± standard errors of means for fecal testosterone metabolite concentration of three breeding males housed in North American zoos of samples collected during various months. *, ** different symbols indicate significant differences ($p < .05$).

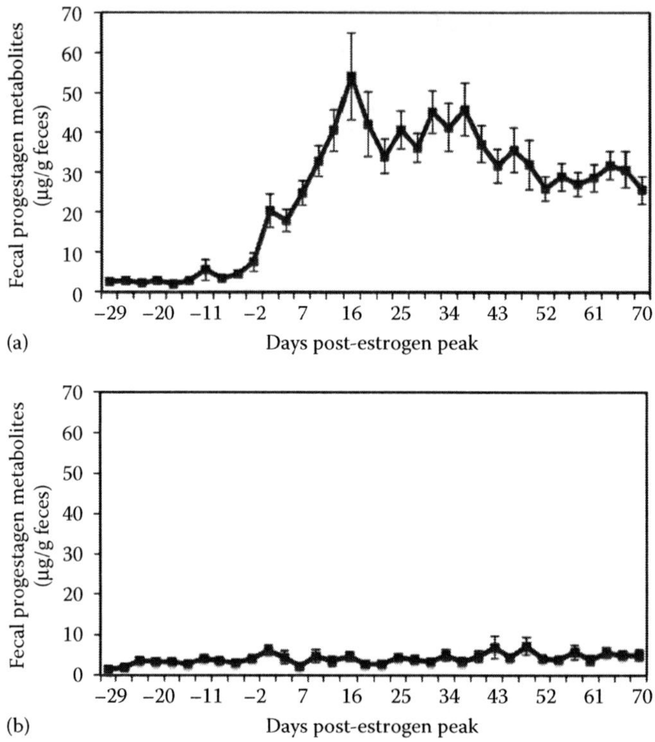

FIGURE 8.2 Longitudinal profiles of fecal progestagen metabolites in paired (a) and singleton (b) females.

of a male. However, there is evidence that maned wolves are induced ovulators, and the presence of a male is required for ovulation (Velloso et al., 1998; Songsasen et al., 2006). Figure 8.2 depicts fecal progestagen metabolite profiles of female maned wolves housed with a male ($n = 30$ reproductive cycles) and those housed alone ($n = 8$ reproductive cycles). Paired individuals exhibit a classical rise in progestagen concentrations following an estrogen peak, while the steroid concentration in singleton females remains at baseline level throughout the breeding season, indicating that ovulation does not occur in the latter.

SEMINAL TRAITS

Compared with other large canids, including the gray wolf (*Canis lupus*; Zindl et al., 2006), red wolf (Goodrowe et al., 1998), coyote (*Canis latrans*; Minter and DeLiberto, 2008), and African wild dog (*Lycaon pictus*; Johnston et al., 2007), which usually ejaculate > 100 million sperm, the maned wolf has a much lower sperm output (Table 8.1). Furthermore, maned wolf ejaculates also contain high proportions of structurally abnormal spermatozoa. Abnormal morphologies include head abnormality (19.1% ± 4.1%), proximal cytoplasmic droplet (9.0% ± 2.5%), distal cytoplasmic droplet (0.6% ± 0.2%), bent midpiece (6.3% ± 1.3%), bent tail (4.0% ± 2.6%), and coiled tail (34.1% ± 4.6%).

Ejaculate traits of wild maned wolves have also been investigated. Evaluation of three ejaculates obtained from two males living in Serra da Canastra National Park, Brazil, indicates that seminal traits of wild wolves are comparable to those of captive individuals (Table 8.1). Ejaculates of wild maned wolves also contained a high proportion of structurally abnormal spermatozoa, including head abnormalities (13.6% ± 4.4%), proximal cytoplasmic droplet (12.2% ± 5.2%), bent tail (25.2% ± 14.0%), and coiled tail (8.4% ± 8.4%). The high proportion of abnormal spermatozoa

TABLE 8.1
Seminal Parameters and Testicular Volume of Captive and Wild Maned Wolves

	Captive Wolves		
Parameters	**3–8 years old (range, *n*)**	**>8 years old (range, *n*)**	**Wild Wolves (range, *n*)**
Volume (ml)	1.8 ± 0.5 (0.1–12.5, 29)[b]	1.3 ± 0.2 (0.1–2.0, 16)[a]	2.1 ± 1.1 (0.09 – 2.3, 3)
pH	7.7 ± 0.1 (6.5–9, 21)	7.2 ± 0.7 (6.5–8.5, 5)	NA
Osmolarity (mOsm)	335.2 ± 7.8 (280–400, 19)	345.0 ± 10.3 (320–370, 4)	NA
Concentration (10^6/ml)	62.6 ± 18.6 (0.2–339, 28)[b]	12.1 ± 3.2 (0.4–23.8, 15)[a]	7.1 ± 3.4 (1.2–13.0, 3)
Total sperm/ejaculate (10^6)	94.5 ± 30.1 (0.1–715, 28)[b]	12.7 ± 3.5 (0.2–50.0, 15)[a]	22.1 ± 14.4 (0.1–49.4, 3)
Motility (%)	63.6 ± 2.9 (50–90, 26)	52.7 ± 5.5 (1–80, 15)	50.0 ± 20.0 (10–70, 3)
Normal morphology (%)	39.6 ± 8.8 (14.4–52.1, 4)	26.9 ± 3.7 (14.3–54.9, 10)	40.6 ± 10.4 (27.7–61.1, 3)
Testicular volume (cm^3)	14.7 ± 1.0 (9.6–25.8, 27)	17.0 ± 1.7 (8.4–38.2, 17)	9.5 ± 0.1 (9.4–9.7, 3)

Note: Due to small samples size, values of wild maned wolves were not included in the analysis. Values are means ± standard errors of means.

NA, not available.

[a,b] Different letters indicate significant difference ($p < .05$).

observed in both captive and wild wolves may be linked to the species' low genetic diversity ($\pi = 0.0013$), resulting from a genetic bottleneck during or at the end of the last glacial maximum (Prates, 2008). Low gene diversity has been associated with poor seminal quality in several carnivore species, including the Florida panther (*Felis concolor coryi*; Facemire et al., 1995), cheetah (*Acinonyx jubatus*; Fitzpatrick and Evans, 2009), black-footed ferret (*Mustela negripes*; Fitzpatrick and Evans, 2009), and gray wolf (Asa et al., 2007). Furthermore, like many wild canid species (Platz et al., 2001), contamination of urine in semen samples obtained by electroejaculation is common in the maned wolf. The presence of urine alters osmolarity and pH of seminal samples, which in turn increases the proportion of sperm with bent and coiled tails and decreases motility (Platz et al., 2001).

Seasonality clearly affects semen production. Analysis of semen samples collected at 4–8-week intervals during two breeding seasons in North America reveals a definitive temporal change in sperm concentration throughout the breeding season (Songsasen and Rodden, 2010). Specifically, semen samples contain low sperm cell numbers early in the breeding season; cell concentration peaks in December and January, followed by an immediate decline.

Age also affects sperm output, although it does not affect seminal quality, that is, motility and morphology (Table 8.1). Males at prime breeding age (3–8 years old) produce significantly higher seminal volume, concentration, and total sperm per ejaculate than older males (>8 years of age) do. A 2-year-old wild wolf produces sperm, although the seminal output (volume: 0.09 ml; concentration: 1.2×10^6 sperm/ml) is much lower than that of an adult individual (volume: 2.3–3.8 ml; concentration: $7.2–13 \times 10^6$ sperm/ml).

FACTORS AFFECTING REPRODUCTIVE SUCCESS

In Situ Populations

Due to their elusive and solitary nature, there is limited information about reproductive success in wild maned wolves. Nevertheless, available data indicate that the mortality rate of wild wolf pups may be high. Dietz (1984) followed four maned wolf pairs and reported that none successfully raised a litter during a 2-year study. Bestelmeyer (2000) assumed ~50% survival of a single litter observed in the Emas National Park, Brazil. Emmons et al. (2012) followed the reproductive life of a female maned wolf (from 3.6 to 8 years of age) living in Noel Kempff Mercado National Park, Bolivia. These investigators reported that the female produced three litters during this period: two were successful (three weaned pups and one juvenile were recorded) and one resulted in the death of all pups 18 days after birth.

To date, there is no information on the causes of pup mortality in the wild. It has been suggested that anthropogenic pressure likely affects the survival of wild wolves (Sollmann et al., 2010). Maned wolves are susceptible to several canine infectious diseases, such as rabies, canine distemper, and canine parvovirus (Fletcher et al., 1979; Rodden et al., 2004). A study conducted in Bolivia has demonstrated that wild maned wolves have been exposed to these infectious diseases (Deem and Emmons, 2005), which are likely transmitted from domestic dogs living in the surrounding villages (Bronson et al., 2008). Furthermore, it has been reported that hematological and blood biochemistry values of wolves living in farmlands or on the border of a protected area differ from those of individuals living inside a national park (May et al., 2009). Wolves living in farmlands also excrete more glucocorticoids than those living in a protected area do, suggesting that individuals living in altered landscapes may suffer from stress associated with human activities (Vynne, 2010; Spercoski et al., 2012). Finally, it has been shown that female maned wolves living outside protected areas excrete less progestagen than individuals ranging inside the park do, suggesting that the former may have lower reproductive success than the latter (Vynne, 2010). Therefore, it appears that direct or indirect contact with humans may compromise the overall health of wild maned wolves, which in turn results in low reproductive success, especially in individuals living outside protected areas.

Ex Situ Populations

Maned wolves breed poorly in captivity, especially in North America (Songsasen and Rodden, 2010). Between 2005 and 2011, the North American population fluctuated between 92 and 79 individuals, which was lower than the target population size of 100 animals (M. Rodden, personal communication). This is mainly due to the failure of paired individuals to successfully reproduce and rear pups. On average, only ~40% of recommended pairings produce pups (M. Rodden, personal communication). Analysis of data shows that of 2629 births, 1052 (40%) pups died within the first 30 days (Holland, 2011). Although the underlying causes of neonatal loss have not been identified, analysis of studbook data showed that pup mortality is largely associated with parental incompetence (Maia and Gouveia, 2002). The survival of pups produced from young (<2 years old) dams is lower ($p<.05$) than that of pups produced from sires of the same age (N. Songsasen, unpublished data). Specifically, 20 of 24 pups (83%) produced from 11 young females died within 30 days of birth, whereas 7 of 20 offspring (35%) produced from young males did so. High pup mortality observed in young females may be due to the lack of maternal skills (Lunn et al., 1994; Thompson et al., 2010) and suboptimal body condition compared to individuals at prime reproductive age (Lunn and Boyd, 1993; Lunn et al., 1994). High pup mortality (72%) was also observed in primiparous females >8 years old, possibly due to compromised uterine integrity. Female cheetahs that are ≥9 years old exhibit a higher incidence of uterine pathology than younger individuals (Crosier et al., 2011). Furthermore, the longer the time interval (in years) from the age a proven female cheetah last gave birth to her death, the more likely that individual expresses uterine hyperplasia (Crosier et al., 2011).

Chronic stress associated with suboptimal management has been indicated as one of the causes of poor reproductive success in carnivore species (Wielebnowski, 1998; Terio et al., 2004). In captive maned wolves, females that successfully raised their young excreted higher progestagen metabolites, especially during the second half of gestation, than individuals that lost pups shortly after birth (Songsasen et al., 2006). Higher fecal corticoid metabolite concentrations during the periovulatory interval were observed in females that lost pups compared to those that successfully raised their young (Songsasen and Rodden, 2010). There appears to be a positive relationship between mean corticoid metabolite concentration in maned wolf pairs and pup mortality (Spercoski, 2007). It has been shown that administration of exogenous cortisol can reduce the concentration of progesterone in several mammalian species (Ogle, 1974; Van Niekerk and Morgenthal, 1982). Therefore, stress associated with captive management may stimulate corticoid excretion that in turn downregulates progesterone production, resulting in suboptimal uterine environment and reduced offspring health.

ASSISTED REPRODUCTIVE TECHNOLOGIES

Assisted "breeding" technologies, including artificial insemination (AI), have proven to be a valuable tool in the genetic management of small, isolated wildlife populations (Pukazhenthi and Wildt, 2004). This technology allows dissemination of genes among geographically isolated populations without the movement of live animals. Additionally, AI with frozen semen allows reinfusion and dissemination of valuable genes even after the death of the sperm donor. Artificial insemination has been successfully used in several carnivore species, including the black-footed ferret (Comizzoli et al., 2009; Howard and Wildt, 2009), cheetah (Howard et al., 1992), and gray wolf (Asa et al., 2006).

To date, assisted reproductive technologies have not been fully developed in the maned wolf. Although semen samples can be routinely obtained from males of prime breeding age, maned wolf sperm appears to be extremely sensitive to cryopreservation procedures (N. Songsasen, unpublished data).

Progress has been made in establishing a method to manipulate the reproductive cycle of female maned wolves. A recent study showed that a gonadotropin-releasing hormone agonist, deslorelin, can be used to induce ovarian activity in females housed with a male (Johnson et al., 2011). Maned wolves receiving a deslorelin implant for 7 days exhibit a rise in progestagen concentration 10 days after the onset of hormone treatment, and the steroid level remains elevated for at least 6 weeks (duration of the study period). In singleton females, although estrogen metabolite concentrations increase after deslorelin treatment, progestagen levels remain at the baseline throughout the study period, indicating that ovulation does not occur in the absence of the male (Johnson et al., 2011). This finding confirms the hypothesis that the maned wolf is, indeed, an induced ovulator.

FUTURE PRIORITIES

In Situ Populations

To date, there is very little information on the survival rate of pups to adulthood in wild maned wolves. It has been suggested that wild maned wolves have a short lifespan, with adult survival rate fluctuating from 0.28 to 0.97 (Sollmann et al., 2010). The survival rate of subadults is estimated to be 0.63 (Sollmann et al., 2010). Available information has indicated that reproductive success may be low in disturbed landscapes (Vynne, 2010). Environmental changes associated with anthropogenic activities likely affect reproductive fitness of most wildlife species, although the impact may vary among taxa (Milligan et al., 2012). Thus, future research should focus on determining the impact of environmental (including climate) change on reproductive success (i.e., pregnancy and pup survival rates) of wild maned wolves. Finally, recent socioecological discoveries, such as evidence that young maned wolves remain in the parental home range beyond the subsequent birth and may help with the care of their younger siblings, warrant additional long-term investigations of social interaction in this fascinating canid.

Ex Situ Populations

During the past two decades, our understanding of maned wolf reproductive biology has vastly increased. Noninvasive hormone monitoring has proven to be a useful tool in determining reproductive mechanisms and potential causes of poor reproductive success in this species, especially in captive environments. Currently, the captive population of this species, especially in North America, is not self-sustaining due to poor reproduction and suboptimal health. Thus, future research should focus on improving reproduction and health of *ex situ* animals, as well as developing assisted breeding techniques, including AI, sperm cryopreservation, and manipulation of ovarian activity in singleton individuals, to enhance genetic management of this population. Furthermore, the findings that this species is an induced ovulator and that females may delay estrus in the absence of a male are fascinating. Future research should focus on elucidating the mechanisms regulating estrus and ovulation in female maned wolves.

ACKNOWLEDGMENTS

The authors are grateful to the AZA Maned Wolf Species Survival Plan for supporting reproductive studies both *ex situ* and *in situ*. We also thank the Morris Animal Foundation and AZA's Conservation Endowment Fund for providing funding for the studies.

REFERENCES

Asa C, Bauman K, Callahan P, Bauman J, Volkmann DH and Jochel W. 2006. Induction of fertile estrus with either natural mating or artificial insemination followed by birth of pups in gray wolves (*Canis lupus*). *Theriogenology* 66:1778–1782.

Asa C, Miller P, Agnew M, Rebolledo JAR, Lindsey SL, Callahan M and Bauman K. 2007a. Relationship of inbreeding with sperm quality and reproductive success in Mexican gray wolves. *Animal Conservation* 10:326–331.

Asa CS, Bauman JE, Coonan TJ and Gray MM. 2007b. Evidence for induced estrus or ovulation in a canid, the island fox (*Urocyon littoralis*). *Journal of Mammalogy* 88:436–440.

Bestelmeyer SV. 2000. Solitary, reproductive, and parental behavior of maned wolves (*Chrysocyon brachyurus*). PhD Thesis, Colorado State University, Fort Collins.

Brady CA. 1981. The vocal repertoires of bush dog (*Speothos venaticus*), crab eating fox (*Cerdocyon thous*), and maned wolf (*Chrysocyon brachyurus*). *Animal Behaviour* 29:649–669.

Bronson E, Emmons LH, Murray S, Dubovi EJ and Deem SL. 2008. Serosurvey of pathogens in domestic dogs on the border of Noel Kempff Mercado National Park, Bolivia. *Journal of Zoo and Wildlife Medicine* 39:28–36.

Comizzoli P, Crosier AE, Songsasen N, Szykman-Gunther M, Howard J and Wildt DE. 2009. Advances in reproductive science for wild carnivore conservation. *Reproduction in Domestic Animals* 44(Suppl):47–52.

Crosier AE, Comizzoli P, Baker T, Davidson A, Munson L, Howard J, Marker LL and Wildt DE. 2011. Increasing age influences uterine integrity, but not ovarian function or oocyte quality, in the cheetah (*Acinonyx jubatus*). *Biology of Reproduction* 85:243–253.

Deem SL and Emmons LH. 2005. Exposure of free-ranging maned wolves (*Chrysocyon brachyurus*) to infectious and parasitic disease agents in the Noel Kempff Mercado National Park, Bolivia. *Journal of Zoo and Wildlife Medicine* 36:192–197.

Dietz JM. 1984. *Ecology and Social Organization of the Maned Wolf* (Chrysocyon brachyurus). Smithsonian Institution Press, Washington, DC.

Dietz JM. 1985. *Chrysocyon brachyurus. Mammal Species* 234:1–4.

Emmons LH, Chávez V, Del Aguila LF, Angulo S and Muir M. 2012a. Ranging patterns. In: Emmons LH (ed.), *The Maned Wolves of Noel Kempff Mercado National Park.* Smithsonian Institution Scholarly Press, Washington, DC, pp. 25–36.

Emmons LH, Chávez V, Del Aguila LF, Augulo S and Muir M. 2012b. Social interactions and reproduction. In: Emmons LH (ed.), *The Maned Wolves of Noel Kempf Mercado National Park.* Smithsonian Institute Scholarly Press, Washington, DC, pp. 57–75.

Facemire CF, Gross TS and Guillette LJ. 1995. Reproductive impairment in the Florida panther—Nature or nurture. *Environmental Health Perspectives* 103:79–86.

Fitzpatrick JL and Evans JP. 2009. Reduced heterozygosity impairs sperm quality in endangered mammals. *Biology Letter* 5:320–323.

Fletcher KC, Eugster AK, Schmidt RE and Hubbard GB. 1979. Parvovirus infection in maned wolves. *Journal of the American Veterinary Medical Association* 175:897–900.

Goodrowe KL, Hay MA, Platz CC, Behrns SK, Jones MH and Waddell WT. 1998. Characteristics of fresh and frozen-thawed red wolf (*Canis rufus*) spermatozoa. *Animal Reproduction Science* 53:299–308.

Gudermuth DF, Concannon PW, Daels PF and Lasley BL. 1998. Pregnancy-specific elevations in fecal concentrations of estradiol, testosterone and progesterone in the domestic dog (*Canis familiaris*). *Theriogenology* 50:237–248.

Holland, R. 2011. *International Studbook 2010 for the Maned Wolf (Chrysocyon brachyurus)*. Zoo Leipzig, Leipzig, Germany.

Howard J and Wildt DE. 2009. Approaches and efficacy of artificial insemination in felids and mustelids. *Theriogenology* 71:130–148.

Howard J, Donoghue AM, Barone MA, Goodrowe KL, Blumer ES, Snodgrass K, Starnes D, Tucker M, Bush M and Wildt DE. 1992. Successful induction of ovarian activity and laparoscopic intrauterine artificial insemination in the cheetah (*Acinonyx jubatus*). *Journal of Zoo and Wildlife Medicine* 23:288–300.

Jácomo ATA, Kashivakura CK, Ferro C, Furtado MM, Astete SP, Tôrres NT, Sollmann R and Silveira L. 2009. Home range and spatial organization of maned wolves in the Brazilian grasslands. *Journal of Mammalogy* 90:150–157.

Johnson AE, Padilla LR, Hope K, Wildt DE and Songsasen N. 2011. Induction of ovarian activity in the maned wolf (*Chrysocyon brachyurus*) using a GnRH-agonist. *Reproduction, Fertility and Development* 23:112–113.

Johnston SD, Ward D, Lemon J, Gunn I, MacCallum CA, Keeley T and Blyde D. 2007. Studies of male reproduction in captive African wild dogs (*Lycaon pictus*). *Animal Reproduction Science* 100:338–355.

Kleiman DG. 1972. Social behavior of the maned wolf (*Chrysocyon brachyurus*) and bush dog (*Speothos vanaticus*): A study in contrast. *Journal of Mammalogy* 53:791–806.

Lunn NJ and Boyd IL. 1993. Effect of maternal age and condition on parturition and the perinatal-period of Antarctic fur seals. *Journal of Zoology* 229:55–67.

Lunn NT, Boyd IL and Croxall JP. 1994. Reproductive performance of female Antarctic fur seals: the influence of age, breeding experience, environmental variation and individual quality. *Journal of Animal Ecology* 63:827–840.

Maia OB and Gouveia AMG. 2002. Birth and mortality of maned wolves *Chrysocyon brachyurus* (Illiger, 1811) in captivity. *Brazilian Journal of Biology* 62:25–32.

Maia OB, Jácomo ATA, Bringel BA, Kashivakura CK, Oliveira CA, Teodoro LOF, Silveira L, et al. 2008. Comparison of serum hormone levels of captive and free-living maned wolves *Chrysocyon brachyurus*. *Brazilian Journal of Medical and Biological Research* 41:176–179.

May JA, Songsasen N, Azevedo FC, Santos JP, Paula RC, Rodrigues FHG, Rodden MD, Wildt DE and Morato RG. 2009. Hematology and blood chemistry parameters differ in free ranging maned wolves (*Chrysocyon brachyurus*) living in Serra da Canastra National Park versus adjacent farmlands, Brazil. *Journal of Wildlife Diseases* 45:81–90.

Melo LFB, Sabato MAL, Magni EMV, Young RJ and Coelho CM. 2007. Secret lives of maned wolves (*Chrysocyon brachyurus* Illiger 1815): As revealed by GPS tracking collars. *Journal of Zoology* 271:27–36.

Milligan SR, Holt WV and Lloyd R. 2009. Impacts of climate change and environmental factors on reproduction and development of wildlife. *Philosophical Transactions of Royal Society B* 364:3313–3319.

Minter LJ and DeLiberto TJ. 2008. Seasonal variation in serum testosterone, testicular volume, and semen characteristics in the coyote (*Canis latrans*). *Theriogenology* 69:946–952.

Ogle TF. 1974. Effects of ACTH on ovarian histochemistry and maintenance of pregnancy in deer mice. *Biology of Reproduction* 11:288–296.

Platz C. 2001. Semen collection in wild canidae using electroejaculation. *The 1st International Symposium on Assisted Reproductive Technology for the Conservation & Genetic Management of Wildlife*, Omaha's Henry Doorly Zoo, 17–18 January, pp. 133–134.

Prates Jr PH. 2008. Diversidade genética e história evolutiva do lobo guará (*Chrysocyon brachyurus*). PhD Thesis, Pontificia Universidade Catolica do Rio Grande Do Sul, Porto Alegre, Brazil.

Pukazhenthi BS and Wildt DE. 2004. Which reproductive technologies are most relevant to studying, managing and conserving wildlife? *Reproduction, Fertility and Development* 16:33–46.

Rodden M, Rodrigues F and Bestelmeyer S. 2004. Maned wolf (*Chrysocyon brachyurus*). In: C Sillero-Zubiri and DW Macdonald (eds), *Canids: Foxes, Wolves, Jackals and Dogs*. IUCN/SSC Canid Specialist Group, Cambridge, UK, pp. 38–43.

Rodden MD, Sorenson LG, Sherr A and Kleiman DG. 1996. Use of behavioral measures to assess reproductive function in maned wolves (*Chrysocyon brachyurus*). *Zoo Biology* 15:565–585.

Sollmann R, Furtado MM, Jácomo ATA, Tôrres NM and Silveira L. 2010. Maned wolf survival rate in central Brazil. *Journal of Zoology* 282:207–213.

Songsasen N and Rodden MD. 2010. The role of the species survival plan in maned wolf *Chrysocyon brachyurus* conservation. *International Zoo Yearbook* 44:1–13.

Songsasen N, Rodden M, Brown JL and Wildt DE. 2006. Patterns of fecal gonadal hormone metabolites in the maned wolf (*Chrysocyon brachyurus*). *Theriogenology* 66:1743–1750.

Spercoski KM. 2007. Corticóides fecais de lobos-guarás (*Chrysocyon brachyurus*, Illiger 1811) de Cativeiro e vida livre. M.Sc. Thesis, Universidade Federal do Paraná, Curitiba, Brazil.

Spercoski KM, Morais RN, Morato RG, de Paula RC, Azevedo FC, May-Júnior J, Santos JP, Reghelin AL, Wildt DE and Songsasen N. 2012. Adrenal activity in maned wolves is higher on farmlands and park boundaries than within protected areas. *General and Comparative Endocrinology* 179(2):232–240.

Terio KA, Marker L and Munson L. 2004. Evidence for chronic stress in captive but not free-ranging cheetahs (*Acinonyx jubatus*) based on adrenal morphology and function. *Journal of Wildlife Diseases* 40:259–266.

Thompson KV, Baker AJ and Baker AM. 2010. Parental care and behavioral development in captive mammals. In: DG Kleiman, KV Thompson and CK Baer (eds), *Wild Mammals in Captivity*. University of Chicago Press, Chicago, pp. 367–385.

Van Niekerk CH and Morgenthal JC. 1982. Fetal loss and the effect of stress on plasma progesterone levels in pregnant thoroughbred mares. *Journal of Reproduction and Fertility* 32(Suppl):453–457.

Velloso A, Wasser SK, Monfort SL and Dietz JM. 1998. Longitudinal fecal steroid excretion in maned wolves (*Chrysocyon brachyurus*). *General and Comparative Endocrinology* 112:96–107.

Vynne C. 2010. Landscape use by wide-ranging mammals of the cerrado. PhD Thesis, University of Washington, Seattle.

Walker SL, Waddell WT and Goodrowe KL. 2002. Reproductive endocrine patterns in captive female and male red wolves (*Canis rufus*) assessed by fecal and serum hormone analysis. *Zoo Biology* 21:321–335.

Wasser SK, Velloso AL and Rodden MD. 1995. Using fecal steroids to evaluate reproductive function in female maned wolves. *Journal of Wildlife Management* 59:889–894.

Wielebnowski N. 1998. Contributions of behavioral studies to captive management and breeding of rare and endangered mammals. In: T Caro (ed.), *Behavioral Ecology and Conservation Biology*. Oxford University Press, New York, pp. 130–162.

Zindl C, Asa CS and Gunzel-Apel A-R. 2006. Influence of cooling rates and addition of Equex pasta on cooled and frozen-thawed semen of generic gray (*Canis lupus*) and Mexican gray wolves (*C. l. Baileyi*). *Theriogenology* 66:1797–1802.

9 Conservation Medicine

Joares Adenilson May Jr. and Paulo Anselmo Nunes Felippe

CONTENTS

Handling and Sampling Maned Wolves in the Wild 109
Chemical Restraint and Anesthesia 110
Examination 111
Hematology and Blood Chemistry Parameters 111
Infectious Diseases 112
Leptospira spp. 114
Leishmania spp. 115
Distemper 116
Canine Parvovirus 117
Rabies 117
Ectoparasites 118
Neospora Caninum and *Toxoplasma Gondii* 119
Dioctophyme Renale 119
Future Research 120
References 122

Maned wolves are a particular species with unique nutritional needs and anatomical, physiological, behavioral, and ecological adaptations, which are relevant to understanding and managing their health status in captivity and in the wild. Understanding how maned wolves live in a natural habitat can provide new techniques to prevent many problems related to captive management (e.g., parasites and infections; nutritional and behavior disorders). As a large carnivore, the maned wolf requires special techniques of physical and chemical restraint for their handling, examination, and collection of biological samples, in both captive and wild situations. Infectious diseases have been identified as a great threat to maned wolves, in captivity and in the wild. Updated knowledge on important viral, bacterial, and parasitological diseases affecting this species will be presented, as well as miscellaneous conditions affecting health within the context of maned wolf populations.

HANDLING AND SAMPLING MANED WOLVES IN THE WILD

Maned wolves are big canids and they bite to defend themselves. Physical restraint is sometimes used to manipulate wolves in captivity for short time periods to avoid the use of drugs. But to capture them in the field, it is necessary to use traps. Physical restraint is then used in conservation projects to collect biological samples and place radio collars on the wolves' necks. Capturing maned wolves is difficult, but a few methods are available.

Dietz (1984) used wooden box traps baited with live chicken to attract wolves. Today metal box traps are used, baited with cooked chicken, sardine, and bacon. (Mattos 2003; May-Júnior et al. 2009). Other methods to capture maned wolves have been tested in Brazil. Edsel Amorim and Joares Adenilson May Jr. (unpublished data) used foot snares to capture maned wolves successfully

FIGURE 9.1 Maned wolf captured by foot snare and moved out with catch pole.

in cerrado areas. This method could provide an improved alternative for the capture of maned wolves (Figure 9.1).

Chemical Restraint and Anesthesia

Drugs such as ketamine, tiletamine, xylazine, detomidine, midazolam, zolazepam butorfanol, and acepromazine have been recommended for use on maned wolves in captivity, taking advantage of their different pharmacological characteristics, depending on the planned procedure (Bush 1996; Gomes 2007; Nunes et al. 2007). Many drug associations such as tiletamine, zolazepam, and acepromazine can be used orally with good results (Gomes 2007). In a wildlife project, May-Júnior et al. (2009) administered tiletamine–zolazepan for chemical restraint using a hand syringe, blowpipe, or cart pole. The wolves responded very quickly, and became relaxed after 4–5 minutes. Each wolf was monitored after the procedure to ensure uneventful recovery within the box trap from anesthesia, before being released into its original area of capture (Figure 9.2).

Following the use of a foot snare as a physical restraint, wolves recovered in open areas without any problem.

Nowadays there are many models of portable anesthesia equipment that are easily transported. Inhalation drugs such as isofluorane and halothane can be used with excellent results, even in field conditions (Nunes et al. 2007).

FIGURE 9.2 Collared maned wolf recovering inside a trap.

Examination

The monitoring of chemical restraint or anesthesia is one of the most important procedures to ensure the physical integrity of the animals and staff. Many complications involving some chemical restraint protocols are described in the literature. Among the most common are hyperthermia, bradycardia, hypotension, and atrioventricular block (Nunes et al. 2007).

A complete clinical examination must be conducted, evaluating all organic traits. Vital signs, including body temperature and heart and respiratory features and rates, should be recorded every 15 minutes during the anesthesia (Figure 9.3). Wolf age can be estimated by physical characters (e.g., weight, scars, coat appearance, and dentition) according to the criteria of Dietz (1984). Based on overall appearance, each wolf can be given a numerical condition score (1–3, largely based on fat covering, with a 3 rating for wolves with the best conformation and slightly detectable fat compared to a 1 rating indicative of poor condition and no fat covering). Wolves also need to be examined for parasitic infestation, and depending on the requirements biological samples can be taken (Figure 9.3).

Hematology and Blood Chemistry Parameters

A prerequisite to examine the sensitivity of any wildlife species to disease is the characterization of baseline species norms and general health status (Karesh et al. 1995). To date, health-related data have been generated for wild maned wolves, including information on hematology, blood chemistry, and incidence of parasitic infestations (Dietz 1984; Mattos 2003; Deem and Emmons 2005). Interestingly, there is a substantial database on health issues associated with maned wolves living *ex situ*—in zoological collections—beginning almost 30 years ago (Fletcher et al. 1979; Bush 1980; Norton 1990; Maia and Gouveia 2002). This includes detailed information on the species' hematology, serum biochemistry (Barbiers and Bush 1994), and overall sensitivity to rabies, canine distemper, and parvovirus pathogens (Fletcher et al. 1979; Norton 1990).

Although maned wolf hematology and serum biochemistry have been evaluated extensively in captive populations in zoos (Barbiers and Bush 1994; Maned Wolf Species Survival Plan 2007), only limited data are available for free-living populations. Dietz (1984), in his original classic studies of this species in the Serra da Canastra National Park, measured hematologic variables from eight wild individuals (five males and three females). Mattos (2003) measured 26 blood traits in 17 wild maned wolves living in São Paulo State. Interestingly, all metrics reported for wild maned wolves fall within a general and similar range to conspecifics held in zoos (Barbiers and Bush 1994). May-Júnior et al. (2009), in Serra da Canastra National Park, Brazil, captured maned wolves and evaluated hematology and serum biochemistry values. All wolves were in good body condition and

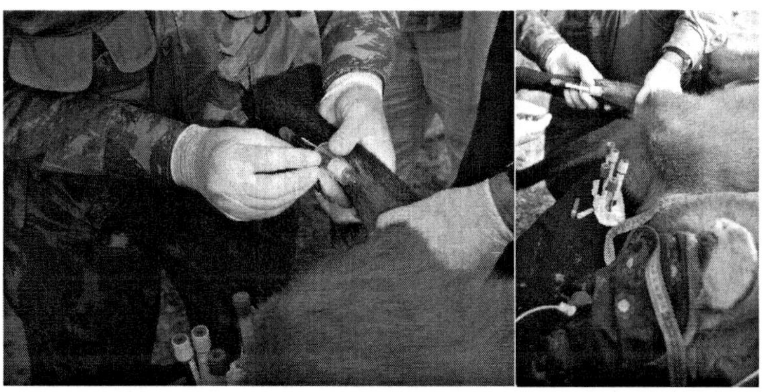

FIGURE 9.3 Collecting a blood sample.

TABLE 9.1
Hematology Values for Male versus Female Maned Wolves Living in the Serra da Canastra National Park (Brazil)

Hematology Trait	Male (n = 12)	Female (n = 16)
Erythrocytes (10^6/mm^3)	5.0 ± 0.2 (4.06–5.65)	4.9 ± 0.1 (4.13–5.87)
Hemoglobin (dl)	13.0 ± 0.4 (10.8–14.4)	12.8 ± 0.3 (10.7–15.4)
Hematocrit (%)	40.6 ± 1.1 (36–44)	39.8 ± 1.0 (33–48)
Mean corpuscular volume (fl)	81.0 ± 1.4 (76–89)	81.3 ± 0.7 (76–86)
Mean corpuscular hemoglobin (/g)	25.8 ± 0.4 (23–28)	26.1 ± 0.3 (24–28)
Mean corpuscular hemoglobin concentration (g/l)	31.9 ± 0.4 (30–34)	32.0 ± 0.3 (30–34)
Leukocytes (10^3/mm^3)	11.8 ± 0.9 (9.2–19.1)	12.4 ± 0.7 (8.6–19.4)
Neutrophils (10^3/mm^3)	8.0 ± 0.8 (5.1–12.9)	8.2 ± 0.7 (3.5–16.1)
Eosinophils (10^3/mm^3)	0.7 ± 0.1 (0–0.9)	0.8 ± 0.1 (0–1.2)
Basophils (10^3/mm^3)	0.007 ± 0.007 (0–92)	0.0 ± 0.0 (0)
Lymphocytes (10^3/mm^3)	2.4 ± 0.5 (1.5–3.8	2.7 ± 1.0 (0.8–9.6)
Monocytes (10^3/mm^3)	0.6 ± 0.2 (0.3–1.3)	0.5 ± 0.1 (0.1–1.1)

Note: Values given as mean ± SEM (range).

appeared overtly healthy. The average body weight of subadults was 22.7 ± 3.1 kg; hematology values were not affected by gender (Table 9.1). Females had higher concentrations of glucose, creatinine phosphokinase, and cholesterol but lower concentrations of potassium than males (Table 9.2).

The influence of cholinesterase activities on genetic, physiological, and pathological conditions is well known (Lepage et al. 1985). Exposure to organophosphate or carbamate insecticides significantly decreases cholinesterase activity (Furlanello et al. 2006). Stress also has been recognized to induce secretion of acetylcholinesterase, a type of cholinesterase that modulates hematopoiesis (Grisaru et al. 2001) and neuronal functions (Small et al. 1996; Nizri et al. 2007), and promotes inflammation-associated thrombopoiesis (Pick et al. 2006). Cholinesterase has also been suggested to play a role in tumorigenesis, especially in hemopoietic malignancies (Soreg et al. 1991). May-Júnior et al. (2009) showed that the average cholinesterase concentration in maned wolves (disregarding location) was within the range of that reported for the domestic dog (3405–6561 U/l; Furlanello et al. 2006). However, when the data were examined on the basis of location, it was discovered that wolves living on farmlands consistently had higher cholinesterase concentrations than those living near the park border or within the park. It was expected that wolves living in farmland would have more chance of exposure to organophosphate/carbonate pesticides, commonly used in coffee farms, than those living inside the park. This in turn would lower cholinesterase activity in the former population (Furlanello et al. 2006). Therefore, variations in cholinesterase activity among wolves living in different locations may be caused by other factors associated with direct or indirect interactions with humans and domestic species (May-Júnior et al. 2009).

INFECTIOUS DISEASES

Over the last decade, there has been growing interest in the relationship between alterations in human-induced habitat availability/quality and the health of wildlife living in or adjacent to those landscapes (Laurenson et al. 1997; Daszak et al. 2000; Cleaveland et al. 2001; Gillin et al. 2002; Lanfranchi et al. 2003). Such encroachments increase the likelihood of direct or indirect contact between humans and their domesticated animals and wildlife that, in turn, increases the risk of disease transmission across the species involved (Lanfranchi et al. 2003). There are now classic examples establishing the legitimacy of such concerns, for instance, the transmission of the rabies pathogen from domestic dogs

TABLE 9.2
Serum Biochemistry Values for Male versus Female Maned Wolves in the Serra da Canastra National Park (Brazil)

Serum Biochemistry Trait	Male (n = 12)	Female (n = 16)
Alanine transaminase (IU/l)	70.1 ± 12.1 (24–278)	66.1 ± 20.1 (32–28)
Aspartate transaminase (IU/l)	48.2 ± 6.1 (22–94)	52.9 ± 5.7 (14–117)
γ-glutamyl transferase (IU/l)	2.8 ± 0.6 (1–4)	3.3 ± 0.5 (0–6)
Glucose (mg/dl)	89.5 ± 6.6[a] (42–119)	115 ± 5.6[b] (71–150)
Urea (mg/dl)	65.4 ± 8.5 (29–142)	57.3 ± 4.0 (31–90)
Creatinine (mg/dl)	1.1 ± 0.0 (0.8–1.4)	1.0 ± 0.0 (0.7–1.3)
Total protein (g/dl)	7.7 ± 0.3 (6.3–9.3)	7.3 ± 0.3 (4–8.8)
Albumin (g/dl)	2.5 ± 0.1 (2–2.9)	2.5 ± 0.1 (2.5–3.5)
Globulin (g/dl)	5.2 ± 0.3 (3.6–6.6)	4.7 ± 0.2 (2.5–6.5)
Total bilirubin (mg/dl)	0.3 ± 0.0 (0.1–0.4)	0.3 ± 0.0 (0.1–0.5)
Creatine phosphokinase (IU/l)	181.2 ± 41.9[a] (79–542)	324.8 ± 38.4[b] (131–624)
Lactate dehydrogenase (IU/l)	200.4 ± 28.0 (79–378)	208.2 ± 35.6 (55–544)
Triglycerides (mg/dl)	38.2 ± 10.0 (4–118)	23.7 ± 3.2 (6–299)
Cholesterol (mmol/l)	4.0 ± 0.2[a] (2.7–5.04)	4.8 ± 0.2[b] (3.64–7.77)
Uric acid (mg/dl)	0.5 ± 0.1 (0.2–69)	0.4 ± 0.1 (0.1–1.2)
Sodium (mmol/l)	143.5 ± 1.1 (135–149)	145.2 ± 1.6 (126–152)
Potassium (mmol/l)	4.9 ± 0.2[a] (4–5.8)	4.4 ± 0.1[b] (4–4.8)
Chloride (mmol/l)	104.5 ± 2.0 (96–117)	108.7 ± 1.8 (98–119)
Calcium (mmol/l)	2.3 ± 0.1 (1.8–2.7)	2.2 ± 0.1 (1.8–3)
Phosphorus (mg/dl)	5.7 ± 0.4 (3.9–8.20)	5.5 ± 0.4 (3.5–8.9)
Cholinesterase	2950.8 ± 226.0 (1932–4705)	3054.5 ± 279.3 (859–5109)

Note: Values given as mean ± SEM (range).
[a,b] Different letters within the same row indicate a difference between groups.

to the Ethiopian wolf (*Canis simensis*) and African wild dog (*Lycaon pictus*), both of which suffered significant population declines (Sillero-Zubiri and MacDonald 1996; Woodroffe et al. 1997). To date, similar studies have not been conducted for any Latin American carnivore. Although so far there has been no evidence of the disease in Brazilian free-ranging maned wolf populations, a recent study in Bolivia has shown that wild individuals of this species have been exposed to common infectious diseases of domestic dogs (*Canis familiaris*) (Deem and Emmons 2005). These include canine distemper virus, canine parvovirus (CPV), rabies virus, canine coronavirus, canine adenovirus, *Leptospira interrogans*, *Toxoplasma gondii*, and *Dirofilaria immitis*. Therefore, the maned wolf in other regions, including Brazil, may well be at risk of multiple diseases.

Several outbreaks of pathogens have caused the decline of wild carnivore populations in recent decades. In addition to the negative impact on wild populations, there is a concern about the transmission of some of these agents to humans and domestic animals. In fact, environmental alterations have resulted in changes in the pathogen–host relationship. Therefore, health monitoring of wild animals is considered an important component in programs for control or eradication of diseases, in the public and animal health programs, and in the decision-making process for the management and conservation of wild species. Compared to habitat loss, hunting, and pollution, the occurrence of disease may appear to be a minor problem for the conservation of wild species. However, the occurrence of pathogens can affect the abundance and distribution of animals, and in recent decades the impact of diseases in wild populations of animals has drawn the attention of conservationists (McCallum and Dobson 1995; Murray et al. 1999; Woodroffe 1999; Cleaveland et al. 2007). Several species of endangered carnivores have experienced declines in their populations due to epidemics caused by infectious diseases, such as outbreaks of distemper in black-footed ferrets (*Mustela nigripes*; Thorne and Williams 1988), lions (*Panthera leo*; Roelke-Parker et. al. 1996), and African wild dogs (*L. pictus*; Alexander et al. 1996) and of rabies in Ethiopian wolves (*C. simensis*; Sillero-Zubiri et al. 1996). These epidemics raised the concern that diseases can significantly increase the mortality of free-living wild animals and highlighted the need to integrate epidemiology in the conservation and management of wild carnivores.

Besides the impact of diseases on wild populations, there is growing concern about the transmission of parasites between humans and wild and domestic animals. In fact, 61% of all human pathogens are classified as zoonoses (Taylor et al. 2001) and about 77% of pathogens in livestock and 91% of domestic carnivore pathogens infect multiple hosts (Haydon et al. 2002). Thus, if these infections can spread widely among human populations once introduced (influenza, HIV, Ebola) and can be transmitted from animals, which are often their reservoirs (rabies, brucellosis, leptospirosis), zoonotic infections may pose a serious risk to socioeconomic and public health. The increase in infectious diseases was caused by changes in ecological interactions between pathogens and hosts (Dobson and Carper 1996; Daszak et al. 2001). Such changes may be natural or anthropogenic in origin, caused by the expansion of human activities and farming frontiers, habitat fragmentation, and pollution, among other factors (Patz et al. 2000). These ecological changes allow increased contact between pathogen species and new host populations and allow for natural selection pressures to act, resulting in the dominance of pathogens that adapt to these new environmental conditions (Daszak et al. 2001). The occurrence of epidemics in wildlife caused by parasites from domestic animals may become increasingly common (Funk et al. 2001).

Populations of wild carnivores can suffer severe declines related to diseases of domestic animals. Other species such as foxes and coyotes, which are well adapted to disturbed habitats, can serve as sources of infection for humans and domestic animals (Aguirre 2009). Additionally, the transmission of pathogens between wild and domestic animals is even more worrying if they are fragmented in environments with low genetic variability and/or exposed to emerging pathogens (McCallum and Dobson 2002; Patz et al. 2004; Travis et al. 2006), which is a common situation for much of the Brazilian wildlife. This interface between the health of humans and wild and domestic animals is embedded in the concept of conservation medicine (Tabor 2002).

Another important aspect is that carnivores can act as "bioaccumulators" of exposure to pathogens. Since they occupy the top of the food web, the consumption of infected hosts results in high rates of infection. Thus, some species of carnivores can be used as sentinels and strategic targets in surveillance programs for the detection of pathogens (Cleaveland et al. 2006a; Aguirre 2009). A major contribution of the effective monitoring of diseases in wild free-living animals is the detection of early stages of the occurrence of new or emerging diseases, some of which may have serious economic and zoonotic implications (Mörner et al. 2002).

Many epidemiological and other ecological factors may influence the distribution of infection in a susceptible population, but the challenge for conservationists and agents of public health is to identify the conditions that lead to an epidemic before it happens. Thus, the study of diseases in wild populations is critical to implementing effective programs for control and eradication of diseases and the development of policies for public and animal health and the management and conservation of wild species (Jorge et al. 2010).

Leptospira spp.

Leptospirosis is a worldwide zoonosis that affects most mammals (Thiermann 1984). These may become carriers, thus contributing to the spread of the disease. Leptospirosis occurs in urban, rural, and wild settings, in endemic and epidemic forms. It is affected by the interactions of environmental factors and different groups of host animals (Vasconcellos 1987). Animals can become infected through intact skin (immersed in water for long periods) and mucosa by contaminated water and food (Levett 2001; Corrêa 2007). Serological surveys in captive wild carnivores in Brazil have demonstrated significant levels of exposure in dogs and cats (Guerra-Neto et al. 2004). Jorge (2008) tested 75 wild carnivores captured in a reserve in the Pantanal of Mato Grosso and 32 were considered positive for exposure (three of eight maned wolves). The detection of positive individuals in all sampled species of wild carnivores in the northern Pantanal (Jorge 2008) indicates that the epidemiology of leptospirosis in the region is quite complex. The greatest impact on populations of wild mammals has been through abortion and reproductive losses (Roth 1972). However, the actual role of the maned wolf in the cycle and the impact of leptospirosis in free-living populations are still unknown.

Leishmania spp.

Leishmaniasis, a complex of diseases caused by parasites of the genus *Leishmania*, is considered to be emerging in some regions and reemerging in others and constitutes a serious public health problem. These diseases are currently endemic in 88 countries and more than 350 million people are at risk, with an estimated incidence of 2 million new cases each year (Desjeux 2004).

In Latin America, the *Leishmania* species cause two diseases with different clinical forms in humans: the American visceral leishmaniasis and cutaneous leishmaniasis. Transmission occurs through insect bites of females of different species of sand flies (Diptera: Phlebotominae), popularly known as "palha" and "tatuquira" mosquitoes, among others (Brazil 2007). Several orders of mammals, especially rodents and wild canids, are the primary reservoir of *Leishmania* spp. (Ashford 2000; Dantas-Torres 2007; Beck et al. 2008).

In Brazil, the pioneering studies of Deane and Deane (1954) described the first reports of natural infection with *Leishmania* spp. in the hoary fox (*Lycalopex vetulus*) and its possible role as a reservoir. Other studies report exposure to *Leishmania* spp. in free-living wild carnivores. Curi et al. (2006) reported the prevalence of anti-*Leishmania* antibodies in wild canids from the Serra do Cipó, Minas Gerais. Of the 21 animals analyzed, two maned wolves and two crab-eating foxes (*Cerdocyon thous*) were positive. In the northern Pantanal, specimens of the maned wolf, crab-eating raccoon, and ocelot were diagnosed using molecular techniques as naturally exposed to *Leishmania braziliensis* (Jorge 2008).

In wild carnivores kept in captivity in a zoo in Belo Horizonte, Minas Gerais, Luppi et al. (2008) reported positive serology for the crab-eating fox, maned wolf, hoary fox, and bush dog (*Speothus venaticus*). Of these, a bush dog and a hoary fox (*Lycalopex vetulus*) developed clinical signs. The first had died and the second was euthanized because of its poor clinical condition and prognosis. At necropsy, both showed characteristic lesions of visceral leishmaniasis confirmed by histopathology and immune histochemistry (Jorge et al. 2010).

The studies cited demonstrate the possible involvement of wild carnivores in the epidemiology of leishmaniasis. However, a detailed study is needed to clarify the roles played by different species of wild carnivores in the epidemiology of leishmaniasis in various regions where it occurs. Little is known about the potential impact of different *Leishmania* species on populations of wild carnivores (Jorge et al. 2010).

Regarding control measures, the Brazilian program for control of visceral leishmaniasis (Brazil 2003) is based on the treatment of human cases, vector control by insecticide use, and control of canine reservoirs (serosurvey of dogs and euthanasia of seropositive dogs) (Luppi et al. 2008). In infected wild animals, several issues must be considered before opting for euthanasia, since some species are endangered, such as the maned wolf, and there is a lack of information on the potential for transmission in various species of carnivores in Brazil. Therefore, understanding the role that the maned wolf plays in the maintenance and expansion of the transmission cycles of *Leishmania* spp. is essential in defining effective measures to control this disease (Jorge et al. 2010).

Distemper

Canine distemper has been known in Europe for at least 200 years (Williams 2001). Its occurrence was described in free-living wild carnivores during the second half of the twentieth century (Helmbolt and Jungherr 1955). It occurs in all families of carnivores and is distributed throughout the world. Close contact between affected and susceptible animals is necessary for transmission to occur, since the virus is inactivated rapidly in the environment by UV irradiation, heat, and drying. Solitary and territorial animals, such as the maned wolf, have less chance of transmitting or contracting the virus. Thus, several factors are important in the epidemiology of canine distemper in a particular region, such as the susceptibility of hosts, the density of sympatric populations of susceptible hosts, and the existence of behavioral aspects that favor intra- and interspecific contact and therefore transmission of the agent (Williams 2001).

Several outbreaks of canine distemper have been reported in free-living carnivores in Africa and North America, some of them causing significant population declines due to increased mortality. In the United States, an outbreak of distemper hit what was considered at the time the last colony of black-footed ferrets (*M. nigripes*) in nature, causing the death of several individuals (Williams et al. 1988). At that time, the species was considered extinct in the wild. Davidson et al. (1992) determined that the virus was responsible for 78% mortality of gray foxes (*Urocyon cinereoargenteus*) in southeastern United States. Another endangered species threatened by this pathogen is the African wild dog (*L. pictus*). The agent has been implicated in significant mortality of individuals of the species in Botswana (Alexander et al. 1996) and Tanzania (Van de Bildt et al. 2002), and the disappearance of the species in the Masai Mara Reserve, Kenya (Alexander and Appel 1994). In the same park, 55% of the lions (*Panthera leo*) tested were seropositive (Kock et al. 1998). In Serengeti, distemper caused the death of about one-third of the population of lions (Roelke-Parker et al. 1996). In some cases, domestic dogs were identified as the most likely source of the agent (Alexander and Appel 1994; Roelke-Parker et al. 1996; Cleaveland et al. 2000).

For carnivorous mammals of South America, publications related to the occurrence of distemper, or even exposure to the agent, are still scarce. Until the late twentieth century, published studies on South American species were restricted to case reports or surveys of the occurrence of the disease in animals kept in captivity. According to Maia and Gouveia (2002), between 1989 and 1998, 9% of the mortality of maned wolves in captivity in Brazil was attributed to infectious diseases. Within this group, distemper was related to 19.4% of deaths. Over the past 10 years, however, a few studies have been published about serologic surveys of maned wolves in the wild. In the Noel Kempf Mercado National Park, Bolivia, two out of four captured maned wolves were seropositive (Deem and Emmons 2005). In Minas Gerais, Curi (2005) serologically tested three maned wolves captured in the Serra do Cipo. All were negative and domestic dogs sampled in the region had 65.7% positivity (46/70). In the northern Pantanal, Jorge (2008) tested 8 maned wolves and 69 domestic dogs for exposure to the distemper virus; 3 maned wolves and 56 domestic dogs tested positive.

The positive results obtained in the serodiagnosis of free-living carnivores in South America indicate that in some regions, these animals were exposed to canine distemper virus or to an antigenically similar agent.

Canine distemper is a common disease in domestic dogs in Brazil and a recent molecular study of this virus shows that it is grouped with the North American, European, and Japanese genotypes, which are newly described (Rosa et al. 2012). It is possible that in some of the studied South American regions the canine distemper virus was initially transmitted from domestic dogs to wild carnivores, as has been suggested in studies in Africa (Alexander and Appel 1994; Roelke-Parker et al. 1996; Cleaveland et al. 2000). A large proportion of positive animals found among domestic dogs in most surveys in our continent reinforces this hypothesis. Consistent with the findings of Courtenay et al. (2001), there is a threat resulting from easy contact between domestic dogs and maned wolves using inhabited rural areas.

Canine Parvovirus

Viruses belonging to the group of parvoviruses infect and cause clinical manifestations in a large proportion of species of wild canids (Cleaveland et al. 2006b) and the CPV can infect felids (Barker and Parrish 2001). In 1978, the first cases occurred in Europe, but quickly spread across the world (Parrish 1990). In 1979, antibodies were detected in coyotes (*Canis latrans*) in the United States and by 1981 most animals of this species in Texas, Idaho, and Utah were seropositive (Thomas et al. 1984).

The parvoviruses are very resistant in the environment. When protected from direct sunlight, high temperatures, and dry weather conditions they can remain viable for months (Gordon and Angrick 1986). When compared with other viral agents such as rabies and distemper they have a relatively long period of infectivity and once introduced into a region, they tend to persist. Due to these characteristics, the parvovirus tends to remain in the endemic occurrence region (Cleaveland et al. 2006b).

The age of affected animals is very important in the pathogenesis, because parvovirus requires dividing cells for replication. Young animals have large numbers of dividing cells and tissues as the myocardium and brain are frequently affected. In adult animals, the intestinal epithelial cells and lymphoid tissues are usually the most affected. Consequently, animals of this age often have gastroenteritis, lymphopenia, and leukopenia (Steinel et al. 2001).

The possible increase in mortality of pups is cited in many studies as a major problem and this may be caused by parvovirus in populations of wild carnivores. In a population of wolves (*Canis lupus*) in Minnesota (United States), sampled and monitored between 1979 and 1990, the proportion of pups captured was lower in years subsequent to years when the prevalence of CPV of sampled animals was higher; as a result a decrease in the rate of population increase was observed (Mech and Goyal 1993). For the African wild dog, Creel et al. (1997) observed that the number of pups in litters was lower in regions where the prevalence of antibodies to CPV was higher.

In relation to the maned wolf in zoos, no records of disease are attributed to CPV (Fletcher et al. 1979; Mann et al. 1980; Maia and Gouveia 2002). In the Noel Kempf Mercado National Park, Bolivia, four captured and serologically tested maned wolves were positive, resulting in 100% positivity in the sample (Deem and Emmons 2005). In Minas Gerais, Curi (2005) serologically tested three maned wolves captured in the Serra do Cipo. All were considered positive. Domestic dogs sampled in the region had 58.6% positivity (41/70). In the northern Pantanal, Jorge (2008) tested 76 carnivores for exposure to CPV: 70 were positive, including seven out of eight maned wolves. Domestic dogs were also tested in the region. Among the 102 animals sampled, 98 were positive.

Diagnosis studies conducted in free-ranging maned wolves indicate exposure to parvovirus, in some cases at quite high levels. These results are not surprising considering the positive results in domestic dogs in all regions sampled and the considerable resistance in the environment characteristic of this virus group (Gordon and Angrick 1986), favoring its persistence in places where it is introduced (Cleaveland et al. 2006b). However, they indicate the need for further studies to find out whether the transmission of this pathogen is increasing the mortality of populations of maned wolves in the regions sampled (Jorge et al. 2010).

Rabies

Rabies is one of the oldest infectious diseases described. It was known in Mesopotamia and Egypt (Megid 2007). The rabies virus affects the central nervous system of all mammalian species (Rupprecht et al. 2001) and causes high levels of mortality (Rupprecht et al. 2001; Megid 2007). The rabies virus belongs to the family Rhabdoviridae and genus *Lyssavirus*, and is transmitted by the saliva of infected bats and terrestrial mammals in episodes of biting (Megid 2007). Understanding rabies in wild animals requires an understanding of the agent–host–environment relationship (Wandeler 1993). The introduction of rabies can bring serious consequences to populations of wild carnivores, especially those that are small and isolated (Gascoyne et al. 1993). Examples of this are outbreaks of the disease in populations of Ethiopian wolves (*C. simensis*; Laurenson et al. 1997;

Sillero-Zubiri et al. 1996), gray wolves (*C. lupus*; Weiler et al. 1995), and African wild dogs (Kat et al. 1996; Hofmeyr et al. 2000).

In Brazil, dogs are the main reservoir in the urban cycle and bats are the main reservoir in the sylvatic cycle. However, studies over the past decade have detected cases of rabies in nonvampire bats, monkeys, and wild canids (Brazil 2006a).

In a survey conducted in five locations of the cerrado and Pantanal biomes, 26 of 211 wild carnivores sampled were considered positive for serodiagnosis, among there were 13 of 91 captured maned wolves (Jorge et al. 2010). In other regions of the country, information on the exposure of maned wolves is lacking or insufficient.

The literature data on the exposure of free-living wild carnivores to the rabies virus in different parts of the planet herein demonstrate that these animals may represent important reservoirs of the agent regionally, but also that their survival is threatened by it. Further studies on the circulation of the virus in free-ranging maned wolves should be encouraged and measures aimed at controlling the occurrence of the agent in wildlife should begin to be planned and tested, based on actions taken successfully in Europe and North America, for example, oral vaccination (Jorge et al. 2010).

Ectoparasites

Belonging to the class Arachnida and order Acarina (Allan 2001), ticks are one of the most important groups of ectoparasites of terrestrial, semiaquatic, and flying vertebrates in all continents of the world, including Antarctica (Labruna et al. 2005). They feed on blood and tissue fluids, causing disturbance and changes such as anemia, skin lesions, and hypersensitivity (Hargis 1990), although their main epidemiological role is as a vector for a wide variety of pathogens (Allan 2001).

Animal populations in nature are regulated by abiotic and biotic factors, including parasites (Borgsteede 1996). Hosts and parasites have evolved in parallel over thousands of years and various ticks depend on the presence of one or more species of host to develop their full life cycle (Grenfell and Gulland 1995; Poulin 1997; Labruna et al. 2005).

With fragmentation of the environment, local extinctions of mammals have been increasingly frequent (Fahrig and Merriam 1994). Thus, many ticks have struggled to develop their life cycle in their specific host, which makes them look for new species to complete their cycle, creating new host–parasite relationships (Poulin 1997). These new relationships can weaken the new host and may even lead to its death (Borgsteede 1996).

Ticks generally spend 90% of their cycle outside the host, which makes them vulnerable to environmental changes. The study of tick species of carnivores may indicate scales of environmental change in certain areas (Labruna et al. 2005). The study of evolutionary ecology of parasite populations helps to understand their role in the population dynamics of their hosts (Grenfell and Gulland 1995), since they may pose a threat to small vulnerable populations of wild animals (Cleaveland et al. 2003) as they become more susceptible to the transmission of pathogens.

Carnivores' need for large areas where they develop their ecological role (Emmons 1999) and contact with various prey species (Oliveira 1994) increase their risk of infection with different kinds of ticks (Labruna et al. 2005). Environmental changes, mainly influenced by humans, can also promote their contact with new species (Borgsteede 1996).

Ticks feeding on a wide variety of domestic and wild vertebrates (Randolph et al. 2003) may transmit to the host various pathogens, for example, *Babesia* spp., *Borrelia burgdorferi*, *Francisella tularensis*, *Rickettsia* spp., and *Cytauxzoon felis* (Röken 1993). Labruna et al. (2005) describe the species of ticks that have been found parasitizing maned wolves, the most common being of the genus *Amblyomma: Amblyomma. ovale*, *Amblyomma cajenennense*, *Amblyomma aureolatum*, *Amblyomma tigrinum*, *Amblyomma parvum*, and *Rhipicephalus (Boophilus) microplus. Rhipicephalus (Boophilus) microplus*'s primary host is livestock, and it was found in maned wolves using grazing areas for cattle. Wolves living near the park border appeared to have higher levels of tick infestation than those inside the park and on farmlands (May-Júnior et al. 2009; Figure 9.4).

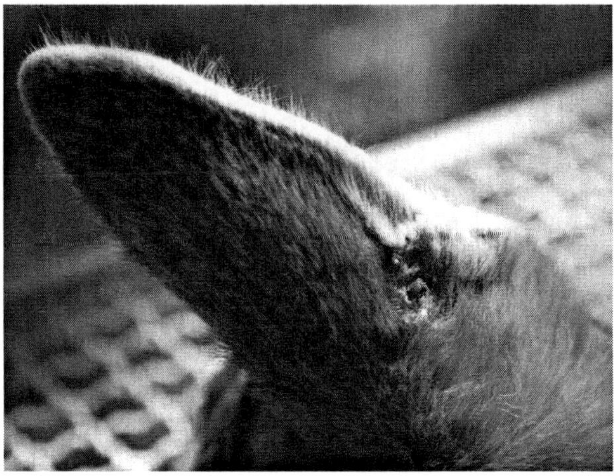

FIGURE 9.4 Ticks on ear of maned wolf.

This may explain the prevalence of wolves on the border with more microcytic red blood cells (seven out of ten wolves captured) and a low mean corpuscular volume. It is possible that the lower mean corpuscular volume of wolves living around park borders compared to those within the park and farms is due to loss of iron associated with blood loss from external parasites or, alternatively, to a nutritional deficiency (Coles 1984; Failace 2003; Bain 2004). Certainly in the case of Brazil these diseases can be important too, especially considering the impact on the populations of different intermediate hosts and the wide range of items in the diet of the maned wolf (Jorge et al. 2010).

Neospora Caninum and *Toxoplasma Gondii*

Neospora caninum and *Toxoplasma gondii* are two apicomplexan parasites that can cause neurologic disease and abortions in their intermediate hosts and less frequently cause disease in their definitive hosts. The definitive hosts of *N. caninum* and *T. gondii* are canids and felids (wild and domestic), respectively, and prevalence in the environment is in part dependent on maintenance of the life cycle through the definitive hosts. Both parasites can be transmitted between domestic and wildlife hosts where their ranges overlap. Prevalence of the parasite in a host population will influence the rate of transmission to other species. Determination of parasite prevalence in populations can provide insight into population fitness and likelihood of disease transmission (Stieve et al. 2010). Studies worldwide have shown the importance of *N. caninum* and *T. gondii* in populations of their intermediate hosts, with high serological prevalence in some species of deer and bovidae and, consequently, wild canids (De Craeye et al. 2011; Stieve et al. 2010; Yu et al. 2009; Jakubek et al. 2007).

Despite the global importance of neosporidiosis and toxoplasmosis for production animals, public health, and ecosystems, no studies have been conducted in Brazil involving the role of free-living maned wolves in the ecoepidemiology of these diseases.

Blood samples were collected from 94 Brazilian wild canids in São Paulo and Mato Grosso States and federal district zoos; 49 (50.5%) wild canids were positive to *T. gondii* and 40 (41.2%) to *N. caninum* (André et al. 2010). In another study, Vitalino et al. (2004) investigated 59 maned wolves originating from six zoos and one ecological reserve of the southeastern and midwestern regions of Brazil. Forty-four (74.6%) had antibodies to *T. gondii* and 5 (8.5%) to *N. caninum*, but there was no positivity in the samples from the ecological reserve.

Dioctophyme Renale

Molin (1860) recognized *Dioctophyme renale* in maned wolves. Maned wolves infested by the parasite have the worms in the right kidney, just like domestic dogs (Matera et al. 1968), and common

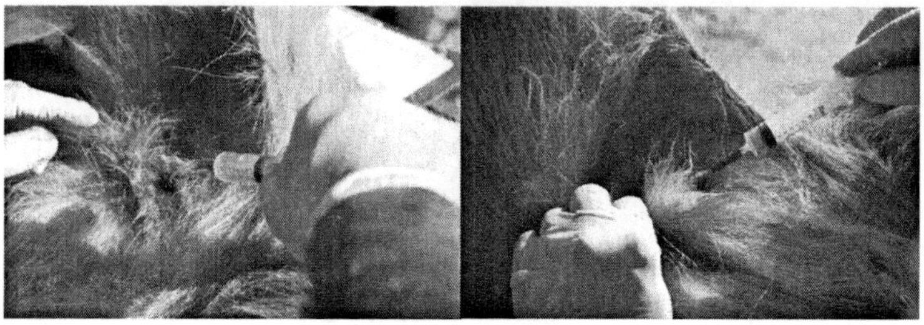

FIGURE 9.5 Collecting urine.

ova are found in the urine of free-ranging maned wolves. *D. renale* is often associated with a hyperplasic right kidney in infected animals (Norton 1990). The parasite cycle in the maned wolf is not completely understood. Research to understand the relationship between the parasite and the host is necessary to understand the risk of the parasitism to maned wolves. In the period 2004–2007, 41 maned wolves were captured in Serra da Canastra National Park, and 24 of them had ova in their urine (Joares Adenilson May Jr., unpublished data; Figure 9.5).

Some animals had ova in the first capture, but none the second or third time. More research is necessary to understand the life cycle of the parasite.

FUTURE RESEARCH

As can be seen in recent studies in Brazil and neighboring countries, free-ranging maned wolves are exposed to pathogens that commonly cause morbidity and mortality in domestic animals and that, historically, have caused mortality in free-living carnivores on other continents. This demonstrates the importance of taking into consideration the diagnosis of exposure to pathogens from wildlife when preparing plans of protected areas (PAs). It also reinforces the need for the implementation of buffer zones (range in the vicinity of PAs that aims to reduce the impact of human activities on the biota protected by it; Brazil, 2000) for areas intended for protection of wildlife, including the implementation of measures aimed at controlling pathogens in domestic animals in these areas. The maned wolf is also exposed to agents that cause diseases in humans and domestic animals. Although the results of studies point to the possibility that these animals act as reservoirs for diseases such as leishmaniasis, rabies, and leptospirosis, it is necessary to assess more thoroughly the actual role of the epidemiological cycle of these diseases, and to conduct studies to determine if they are important in the transmission of these agents. Throughout this review, we have noted that the amount of information on the exposure of maned wolves to infectious and parasitic agents has increased considerably in recent years. We are approaching a point where the data produced will form the basis of effective management measures for the conservation of maned wolves and measures to control the movement of pathogens of public health importance. Obviously, studies involving surveys of exposure to pathogens continue, and will continue for a long time, having great importance in our continent, because there are still many species of host organisms and areas to be covered. But researchers working in this area should remember that the aim is to apply this knowledge to the control of movement of pathogens of importance to biodiversity conservation and public health.

Unfortunately, in Brazil there are no national data on road kill of wildlife or those captured or sighted in urban areas. In the case of maned wolves, some data obtained from studies involving specific populations are a cause for concern (Prada 2004).

In a brief survey in the local press about the number of animals that were captured in southeastern Brazilian cities, 51 occurrences were reported in the years 2010 and 2011, even in cities that are not close to the cerrado, previously considered as the habitat of the species in Brazil. Some cases of maned wolves captured in coastal regions were also reported (staff survey). The cerrado is one of the world's biodiversity hotspots. In the last 35 years, more than 50% of its approximately 2 million km^2 has been transformed into pasture and agricultural lands planted with cash crops (Klink and Machado 2005).

More specifically, within 100 km of São Paulo (Brazil's largest city and capital of the state), we captured eight maned wolves within urban centers, and three of these were translocated and monitored with a radio collar. A curious case was the discovery of a female maned wolf with a pup living inside a stormwater drain in Jundiai (50 km from São Paulo), which could be seen several times a day (unpublished data).

Given the above, certainly further studies of maned wolf population dynamics in Brazil need to be done, especially with regard to their occupation of other biomes and periurban and urban areas of Brazil.

Major histocompatibility complex (MHC) is a group of genes that encode proteins that are related with immune binding and present antigens to T-cells (defense cells, immune recognition). They are important to immune surveillance and host–parasite interactions (Ujvari and Belov 2011). Analyzing DNA from felids, researchers showed different lineages of MHC related to MHC from domestic cats, and in cheetahs (*Acynonix jubatus*) results showed high similarity between individuals or the same genotype (Drake et al. 2004). Measuring levels of polymorphism in these genes (MHC) can provide indirect measures of the immunological fitness of populations. Understanding MHC diversity in maned wolves is very important to knowing the real significance of diseases for its conservation, and could indicate the necessity of *ex situ* management.

Among the infectious diseases, visceral leishmaniasis is very important. It is a reemerging disease in Brazil (1600 municipalities with autochthonous transmission), the country which holds 90% of the leishmaniasis cases diagnosed in Latin America. In recent years, the disease's pattern of transmission has changed in urban areas, as previously occurred in rural areas (Brazil 2006b). Domestic and wild canids are reservoirs of the agent, thus both free-living and captive maned wolves could be participating in the cycle, making it relevant not only for public health but also for species conservation. Certainly this and other aspects of the disease ecoepidemiology deserve further study in Brazil.

The impact of rabies in populations of wild canids in Brazil is also a concern, since there are recent reports of disease transmission to humans by these animals, especially in the northeast of Brazil. Besides the aspects related to the lethality of the disease itself, one of the rabies viruses isolated in *C. thous* in Ceará State (northeast Brazil) was reactive with monoclonal antibodies that identify the rabies virus circulating in hematophagous bats of Colombia. This may indicate a large area of disease transmission and therefore an ecoepidemiology more complex than initially expected. Therefore, this disease could be having a significant impact on the population of wild canids in South America and could even become a serious public health problem (Batista et al. 2007).

RNA virus genomes, especially those of the Paramyxoviridae family, have a high rate of evolution. The RNA polymerase misses one nucleotide in 10^3 to each copy; this means that after the replication process different viral genotypes are obtained (*quasiespecies* concept) (Flint et al. 2004). This feature provides this viral family with a great plasticity to adapt to new hosts. The distemper virus, which belongs to this family (discussed earlier in this chapter), has shown such plasticity and was recently isolated in noncarnivorous species such as the primates *Macaca fuscata* (Yoshikawa et al. 1989) and *Macaca mulatta* in China (Qiu et al. 2011). This family gained new importance in human and animal health and consequently in conservation medicine when it was associated with the emergence of two new paramyxoviruses, the Hendra (Australia) and Nipah (Malaysia) viruses, from reservoir animals (horses, pigs, and bats), which reached the human species with a high fatality rate.

REFERENCES

Aguirre, A.A. 2009. Wild canids as sentinels of ecological health: A conservation medicine perspective. *Parasites & Vectors*, 2: S7.

Alexander, K.A. & Appel, M.J.G. 1994. African wild dogs (*Lycaon pictus*) endangered by a canine distemper epizootic among domestic dogs near the Masai Mara National Reserve, Kenya. *Journal of Wildlife Diseases*, 30: 481–485.

Alexander, K.A., Kat, P.W., Munson, L.A., Kalake, A. & Appel, M.J.G. 1996. Canine distemper-related mortality among wild dogs (*Lycaon pictus*) in Chobe National Park, Botswana. *Journal of Zoo Wildlife Medicine*, 27: 426–427.

Allan, S.A. 2001. Ticks (Class Arachnida: Ordem Acarina). In: W.M. Samuel, M.J. Pybs & A.A. Kocan (eds), *Parasitic Diseases of Wild Mammals*, pp. 72–106. Iowa State University Press, Ames.

André, M.R., Adania, C.H., Teixeira, R.H., Silva, K.F., Jusi, M.M., Machado, S.T., de Bortolli, C.P., et al. 2010. Antibodies to *Toxoplasma gondii* and *Neospora caninum* in captive neotropical and exotic wild canids and felids. *Journal of Parasitology*, 96(5): 1007–1009.

Ashford, R.W. 2000. The leishmaniases as emerging and reemerging zoonoses. *International Journal for Parasitology*, 30: 1269–1281.

Bain, B.J. 2004. *Células Sanguíneas: Um Guia Prático*, 3rd edn, Editora Artmed, São Paulo, Brazil.

Barbiers R. & Bush, M. 1994. Medical management of maned wolves. In: N. Fletchall, M. Rodden & S. Taylor (eds), *Husbandry Manual for the Maned Wolf* (Chrysocyon brachyurus), pp. 52–66. John Ball Zoo, Grand Rapids, MI.

Barker, I.K. & Parrish, C.R. 2001. Parvovirus infections. In: E.S. Williams & I.K. Barker (eds), *Infectious Diseases of Wild Mammals*, 3rd edn, pp. 3–36. Iowa State University Press, Ames.

Batista, H.B.C.R., Franco, A.C. & Roehe, P.M. 2007. Raiva: Uma breve revisão. *Acta Scientiae Veterinae*, 35(2): 125–144.

Beck, A., Beck, R., Kusak, J., Gudan, A., Martinkovic, F., Artukovic, B., Hohsteter, M., Huber, D., Marinculic, A. & Grabarevic, Z. 2008. A case of visceral leishmaniasis in a gray wolf (*Canis lupus*) from Croatia. *Journal of Wildlife Diseases*, 44: 451–456.

Borgsteede, F.H.M. 1996. The effect of parasites on wildlife. *The Veterinary Quarterly*, 18: S138– S140.

Brazil. 2000. Convenção sobre diversidade biológica. Ministério do Meio Ambiente. Editora do Ministério do Meio Ambiente, Brasília. Série Biodiversidade, no. 1.

Brazil. 2003. Manual de vigilância e controle da leishmaniose visceral. Ministério da Saúde, Brasília.

Brazil. 2006a. Doenças transmissíveis: Raiva humana. Ministério da Saúde, Brasília. http://portal.saude.gov.br/portal/svs/visualizar_texto.cfm?idtxt=21906. (Accessed on 07/12/2009).

Brazil. 2006b. Manual de vigilância e controle da leishmaniose visceral. Ministério da Saúde, Secretaria de Vigilância em Saúde, Departamento de Vigilância Epidemiológica. Editora do Ministério da Saúde, Brasília.

Brazil. 2007. Manual de vigilância da leishmaniose tegumentar americana, 2nd edn. Ministério da Saúde, Brasília.

Bush, M. 1980. Medical management of maned wolves (*Chrysocyon brachyurus*). In: *Proceedings: American Association of Zoo Veterinarians*, pp. 131–132.

Bush, M. 1996. Methods of capture, handling and anesthesia. In: D.G. Kleiman, M.E. Allen, K.V. Thompson & S. Lumpkin (eds), *Wild Mammals in Capitivity: Principles and Techniques*, pp. 25–40. 1st edn. The University of Chicago Press, Chicago.

Cleaveland, S., Appel, M.G.J., Chalmers, W.S.K., Chillingworth, C., Kaare, M. & Dye, C. 2000. Serological and demographic evidence for domestic dog as a source of canine distemper virus infection for Serengeti wildlife. *Veterinary Microbiology*, 72: 217–227.

Cleaveland, S., Hess, G.R., Dobson, A.P., Laurenson, M.K., McCallum, H.I., Roberts, M.G. & Woodroffe, R. 2001. The role of pathogens in biological conservation. In: P.J. Hudson, A. Rizzoli, B.T. Grenfell, H. Heesterbeek, A.P. Dobson (eds), *The Ecology of Wildlife Diseases*, pp. 139–150. Oxford University Press, Oxford, UK.

Cleaveland, S., Meslin, F.X. & Breiman, R. 2006a. Dogs can play useful role as sentinel hosts for disease. *Nature*, 440: 605.

Cleaveland, S., Laurenson, K., Funk, S. & Packer, C. 2006b. Impact of viral infections in wild carnivore populations. In: R.G. Morato, F.H.G. Rodrigues, E. Eizirik, P.R. Mangini, F.C.C. Azevedo & J. Marinho-Filho (orgs.) (eds), *Manejo e conservação de carnívoros neotropicais*, pp. 326–349. IBAMA, São Paulo.

Cleaveland, S., Mlengeya, T., Kaare, M., Haydon, D., Lembo, T., Laurenson, M.K. & Packer, C. 2007. The conservation relevance of epidemiological research into carnivore viral diseases in the Serengeti. *Conservation Biology*, 21: 612–622.

Coles, E. 1984. *Patologia clínica veterinária*, 3rd edn. Editora Manole, São Paulo.
Corrêa, S.H.R. 2007. Leptospirose. In: Z.S. Cubas, J.C.R. Silva & J.L. Catão-Dias (eds), *Tratado de animais selvagens, medicina veterinária*, pp. 736–741. Roca, São Paulo.
Courtenay, O., Quinnell, R.J. & Chalmers, W.S.K. 2001. Contact rates between wild and domestic canids: No evidence of parvovirus or canine distemper virus in crab-eating foxes. *Veterinary Microbiology*, 81: 9–19.
Creel, S., Creel, N.M., Munson, L., Sanderlin, D. & Appel, M.J. 1997. Serosurvey for selected viral diseases and demography of African wild dogs in Tanzania. *Journal of Wildlife Diseases*, 33: 823–832.
Cross, M.L., Buddle, B.M. & Aldwell, F.E. 2007. The potential of oral vaccines for disease control in wildlife species. *The Veterinary Journal*, 174: 472–480.
Curi, N. 2005 Avaliação do estado de saúde e do risco de transmissão de doenças entre canídeos (Mammalia, Carnívora) silvestres e domésticos na região da Serra do Cipó, Minas Gerais: Implicações para a conservação. Dissertação de mestrado. Programa de pós-graduação em zoologia de vertebrados. Pontifícia Universidade Católica de Minas Gerais, Belo Horizonte, Brazil.
Curi, N.H.A., Miranda, I. & Talamoni, S.A. 2006. Serologic evidence of leishmania infection in free-ranging wild and domestic canids around a Brazilian national park. *Memórias do Instituto Oswaldo Cruz*, 101: 99–101.
Dantas-Torres, F. 2007. The role of dogs as reservoirs of Leishmania parasites, with emphasis on *Leishmania (Leishmania) infantum* and *Leishmania (Viannia) braziliensis*. *Veterinary Parasitology*, 149: 139–146.
Daszak, P., Cunningham, A.A. & Hyatt, A.D. 2000. Emerging infectious diseases of wildlife threats to biodiversity and human health. *Science*, 287: 443–449.
Daszak, P., Cunningham, A.A. & Hyatt, A.D. 2001. Anthropogenic environmental change and the emergence of infectious diseases in wildlife. *Acta Tropica*, 78: 103–116.
Davidson, W.R., Nettles, V.F., Hayes, L.E., Howerth, E.W. & Couvillion, C.E. 1992. Diseases diagnosed in gray foxes (*Urocyon cinereoargentateus*) from the southeastern United States. *Journal of Wildlife Diseases*, 28: 28–33.
Deane, L.M. & Deane, M.P. 1954. Encontro de leishmanias nas vísceras e na pele de uma raposa, em zona endêmica de calazar, nos arredores de Sobral, Ceará. *Hospital*, 45: 419–421.
De Craeye, S., Speybroeck, N., Ajzenberg, D., Dardé, M.L., Collinet, F., Tavernier, P., Van Gucht, S., Dorny, P. & Dierick K. 2011. *Toxoplasma gondii* and *Neospora caninum* in wildlife: Common parasites in Belgian foxes and Cervidae? *Veterinary Parasitology*, 178(1–2): 64–69.
Deem, S.L. & Emmons, L.H. 2005. Exposure of free-ranging maned wolves (*Chrysocyon brachyurus*) to infectious and parasitic disease agents in the Noël Kempff Mercado National Park, Bolivia. *Journal of Zoo and Wildlife Medicine*, 36: 192–197.
Desjeux, P. 2004. Leishmaniasis: Current situation and new perspectives. *Comparative Immunology Microbiology & Infectious Disease*, 27: 305–318.
Dietz, J.M. 1984. Ecology and social organization of the maned wolf (*Chrysocyon brachyurus*). *Smithsonian Contributions to Zoology*, 392: 1–51.
Dobson, A.P. & Carper, E.R. 1996. Infectious diseases and human population history. *Bioscience*, 46: 115–126.
Drake, G.J.C., Auty, H.K., Ryvar, R. Ollierwe, W.E.R., Kitchenerac, A.C., Freemanar, A.R. & Radford, A.D. 2004. The use of reference strand-mediated conformational analysis for the study of cheetah (*Acinonyx jubatus*) feline leucocyte antigen class II DRB polymorphisms. *Molecular Ecology*, 13: 221–229.
Emmons, L.H. 1999. *Neotropical Rainforest Mammals, A Field Guide*, 2nd edn. The University of Chicago Press.
Fahrig, L. & Merriam, G. 1994. Conservation of fragmented populations. *Conservation Biology*, 8: 50–59.
Failace, R. 2003. *Hemograma, manual de interpretação*, 4th edn. Editora Artmed, Porto Alegre.
Fletcher, K.C., Eugster, A.K., Schmidt, R.E. & Hubbard, G.B. 1979. Parvovirus infection in maned wolves. *Journal of the American Veterinary Medical Association*, 175: 897–900.
Flint, S.J., Enquist, L.W., Krug, R.M., Racaniello, A.R. & Skalka, A.M. 2004. *Principles of Virology Molecular Biology, Pathogenesis, and Control of Animal Viruses*, 2nd edn, pp. 761–762. ASM Press, Washington.
Funk, S.M., Fiorello, C.V., Cleaveland, S. & Gompper, M.E. 2001. The role of disease in carnivore ecology and conservation. In: J.L. Gittleman, S.M. Funk, D. Macdonald & R. Wayne (eds), *Carnivore Conservation, Conservation Biology 5*, pp. 443–466. Cambridge University Press, Cambridge, UK.
Furlanello, T., Simonato, G., Caldin, M., De Lorenzi, D., Lubas, G., Bernardini, D. & Solano-Gallego, L. 2006. Validation of an automated spectrophotometric assay for the determination of cholinesterase activity in canine serum. *Veterinary Research and Communication*, 30: 723–733.
Gascoyne, S.C., Laurenson, M.K., Lelo, S. & Borner, M. 1993. Rabies in African wild dogs (*Lycaon pictus*) in the Serengeti region, Tanzania. *Journal of Wildlife Diseases*, 29: 396–402.

Gillin, C.M., Tabor, G.M. & Aguirre, A.A. 2002. Ecological health and wildlife disease management in national parks. In: A.A. Aguirre, R.S. Ostfeld, G.M. Tabor, C. House & M.C. Pearl (eds), *Conservation Medicine, Ecological Health in Practice*, pp. 253–264. Oxford University Press, New York.

Gomes, M.S. 2007. Carnivora (lobo-guará, cachorro-do-mato, raposa-do-campo). In: Z.S. Cubas, J.C.R. Silva & J.L. Catão-Dias (eds), *Tratado de Animais Selvagens*, pp. 492–504. Roca, São Paulo.

Gordon, J.C. & Angrick, E.J. 1986. Canine parvovirus: Environmental effects on infectivity. *American Journal of Veterinary Research*, 47: 1464–1467.

Grenfell, B.T. & Gulland, F.M.D. 1995. Introduction: Ecological impact of parasitism on wildlife host populations. *Parasitology*, 111: S3–S14.

Grisaru, D., Deutsch, V., Shapira, M., Pick, M., Sternfeld, M., Melamed-Book, N., Kaufer, D., et al. 2001. ARP, a peptide derived from the stress-associated acetylcholinesterase variant, has hematopoietic growth promoting activities. *Molecular Medicine*, 7: 93–105.

Guerra-Neto, G., Girio, R.J.S., Andrade, T.M., Koproski, L.P., Moraes, W. & Santos, L.C. 2004. Ocorrência de anticorpos contra *Leptospira* spp. em felídeos neotropicais pertencentes ao criadouro de animais silvestres da Itaipu Binacional e ao Zoológico Municipal Bosque Guarani, Foz do Iguaçu, Estado do Paraná. *ARS Veterinária*, 20: 75–80.

Hargis, A.M. 1990. Sistema Tegumentar. In: R.G. Thomson (ed.), *Patologia Veterinária Especial*, pp. 1–77. Editora Manole, São Paulo.

Haydon, D.T., Cleaveland, S., Taylor, L.H. & Laurenson, M.K. 2002. Identifying reservoirs of infection: A conceptual and practical challenge. *Emerging Infection Diseases*, 8: 1468–1473.

Helmbolt, C.F. & Jungherr, E.L. 1955. Distemper complex in wild carnivores simulating rabies. *American Journal of Veterinary Research*, 16: 463–469.

Hofmeyr, M., Bingham, J., Lane, E.P., Ide, A. & Nel, L. 2000. Rabies in African wild dogs (*Lycaon pictus*) in the Madikwe Game Reserve, South Africa. *Veterinary Record*, 146: 50–52.

Jakubek, E.B., Farkas, R., Pálfi, V. & Mattsson, J.G. 2007. Prevalence of antibodies against *Toxoplasma gondii* and *Neospora caninum* in Hungarian red foxes (*Vulpes vulpes*). *Veterinary Parasitology*, 144(1–2): 39–44.

Jorge, R.S.P. 2008. Caracterização do estado sanitário dos carnívoros selvagens da RPPN SESC Pantanal e de animais domésticos da região. Tese de Doutorado. Programa de Pós-Graduação em Epidemiologia Experimental e Aplicada às Zoonoses da Faculdade de Medicina Veterinária e Zootecnia da Universidade de São Paulo, São Paulo, Brasil.

Jorge, R.S.P., Rocha F.L., May-Júnior, J.A. & Morato, R.G. 2010. Ocorrência de patógenos em carnívoros selvagens brasileiros e suas implicações para a conservação e saúde pública. *Oecologia Australis*, 14(3): 686–710.

Karesh, W.B., Hart, J.A., Hart, T.B., House, C., Torres, A., Dierenfeld, E.S., Braselton, W.E., Puche, H. & Cook, R.A. 1995. Health evaluation of five sympatric duiker species (*Cephalophus spp.*). *Journal of Zoo and Wildlife Disease*, 26: 485–502.

Kat, P.W., Alexander, K.A., Smith, J.S., Richardson, J.D. & Munson, L. 1996. Rabies among African wild dogs (*Lycaon pictus*) in the Masai Mara, Kenya. *Journal of Veterinary Diagnostic Investigation*, 8: 420–426.

Klink, C.A. & Machado, R.B. 2005. A conservação do Cerrado Brasileiro. *Megadiversidade*, 1(1).

Kock, R., Chalmers, W.S.K., Mwanzia, J., Chillingworth, C., Wambua, J., Coleman, P.G. & Baxendale, W. 1998. Canine distemper antibodies in lions of the Masai Mara. *Veterinary Record*, 142: 662–665.

Labruna, M.B., Jorge, R.S.P., Sana, D.A., Jacomo, A.T.A., Kashivakura, C.K., Furtado, M.M., Ferro, C., et al. 2005. Ticks (Acari: Ixodida) on wild carnivores in Brazil. *Experimental and Applied Acaralogy*, 36: 149–163.

Lanfranchi, P., Ferroglio, E., Poglayen, G. & Guberti, V. 2003. Wildlife veterinarian, conservation and public health. *Veterinary Research Communications*, 27: 567–574.

Laurenson, K., Shiferaw, F. & Sillero-Zubiri, C. 1997. Disease, domestic dogs and the Ethiopian wolf: Current situation. In: C. Sillero-Zubiri & D. Macdonald (eds), *The Ethiopian Wolf: Status and Conservation Action Plan*, pp. 32–40. IUCN/SSC Canid Specialist Group, Gland, Switzerland.

Laurenson, M.K., Mlengeya, T., Shiferaw, F. & Cleaveland, S. 2005. Approaches to disease control in domestic canids for the conservation of endangered wild carnivores. In: S.A. Osofsky, S. Cleaveland, W.B. Karesh, M.D. Kock, P.J. Nyhus, L. Starr & A. Yang (eds), *Conservation and Development Interventions at the Wildlife/Livestock Interface: Implications for Wildlife, Livestock, and Human Health*, pp. 141–146. IUCN Publications Services Unit, Gland, Switzerland and Cambridge, UK.

Lepage, L., Seschiele, F., Gueguen, R. & Siest, G. 1985. Total cholinesterase in plasma: Biological variations and reference limits. *Clinical Chemistry*, 31: 546–550.

Levett, P.N. 2001. Leptospirosis. *Clinical Microbiology Reviews*, 14: 296–326.

Luppi, M.M., Malta, M.C., Silva, T.M., Silva, F.L., Motta, R.O., Miranda, I., Ecco, R. & Santos, R.L. 2008. Visceral leishmaniasis in captive wild canids in Brazil. *Veterinary Parasitology*, 155: 146–151.

Maia, O.B. and Gouveia, A.M.G. 2002. Birth and mortality of maned wolves *Chrysocyon brachyurus* (Illiger, 1811) in captivity. *Brazilian Journal of Biology*, 62: 25–32.

Maned Wolf Species Survival Plan. 2007. *Maned Wolf Husbandry Manual*. Maned Wolf Species Survival Plan of Association of Zoos and Aquariums, USA.

Mann, P.C., Bush, M., Appel, M.J.G., Beehler, B.A. & Montali, R.J. 1980. Canine parvovirus infection in South American canids. *Journal of the American Veterinary Medical Association*, 177: 779–783.

Matera, E.A., Saliba, A.M. & Matera, A. 1968. The occurrence of dioctophymiasis in the maned wolf *Chrysocyon brachyurus*. *International Zoo Yearbook*, 8: 24–26.

Mattos, P.S.R. 2003. Epidemiologia e genética populacional do lobo-guará (*Chrysocyon brachyurus*, Illiger, 1915) (Carnívora, Canidae) na região Nordeste do estado de São Paulo. Master Dissertation, University of São Carlos, São Carlos, São Paulo, Brazil.

May-Júnior, J.A., Songsasen, N., Azevedo, F.C., Santos, J.P., Paula, R.C., Rodrigues, F.H.G., Rodden, M.D., Wildt, D.E. & Morato, R.G. 2009. Hematology and blood chemistry parameters differ in free-ranging maned wolves (*Chrysocyon brachyurus*) living in the Serra da Canastra National Park versus adjacent farmlands, Brazil. *Journal of Wildlife Diseases*, 45(1): 81–90.

McCallum, H. & Dobson, A. 1995. Detecting disease and parasite threats to endangered species and ecosystems. *Trends in Ecology and Evolution*, 10: 190–194.

McCallum, H. & Dobson, A. 2002. Disease, habitat fragmentation and conservation. *Proceedings of the Royal Society of London*, 269: 2041–2049.

Mech, L.D. & Goyal, S.M. 1993. Canine parvovirus effect on wolf population change and pup survival. *Journal Wildlife Diseases*, 29: 330–333.

Megid, J. 2007. Raiva. In: Z.S. Cubas, J.C.R. Silva & J.L. Catão-Dias (orgs.) (eds), *Tratado de animais selvagens—Medicina veterinária*, pp. 785–798. Editora Roca, São Paulo.

Molin, R. 1860. Uma monografia del genera Spiroptera. *Sitzungsberichte der Kaiserlichen Akademie der Wissenschaften in Wien*, 38: 911–1005.

Mörner, T., Obendorf, D.L. & Artois, M. 2002. Surveillance and monitoring of wildlife diseases. *Revue Scientifique et Technique de l'Office International des Epizooties*, 21: 67–76.

Murray, D.L., Kapke, C.A., Everman, J.J. & Fuller, T.K. 1999. Infectious disease and the conservation of free-ranging large carnivores. *Animal Conservation*, 2: 241–254.

Nizri, E., Wirguin I. & Brenner, T. 2007. The role of cholinergic balance perturbation in neurological diseases. *Drug News and Perspective*, 20: 412–429.

Norton, T.M. 1990. Medical management of maned wolves. In: R.C. Cambre, J.E. Whitney, C.L. Garrett, M. Zoller & J. Nordstrom (eds), *Proceedings: American Association of Zoo Veterinarians*, pp. 61–63, American Association of Zoo Veterinarians, Denver, CO.

Nunes, A.L.V., Cruz, M.L. & Cortopassi, S.R.G. 2007. Anestesiologia. In: Z.S. Cubas, J.C.R. Silva & J.L. Catão-Dias. *Tratado de Animais Selvagens*, pp. 1040–1067. ROCA, São Paulo.

Oliveira, T.G. 1994. *Neotropical Cats: Ecology and Conservation*. EDUFMA, São Luís, MA.

Parrish, C.R. 1990. Emergence, natural history, and variation of canine, mink, and feline parvoviruses. *Advances in Virus Research*, 38: 403–450.

Patz, J.A., Graczyk, T.K., Geller, N. & Vittor, A.Y. 2000. Effects of environmental change on emerging parasitic diseases. *International Journal of Parasitology*, 30: 1395–1405.

Patz, J.A., Daszak, P., Tabor, G.M., Aguirre, A.A., Pearl, M., Epstein, J., Wolfe, N.D., et al. 2004. Unhealthy landscapes: Policy recommendations on land use change and infectious disease emergence. *Environmental Health Perspectives*, 112: 1092–1098.

Pick, M., Perry, C., Lapidot, T., Guimaraes-Sternberg, C., Naparstek, E., Deutsch, V. & Soreq, H. 2006. Stress-induced cholinergic signaling promotes inflammation-associated thrombopoiesis. *Blood*, 107: 3397–3406.

Poulin, R. 1997. Species richness of parasite assemblages: Evolution and patterns. *Annual Reviews Ecological Systems*, 28: 341–358.

Prada, C.S. 2004. Atropelamento de vertebrados silvestres em uma região fragmentada do Nordeste do estado de São Paulo: Quantificação impacto e análise de fatores envolvidos. Dissertation Presented in the Postgraduate Program in Ecology and Natural Resources of the Center for Biological and Health Sciences, Federal University of São Carlos, São Carlos.

Qiu, W., Zheng, Y., Fan, Q., Liu, H., Zhang, F., Wang, W. & Liao, G. 2011. Canine distemper outbreak in rhesus monkeys, China. *Emerging Infectious Diseases*. 17(8).

Randolph, S.E., Chemini, C., Furlanello, C., Grnchi, C., Hails, R.S., Hudson, P.J., Jones, L.D., et al. 2003. The role of pathogens in biological conservation. In: P.J. Hudson, A. Rizzoli, B.T. Grenfell, H. Heesherbeek & A.P. Dobson (eds), *The Ecology of Wildlife Diseases*, pp. 139–150. Oxford University Press, Oxford, UK.

Roelke-Parker, M.E., Munson, L., Packer, C., Kock, R., Cleaveland, S., Carpenter, M., O'Brien, S.J., et al. 1996. A canine distemper virus epidemic in Serengeti lions (*Panthera leo*). *Nature*, 379: 441–445.

Röken, B.O. 1993. Parasitic diseases of carnivores. In: M.E. Fowler (ed.), *Zoo & Wild Animal Medicine: Current Therapy 3*, pp. 399–404. W.B. Saunders Company, Philadelphia, PA.

Rosa, G.N., Domingues, H.G., Santos, M.M.A.B., Felippe, P.A.N., Spilki, F.R. & Arns, C.W. 2012. Molecular detection and phylogenetic analysis of the gene H from canine distemper virus isolates circulating at the municipality of Campinas, São Paulo. *Pesquisa Veterinária Brasileira*, 32(1): 72–77.

Roth, E.E. 1972. Leptospirosis. In: J.W. Davis, L.H. Karstad & D.O. Trainer (eds), *Enfermidades infecciosas de los mamíferos salvajes*, pp. 356–368. Editora Acribia, Barcelona.

Rupprecht, C.E., Stöhr, K. & Meredith, C. 2001. Rabies. In: E.S. Williams, I.K. Barker (eds), *Infectious Diseases of Wild Mammals*, 3rd edn, pp. 3–36. Iowa State University Press, Ames.

Sillero-Zubiri, C. & MacDonald, D. (eds). 1996. *The Ethiopian Wolf. Status and Conservation Action Plan.* IUCN/SSC Canid Specialist Group, Gland.

Sillero-Zubiri, C., MacDonald, D.W. & King, A.A. 1996. Rabies and mortality in Ethiopian wolves (*Canis simensis*). *Journal of Wildlife Diseases*, 32: 80–86.

Small, D.H., Michaelson, S. & Sberna, G. 1996. Non-classical actions of cholinesterase: Role in cellular differentiation, tumorigenesis and Alzheimer's disease. *Neurochemistry International*, 28: 453–483.

Soreg, H., Lapidot-Lifson, Y. & Zakut, H. 1991. A role for cholinesterase in tumorigenesis. *Cancer Cells*, 3: 511–516.

Steinel, A., Parrish, C.R., Bloom, M.E. & Truyen, U. 2001. Parvovirus infections in wild carnivores. *Journal of Wildlife Diseases*, 37: 594–607.

Stieve, E., Beckmen, K., Kania, S.A., Widner, A. & Patton, S. 2010. *Neospora caninum* and *Toxoplasma gondii* antibody prevalence in Alaska wildlife. *Journal of Wildlife Diseases*, 46(2): 348–55.

Tabor, G.M. 2002. Defining conservation medicine. In: A.A. Aguirre, R.S. Ostfeld, G.M. Tabor, C. House & M.C. Pearl (eds), *Conservation Medicine: Ecological Health in Practice*, pp. 8–16. Oxford University Press, New York.

Taylor, L.H., Latham, S.M. & Woolhouse, M.E.J. 2001. Risk factors for human disease emergence. *Philosophical Transactions of the Royal Society of London. Series B, Biological. Science*, 356: 983–989.

Thiermann, A.B. 1984. Leptospirosis: Current developments and trends. *Journal of the American Veterinary Medical Association*, 184: 722–725.

Thomas, N.J., Foreyt, W.J., Evermann, J.F., Windberg, L.A. & Knowton, F.F. 1984. Seroprevalence of canine parvovirus in wild coyotes from Texas, Utah, and Idaho (1972 to 1983). *Journal of the American Veterinary Medical Association*, 185: 1283–1287.

Thorne, E.T & Williams, E.S. 1988. Disease and endangered species: The black-footed ferret as a recent example. *Conservation Biology*, 2: 66–74.

Travis, D.A., Hungerford, L., Engel, G.A.M & Jones-Engel, L. 2006. Disease risk analysis: A tool for primate conservation planning and decision making. *American Journal of Primatology*, 68: 855–867.

Ujvari, B. & Belov, K. 2011. Major histocompatibility complex (MHC) markers in conservation biology. *International Journal of Molecular Sciences*, 12: 5168–5186.

Van de Bildt, M.W.G., Kuiken, T., Visee, A.M., Lema, S., Fitzjohn, T.R. & Osterhaus, A.D.M.E. 2002. Distemper outbreak and its effect on African wild dog conservation. *Emerging Infectious Diseases*, 8: 211–213.

Vasconcellos, S.A. 1987. O papel dos reservatórios na manutenção da leptospirose na natureza. *Comunicações Científicas da Faculdade de Medicina Veterinária e Zootecnia da Universidade de São Paulo*, 11: 17–24.

Vitalino, S.N., Silva, D.A., Mineo, T.W., Ferreira, R.A., Bevilacqua, E. & Mineo, J.R. 2004. Seroprevalence of *Toxoplasma gondii* and *Neospora caninum* in captive maned wolves (*Chrysocyon brachyurus*) from southeastern and midwestern regions of Brazil. *Veterinary Parasitology*, 122(4): 253–260.

Wandeler, A.I., Capt, S., Gerber, H., Kappeler, A. & Kipfer, R. 1988. Rabies epidemiology, natural barriers and fox vaccination. *Parasitologia*, 30: 53–57.

Weiler, G.J., Garner, G.W. & Ritter, D.G. 1995. Occurrence of rabies in a wolf population in northeastern Alaska. *Journal of Wildlife Diseases*, 31: 79–82.

Williams, E.S. 2001. Canine distemper. In: E.S. Williams & I.K. Barker (eds), *Infectious Diseases of Wild Mammals*, 3rd edn, pp. 50–59. Iowa State University Press, Ames.

Williams, E.S., Thorne, E.T., Appel M.J.G. & Brlitsky D.W. 1988. Canine distemper in black-footed ferrets (*Mustela nigripes*) from Wyoming. *Journal of Wildlife Diseases*, 24: 385–398.

Woodroffe, R. 1999. Managing disease threats to wild mammals. *Animal Conservation*, 2: 185–193.

Woodroffe, R., Cleaveland, S., Courtenay, O., Laurenson, K. & Artois, M. 1997. *The African Wild Dog: Status, Survey and Conservation Action Plan.* IUCN/SSC Canid Specialist Group, IUCN, Gland.

Yoshikawa, Y., Ochikubo, F., Matsubara, Y., Tsuruoka, H., Ishii, M. & Shirota, K. 1989. Natural infection with canine distemper virus in a Japanese monkey (Macaca fuscata). *Veterinary Microbiology*, 20: 193–205.

Yu, X., Chen, N., Hu, D., Zhang, W., Li, X., Wang, B., Kang, L., Li, X., Liu, Q. & Tian K. 2009. Detection of *Neospora caninum* from farm-bred young blue foxes (*Alopex lagopus*) in China. *Journal of Veterinary Medical Science*, 71(1): 113–115.

Part II

Case Studies and Conservation Programs

Drawing of maned wolf by Marcelo Pinto.

10 Conservation of a Population of Maned Wolves, *Chrysocyon brachyurus*, on a Small Reserve in the Cerrado

Flávio H.G. Rodrigues, Arnaud L.J. Desbiez, Raquel L.S.C. Grando, Ana C.R. Lacerda, Adriani Hass, and Wesley R. Silva

CONTENTS

Introduction 131
Study Area 132
Material and Methods 133
VORTEX Simulation Model 134
  VORTEX Baseline Model Parameters 134
    Supplementation 136
    Decrease in Road Mortality 136
Results 136
  Home Ranges and the Use of External Areas 136
  Population and Production Estimate 138
  Diet 139
  Causes of Mortality 139
  Baseline Model 140
  Alternative Futures 140
  Decrease of Road Mortality 140
Discussion 142
  Small Reserves, Large Home Ranges 142
  Small Reserves, Small Populations 143
  Mortality Factors and Direct Threats to the Population 144
  Viability of the Population 145
  Next Steps: Small Reserves and the Search for Solutions for the Conservation of Cerrado 145
Acknowledgments 148
References 148

INTRODUCTION

The maned wolf (*Chrysocyon brachyurus* Illiger) is widely distributed in open formations in South America. However, it has disappeared from some regions to the south of its distribution, mainly due to a loss of natural habitat (Queirolo et al. 2011). The main region in which the maned wolf is currently found is the Brazilian cerrado, where it is frequently found in practically all the protected areas (Rodrigues and Oliveira 2006). However, the cerrado is one of the most endangered environments in the world and roughly 80% of it has been altered due to agriculture and cattle farming, hydroelectric power generation and urbanization (Myers et al. 2000; Mittermeier et al. 2004), leaving the landscape highly fragmented. As a result, the maned wolf is considered endangered (Vulnerable) in Brazil (Machado et al. 2008) and Near Threatened by the IUCN. Only 2.5% of the total area of the cerrado is made up of protected areas (Machado et al. 2004) and the majority of these are isolated and not large enough to sustain viable populations of large predators such as the maned wolf, jaguar (*Panthera onca*), and cougar (*Puma concolor*) (Rodrigues and Oliveira 2006). In addition, reserves near urban areas are under even greater pressure from invasion by exotic plants and animals, hunting, fire, road kills, and isolation of populations. Therefore, only a fraction of the original fauna can be maintained in reserves of this type.

Within this scenario of intense degradation and fragmentation of the cerrado habitat, it is important to understand the situation faced by maned wolves in small fragments. Although the maned wolf is still common in the cerrado biome, we know that the stress levels of individuals living within a mosaic of natural areas and farms are higher than those of individuals living in conservation areas (Spercoski 2007) and that populations of maned wolves that inhabit fragmented ranges suffer even greater impacts directly related to human activity, such as being killed on roads (Vieira 1996; Rodrigues et al. 2002; Paula et al. 2007; Rodden et al. 2004) and are reduced to preying on domestic birds (Dietz 1984), among other animals. Animals rescued after falling victim to these impacts (road accidents or captured on farms, for example) are constantly released into protected areas using no criteria whatsoever and with no monitoring of individual animals to learn about their fate. On the one hand, this might be an additional negative impact (transmission of diseases, competition with resident individuals, for example), but it may also increase the viability of isolated populations, in the event that these animals remain and reproduce in the release area. It is therefore necessary to assess the behavior of translocated animals in order to learn whether the releases conducted are effective and whether they can serve as a management tool. Microsatellite data indicate that there is no structure pattern for the species in Brazil (Fontoura-Rodrigues 2010), However, since the fragmentation process is recent, in evolutionary terms, it is possible that the populations may already be fragmented, even though this process has not yet been detected in any genetic analyses.

Despite a significant presence of maned wolves in cerrado protected areas, many of these populations may not be large enough to be viable (Rodrigues and Oliveira 2006). However, these small reserves may be essential to maintaining biodiversity in highly fragmented regions, such as the cerrado, as they may serve to improve connectivity between populations in different fragments, providing the necessary conditions for the maned wolf to continue living in the cerrado. It is therefore of fundamental importance to analyze the conditions of the species' survival in these small protected areas.

Computer modeling is a valuable and versatile tool for assessing the risk of decline and extinction of wildlife populations. Complex and interacting factors that influence population persistence and health can be explored, including natural and anthropogenic causes. Models can also be used to evaluate the effects of alternative management strategies to identify the most effective conservation actions for a population or species and to identify research needs. Such an evaluation of population persistence under current and varying conditions is commonly referred to as a population viability analysis (PVA).

The purpose of this work is to assess the viability of the population of maned wolves at the Águas Emendadas Ecological Station, which is a small reserve located near an urban center in the Distrito Federal in central Brazil, and to discuss the role of small reserves in the conservation of the species.

STUDY AREA

The study was conducted at the Águas Emendadas Ecological Station (ESECAE), located in the northern part of the Distrito Federal in central Brazil (Figure 10.1). The station covers an area

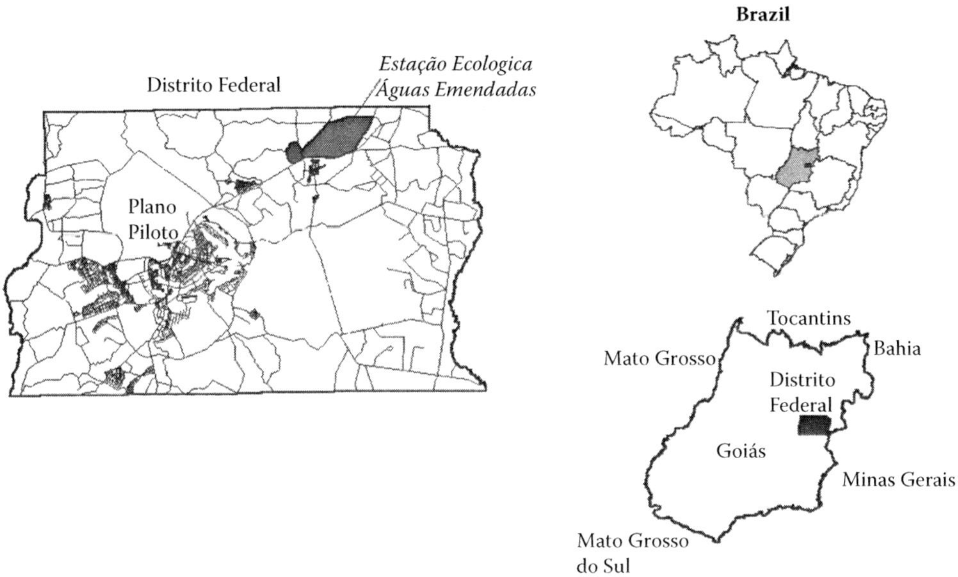

FIGURE 10.1 Location of Estação Ecológica de Águas Emendadas (Águas Emendadas Ecological Station), Federal District, Brazil.

of roughly 104 km^2 and is the drainage divide between two large Brazilian drainage basins: the Amazon and the Paraná, joined in a common source (which is the reason for the name Águas Emendadas, which means "joined waters" in English). The vegetation is typical of the cerrado, with various features: grasslands, *veredas*, riparian forests, *cerradão*, and *cerrado sensu strict* (Silva-Júnior and Felfili 1996). Two distinct seasons can be identified: a dry season between April and September and a rainy season between October and March. ESECAE is located beside the city of Planaltina (roughly 230,000 inhabitants) and is composed of a main polygon and an adjacent area, where Bonita Lake is located. The main polygon is bordered by four highways: three maintained by the federal district (DF 205, DF 128, and DF 345) and one federal highway (BR 020), with only the DF 205 highway being unpaved. DF 128 cuts across a small portion of the station, splitting the main polygon area from Bonita Lake. The city of Planaltina extends to the edge of the station's fence near Bonita Lake and many small farms are located along the DF 205 and DF 128 highways and are dedicated to small-scale subsistence agriculture, pastureland, and chicken farming. These small farms are the cause of domestic cats and dogs entering the station. The process of native vegetation degradation around ESECAE is accelerating. In 1987, 41.2% of the surrounding area was covered in natural vegetation, but by 1996, this proportion had fallen to 35.8% (Machado et al. 1998). During this period, the area being used for agriculture increased by 76.7%, while urban areas increased by 95.2% (Machado et al. 1998). The areas that present the greatest risk for ESECAE, according to the assessment of these authors, are the city of Planaltina (DF), the agricultural area to the east of this city, the area around Bonita Lake, and the city of Planaltina de Goiás (GO) (77,000 inhabitants), located 5 km to the north of the station. Even with these problems, ESECAE is one of the most important protected areas in the Distrito Federal and shelters various species of vertebrate, including many listed as endangered (Brandão and Araújo 1998; Bagno 1998; Marinho-Filho et al. 1998).

MATERIAL AND METHODS

The study was conducted between April 1997 and November 1999. In order to monitor the movements of the maned wolves and to estimate their use of different areas, we captured these animals in a cage trap with an independently triggered guillotine-style door, placed at strategic locations in ESECAE and baited with cooked chicken heads or raw beef. When a wolf was captured, we anesthetized it, measured it, took blood samples, marked it with a numbered tag, and fitted it with a radio collar. The animal was released only after it had completely recovered from the anesthesia. Also, we took advantage of the arrival of three wolves at the Brasília Zoo from the Distrito Federal region and surroundings, which were to be released into ESECAE, to assess the translocation of individuals as a possible management tool. These wolves were subjected to the same identification procedures (tagged and fitted with a radio collar), biometric measurements, and sample collection that were used with the wolves caught in traps.

The wolves were monitored using radio telemetry, with triangulation using ground-based azimuths from known coordinates, or by following the signal until the animal was sighted. In parallel, all wolf sightings (tagged or not) in ESECAE were recorded. The locations were plotted on a map in order to track the movements of the wolves in ESECAE and the surrounding areas. We grouped together all the locations (identified via telemetry or by sighting) for each individual and calculated its home range, using the minimum convex polygon (MCP) method.

In order to determine whether the maned wolves were preying on domestic animals, we collected fecal samples during the period from November 1996 to August 1999 along all the internal and bordering roads and highways at ESECAE. The samples were washed in running water using sieves and checked for the presence of food elements (see Rodrigues et al. [2007] for details of the food categories utilized and the general diet analysis).

The roads surrounding ESECAE were checked two or three times a week in order to quantify instances of maned wolves being killed on roads by vehicles. The wolves that were found dead along these roads were collected and, when possible, a necropsy was performed on the animals to determine their general state of health when they died. The roads along the borders of ESECAE were monitored from December 1997 to October 2000. There were no recorded instances of wolves dying from causes other than road kills.

VORTEX SIMULATION MODEL

The simulation software program VORTEX (v9.57) was used to examine the viability of maned wolf populations. VORTEX is a Monte Carlo simulation of the effects of deterministic forces as well as demographic, environmental, and genetic stochastic events on wild populations. VORTEX models population dynamics as discrete sequential events that occur according to defined probabilities. The program begins by creating individuals to form the starting population and progressing through life cycle events (e.g., births, deaths, dispersal, catastrophic events), typically on an annual basis. Events such as breeding success, litter size, sex at birth, and survival are determined based on designated probabilities. Consequently, each run (iteration) of the model gives a different result. By running the model hundreds of times, it is possible to examine the probable outcome and range of possibilities. For a more detailed explanation of VORTEX and its use in population viability analysis, see Lacy (1993, 2000) and Miller and Lacy (2003).

VORTEX Baseline Model Parameters

The baseline model developed during the maned wolf population and habitat viability assessment (PHVA) (Paula et al. 2007) was adapted to reflect the population of maned wolves at Águas Emendadas, where we conducted research from 1996 to 2003. Therefore, although some of the data entered into the model were based on best guesses, these guesses were based on years of experience from the Águas Emendadas maned wolf population and are a good illustration of how the general model created during the workshop can be adapted to reflect the demographics of a known population.

The population characteristics presented in Table 10.1 were accepted as realistic for the Águas Emendadas maned wolf population. Details about the meaning and significance of the different

TABLE 10.1
Population Parameter Values of the Águas Emendadas Maned Wolf Population Loaded in the VORTEX Software

Number of populations	1
Initial population size	30
Carrying capacity	30
Dispersal among populations	None
Inbreeding depression?	Yes
Percentage of the effect of inbreeding due to recessive lethal alleles	50
Breeding system	Long-term monogamy
Age of first reproduction (♀/♂) (years)	2/2
Maximum age of reproduction	12
Annual percentage of adult females reproducing (SD)	70 (10)
Density-dependent reproduction?	No
Maximum litter size	5
Distribution of Offspring per Female per Year (in %)	
1 offspring	21
2 offspring	24
3 offspring	28
4 offspring	19
5 offspring	8
Overall offspring sex ratio	0.5
Adult males in breeding pool	100%
% mortality age 0–1, ♀/♂ (SD)	60 (10)
% mortality age 1–2, ♀/♂ (SD)	60 (10)
% mortality age after 2, ♀/♂ (SD)	10 (20)
Catastrophe	None
Harvest	2% of adults
Supplementation	None

parameters entered into the model for maned wolves can be found in the PHVA VORTEX report (Paula et al. 2007).

In captivity, maned wolves may live up to 16 years (Rodden et al. 2004). Life expectancy of maned wolves is approximately 10–12 years in the wild (based on age estimates of wild animals in this study). The population was modeled for 100 years (approximately 15 generations) so that long-term population trends could be observed and yet not so far into the future that results become too uncertain.

As alternative scenarios, two conservation measures were suggested: (1) supplementation of the population through the translocation of maned wolves from a neighboring population and (2) decreasing the mortality of maned wolves as a result of road kills by 50%. Results reported for each VORTEX modeling scenario include the following:

rs (SD): The mean rate of stochastic population growth or decline and standard deviation, demonstrated by the simulated populations, averaged across years and iterations, for all simulated populations that are not extinct. This population growth rate is calculated for each year of the simulation, prior to any truncation of the population size due to the population exceeding the carrying capacity.

P(S)100: Probability of the population surviving and not falling below the critical "quasi-extinction" threshold of 10 individuals after 100 years, determined by the proportion of

TABLE 10.2
Mortality Rates of Maned Wolf Populations from Águas Emendadas with a 50% Reduction in Mortality due to Road Kill

	Baseline	50% Decrease in Road Kill
% mortality age 0–1, ♀/♂ (SD)	60 (10)	45 (10)
% mortality age 1–2, ♀/♂ (SD)	60 (10)	35 (5)
% mortality age after 2, ♀/♂ (SD)	10 (2)	10 (2)

500 iterations within a given scenario that fall below the threshold within the given time frame.

T(E): The average time to population extinction, in 100 years.

N100 (SD): The mean (standard deviation) population size at the end of the simulation, averaged across all simulated populations, including those that are deemed extinct.

GD100: The gene diversity or expected heterozygosity of the extant populations, expressed as a percentage of the initial gene diversity of the population. The fitness of individuals usually declines proportionately with gene diversity.

Supplementation

Translocation of wild maned wolves from adjacent populations to Águas Emendadas was modeled. Two scenarios were analyzed.

- The supplementation of one pair of subadult maned wolves every 2 years for 50 years
- The supplementation of two pairs of subadult maned wolves every year for 50 years

We chose to model the translocated animals as subadults. This decision was made because we believe that translocated animals will probably travel to establish a territory and therefore are more likely to have a mortality rate similar to that of subadults rather than adults. Furthermore, translocated animals are not likely to have pups in the first year after translocation since they have to establish a territory and find a mate.

Therefore, the animals supplementing the population will probably respond to the same parameters as a subadult even if they are translocated as adults.

Decrease in Road Mortality

The baseline model also includes mortality rates due to natural causes (see the PHVA model in Paula et al. [2007] for more detail). Therefore, a 50% reduction in road-related deaths leads to the mortality rates presented in Table 10.2.

RESULTS

Home Ranges and the Use of External Areas

We employed a set of 422 traps*night to capture the maned wolves. A total of seven individuals (four males and three females) were captured on 15 different occasions.

The home ranges of the maned wolves were large, between 4.43 and 104.90 km^2 (Table 10.3). If we consider only animals for which a minimum of ten locations were recorded, the average size of the home ranges was 56.95 ± 34.30 km^2. All seven maned wolves monitored utilized areas outside ESECAE (Figure 10.2), in areas varying from pastures and *Pinus* plantations to small farms and

TABLE 10.3
Home Range Sizes of Maned Wolves at Águas Emendadas Ecological Station and Proportion of Area outside the Protected Area

Individual	Months of Monitoring	Number of Locations	Home Range (ha)	% outside ESECAE
75	19	150	10,490	40.6
81	8	45	3,952	14.6
81b	8	44	2,344	24.6
92	3	15	1,553	49.5
94	2	8	443	6.1
84	5	21	7,522	42.1
89	24	109	4,960	35.1
40	2	5	582	<1.0

areas with natural vegetation. The proportion of the area used outside the station was high, with the exception of the animals monitored for only a short time and Female #81 (Table 10.3). Male #89 frequently visited a eucalyptus plantation and stayed there for a few days. We monitored two litters of Female #75 in 1997 and 1998. In both cases the females gave birth in a pasture area outside ESECAE and the pups remained in this area for a month and a half the first time and up to 3 months the second time.

The three wolves that were translocated to ESECAE did not remain in the area. The first, an adult female, remained near the release area for 3 days before disappearing. A great effort was made to locate her, including flights over the region in a helicopter, but she was not successfully located. The second, an adult male, crossed the entire station in 3 days and on the fourth day he was found outside ESECAE, after which we lost contact with him also. The third animal, an adult female, disappeared on the second day after being released. Even though 100% of the results of the translocations realized were negative, we modeled an alternative scenario with translocations serving as a management alternative, since different release strategies may increase the chance of translocated animals establishing themselves.

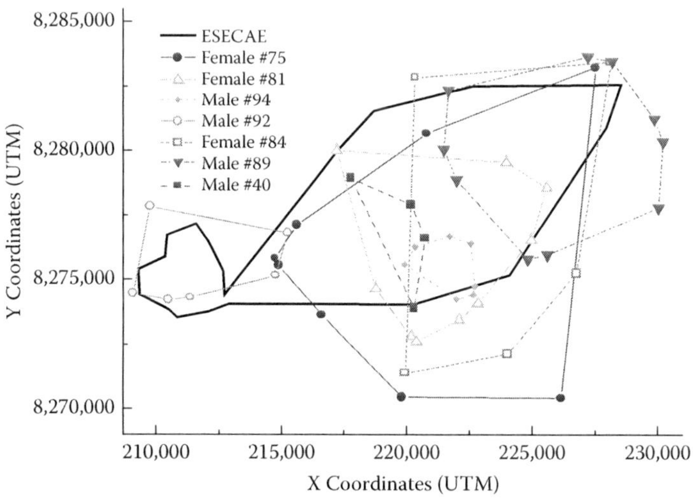

FIGURE 10.2 Home range of seven maned wolves monitored at Águas Emendadas Ecological Station. The x and y axes correspond to the UTM coordinates of the study area.

Population and Production Estimate

Based on the location of the wolves monitored by radio and sightings of other untagged animals, we estimated the minimum population of wolves that utilize ESECAE as part of their home range. In the Bonito Lake region, in the extreme southwest of the station, we captured an adult male (#92), which occupied the lake region and adjacent external areas. Another animal, not tagged, was seen and two wolves were heard vocalizing in the area. Thus, at least two animals inhabited the lake area, probably Male #92 and a female. Female #75 was captured in the southeastern part of the station and occupied a large area inside and outside the station (Figure 10.2), though her activity was concentrated in the southeastern area (Figure 10.3). At least one male inhabited the same area and was seen walking together with her during the period between June and September 1997, when Female #75 had pups. Male #89 occupied the northeastern area of the station (Figure 10.3). In the same area, we found traces of an adult and a pup and sighted a pup in August 1998, which were not Female #75 and her pups, which were on the other side of the station. Therefore, there are at least two more pairs occupying the area: one predominantly in the northeast and another in the southeast. Female #84 overlapped her home range with that of Male #89, often moving quite close to each other, and when Female #75 died, Female #84 occupied her home range. However, since we were only able to monitor Female #84 for a short time prior to the death of Female #75, we were not able to definitively determine that their home ranges did not previously overlap. Female #85 is probably the one that formed the pair with Male #89 or a descendant of this pair. The western part of the main polygon in ESECAE was also occupied by maned wolves, though we were not able to capture any animals in this region.

Wolves were observed a few times in this region, including two different animals sighted on the same night, and tracks and feces were constantly present. In Figure 10.3, we can see that none of the tagged wolves occupied the northwestern area of the station and the southwestern part of the main polygon was the limit of the home ranges of two individuals, though none of them concentrated their activities in this area. At least one pair and perhaps two (one in the southwest of the main polygon and the other in the northwest) occupied this area. Male #40 was excluded from the analysis, since few records of his location were available to gain a reasonable idea of his home range, though it is likely that this wolf was the male that occupied the same area as Female #75 or one of the males that inhabited the western region of ESECAE. Male #94 and Female #81 were also excluded, since they were young individuals, the pups of Female #75, and had not yet dispersed to their future home ranges. Thus, between four and five adult pairs of maned wolves inhabit ESECAE or part of it.

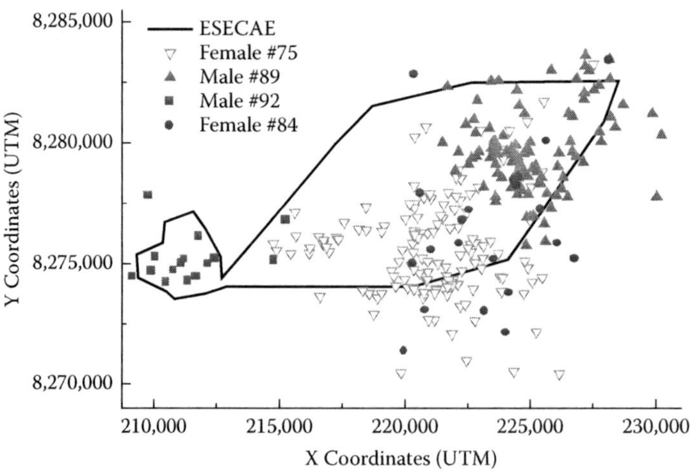

FIGURE 10.3 Locations of four maned wolves monitored at Águas Emendadas Ecological Station. The *x* and *y* axes correspond to the UTM coordinates of the study area.

The sizes of the maned wolf litters recorded in ESECAE were between two and three pups. During the study, Female #75 gave birth two times: in June 1997 and June 1998; in both instances she produced two pups (at least this is what we believe, since we were not able to observe the litters as soon as they were born). When she died, in June of 1999, she was pregnant with three embryos. A young female killed by a vehicle in June 2000 was pregnant with two pups and was very close to delivery. Considering the number of pups per litter and one litter per female per year, on average between eight and ten pups are born per year in ESECAE and the surrounding areas.

Diet

We analyzed 328 maned wolf fecal samples and identified 901 food items: 40% were animal in origin and 60% vegetation. Twenty species of fruits were recorded, with wolf's fruit (*Solanum lycocarpum*) being the main fruit consumed, found in 74.1% of the samples. Among the items of animal origin, we identified at least 28 species of vertebrates. Small mammals (Rodentia + Didelphimorphia), representing 16.7% of the total items, formed the most important category among the items of animal origin, followed by birds (10.1%) and armadillos (6.7%). The only domestic animals found were chickens, which were found in only two samples (0.6% of the samples, representing 0.2% of the total items).

Causes of Mortality

During 35 months of monitoring the roads bordering ESECAE, we found 13 maned wolves killed on the roads, resulting in an average of 0.4 wolves hit per month or 4.5 wolves hit per year. This figure may still be low, as some wolves may be hit but be able to leave the road, dying later undetected. Most of the wolves hit were young, up to 1 year old and, thus, had not yet had a chance to reproduce (Table 10.4). The number of wolves hit was greater on the BR 020 and DF 345 highways, where there is a greater flow of traffic, and less on DF 128 and DF 205. This last highway, on which only one hit wolf was recorded, has the lowest traffic flow and is the only one that is not paved, which

TABLE 10.4
Maned Wolves Found Road Killed in the Borders of Águas Emendadas Ecological Station

Time	Sex	Age (months)	Road
December/97	Male	Adult	DF 205
May/98	Female	12	DF 128
July/98	Female	12	DF 128
August/98	Male	3	BR 020
November/98	?	?	DF 345
June/99	Female	Adult[a]	BR 020
June/99	Female	?	DF 345
April/00	?	12	DF 345
May/00	?	?	DF 345
May/00	?	?	DF 345
June/00	Female	12[b]	BR 020
September/00	?	5	BR 020
October/00	?	5	BR 020

[a] Pregnant with three pups.
[b] Pregnant with two pups.

means vehicles travel along it more slowly. On the other highways, the speed limit of 80 km/h is rarely adhered to and it is common to observe vehicles traveling at higher speeds. The majority of wolves are hit during the dry season, especially in May and June.

Wolves were also killed by farmers from small farms that border ESECAE, supposedly because the wolves prey on domestic animals such as chickens and ducks, according to reports made by some residents in the area. However, as this is an illegal activity, this type of information is not readily available. Apparently, the culling of maned wolves was a less important factor in mortality than road kills, since we did not record mortalities of this type in any of the tagged animals.

Baseline Model

The baseline model results in a negative deterministic growth rate (r_{det}): $r_{det} = -0.032$ ($\lambda = 0.969$). This means that the maned wolf population is decreasing each year, even in the absence of stochastic events, and will therefore eventually become extinct (probability of extinction = 1). The stochastic growth rate (r_{stoch}) is −0.083 (SD = 0.259). The mean population size of the Águas Emendadas maned wolf population is shown in Figure 10.4. Generation time (the average age of reproduction) is 5.6 years for both males and females. Adult sex ratio of adult males to adult females is 1. Stable age distribution is presented in Table 10.5.

Alternative Futures

Supplementation with a pair of maned wolves, whether done every year or every 2 years, did not ensure the survival of the Águas Emendadas maned wolf population. In fact, the stochastic growth rate remained negative (Table 10.6). The increase in the rate of supplementation did increase the average time to population extinction (Figure 10.5). Therefore, supplementation can delay extinction. However, supplementation does not change the deterministic characteristics of the model population and therefore is not an appropriate long-term conservation solution for this population. However, supplementation may be a useful strategy for preventing the immediate extinction or increasing the genetic diversity of the population.

Decrease of Road Mortality

The results from a 50% reduction of wolves killed on roads are extremely encouraging and show that the population of maned wolves is likely to persist over the next 100 years. The stochastic

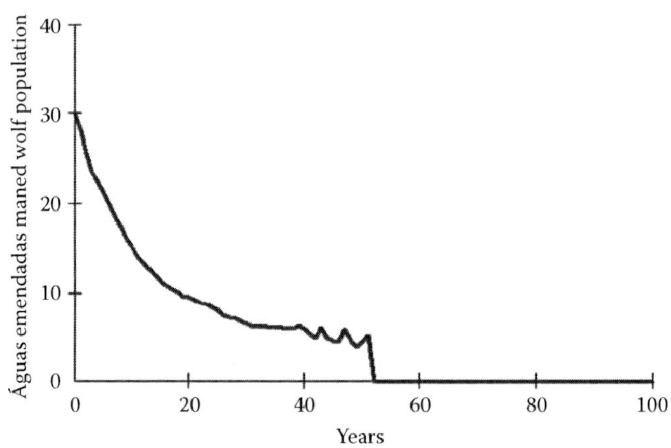

FIGURE 10.4 Mean population size of the Águas Emendadas maned wolf population at the end of the simulation, averaged across all simulated populations (500).

TABLE 10.5
Stable Age Distribution of the Águas Emendadas Maned Wolf Population

Age Class	Females	Males
0	0.202	0.202
1	0.083	0.083
2	0.034	0.034
3	0.03	0.03
4	0.027	0.027
5	0.023	0.023
6	0.02	0.02
7	0.018	0.018
8	0.016	0.016
9	0.014	0.014
10	0.012	0.012
11	0.011	0.011
12	0.009	0.009

TABLE 10.6
Impact of Supplementation on the Águas Emendadas Maned Wolf Population

	r_{stoch}	P(S)	TE	N100	Genetic Diversity
No supplementation	–0.083	0	23.06	0	0
One pair supplemented every 2 years	–0.029	0	43.94	0	0
One pair supplemented every year	–0.023	0.002	65.33	3	0

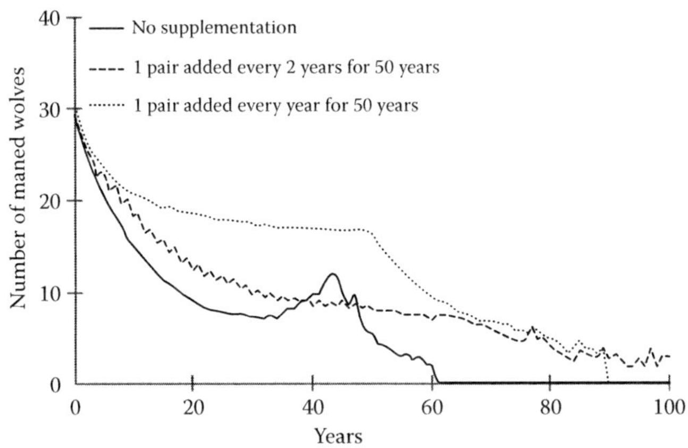

FIGURE 10.5 Mean population size of the Águas Emendadas maned wolf population with and without supplementation at the end of the simulation, averaged across all simulated populations (500).

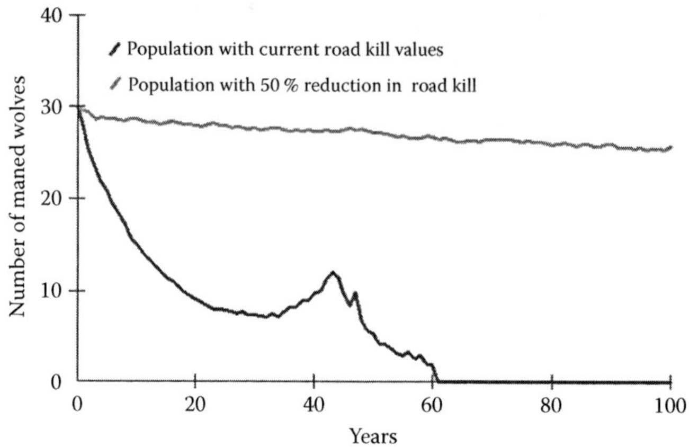

FIGURE 10.6 Mean population size of the Águas Emendadas maned wolf population with a decrease in road kill at the end of the simulation, averaged across all simulated populations (500).

TABLE 10.7
Impact of a 50% Decrease in Road Kill on the Águas Emendadas Maned Wolf Population

	r_{stoch}	P(S)	TE	N100	SD N100	Genetic Diversity
Current estimates	–0.083	0	23.06	0	0	0
Reduction of 50% of road kill	0.158	0.932	76.21	25.75	6.09	0.4996

growth rate (r_{stoch}) is 0.158, enabling the population to grow when below carrying capacity. The probability of the wolves surviving for 100 years is almost 1, and the mean population size at 100 years is 26 wolves. These results are presented in Figure 10.6 and Table 10.7. However, almost half the genetic diversity of the population will have been lost. The low retention of gene diversity is in part due to the very low initial population size and to the fact that we modeled a closed population in which no immigration of unrelated animals is included. However, it is hoped that deaths on roads will decrease and unrelated animals from other populations will enter the Águas Emendadas population. If this does not happen and the population remains closed to immigration from outside the reserve, supplementation of the population with unrelated animals will be necessary to maintain genetic diversity levels.

DISCUSSION

Small Reserves, Large Home Ranges

The wolves' home ranges inside ESECAE were very large, 56.95 ± 34.30 km^2, based on those of individuals for which there were an arbitrary minimum number of ten localizations. The average home ranges of maned wolves found in other studies varies from 25.23 to 80.18 km^2 (de Azevedo 2008; Dietz 1984; Carvalho and Vasconcellos 1995; Jácomo et al. 2009; Mantovani 2001; Rodrigues 2002). Based on the sizes of the home ranges occupied by the maned wolves in ESECAE, it is obvious that the size of the station is not sufficient to support a stable population of maned wolves within its boundaries. For example, at ESECAE the home of Female #75 alone was already larger than the station itself and the other individuals for which there were at least ten localizations occupied home ranges equivalent to 15%–48% of the total area of the unit. For this reason, all the wolves we monitored utilized, to a greater or lesser degree, areas outside ESECAE. The wolves that occupied smaller areas outside ESECAE (Table 10.3) are those that were located few times, and with an increase in the number of localizations, the proportion of the home range outside the station would tend to increase.

Though activity is common in areas altered by man, including fundamental activities such as bearing pups, the natural area represented by ESECAE is fundamental to the maintenance of the wolves, since the greater portion of their home ranges was within the protected area and it is within the reserve that they found their main food sources, whether fruit in general (except wolf's fruit, *S. lycocarpum*, which is common in pastures), small mammals, or armadillos (Rodrigues et al. 2007).

Small Reserves, Small Populations

The large size of the home ranges, associated with the low level of overlap of these ranges, does not permit a high population density and on a small reserve the size of the population will certainly be small. The five resident pairs at ESECAE are certainly not enough to maintain a viable population for very long in a state of isolation. These five pairs along with the young wolves that have not yet left their mothers would result in a total maximum population of approximately 30 individuals, which is why this number was used as the initial size in the models. The population of maned wolves at ESECAE today may experience a gene flow with other larger populations, such as that in the military area in Formosa (GO), to the northeast of the station, and that in the Brasília National Park and Cafuringa Environmental Protection Area (DF), to the southwest, both approximately 20 km away. However, the accelerated process of degradation in the areas surrounding the Águas Emendadas Station, especially as a result of urbanization (Machado et al. 1998), may, in a few years, sever these links with other natural areas. Loss of habitat and fragmentation may reduce the wild population, which may result in local extinctions and genetic problems caused by inbreeding (Gaines et al. 1997). Carnivores have a high dispersal rate and disperse over great distances, which is why the gene flow between populations of species with high mobility tends to be high and the difference between populations tends to be relatively small (Wayne and Koepfli 1996), which has already been observed in maned wolves (Fontoura-Rodrigues 2010; Lion et al. 2011). On the other hand, even in populations of species that are highly mobile, significant levels of genetic structure may exist if their distribution has been fragmented by recent alterations to their habitat or by prehistoric barriers (O'Brien et al. 1987; Wayne et al. 1992; Roelke et al. 1993; Roy et al. 1994). Even though some species may survive with a low level of genetic diversity with no great problems, the effects of the loss of heterozygosity in some populations are drastic, including reduced viability of sperm, a high rate of cryptorchidism in males, high frequency of cardiac defects and diseases, and reductions in the growth rate of pups, in the survival rate, and in successful breeding (Fergus 1991; Barone et al. 1994; O'Brien 1996; Roelke et al. 1993). In Brazil, there is still little evidence of genetic problems involving small populations, though the first case of cryptorchidism was recently reported involving wild felines and this is thought to be related to habitat fragmentation (Michalski et al. 2001). The population of wolves at ESECAE does not differ from other populations in the Distrito Federal region (Lion et al. 2011) and, in a general sense, genetic structure has not been found in wolves from different Brazilian populations (Fontoura-Rodrigues 2010). However, it may only be a question of time as there may be isolated populations that simply have not yet had time to develop such differences.

In addition to the genetic problems that can ensue as a function of a small population, demographic factors may eliminate a small population even before the loss of genetic variability becomes a problem (Lande 1998). In small populations, for example, the chance of finding sexual partners is diminished and random variations may result in deviations in the gender ratio in such a manner that fewer descendants can be produced from the same sized population. The chances of extinction due to catastrophe or to random factors are also much greater (Shaffer 1981). In an isolated setting, the small population of wolves that inhabit ESECAE may become extinct in a short time due to genetic factors, demographic factors, or both.

Mortality Factors and Direct Threats to the Population

In addition to the direct threats inherent in a small population, the ESECAE maned wolves are subject to other external threats, which are important causes of mortality and, consequently, of a reduction in the size of the population. Maned wolves are frequently killed on roads in the areas surrounding ESECAE. As the animals leave the station, they are constantly faced with the risk of crossing the highways that border the unit. The majority of the wolves hit were young animals, up to 1 year old, which, therefore, had not yet mated or were about to reproduce for the first time. A lack of experience with highways is probably one of the factors that lead to high mortality among young animals. However, animals accustomed to crossing highways, such as Females #75 and #81, also perished as a result of road kills. The average of 4.5 wolves killed on roads per year represents a third of the annual production of pups (considering litters of between two and three pups) and therefore is certainly the main cause of death for the population of wolves at ESECAE. The situation is even more critical on small reserves, since in addition to the smaller populations, the core area of the reserve is not sufficient to house the entire home range of some individuals and, as a result, practically all of them have to cross highways as part of their regular activities. The BR 020 highway is also where the highest number of vehicles is to be found. The paved highways have a much higher impact, due to both the higher than normal traffic levels and, mainly, the higher speeds at which vehicles travel on them.

The majority of road kills on the highways that border ESECAE and other locations could be prevented. Logically, there are situations in which an animal crosses a highway unexpectedly and it is impossible to avoid an accident. This usually occurs when the vegetation along the side of the highway is quite high, obscuring animals' and drivers' vision. But speeding is also one of the main causes of wild animals being hit. Speeding makes it impossible for the driver to swerve to avoid the animal, stop, or slow down to wait for the animal to cross the road, thus avoiding hitting the animal. Also, a large number of the animals are hit on purpose or due to a lack of concern on the part of motorists. Drivers, especially truck drivers, often do not attempt to avoid hitting wild animals and, even worse, often intentionally run the animals over, as has already been observed with two ferrets (*Galictis* sp.) on the BR 020 highway (M.A. Bagno, pers. comm.), obtaining from this a strange and sinister satisfaction. Among mammals, carnivores are hit most often (Rodrigues et al. 2002; Silveira 1999; Vieira 1996). It is likely that carnivores' susceptibility to road kills is due to the fact that they are highly mobile and tend to eat the carcasses of other animals hit along roads, and are thus also at risk of dying the same way. Among carnivores, the maned wolf is one of the most susceptible to being hit (Rodrigues et al. 2002; Silveira 1999).

Highways are not recommended as borders for preservation areas, due to the danger of animals being killed on roads, the greater incidence of fires entering the reserve and because they make the reserve vulnerable to unauthorized access. Despite this, a large number of Brazilian protected areas are bordered by highways, paved or not, and in some cases highways run through the units.

Although we do not have any idea of the frequency with which it occurs, the killing of animals by farmers and local residents may also be an important risk factor. All the wolves monitored using radiotelemetry have home ranges that extend, at least in part, beyond the borders of the station, meaning that contact with humans is intense in the areas surrounding ESECAE. Since the vast majority of farmers raise chickens, which are normally left free to roam the property, it is to be expected that these will be predated by wolves and other wild carnivores. Some of the farmers accuse wolves of killing their chickens, though we observed only two samples with the remains of chickens, indicating that the consumption of domestic animals is infrequent and of little importance to the maned wolves. In other studies, the frequency of chicken remains in the feces of maned wolves is also very low (Dietz 1984; Motta-Júnior et al. 1996), though there have been reports of wolves killing many chickens in a single night, and, for small farmers, the impact could be significant.

Another source of risk for the population of maned wolves at ESECAE is access by domestic animals, especially dogs (*Canis familiaris*). The domestic dog is a cosmopolitan plague that interacts with native species via predation, competition for limited resources, and the introduction of diseases, resulting in serious damage to wild fauna (Artois 1997; May and Norton 1996). The main impact dogs have on reserves is predation on native species (Yanes and Suárez 1996; Boitani 1983; Barnett and Rudd 1983; Kruuk and Snell 1981; Causey and Cude 1978, 1980; Iverson 1978; Lowry and MacArthur 1978; Scott and Causey 1973). Domestic dogs tend to form packs and in this situation they can prey on large animals, including wolves. In Brazil, the presence of domestic dogs in protected areas near urban areas is frequent. At ESECAE, many cases of attacks by packs of dogs on wild animals have been reported (Marinho-Filho et al. 1998). In Brasília National Park, it is estimated that the most frequent cause of death of native fauna may be attacks by dogs (Horowitz 1992; Lacerda et al. 2009) and the presence of domestic dogs is one of the main management problems (Funatura/IBAMA 1998). In this park, the maned wolf seems to avoid areas with dog tracks (Lacerda et al. 2009). The presence of domestic dogs can result in other types of harm, such as competition for resources with wild predators and the transmission of diseases (Rodden et al. 2004). In captivity, maned wolves are susceptible to many canine diseases caused by viruses, including canine distemper, parvovirus, rabies, and adenovirus (Rodden et al. 2004; Mann et al. 1980; Fletcher et al. 1979). As disease treatment and prevention for domestic animals by low-income owners are poor, domestic dogs that inhabit areas near ESECAE may be a significant source of diseases that threaten wolves and other wild carnivores at the station.

The problem of access by domestic animals is not easy to solve, since the human population (and with it that of domestic animals) is constantly increasing around the station and it is practically impossible to prevent access by these animals to the conservation unit. The routine elimination of dogs that have established themselves in various reserves, however, has had little effect in the face of the increase in the canine population in these areas. It is estimated that in 1997 alone a total of 700 dogs were culled (by shooting) in the Brasília National Park (Lindberg 1998). The minimum that should be undertaken, however, is a vaccination program and sanitary monitoring of the animals, which would produce benefits for the wild animals and for the local human population.

Viability of the Population

The simulations conducted as part of the VORTEX program indicate that the population of wolves at ESECAE will not last for very long if the wolves continue to be killed by the roads and remain isolated from other populations. The addition of new individuals has not substantially altered this scenario and the chances are high that the population will go extinct. However, a 50% reduction in vehicle-related mortality could make the population viable in the long term. This is the main mortality factor among the population and it is crucial that it be given due priority. A 50% reduction in the current rate is an ambitious goal, however, and it involves a series of factors, which are discussed below.

Next Steps: Small Reserves and the Search for Solutions for the Conservation of Cerrado

The cerrado is one of 25 "hotspots" in the world and roughly 80% of the cerrado can no longer be found in its natural state (Myers et al. 2000). The portion of the cerrado covered by protected areas is small (Machado et al. 2004), and only a few areas are left that are large enough to preserve viable populations of large mammals. Within this context, small reserves are fundamental to the effort to conserve what is left of this important biome. However, in order to maximize the biodiversity that can be conserved in these units, they must be managed, in order to minimize the anthropic pressures to which they are subjected. Animals that need large areas, such as large predators, used to be distributed over a vast and contiguous stretch of land, sometimes over entire continents, and they are now confined to small remaining habitats. Small reserves, most of the time, are not capable of sheltering the entire community of fauna in the region. For some species, even the largest reserves present this restriction. At Águas Emendadas, we can still find a large portion of the original mammal fauna, though some species are now very rare, such as the giant armadillo, *Priodontes maximus*; the collared peccary, *Pecari tajacu*; and the cougar, *Puma concolor* (Marinho-Filho et al. 1998). This situation is common among the majority of small reserve areas in the cerrado. Within this scenario of small fragments still available for conservation and the need for large areas in order to maintain viable populations of some species, it is of paramount importance that conservation efforts are focused on the area of landscape management. Connectedness is an important factor that must be managed in this landscape. The solution for the conservation of the fauna in these small reserves may lie in a system of connected protected areas that can thus form metapopulations. To do this, there must be a dispersal pattern between the reserves, whether natural (corridors or dispersal via a partially permeable matrix) or managed, with transfers of individual animals. The maned wolf is a species that tolerates human presence well, insofar as it makes use of pastures and cleared areas. This type of behavior certainly increases the species' capacity for dispersal, which does not depend solely on natural areas as crossing points. It is likely that the abundance of wolf's fruit plants (the most common food item in the diet of *C. brachyurus*) in altered areas, especially in pastures, is one of the factors that allow the species to utilize these areas for their dispersal, since these animals are guaranteed of finding food during the crossing (Rodrigues et al. 2002). Despite being an endangered species (Machado et al. 2008), the maned wolf, in some areas, is expanding its distribution, taking advantage of the transformation of forests into open areas in the Zona da Mata region of Minas Gerais state (Dietz 1984). In these expansion areas, wolf's fruits (*S. lycocarpum*) were present in 77% of the samples and were responsible for 43% of the total biomass ingested and 98% of the fruit biomass ingested (Santos et al. 2003), which reinforces the idea that the colonization of open areas by wolf's fruit is an important factor that allows anthropic areas to be crossed by maned wolves.

However, the isolation process is increasing daily. The two cities closest to ESECAE have been expanding very rapidly, as have agricultural areas. Between 1987 and 1996, the urban areas around ESECAE expanded by 95.2% and agricultural areas grew by 76.7%; as a result, there was a significant reduction in natural areas and pastures (Machado et al. 1998). In a state of isolation, in which the dispersal of individuals is unlikely or insufficient, the translocation of individuals is an alternative for reducing the effects of the geographical isolation of small populations. The exchange of individuals between populations is an option for increasing genetic viability, restoring a population, or increasing population sizes. This type of management has been suggested as the only means of reversing the consequences of genetic depression in the Florida panther, for example (Roelke et al. 1993). However, translocations may not meet with success due to three problems related to animal behavior following release: first, some animals tend to return to their original home ranges (Rogers 1988); second, even when they cannot return to their original home ranges, animals may not remain in the release area and may establish themselves in other locations (Fritts et al. 1984); and, third, the survival rate may be low (O'Brien and McCullogh 1985). Translocations of mammals in Brazil, including carnivores, are fairly common with animals from fauna rescue programs at large construction projects or animals caught or captured for other reasons. Very little of this work is documented or monitored in terms of the success of these translocations though. Rodrigues and Marinho-Filho (1999) monitored small felines (*Herpailurus yagouaroundi* and *Leopardus tigrinus*) translocated in the region around the Serra da Mesa Hydroelectric Plant and concluded that the individuals left the release area immediately, indicating that translocations with the goal of genetic exchange would not be successful with these species. The three maned wolves translocated to ESECAE also did not remain in the area, leaving the range of the radio receivers within a few days. We do not have any information on whether the animals survived, but it is likely, based on the trajectory that was calculated before their signals were lost, that they did not return to their original home ranges. Therefore, the translocation of maned wolves, at least in the manner in which it was carried out, is not an effective management tool. The release method may influence whether the animal remains in the release area and its survival. The majority of translocations involved "hard releases," in which the animal is simply released into a specific location. "Soft release" involves holding the animal in the release area for a period of time before releasing it (Kleiman 1989). Soft releases promote a greater connection with the release area and reduce postrelease movements (Fritts et al. 1984). It is possible that the use of other release methods may increase the success of maned wolf translocations. Another factor that may have influenced the success of the translocations was the presence of other adult animals in the area. Maned wolves use scent markings to mark their territory, and encountering such markings may inhibit other animals' willingness to remain in the area. The translocation of younger animals may also be a useful strategy, since they would not have established home ranges in other areas. There is still a need for specific studies of maned wolf dispersal patterns (heading, distance, sex of the animals that disperse, etc.), in order to better understand which conservation strategies should be employed. However, even with the soft release of younger individuals, which may increase the chances that translocated animals will remain in the area, this measure would not be effective for increasing the viability of the wolf population at ESECAE, according to the VORTEX program models, which indicate that the chances of the animals remaining would not be significantly changed.

On the contrary, the chances of a population remaining would increase substantially, becoming viable in a period of 100 years, if the number of animals killed on roads was reduced by 50%. Road kills are a major problem for the Águas Emendadas population. Each year at least 15% of the population is killed by cars. The objective of reducing by half the number of animals killed on roads is ambitious and requires enormous effort from different segments of society, from the government to the population itself. Changing the behavior of drivers is a major challenge. The first mitigation measure that should be adopted is a reduction in the speed limit on the stretches of highway that border reserves to a speed that will give drivers time to avoid an accident. It is therefore essential that speed control measures be adopted, such as radar equipment, electronic barriers, and obstacles, as other enforcement measures would not be respected, as is the case today. Another essential measure is the periodical clearing of the sides of highways, cutting down herbaceous vegetation. This would improve visibility for drivers and animals, allowing drivers to see the animals before they cross the highway and allowing animals to see cars at a distance and choose not to cross at that time. These measures must be adopted along the four highways that surround ESECAE, especially on the DF 345 and BR 202 highways, where the majority of accidents occur. In parallel to speed reduction measures and management of vegetation along the highways, driver education campaigns should be undertaken, with road signs warning of wild animal crossings in the area, explanatory pamphlets, and other informative materials. Signs were installed along the paved highways following the collection of the data presented here and a reduction has been recorded in the number of road kill incidents (Alcântara 2006) in comparison with the period up until 2002 (Rodrigues et al. 2002).

It is not known, however, if this is due to the signs or to a gradual reduction in the abundance of fauna on the reserve (Alcântara 2006). It is certain that the number of maned wolves killed on the roads has not changed in the years following the present study (FHGR, personal observation). One possibility that could be quite effective is making this issue part of the obligatory driver education courses required to obtain a driving license.

In addition to the legislative and warning measures discussed above, other management measures could be adopted. Some animals utilize tunnels to cross highways, but we do not know if maned wolves utilize these routes. Therefore, the utilization of fences and tunnels should be the subject of studies that assess their efficacy as passages for certain animals and whether they limit animals' natural movement. Fitting maned wolves with retroreflective colored collars that are picked up by car headlights could be a short-term solution for reducing mortality rates due to vehicles. Furthermore, such a project could garner sufficient media attention to serve as an efficient educational tool regarding maned wolves throughout Brazil. In a project in Zimbabwe, the Painted Dog Conservation Project (PDC), painted dogs were fitted with retroreflective colored collars that are picked up by car headlights. The project claims that this has reduced mortality rates by 50%

(G. Rasmussen, pers. comm.). This, however, requires constant efforts to capture animals and fit them with radio collars. It is essential that the most appropriate preventive measures are assessed and tested.

Educational actions are essential for the conservation of wild animals in areas with human populations. Discussing issues inherent to the theme with the community could help them to recognize the importance of natural resources to their own well-being and survival. Following that, some simple measures that would benefit both native fauna and people could be discussed and proposed. For example, the construction of chicken coops, where chickens could be sheltered at night, would be sufficient to prevent predation and eliminate the killing of wolves and other predators for this reason. Caring for the health of dogs and other domestic animals would prevent the transmission of diseases to their owners.

Even though the dynamics of these small reserves and the biology of the majority of the species there are poorly understood, and even considering that a more detailed understanding is important in order to outline more precise management strategies, impact mitigation actions are urgently needed and must be implemented immediately, for the conservation of both maned wolves and other vertebrates. In the case of ESECAE, the main tasks are to maintain gene flow with other populations and to reduce mortality along highways.

ACKNOWLEDGMENTS

We thank Luciano Corrêa, former director of Águas Emendadas Ecological Station, for logistical support and friendship. We thank Jader Marinho-Filho for lending the traps, Fernanda Vinci dos Santos for assisting in capturing and anesthetizing animals, and both also for their friendship. We thank Fundação O Boticário de Proteção à Natureza (Fundação Grupo Boticário) for financial support and CAPES for the PhD scholarship to FHGR.

REFERENCES

Alcântara, T. B. 2006. Evolução dos índices de atropelamento de vertebrados silvestres nas rodovias do entorno da Estação Ecológica de Águas Emendadas, DF, Brasil, e eficácia de medidas mitigadoras. Dissertação (Mestrado em Ecologia), Universidade de Brasília.

Artois, M. 1997. Managing problem wildlife in the "Old World": A veterinary perspective. *Reproduction, Fertility and Development* 9: 17–25.

de Azevedo, F. C. 2008. Área de vida e organização espacial de lobos-guará (*Chrysocyon brachyurus*) na região do Parque Nacional da Serra da Canastra, Minas Gerais, Brasil. Dissertação de Mestrado. Universidade Federal de Minas Gerais, Belo Horizonte, Minas Gerais.

Bagno, M. A. 1998. As aves da Estação Ecológica de Águas Emendadas. In: Marinho-Filho, J. S., F. H. G. Rodrigues & M. M. Guimarães (eds), *Vertebrados da Estação Ecológica de Águas Emendadas*, pp. 22–33. SEMATEC/IEMA, Brasília, DF.

Barnett, B. D. & R. L. Rudd. 1983. Feral dogs of the Galapagos Islands: Impact and control. *International Journal for the Study of Animal Problems* 4: 44–58.

Barone, M. A., M. E. Roelke, J. Howard, J. L. Brown, A. E. Anderson & D. E. Wildt. 1994. Reproductive characteristics of male Florida panthers: Comparative studies from Florida, Texas, Colorado, Latin America and North American Zoos. *Journal of Mammalogy* 75(1): 150–162.

Boitani, L. 1983. Wolf and dog competition in Italy. *Acta Zoologica Fennica* 174: 259–264.

Brandão, R. A. & A. F. B. Araújo. 1998. A herpetofauna da Estação Ecológica de Águas Emendadas. In: Marinho-Filho, J. S., F. H. G. Rodrigues & M. M. Guimarães (eds), *Vertebrados da Estação Ecológica de Águas Emendadas*, pp. 9–21. SEMATEC/IEMA, Brasília.

Carvalho, C. T. & L. E. M. Vasconcellos. 1995. Disease, food and reproduction of the maned wolf—*Chrysocyon brachyurus* (Illiger) (Carnivora, Canidae) in southeast Brazil. *Revista Brasileira de Zoologia* 12(3): 627–640.

Causey, M. K. & C. A. Cude. 1978. Feral dog predation of the gopher tortoise, *Gopherus polyphemus* (Reptilia, Testudines, Testudinidae) in southeast Alabama. *Herpetological Review* 9: 94–95.

Dietz, J. M. 1984. Ecology and social organization of the maned wolf. *Smithsonian Contributions to Zoology* 392: 1–51.

Fergus, C. 1991. The Florida panther verges on extinction. *Science* 251: 1178–1180.

Fletcher, K. C., A. K. Eugster, R. E. Schmidt & G. B. Hubbard. 1979. Parvovirus infection in maned wolves. *Journal of the American Veterinary Medical Assocociation* 175: 897–900.

Fontoura-Rodrigues, M. L. 2010. Diversidade genética e estrutura populacional do lobo-guará (*Chrysocyon brachyurus*). Master thesis in zoology. Pontifícia Universidade Católica do Rio Grande do Sul.

Fritts, S. H., W. J. Paul & L. D. Mech. 1984. Movements of translocated wolves in Minnesota. *Journal of Wildlife Management* 48: 709–721.

Funatura/IBAMA. 1998. *Plano de manejo do Parque National de Brasilia*—Revisão, Vol. 1. Funatura/IBAMA, Brasília.

Gaines, M. S., J. E. Diffendorfer, R. H. Tamarin & T. S. Wittam. 1997. The effects of habitat fragmentation on the genetic structure of small mammal populations. *Journal of Heredity* 88(4): 294–304.

Horowitz, C. 1992. *Plano de manejo do Parque Nacional de Brasília: Avaliação da metodologia de planejamento adotada*, Execução e Resultados Alcançados no decênio *79/89*. Tese de Mestrado. Universidade de Brasília, Brasília.

Iverson, J. B. 1978. The impact of feral cats and dogs on populations of the West Indian rock iguana, *Cyclura carinata. Biological Conservation* 14: 63–73.

Jácomo, A. T. A., C. K. Kashivakura, C. Ferro, M. M. Furtado, S. P. Astete, N. M. Torres, R. Sollmann & L. Silveira. 2009. Home range and spatial organization of maned wolves in the Brazilian grasslands. *Journal of Mammalogy* 90(1): 150–157.

Kleiman, D. G. 1989. Reintroduction of captive mammals for conservation: Guidelines for reintroducing endangered species into the wild. *Bioscience* 39: 152–161.

Kruuk, H. & H. Snell. 1981. Prey selection by feral dogs from a population of marine iguanas (*Amblyhynchus cristatus*). *Journal of Applied Ecology* 18: 197–204.

Lacerda, A. C. R., W. M. Tomás & J. Marinho-Filho. 2009. Domestic dogs as an edge effect in the Brasília National Park, Brazil: Interactions with native mammals. *Animal Conservation* 12: 477–487.

Lacy, R. C. 1993. VORTEX: A computer simulation model for population viability analysis. *Wildlife Research* 20: 45–65.

Lacy, R. C. 2000. Structure of the VORTEX simulation model for population viability analysis. *Ecological Bulletins* 48: 191–203.

Lande, R. 1998. Genetics and demography in biological conservation. *Science* 241: 1455–1460.

Lindberg, S. M. 1998. Cães ferais do Parque Nacional de Brasília: Uma séria ameaça a fauna. In: *Plano de Manejo do Parque Nacional de Brasília*. IBAMA/FUNATURA, Brasília.

Lion, M. B., A. A. Garda, J. S. Marinho Filho, F. H. G. Rodrigues & E. Eizirik. 2011. Conservation genetics of maned wolves in a highly impacted area of the Brazilian Cerrado biome. *Genetica ('s-Gravenhage)* 139(3): 369–381.

Lowry, D. A & K. L. MacArthur. 1978. Domestic dogs as predators on deer. *Wildlife Society Bulletin* 6(1): 38–39.

Machado, A. M. B., G. M. Drummond & A. P. Paglia. 2008. *Livro vermelho da fauna Brasileira ameaçada de extinção*. 1st edn. MMA; Fundação Biodiversitas, Brasília.

Machado, R. B., L. M. S. Aguiar, C. A. Bianchi, R. L. Vianna, A. J. B. Santos, C. H. Saito & J. F. Timmers. 1998. Áreas de risco no entorno de Unidades de Conservação: Estudo de caso da Estação Ecológica de Águas Emendadas, Planaltina. In: Marinho-Filho, J. S., F. H. G. Rodrigues & M. M. Guimarães (eds), *Vertebrados da Estação Ecológica de Águas Emendadas*, pp. 64–75. SEMATEC/IEMA, Brasília.

Machado, R. B., L. M. S. Aguiar, M. B. Ramos-Neto, F. H. G. Rodrigues, A. Hass & F. G. Aquino. 2004. *Atlas de conservação da natureza brasileira—Unidades Federais*. 1st edn, Vol. 1. Editora Metalivros, São Paulo.

Mann, P. C., M. Bush, M. Appel, B. A. Beehler & R. J. Montali. 1980. Canine parvovírus infection in South American canids. *Journal of the American Veterinary Medical Assocication* 177: 779–783.

Mantovani, J. E. 2001. Telemetria convencional e via satélite na determinação da área de vida de três espécies de carnívoros da região nordeste do Estado de São Paulo. 133f. Tese de Doutorado, Universidade Federal de São Carlos, São Carlos.

Marinho-Filho, J. S., F. H. G. Rodrigues & M. M. Guimarães. 1998. Mamíferos da Estação Ecológica de Águas Emendadas. In: Marinho-Filho, J. S., F. H. G. Rodrigues & M. M. Guimarães (eds), *Vertebrados da Estação Ecológica de Águas Emendadas*, pp. 34–63. SEMATEC/IEMA, Brasília.

May, S. A & T. W. Norton. 1996. Influence of fragmentation and disturbance on the potential impact of feral predators on native fauna in Australian forest ecosystems. *Wildlife Research* 23: 387–400.

Michalski, F., L. Cullen Jr, T. G. Oliveira, J. R. Crawshaw & G. Peter. 2001. Felid cryptorchidism in Atlantic Forest fragments: Evidence of inbreeding in free-ranging populations? *Mammalia (Paris)* 65(4): 548–550.

Miller, P. S. & R. C. Lacy. 2003. *VORTEX: A Stochastic Simulation of the Extinction Process*. Version 9 User's Manual. Conservation Breeding Specialist Group (SSC/IUCN), Apple Valley, MN.

Mittermeier, R. A., P. R. Gil, M. Hoffman, J. Pilgrim, T. Brooks, C. G. Mittermeier, L. Lamoreux & G. A. B. Fonseca. 2004. Hotspots revisited: Earth's biologically richest and most endangered terrestrial ecoregions. CEMEX & Agrupacion Sierra Madre, Mexico City.

Motta-Júnior, J. C., S. A. Talamoni, J. A. Lombardi & K. Simokomaki. 1996. Diet of the maned wolf, *Chrysocyon brachyurus*, in central Brazil. *Journal of Zoology (London)* 240: 277–284.

Myers, N., R. A. Mittermeier, C. G. Mittermeier, G. A. B. Fonseca & J. Kent. 2000. Biodiversity hotspots for conservation priorities. *Nature* 403: 853–858.

O'Brien, S. 1996. Conservation genetics of the Felidae. In: Avise, J. C. & J. L. Hamrick (eds), *Conservation Genetics: Case Histories from Nature*, pp. 50–74. Chapman and Hall, New York.

O'Brien, M. K. & D. R. McCullogh. 1985. Survival of black-tailed deer following relocation in California. *Journal of Wildlife Management* 49: 115–119.

O'Brien, S. J., J. S. Martenson, C. Packer, L. Herbst, V. de Vos, P. Joslin, J. Ott-Joslin, D. E. Wildt & M. Bush. 1987. Biochemical genetic variation in geographic isolates of African and Asian lions. *National Geographic Research* 3: 114–124.

Paula, R. C. P. Medici & R. G. Morato. 2007. *Plano de ação para a conservação do lobo-guará*. 1st edn, Vol. 1. IBAMA, Brasília.

Queirolo, D., J. R. Moreira, L. Soler, L. H. Emmons, F. H. G. Rodrigues, A. A. Pautasso, J. L. Cartes & V. Salvatori. 2011. Historical and current range of the Near Threatened maned wolf *Chrysocyon brachyurus* in South America. *Oryx (Oxford)* 45: 296–303.

Rodden, M., F. H. G. Rodrigues & S. Bestelmeyer. 2004. Maned wolf (*Chrysocyon brachyurus*). In: Sillero-Zubiri, C., M. Hoffmann & D. W. Macdonald. (Org.) (eds), *Canids: Foxes, Wolves, Jackals and Dogs. Status Survey and Conservation Action Plan*, pp. 38–44. IUCN/SSC Canid Specialist Group, Gland.

Rodrigues, F. H. G. 2002. Biologia e conservação do lobo-guará na Estação Ecológica de Águas Emendadas. 105f. Tese de doutorado, Universidade Estadual de Campinas, Campinas, SP.

Rodrigues, F. H. G. & J. S. Marinho-Filho. 1999. Translocation of two species of small wild cats in central Brazil: A preliminary report. *Cat News* 30: 28–28.

Rodrigues, F. H. G., & T. G. Oliveira. 2006. Unidades de conservação e seu papel na conservação de carnívoros brasileiros. In: Morato, R. G., F. H. G. Rodrigues, E. Eizirik, P. R. Mangini, F. C. C. Azevedo & J. S. Marinho-Filho. (Org.) (eds), *Manejo e conservação de carnívoros neotropicais: I Workshop de pesquisa para a conservação*, pp. 97–110. IBAMA, São Paulo.

Rodrigues, F. H. G., A. Hass, L. M. Rezende, C. S. Pereira, C. F. Figueiredo, B. F. Leite & F. G. F. Franca. 2002. Impacto de rodovias sobre a fauna da Estação Ecológica de Águas Emendadas, DF. In: Anais do III Congresso Brasileiro de Unidades de Conservação, pp. 585–593.

Rodrigues, F. H. G., A. Hass, A. C. R. Lacerda, R. L. S. C. Grando, M. A. Bagno, A. M. R. Bezerra & W. R. Silva. 2007. Feeding habits of the maned wolf (*Chrysocyon brachyurus*) in the Brazilian Cerrado. *Mastozoologia Neotropical* 14: 37–51.

Roelke, M. E., J. S. Martenson & S. O'Brien. 1993. The consequences of demographic reduction and genetic depletion in the endangered Florida panther. *Current Biology* 3(6): 340–350.

Rogers, L. L. 1988. Homing tendencies of large mammals: A review. In: Nielsen, L. & R. D. Brown (eds), *Translocations of Wild Animals*, pp. 76–91. Humane Society, Milwaukee: Wisconsin.

Roy, M. S., E. Geffen, D. Smith, E. Ostrander & R. K. Wayne. 1994. Patterns of differentiation and hybridization in North American wolf-like canids reveled by analysis of microsatellite loci. *Molecular Biology and Evolution* 11: 553–570.

Santos, E. F., E. Z. Setz & N. Gobbi. 2003. Diet of the maned wolf (*Chrysocyon brachyurus*) and its role in seed dispersal on a cattle ranch in Brazil. *Journal of Zoology* 260: 203–208.

Scott, M. D. & K. Causey. 1973. Ecology of feral-dogs in Alabama. *Journal of Wildlife Management* 37(3): 253–265.

Shaffer, M. L. 1981. Minimum population sizes for species conservation. *BioScience* 31: 131–134.

Silva-Júnior, M. C. & J. M. Felfili. 1996. A vegetação da Estação Ecológica de Águas Emendadas. Secretaria de Meio Ambiente e Tecnologia do Distrito Federal, Brasília, DF.

Silveira, L. 1999. Ecologia e conservação dos mamíferos carnívoros do Parque Nacional das Emas, Goiás. Tese de mestrado. Universidade Federal de Goiás, Goiás.

Spercoski, K. M. 2007. Monitoramento a longo prazo da excreção de corticóides fecais em lobos-guarás (*Chrysocyon brachyurus*, Illiger 1811) de cativeiro e vida livre: Uma contribuição para o manejo e a conservação da espécie. Master thesis in Cell and Molecular Biology. Universidade Federal do Paraná, UFPR, Brazil.

Vieira, E. M. 1996. Highway mortality of mammals in central Brazil. Ciência Cultura. *Journal of the Brazilian Association for the Advancement of Science* 48(4): 270–272.

Wayne, R. K. & K. P. Koepfli. 1996. Demographic and historical effects on genetic variation of carnivores. In: Gittleman, J. L. (ed.), *Carnivore Behavior, Ecology and Evolution*, vol. 2, pp. 453–484. Cornell University Press, Ithaca.

Wayne, R. K., N. Lehman, M. W. Allard & R. L. Honeycutt. 1992. Mitochondrial DNA variability of the gray wolf: Genetic consequences of population decline and habitat fragmentation. *Conservation Biology* 6: 559–569.

Yanes, M. & F. Suárez. 1996. Incidental nest predation and lark conservation in an Iberian semiarid shrub-steppe. *Conservation Biology* 10: 881–887.

11 A Case Study
Diet of Maned Wolf and Its Relationship to Seed Dispersal at a Cattle Ranch in Southeastern Brazil

Eliana Ferraz Santos, Nivar Gobbi, and Paulo Anselmo Nunes Felippe

CONTENTS

Introduction 153
Methods 154
Study Area 154
Experimental Design and Specific Procedures 154
Statistical Analysis 155
Results 155
Discussion 159
Acknowledgments 161
References 161

INTRODUCTION

The Genus *Chrysocyon* comprises a sole species, *Chrysocyon brachyurus* (Illiger, 1811). It is the largest canid of South America, and is well adapted to meadows and savanna habitats. It has long legs, a long head and muzzle, and large ears. Its geographic range includes grasslands of central Brazil, eastern Bolivia, Paraguay, north Argentina and Uruguay (Cabrera and Yepes, 1960; Langguth, 1975). The maned wolf ranges long distances searching for food, scent-marking with urine and scats, even outside the reproductive season (Dietz, 1984). As a typical omnivore, the maned wolf's diet is typical and diverse, including rodents, arthropods, reptiles, birds, fruits, and armadillos (Carvalho, 1976; Dietz, 1984; Jácomo, 1995). The diet of canids can vary seasonally throughout the year and vegetable matter is very important for some species (including the maned wolf, indicating that this species is a trophic generalist; see Dietz, 1984; Jácomo, 1995) at certain times (Nowak, 1991). Within this context, it would be interesting to analyze the maned wolf's diet and its seasonal variation in different places to compare the results and identify possible variation among different regions. These possible differences can be fundamental for the maintenance of this mammal species, which is endangered in Brazil (Coimbra Filho, 1968).

Wolf's fruit (*Solanum lycocarpum*, Solanaceae) is the main item of the maned wolf's diet, thus its nickname (Carvalho, 1976). It is widespread in Brazil, generally in sandy soils and sunny habitats (Leitão Filho, 1975). Solanaceae fruits are known for having glycoalkaloids. Silveira (1969) suggested that these substances could inhibit kidney infection by *Dioctophyma renale*, prevalent in maned wolves. In an experiment carried out by Lombardi and Motta-Junior (1993), the germination of seeds of wolf's fruit was more successful in the samples grown in disturbed cerrado areas than in those grown in more closed cerrado. According to these authors, the maned wolf was found to be able to disperse the seeds of *S. lycocarpum*.

Dispersal in plants is the transport of seeds away from the mother plant. The retrieval or output of diaspores (units that are dispersed, fruits and seeds) of the mother plant can be performed by animals (zoochory) (Morellato and Leitão-Filho, 1995). Studies in different animal species show that predation and seed dispersal by frugivorous animals are very important in plant–animal interactions, with implications for the evolutionary diversity of plant species (Hladik and Hladik, 1969; Mckey, 1975; Janzen, 1977, 1983; Cant, 1979; Lieberman et al., 1979; Mittermeier and van Roosmalen, 1981; Muskin and Fischgrund, 1981; Leighton and Leighton, 1982, 1983; Pianka, 1982; Janson, 1983; van Roosmalen, 1985; Terborgh, 1986). Although maned wolves more frequently consume ripe wolf's fruit (Kleiman, 1977), unripe wolf's fruit is also eaten, and its seeds appear intact and clean in the scats (Carvalho, 1976). So, since wolf's fruit is the main item of the maned wolf's diet and the seeds seem intact after being eliminated by the maned wolf, can this mammal species efficiently disperse wolf's fruit?

Here we aimed to assess the diet of the maned wolf on a ranch in southeast Brazil to compare the food items used by this species in this region with data from other studies and analyze the relationships between its diet and seasons. Furthermore, we evaluated the effect of the passage of the wolf's fruit seeds through the digestive system of the maned wolf on the success of germination of these seeds.

METHODS

Study Area

We conducted our study on a private dairy cattle ranch (São Luís Farm), Matias Barbosa County, Minas Gerais State, Zona da Mata Mineira region. It is one of the largest (566 ha) and oldest properties in the region. The topography is heterogeneous, from plain to mountainous. The climate is subtropical, rainy in summer (November to April), and dry in winter (May to October) (EMBRAPA, 1996). Vegetation is mainly pastureland, wet lowlands, and forest.

Experimental Design and Specific Procedures

Maned wolf (*C. brachyurus*) feeding habits were determined by monthly collection of feces ($N = 150$) in São Luís and neighboring farms (from January to December 1996). With this method we could gather a large number of samples, following monthly and seasonal variation, without interfering with the study population (Don Caster et al., 1990). During our monthly stay (ca. 5 days) on the farm, we searched all trails and dirt roads by foot, crossing all types of vegetation. To standardize the sampling effort, we explored the same routes every month, carefully inspecting trail intersections, rocks, and earth and termite mounds, which are the preferred sites for maned wolf scat depositions (Dietz, 1984). Scats were gathered in labeled plastic bags.

The scats were identified as from maned wolf by size/diameter, presence of its hair, presence of nearby tracks, deposition place, and characteristic odor. Scats were washed in tap water over two fine-mesh screens (0.5 and 1 mm) to recover scutes/scales, feathers, bones, hair, teeth, claws, and seeds, among other items, and dried. Food items were grouped in six categories: grasses, fruits, arthropods, reptiles, birds, and mammals. Wolf's fruit (*S. lycocarpum*) was analyzed separately, due to its importance in the maned wolf diet. Seed identification was helped by a seed reference collection from the Seed Bank of the Integral Technical Assistance Coordination of Campinas (CATI). The importance of each food item was determined by relative occurrence frequency or the percentage of the number of scats with the item by the sum of occurrences of all items (Deblase and Martin, 1981). Biomass was determined by the average weight of each item (*in natura*) found in scats. Relative biomass was calculated by biomass of the item divided by the sum of biomass from all items in scats (Motta-Junior et al., 1996; Juarez, 1997).

To analyze the impact of the passage of wolf's fruit seeds through the gut of the maned wolf on the success of seed germination, we conducted an experiment with two seed germination trials with wolf's fruit (one trial in March and one in September). For each trial, we had two treatments: 100 seeds (50 in the dry season and 50 in the rainy season) from more than 20 maned wolf scats and 100 seeds from more than 10 ripe wolf's fruits gathered during 1996. The latter treatment acted as a control group. Seeds were sown in two plastic trays in mineral substrate and inspected daily for 30 days, to determine germination (when a root appeared).

Statistical Analysis

We compared plant biomass with animal biomass of the food items detected with Goodman's test (1965; within multinomial). Frequencies of occurrence among wolf's fruit and other plants were compared by Goodman's test (1965). For seasonal analysis, we considered months with <100 mm of pluviometrical precipitation as dry. We used a G-test for independence with Yates correction to compare the presence and absence of food items in maned wolf scats between seasons. Species and morphospecies richness in the wolf diet from Matias Barbosa (present study), Água Limpa (Motta-Junior et al., 1996), and Jaborandi (Juarez, 1997) were compared by the rarefaction method of species richness estimation (Krebs, 2000). Nine species reported by Motta-Junior et al. (1996) were grouped with related taxa to correspond to our classification of food items in this analysis. We used Goodman's test (1964; among multinomial) to compare the numbers of germinated seeds between treatments in each trial. We fixed $\alpha = 0.05\%$.

RESULTS

From 3 to 19 scats were collected monthly, totaling 68 scats in the wet season and 82 in the dry season. From these, we identified 29 food items: 18 animal and 11 plant items (Table 11.1). In terms of biomass, animal items represented significantly more biomass (57%) than plant (43.1%) items (Table 11.1; Goodman's test, 1965; within multinomial; $p < .05$).

Wolf's fruit was the most frequent food item among plant items (Goodman's test, 1965; within multinomial; $p < .05$) (Table 11.1). Our personal observations revealed that wolf's fruit availability is markedly seasonal (more fruits ripen in summer than in winter). Other fruit species in the diet were bell peppers (*Capsicum annuum*), apple of Peru (*Nicandra physaloides*), Indian nightshade (*Solanum aculeatissimum*), potato tree (*Solanum erianthum*), winter cherry (*Solanum pseudocapsicum*) (all Solanaceae), coffee (*Coffea arabica*, Rubiaceae), and papaya (*Carica papaya*, Caricaceae). Except for *S. erianthum* (found in two scats) and *S. aculeatissimum* (found in eight scats), all fruit species appeared in one scat each. Guava (*Psidium guajava*), bell peppers, coffee, and papaya are relicts of old gardens.

From animal item occurrences, mammals were, significantly, the most frequent food items, followed by insect and bird food items (Figure 11.1; Goodman's test, 1965; within multinomial; $p < .05$). Fish and reptiles were the least frequent food items (Figure 11.1; Goodman's test, 1965; within multinomial; $p < .05$). With respect to biomass, mammals represented, significantly, the largest biomass of food items, followed by reptiles (Table 11.2; Goodman's test, 1965; within multinomial; $p < .05$). Birds and fish were, significantly, third and fourth in respect of biomass amount, respectively (Table 11.2; Goodman's test, 1965; within multinomial; $p < .05$). Insects represented the least biomass of food items (Table 11.2; Goodman's test, 1965; within multinomial; $p < .05$). The pit viper (*Bothrops jararaca*) appeared in three scats: one in May and two in July. At least two individuals were consumed. All bird remains were of Passeriformes. One egg of a *Crotophaga ani* (Cuculidae) also appeared in the scats.

TABLE 11.1
Contents of Maned Wolf Scats at a Dairy Cattle Ranch in Matias Barbosa County, Minas Gerais, Brazil

Food Item	Occurrence (no. of scats)	Frequency (%)	Biomass (g)	Biomass (%)
Grass, Graminae	68	17.1	—	—
Papaya (*Carica papaya*), Caricaceae	1	0.3	600	0.49
Guava (*Psidium guajava*), Myrtaceae	7	1.8	280	0.23
Joá-bravo (*Solanum aculeatissimum*), Solanaceae	8	2.0	80	0.07
Wolf's fruit (*S. lycocarpum*), Solanaceae	116[a]	29.2	52,200	42.29
Other fruits	14	3.5	—	—
Plant subtotal	**214**	**53.9**	**53,160[a]**	**43.07**
Beetles, Coleoptera, n.i.	16	4.0	32	0.03
Termites, Isoptera, n.i.	11	2.8	1.29	0.001
Cricket, Acrididae	4	1.0	4	0.003
Cricket, Gryllidae	4	1.0	12	0.01
Katydid, Tettigonidae	6	1.5	30	0.02
Other insects	7	1.8		
Tilapia (*Tilapia* sp.), Cichlidae	1	0.3	550	0.45
Pit viper (*Bothrops jararaca*), Viperidae	3	0.8	1,950	1.6
Glass lizard (*Ophiodes striatus*), Aguidae	1	0.3	110	0.09
Skink (*Mabouya* spp.), Scincidae	3	0.8	30	0.02
Birds n.i. (Small-bodied Passeriformes)	44	11.0	860	0.7
White-eared opossum (*Didelphis albiventris*), Didelphidae	2	0.5	1,380	1.12
Grass mouse (*Akodon* spp.), Muridae	9	2.3	193.5	0.16
Rice rat (*Oryzomys* spp.), Muridae	12	3.0	540	0.44
Murid mice, n.i.	33	8.3	990	0.80
Paca (*Cuniculus paca*), Dasyproctidae	2	0.5	8,000	6.50
Long-nosed armadillo (*Dasypus* spp.), Dasypodidae	22	5.5	41,800	33.90
Yellow armadillo (*Euphractus sexcinctus*), Dasypodidae	3	0.8	13,800	11.18
Animal subtotal	**183**	**46.2**	**70,282.8[b]**	**57.02**
Total (plant + animal items)	**397**	**100.0**	**123,442.8**	**100.00**

n.i., not identified.

[a] Significant differences among wolf's fruit and other plant items. Different letters indicate statistical differences between plant and animal biomass values (Goodman's test, 1965; within multinomial; $p < .05$).

The presence of wolf's fruit (*S. lycocarpum*) and insects in scats was positively correlated to rain, thus they were more prevalent in the rainy season (rs = 0.07, $p < .05$ and rs = 0.62, $p < .001$, respectively) (Table 11.3; G-test for independence with Yates correction; $p < .05$). Mammals were negatively correlated with rain (rs = −0.87, $p < .001$), thus they were more prevalent in the dry season (Table 11.3; G-test for independence with Yates correction; $p < .05$). Birds appeared constantly during the entire study period, not prevailing in any season (Table 11.3; G-test for independence with Yates correction; $p < .05$).

Species or morphospecies richness found in the maned wolf diet in Matias Barbosa (present study, 28 species in 150 scats) was lower than that found in Água Limpa (34 species in 104 scats; Motta-Junior et al., 1996) and Jaborandi (30 spp. in 70 scats; Juarez, 1997) (Figure 11.2).

In our experiment with wolf's fruit seeds, the seeds' germination was significantly higher in seeds that were recovered from maned wolf scats than those collected from ripe fruits, in both trials (Figure 11.3; Goodman's test, 1964; among multinomial; $p < .05$). We also observed that the

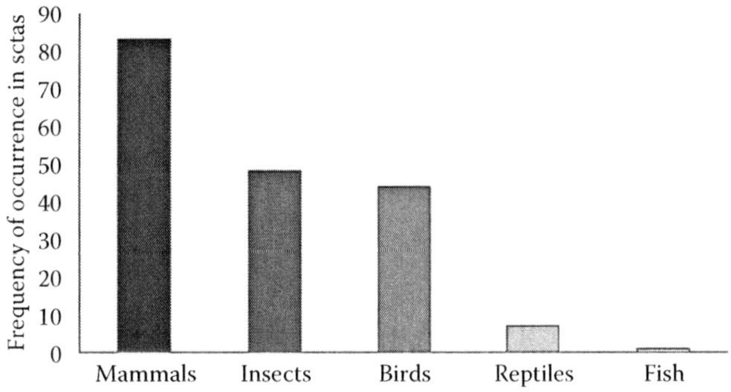

FIGURE 11.1 Frequency of occurrence of animal food items in maned wolf diet at a dairy cattle ranch in Matias Barbosa County, Minas Gerais, Brazil. Different letters indicate statistical differences among animal food items (Goodman's test, 1965; within multinomial; $p < .05$).

TABLE 11.2
Frequency of Biomass of Animal Food Items in Maned Wolf Diet at a Dairy Cattle Ranch in Matias Barbosa County, Minas Gerais, Brazil

Animal Food Items	Biomass (g)
Mammals	66,703.5[a]
Birds	860[c]
Reptiles	2,090[b]
Fish	550[d]
Insects	79.3[e]
Total	**70,282.8**

Different letters indicate statistical differences among animal food items (Goodman's test, 1965; within multinomial; $p < .05$).

TABLE 11.3
Seasonal Variation of Food Items in Maned Wolf Scats at a Dairy Cattle Ranch in Matias Barbosa County, Minas Gerais, Brazil

	No. of Scats (%)			
Item	**Dry Season ($n = 82$)**	**Rainy Season ($n = 68$)**	**G**	***p***
Wolf's fruit	52 (63)	64 (94)	20.29	.000*
Mammals	53 (65)	22 (32)	14.47	.000*
Birds	23 (28)	17 (25)	0.06	.814
Insects	9 (11)	36 (53)	30.32	.000*

[a] Indicates differences extremely significant ($p < .001$) between dry season and rainy season for each food item (G-test for independence with Yates correction; $p < .05$)

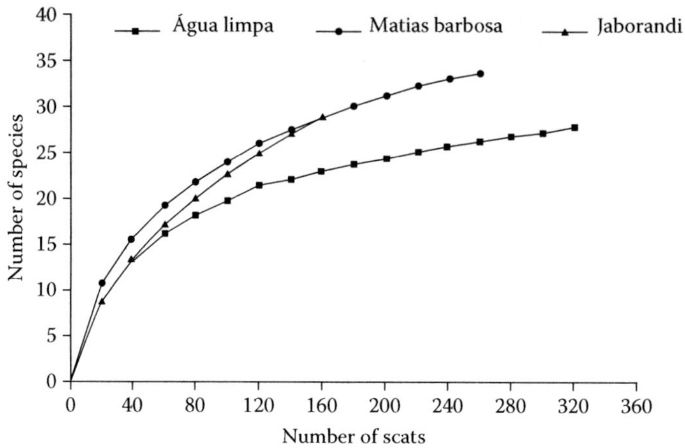

FIGURE 11.2 Rarefaction curves for number of species in scats of maned wolf for Matias Barbosa (present study), Água Limpa (Data from Motta-Junior, J.C., et al., *Journal of Zoology London*, 240, 277–284, 1996), and Jaborandi (Data from Juarez, K.M., Dieta, Uso de Habitat e Atividade deTrês Espécies de Canídeos Simpátricas do Cerrado. Dissert. Mestr.; UNB, Brasília, DF, 1997).

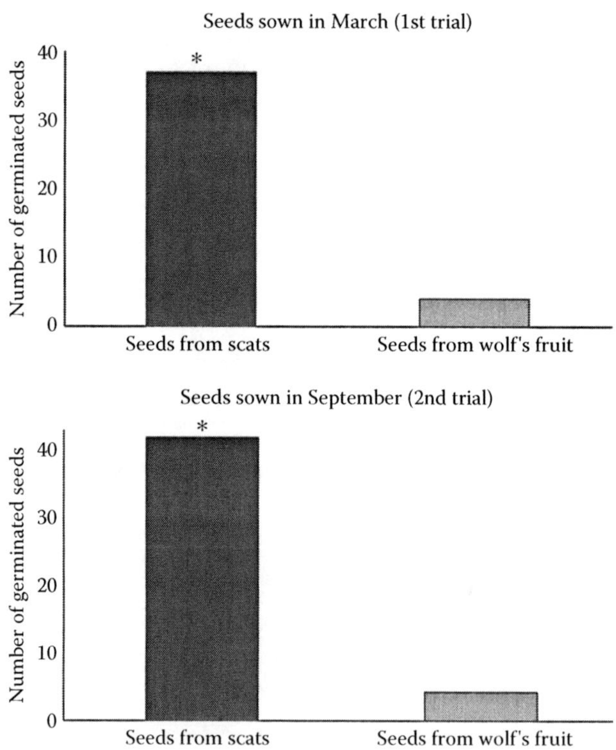

FIGURE 11.3 Germinated seeds in two different treatments: seeds recovered from maned wolf scats (100 seeds sown) and seeds collected from wolf's fruit (100 seeds sown). *Significant differences between treatments in each trial (Goodman's test, 1964; among multinomial; $p < .05$).

germination of seeds from maned wolf scats was faster than the germination of seeds collected from ripe wolf's fruit (personal observation).

DISCUSSION

In the maned wolf's diet, animal items represented more biomass than plant items (Table 11.1) and among animal items, mammals were the most frequent food items, followed by insect and bird food items (Figure 11.1). With respect to biomass, mammals represented the most biomass of food items, followed by reptiles (Table 11.2). These data indicate that animal food items are very important in the maned wolf's diet; further, mammals seem to be the main animal food item in the area studied here. Wolf's fruit (*S. lycocarpum*) was the most frequent food item among plant items (Table 11.1), constituting the basis for the diet in our study. The presence of wolf's fruit (*S. lycocarpum*) and insects in scats was more prevalent in the rainy season, while mammals were more prevalent in the dry season (Table 11.3), indicating that the maned wolf's diet varies throughout the year. The species or morphospecies richness of the maned wolf's diet in Matias Barbosa (present study) was lower than that found in other studies (see Figure 11.2), indicating a loss of species richness in the maned wolf's diet in Matias Barbosa. Moreover, data from our experiment with wolf's fruit seeds showed that the passage of wolf's fruit seeds through the maned wolf's gut considerably improves the seeds' germination (Figure 11.3).

Mammal food items were the most frequent (Figure 11.1) and represented the most biomass (Table 11.2). However, although birds and insects were the second most frequent (Figure 11.1), their biomass was lower than that of reptiles (Table 11.2). In other words, the most frequent item does not always represent the most biomass. Thus, biomass was useful to quantify the importance of items in the diet of the maned wolf, indicating that, in Matias Barbosa, animal food items were more important than plant food items in constituting the amount of food eaten (Table 11.1). Moreover, among animal food items, mammals and reptiles represented the most biomass, although mammals, birds, and insects were the most frequent.

Wolf's fruit was the basis of the maned wolf diet, and was the most frequent item detected among plant food items (Table 11.1). Wolf's fruit starts fruiting more intensively from August to October. These fruits are ripe in the rainy season (personal observation), when they were consumed more (Table 11.3). Wolf's fruit provides water and carbohydrates, and it has higher protein content than other cerrado fruits, which should explain its high consumption in the present study. In Matias Barbosa, as in other study sites, the maned wolf searches actively for wolf's fruit, consuming it year round (Carvalho, 1976; Dietz, 1984; Azevedo and Gastal, 1996; Motta-Junior et al., 1996; Aragona and Setz, 2001). Other wild fruits were more abundant in the ranch studied here in the rainy season, but this did not mean higher consumption by the maned wolf.

The fact that the presence of wolf's fruit (*S. lycocarpum*) and insects in scats was significantly more frequent in the rainy season and mammals were significantly more frequent in the dry season (Table 11.3) shows that the maned wolf's diet varies throughout the year. These data corroborate with those in the literature. The diet of canids, including the maned wolf, can vary seasonally throughout the year and vegetable matter is very important for some species (indicating that the maned wolf is a trophic generalist species; see Dietz, 1984; Jácomo, 1995) at certain times (Nowak, 1991). This variation of diet as a function of seasons should be related to variation of food item availability. It is likely that wolf's fruit and insects are more abundant in the rainy season and that mammals are more abundant in the dry season.

In Matias Barbosa, as in other natural and artificial landscapes, the maned wolf (*C. brachyurus*) had a generalist diet comprising wolf's fruit, small rodents, armadillos, birds, reptiles, and some insects, which are highly available items in many places. Among mammals, rodents were also the most frequent item in Emas (23%; Jácomo, 1995), Água Limpa (22%; Motta-Junior et al., 1996), Jaborandi (32%; Juarez, 1997), and Ibitipoca (38%; Aragona and Setz, 2001), all higher than in Matias Barbosa (14%; present study, see Table 11.4). By biomass, armadillos were more

TABLE 11.4
Major Food Items in Scats of Maned Wolf by Frequency at a Dairy Cattle Ranch in Matias Barbosa (Present Study) Compared to Cerrado and Upland Meadow Studies

	Matias Barbosa (n = 150)	Água Limpa[a] (n = 105)	Emas[b] (n = 645)	Jataí[c] (n = 191)	Canastra[d] (n = 740)	Jaborandi[e] (n = 70)	Ibitipoca[f] (n = 141)
Arthropods	12.1	2.0	1.8	4.9	5.9	3.7	13.8
Lizards	1.1	2.6 (4)	1.1	1.1 (2)	0	0.5 (2)	0.1
Snakes	0.8	0	2.7	1.7 (1)	0.3	1.0 (1)	0.5
Birds	10.9 (1)	13.8 (6)	14.0	6.2 (2)	11.5	8.3 (4)	3.5
Rabbit	0	0	0	2.0	1.1	0	0
Rodents	14	22.0 (7)	23.5	18.3	25.5*	32 (22)	38
Armadillos	6.3 (45)	9.2 (30)	3.9	2.8 (27)	3.1	1.5 (13)	13.3
Wolf's fruit	29.3	25.7	19.6	21.5	32.6	31.9	26.2

Source: (a) Motta-Junior, J.C., et al., *Journal of Zoology London*, 240, 277–284, 1996; (b) Jácomo A.T.A., Nicho Alimentar do Lobo-guará (*Chrysocyon brachyurus*, Illiger, 1811) no Parque nacional das Emas, Goiás. Monografia; Univ. Católica de Goiás, Goiania, 1995; (c) Motta-Junior, J.C. 2000. *Estação Ecológica de Jataí*, São Carlos, RIMA Editora, 2000; (d) Dietz, J.M., *Smithsonion Contributions to Zoology*, 1984; (e) Juarez, K.M., Dissert. Mestr.; UNB, Brasília, DF, 1997; (f) Aragona, M. & Setz, E.Z.F., *Journal of Zoology* 254, 131–136, 2001.

Note: Values in parentheses are biomass (%).

*All small mammals.

important in the maned wolf diet in Matias Barbosa (45%) and elsewhere (30%; Motta-Junior et al., 1996), although in Jaborandi rodents continue to be the main item (13% vs. 22%; Juarez, 1997; see Table 11.4).

Jararaca (*Bothrops jararaca*) is a new item in the maned wolf's diet, although Neuwied's lancehead (*Bothrops neuwiedii*) was already reported by Emas (Jácomo, 1995). Although sparse in the maned wolf's diet, reptiles account for a higher biomass than in Jaborandi (lizards 2% and snakes 1%; Juarez, 1997) and lower than in Água Limpa (4% and 0%, respectively; Motta-Junior et al., 1996; see Table 11.4). Bird frequency and biomass (10.9% and 0.7%) were also intermediate (8.3% and 3.5%, Juarez, 1997; 13.8% and 6.2%, Motta-Junior et al., 1996; see Table 11.4). Avian eggshells were also observed in Canastra (Dietz, 1984) and Ibitipoca (Aragona and Setz, 2001), but none were from domestic stock. The identification of *Dasypus* species here, as elsewhere (Jácomo, 1995; Motta-Junior et al., 1996), shows that the maned wolf hunts either smaller more open field dwellers or larger forest species (see Emmons and Feer, 1997). Although rabbit latrines were common (personal observation), no rabbits appeared in the maned wolf's diet in Matias Barbosa (present study, but see Dietz, 1984 and Motta-Junior, 2000 in Table 11.4). In Emas, Água Limpa, Jaborandi, and Ibitipoca, rabbits were not consumed either (Jácomo, 1995; Motta-Junior et al., 1996; Juarez, 1997; Aragona and Setz, 2001; see Table 11.4). In Canastra, the maned wolf attacked poultry with higher frequency between September and November, when litters are born in the wild (Dietz, 1984). We observed tracks of young maned wolves in March and June in Matias Barbosa, but there was no sign of wolf predation on domestic stock. Wolf tracks were never observed approaching domestic animals in the pasture either. However, ranch people referred to a foraging wolf near the cattle, with no sign of disturbance by the cattle.

Although large, the solitary maned wolf captures prey smaller than other canids that cooperate in hunts (Moehlman, 1989). At Matias Barbosa, the largest prey was *Cuniculus paca* (6–13 kg), which is smaller than that found at other sites (collared anteater, Dietz, 1984; juvenile Pampas deer, Jácomo, 1995). Giant anteater (*Myrmecophaga tridactyla*) remains found in scats at Canastra were considered to have been scavenged from roadkill (Dietz, 1984). Mammals heavier than 5 kg appear with low frequency (0.7% and 0.5%, Motta-Junior et al., 1996; Juarez, 1997, respectively).

In our study, species/morphospecies richness that was found in the maned wolf diet was lower than that found in Água Limpa (Motta-Junior et al., 1996) and Jaborandi (Juarez, 1997) (see Figure 11.2). This fact indicates a loss of species richness in the maned wolf's diet in Matias Barbosa when compared to that found in other studies. Although our frugivory results are at the lower end of the diversity scale (9–33 spp.), as expected for a heavily farmed region, no previous study registered as many species of Solanaceae as ours. Thus, the effect of a loss of dietary richness and increase of Solanaceae items of food on the survival of maned wolves needs to be further investigated.

Solanum, like many other Neotropical genera, has predominantly zoocorous fruits (Albuquerque, 2001). So, the fact that significantly more wolf's fruit seeds germinated after passing through the maned wolf gut (see Figure 11.3) indicates that the maned wolf is a very important species for dispersal of wolf's fruit seeds. Both higher and accelerated seed germination qualify the maned wolf as an important wolf's fruit seed disperser in Matias Barbosa, because they carry large quantities of seeds (ca. 300 seeds, personal observation) and take them far, the whole year round. Our results differ from published experiments conducted in open cerrado, where 67% of seeds germinated in each treatment (Lombardi and Motta-Junior, 1993). In Canastra, Água Limpa, and Jaborandi (Dietz, 1984; Motta-Junior et al., 1996; Juarez, 1997; respectively), highly used substrates for scat deposition differ in quality: termitaria are nutrient rich but rocks present no soil, though this bush also grows in poor soils. Human landscape modifications in Matias Barbosa have made new substrates available (kaolin mines, gravel mounds, road banks). This fact could have favored the germination and development of wolf's fruit (*S. lycocarpum*), increasing fruit availability in the region.

In conclusion, the highly modified landscape of Matias Barbosa farm is responsible for some differences in the wolf's diet in our study, but the most consumed items (wolf's fruit, armadillos, rodents, birds) are the same as in other studies (Dietz, 1984; Aragona and Setz, 2001; Motta-Junior et al., 1996; Juarez, 1997). Availability of these foods in our study site is probably important for the survival and reproduction of this species. Moreover, as the germination tests on seeds from wolf's fruit show that passing through the digestive system of the maned wolf increases viability, this mammal species seems to be very important for the dispersal of wolf's fruit seeds in Matias Barbosa. Future studies should analyze the impact of the loss of dietary richness, compared with other studies, and the increase of Solanaceae items of food on the survival of the maned wolf.

ACKNOWLEDGMENTS

We acknowledge the Brazilian National Science Foundation (CNPq) for a fellowship to E. F. Santos. We are grateful to Mr. Fernando C. Veiga Neto for authorization and encouragement to work on his farm and the use of farm facilities. We thank M. DeVivo and P. Crawshaw for discussions, J. C. Motta-Junior and W. R. Silva for providing manuscript reviews, and Giselda Person for review.

REFERENCES

Albuquerque, L.B. 2001. Polinização e dispersão de sementes em solanáceas neotropicais. PhD Thesis, Campinas State University, São Paulo.

Aragona, M. & Setz, E.Z.F. 2001. Diet of the maned wolf, *Chrysocyon brachyurus* (Mammalia: Canidae), during wet and dry seasons at Ibitipoca State Park, Brazil. *Journal of Zoology* 254: 131–136.

Azevedo, F.C.C. & Gastal, M.L.A. 1996. Análise dos hábitos alimentares de lobo-guará (*Chrysocyon brachyurus*), na APA do Gama/Cabeça do Veado—DF. 3 Congresso de ecologia do Brasil. Departamento de Ecologia—Universidade de Brasília, p. 211.

Cabrera, A. & Yepes J. 1960. *Mamiferos Sud Americanos*, 2nd edn, v2. Buenos Aires, Ediar Editores.

Cant, J. 1979. Dispersal of *Stemmadenia donnell-smithii* by birds and monkeys. *Biotropica* 13: 70–76.

Carvalho, C.T. 1976. Aspectos faunísticos do Cerrado—O lobo-guará (Mammalia—Canidae). Boletim Técnico Inst. Florestal, São Paulo, 21: 1–18.

Coimbra Filho, A.F. 1968. Notes on the reproduction and diet of Azara's fox *Cerdocyon thous* and the hoary fox *Dusicyon vetulus* at Rio de Janeiro Zoo. *International Zoo Yearbook* 6: 168–169.

Deblase, A.F. & Martin, R.E. 1981. *A Manual of Mammalogy with Keys to Families of the World,* 2nd edn. Dubuque, Wm. C. Brown Company.

Dietz, J.M. 1984. Ecology and social organization of the maned wolf (*Chrysocyon brachyurus*). *Smithsonion Contributions to Zoology,* n. 392.

EMBRAPA. 1996. Levantamento semidetalhado de solos da área do Centro Nacional de Pesquisa de Gado de Leite, Coronel Pacheco, MG. Boletim Técnico. n. 76, Rio de Janeiro.

Emmons, L.H. & Feer, F. 1997. *Neotropical Rainforest Mammals: A Field Guide*, 2nd edn. University of Chicago Press.

Goodman, L.A. 1964. Simultaneous confidence intervals for contrasts among multinomial populations. *Annals of Mathematical Statistics, Hayward* 35: 716–725.

Goodman, L.A. 1965. On simultaneous confidence intervals for multinomial proportions. *Techometrics* 7(2): 247–254.

Hladik, A. & Hladik, C.M. 1969. Rapports trophiques entre vegetation at primates dans la foret de Barro Colorado (Panamá). *Terre et Vie* 116: 25–117.

Jácomo, A.T.A. 1995. Nicho alimentar do lobo-guará (*Chrysocyon brachyurus*, Illiger, 1811) no Parque Nacional das Emas, Goiás. Monografia; Univ. Católica de Goiás, Goiania, p. 29.

Janson, C.H. 1983. Adaptation of fruit morphology to dispersal agents in a neotropical forest. *Science* 219: 187–189.

Janzen, D.H. 1977. *Ecologia vegetal nos trópicos*. São Paulo, E. P. U., EDUSP.

Janzen, D.H. 1983. Dispersal of seeds by vertebrate guts. In: Futuyma, D.J. & Slatkin, M. (eds), *Coevolution*. Sunderland, MA, Sinauer, pp. 233–261.

Juarez, K.M. 1997. Dieta, uso de habitat e atividade de três espécies de canídeos simpátricas do Cerrado. Dissert. Mestr.; UNB, Brasília, DF, p. 58.

Kleiman, D.G. 1977. Monogamy in mammals. *The Quarterly Review of Biology* 52: 39–69.

Krebs, C.J. 2000. *Ecological Methodology*, 2nd edn. Menlo Park, CA, Benjamin/Cummings.

Langguth, A. 1975. South American ecology and evolution in the canids. In: Fox, M.W. (ed.), *The Wild Canids*. New York, Van Nostrand Reinhold, pp. 192–206.

Leighton, M. & Leighton, D.R. 1982. The relationships of size of feeding aggregate to size of food patch: Howler monkeys (*Alouatta palliata*) feeding in *Trichilia cipo* fruit trees on Barro Colorado Island. *Biotropica* 14: 81–90.

Leighton, M. & Leighton, D.R. 1983. Vertebrate responses to fruiting seasonality within a Bornean rain forest. In: Sutton, S.L., Whitmore, T.C. & Chadwick, A.C. (eds), *Tropical Rain Forest: Ecology and Management*. Oxford, Blackwell Scientific, pp. 181–196.

Leitão Filho, H.F. 1975. *Plantas invasoras de culturas no Estado de São Paulo*. SP, AGIPLAN/Ministério da Agricultura, v II, 1975.

Lieberman, D., Hall J.B., Swaine M.D. & Lieberman M. 1979. Seed dispersal by baboons in Shai Hills, Ghana. *Ecology* 60: 65–75.

Lombardi, J.A. & Motta-Junior, J.C. 1993. Seed dispersal of *Solanum lycocarpum* St. Hil. (Solanaceae) by the maned wolf, *Chrysocyon brachyurus* Illiger (Mammalia, Canidae). *Ciência e Cultura* 45(2): 126–127.

McKey, D. 1975. The ecology of coevolved seed dispersal systems. In: Gilbert, L.E. & Raven, P.H. (eds), *Coevolution of Animals and Plants*. Austin, University of Texas Press, pp. 159–191.

Mittermeier, R.A. & van Roosmalen, M.G.M. 1981. Preliminary observations on habitat utilization and diet in eight Surinam monkeys. *Folia Primatologica* 36: 1–39.

Moehlman, P.D. 1989. Intraspecific variation in canids social systems. In: Gittleman, J.L. (ed.), *Carnivore Behavior, Ecology, and Evolution*. New York, Cornell University, pp. 143–163.

Morellato, L.P.C. & Leitão-Filho, H.F. 1995. *Ecologia e preservação de uma floresta tropical urbana*. Campinas, Editora da UNICAMP, p. 136.

Motta-Junior, J.C. 2000. Variação temporal e seleção de presas na dieta do lobo-guará, *Chrysocyon brachyurus* (Mammalia: Canidae), na Estação Ecológica de Jataí, Luiz Antônio, SP. In: *Estação Ecológica de Jataí* (JE Santos and JSR Pires, eds), Vol. I. São Carlos, RIMA Editora, pp. 331–346.

Motta-Junior, J.C., Talamoni, S.A., Lombardi, J.A. & Simokomaki, K. 1996. Diet of maned wolf, *Chrysocyon brachyurus*, in central Brazil. *Journal of Zoology, London* 240: 277–284.

Muskin, A. & Fischgrund, A.J. 1981. Seed dispersal of *Stemmadenia* (Apocynaceae) and sexually dimorphic feeding strategies by *Ateles* in Tikal, Guatemala. *Biotropica* 13: 78–80.

Nowak, R.M. 1991. *Walker's Mammals of the World*, 5th edn, Vol. II. Hopkins University, Baltimore, p. 1629.
Pianka, E. 1982. *Ecologia evolutiva*. Editora Omega, Barcelona, pp. 137–288.
van Roosmalen, M.G.M. 1985. Habitat preferences, diet, feeding strategy and social organization of the black spider monkey (*Ateles paniscus paniscus*) (LINNAEUS 1758) in Surinam. *Acta Amazonica* 15(3/4): 1–238.
Silveira, E.K.P. 1969. O lobo-guará (*Chrysocyon brachyurus*) possível ação inibidora de certas solanáceas sobre o nematóide renal. *Vellozia* 7: 58–60.
Terborgh, J. 1986. Keystone plant resources in the tropical forest. In: Soulé, M.E. (ed.), *Conservation Biology 2*. Sinauer, Sunderland, MA, pp. 330–344.

12 Agricultural Expansion and the Future of the Maned Wolf

Carly Vynne

CONTENTS

Materials and Methods 167
Study Site 167
Sampling 167
Distribution across the Landscape 167
Hormone Extraction and Processing 168
Statistical Analysis 168
Results 170
Distribution across the Landscape Matrix 170
Glucocorticoids 171
Thyroid Hormone 171
Progesterone 171
Discussion 173
Acknowledgments 174
References 175

The maned wolf, which is practically endemic to the cerrado, persists today in what is a largely transformed and fragmented landscape. More than half the cerrado's 2 million km^2 has been converted into pasture, cropland, and other uses in the past 40 years and both the total amount and annual rate of clearing in the cerrado are greater than those in the Brazilian Amazon (Klink and Machado 2005). Less than 4% of the maned wolves' global range overlaps with an existing or proposed Categories I–III nature reserve (calculated in a geographic information system (GIS) with data from the International Union for the Conservation of Nature [IUCN 2009] and IUCN & the World Conservation Monitoring Centre [UNEP-WCMC 2010]). Incentives for Brazil to expand its area of land under cultivation for bioethanol production are placing additional pressures on the region, which has traditionally been a major producer of soybeans. Determining the occurrence of the maned wolf outside the protected areas and their likelihood of persisting in an agricultural complex is of utmost priority in planning for the maned wolves' long-term conservation (Rodden et al. 2004).

While the landscape change has been extensive, Brazilian legislation requires that landowners in the cerrado maintain 20% of their land in its natural state as "legal reserve" (Klink and Machado 2005), and together with requirements for riparian protection the amount of private land that is federally mandated for protection is approximately 30%. The habitat fragments and corridors that remain on private lands may be responsible for the continued presence of large and wide-ranging mammals outside the protected areas. Alternatively, and because clearing has been relatively recent, the region may be experiencing an extinction debt (Brooks et al. 1999; Carroll et al. 2004). World food demand and global energy use are placing increased pressure on land use within Brazil, and the country has recently sought to weaken measures protecting habitat on private lands (The Forest

Code, Brazil 1965). The ability of maned wolves to tolerate the expansion of croplands throughout the range is thus likely to be linked to their future outlook.

We have previously shown that maned wolves appear attracted to croplands, and that while open natural grasslands were a preferred habitat type for maned wolves, when the wolves were near (<5 km) croplands they used them extensively (Vynne et al. 2011a). Maned wolves also strongly avoided closed-canopy habitat (Vynne et al. 2011a). The avoidance of closed canopy is noteworthy because (a) the current practice of leaving predominately forests and woodlands in conservation on private lands is likely to benefit maned wolves less than the conservation of grasslands would and (b) cropland expansion in the cerrado typically removes grasslands and leaves taller vegetation intact. Although maned wolves appear to select proximity to croplands when available (Vynne et al. 2011a), we suspected that this use of croplands may have implications for maned wolf physiological health.

Physiological markers may provide a window into the cost of making a living in areas where there is anthropomorphic disturbance or land use change. Habitat disturbance may influence the physiological status of wildlife and provoke stress, reduce reproductive output, and affect nutritional health. To understand how landscape alterations, particularly to croplands, influence physiological health in maned wolves, we measured fecal hormone metabolites in scat samples from free-ranging maned wolves in and around a federal nature reserve in Brazil. Specifically, we measured fecal glucocorticoids, thyroid hormone, and progestins. Glucocorticoids are known to be important chemical messengers in orchestrating numerous physiological functions but at chronically elevated levels these can be detrimental to individuals, resulting in increased susceptibility to disease, reproductive suppression, and even death (Cabezas et al. 2007; Pride 2005; Wingfield and Ramenofsky 1999). Thyroid hormones have profound influences on metabolism, heart rate, blood pressure, nutritional physiology, brain development, and body temperature regulation (Oppenheimer 1999) and are particularly responsive to nutritional deficits; they lower metabolism and allow the body to conserve energy during a nutritional emergency (Eales 1988). Fecal progestin metabolite concentrations have been shown to track sexual receptivity, pregnancy, and probability of pup survival in captive maned wolves (Songsasen et al. 2006; Velloso et al. 1998; Wasser et al. 1995).

We expected to be able to measure the influence of environmental disturbances on maned wolves through feces because radiolabeled infusion studies have shown that maned wolves are predominantly fecal excretors of steroid hormones (Brown et al. 1994, 1997; Velloso et al. 1998; Wasser et al. 1995). Furthermore, adrenocorticotropic hormone (ACTH) challenge studies of fecal glucocorticoids produced biologically meaningful results in related taxa (gray wolves, Sands and Creel 2004; spotted hyena, Goymann et al. 1999). To understand the influence of landscape use on maned wolf physiology, we correlated hormone levels with environmental and landscape variables including distance to the park and other available natural habitat patches, amount of natural vegetation, distance to roads, amount of cropland, and crop height.

Maned wolves are highly territorial and maintain high site fidelity to capture location, even when found on farmlands (May-Júnior et al. 2009). From a previous report in our study area and others (Vynne et al. 2011a,b; Dietz 1984; Jácomo et al. 2009; Sollmann et al. 2010; Trolle et al. 2006), we expected that the landscape matrix outside the park provided an important habitat for maned wolves. We predicted that the maned wolf scat sample recovery per unit effort would be similar when surveying outside the park to when surveying inside the park, but that it would drop in areas far from the park or where there was little remnant vegetation. Also, we expected to find that the matrix supported individuals with established home ranges and perhaps breeding pairs, and that this would be supported by finding both male and female maned wolves throughout the landscape matrix during all surveyed months of the year.

We predicted, however, that this permeable matrix would influence maned wolf physiology. Specifically, we expected that fecal glucocorticoid levels would be higher outside the park, near roads, and during harvest activity since exposure to human disturbance has been shown to elevate glucocorticoid levels in this system (Pereira et al. 2006) and in related species (Creel et al. 2002; Baker et al. 2013). Maned wolves are attracted to croplands, presumably for foraging on rodents

(Vynne 2010, Vynne et al. 2011a), but croplands are subject to pulses in resource availability, and so we also expected that thyroid hormone levels would be either lower or more variable for individuals outside the park. We also expected that wolves living in areas containing <30% of remaining natural habitat would have higher circulating levels of physiological stress (measured as higher cortisol metabolites in samples) and that females would be more likely to experience reproductive suppression (lower progesterone metabolites in samples). We examined these hypotheses against the alternative that matrix composition factors do not have long-term downstream effects; rather, maned wolves habituate to these conditions and maintain relatively low glucocorticoids and high levels of thyroid and progestins.

MATERIALS AND METHODS

Study Site

The 1320 km^2 Emas National Park is a federal reserve located in the southwest of Goiás State, Brazil (S 18°, W 52°). The park protects typical vegetation of the cerrado biome, a biologically rich grassland–woodland mosaic that covers 21% of Brazil (Klink and Machado 2005). Public visitation is restricted to a few locations regularly accompanied by park guides (~15% of the park), hunting is not allowed, and access to the remaining areas is only authorized for guides and researchers. This park is surrounded by croplands (44% of area outside the park), cattle pasture (25%), and habitat remnants, principally closed-canopy woodlands and riparian forests (31%; Vynne et al. 2011b). The main crops in the region include soy, corn, cotton, and sugarcane, and there is a heavy flow of tractors, cars, and trucks on the roads bordering both the western and southern edges of the park during harvest months. The maned wolf density estimates of ~5 wolves/100 km^2 inside Emas National Park put the population at 60–70 individuals, which is alone unlikely to be sustainable over the long term (Silveira et al. 2010).

Sampling

Specially trained detection dogs were employed to locate scats from maned wolves between August 2004 and April 2008. Dogs were trained according to the methods described in Wasser et al. (2004). Sampling sites included 65 5 × 5 km search quadrats, which were distributed throughout the park and the surrounding areas and visited by different teams throughout each sampling season. The quadrats were predetermined to distribute surveys across the study area, ensuring that sampling would maximize the different vegetation types and land uses represented without biasing toward any particular microsites. Sampling was evenly split between park and nonpark sites. The actual survey path for each of 415 visits was recorded with handheld GPS units set to mark a waypoint every 30 m, and the total distance walked was 3170 km. A detailed description of the survey design and the results of detection dog surveys are available in Vynne et al. (2011b).

When a sample was located, the GPS position and data on the habitat and sample condition were recorded. We noted factors that could affect sample DNA and hormone quality and did not collect samples that were both odorless (to the human observer) and formless (e.g., consisted only of a scattering of undigested animal material or seeds; Vynne et al. 2011c). Otherwise, a plum-sized portion of the sample was preserved for fecal DNA extraction and the remainder of the sample was then stored in a plastic bag (Millspaugh and Washburn 2004; Wasser et al. 1996) and retained for subsequent diet and hormone analysis. Samples were kept on ice packs in cooler bags and then frozen upon returning from the field until shipment (on dry ice) to the United States.

Distribution across the Landscape

Only samples that amplified for maned wolf mitochondrial DNA were included in hormone data analysis (Vynne 2010). Analysis of the distribution of samples across the landscape was done by classifying distribution of samples using a GIS and then standardizing the sampling catch effort per kilometer walked. We classified samples as either being inside or outside the park boundary, as well as being within 15 km of the park boundary (<15 km from park) or further than 15 km from the park (>15 km from park), since 15 km is the average straight line daily movement distance of maned wolves in our study area (calculated from unpublished data collected by the Jaguar Conservation Fund). To determine the proportion of area searched in <30% natural habitat, we generated 20,000 random points in the GIS and restricted these to our presumed search area, a 30 m buffer on each side of our track logs. We used the neighborhood function in the GIS to sample the amount of remnant vegetation within 3 km^2 of all samples and random points. We then calculated the proportion of area where maned wolves were found relative to survey effort for areas containing <30% or >30% natural vegetation types. To determine if gender influences maned wolves' use of roads, we used chi-square contingency table analysis to compare observed versus expected values (Zar 1999).

Hormone Extraction and Processing

We prepared samples for hormone assays by freeze-drying for 48 hours or until all moisture was removed from the sample. We sifted dried samples through steel-mesh colanders to remove nonfecal materials such as seeds, fruit pulp, bones, and hair. We recorded pre- and postfreeze dry weights to determine if the percentage of moisture in a sample influenced hormone levels (Millspaugh and Washburn 2004), and samples were expressed as hormone concentrations per gram dry mass since desiccation had previously been shown to control for the dietary effects on hormone excretion rates (Wasser et al. 1993). For steroid extractions, 2.0 mL of 90% ethanol was added to 0.2 g freeze-dried and thoroughly homogenized fecal powder (unpublished data), and for thyroid extractions 15 ml of 70% ethanol was added to 0.1 g of freeze-dried and thoroughly homogenized fecal powder (Wasser et al. 2010). The fecal powder–ethanol mixture was then vortexed on a multitube pulsing vortexer for 30 min at 1 pulse/sec (Glas-Col, Terre Haute, IN) and then centrifuged for 20 min at 2200 rpm. We then pipetted off the supernatant and stored it in an airtight tube at −20°C (Wasser et al. 2010).

Before assaying, fecal extracts were diluted in phosphate-buffered saline (1:120 for glucocorticoids, 1:20 for thyroid, and 1:240 for progesterone). Fecal glucocorticoid and thyroid hormones were assayed with commercial [^{125}I] assay kits (MP Biomedicals, Costa Mesa, CA). The fecal glucocorticoid assay was previously used by Cummings et al. (2007) to study the relationship between environmental enrichment and fecal glucocorticoids in captive maned wolves and we used high-performance liquid chromatography (HPLC) analysis to validate the fecal cortisol metabolite assay in maned wolves by comparing the excretion profile to that described for wild baboons (Wasser et al. 2000) using the same extraction method, antibody, and HPLC gradient.

The thyroid hormone assay was validated for recovery by our laboratory (Wasser et al. 2010). Fecal progesterone was assayed with an in-house 3 H radioimmunoassay using antibody CL# 425 (C. Munro, University of California at Davis, Davis), which was previously validated for captive maned wolves (Wasser et al. 1995; Velloso et al. 1998). All samples were assayed in duplicate, with NSBs and blanks in quadruplicate, a full standard curve in duplicate, and both manufacturer and study low and high controls in duplicate. Any samples falling outside the 15%–85% bound or >10% CV between duplicates were reassayed. All assays showed excellent parallelism and accuracy for maned wolf feces, and intra- and interassay variation was <12%.

Statistical Analysis

To test the hypothesis that location relative to the park, landscape composition, and disturbance variables (harvest activity, proximity to roads) influence stress, nutritional physiology, and reproductive physiology, we used general linear models (GLMs) to predict glucocorticoid, thyroid, and progestin levels, respectively. We expected that higher glucocorticoid levels would be associated with the dry season, breeding, harvest activity, lower crop height, and areas of greater disturbance (outside park, further from natural habitat, more cropland, less natural habitat, closer to roads). We expected that thyroid hormone would be highest (indicating better nutritional status) during the wet season and when crops were highest (when crops are fruiting and rodents are presumed to be most abundant). We expected both thyroid and progesterone hormone levels to be highest in areas associated with less disturbance (inside park, nearer to natural habitat, and in areas with less cropland, more natural habitat, and further from roads). Because we also suspected that the croplands were likely to experience seasonal pulses in rodent availability, we used Bartlett's test for equal variances (Zar 1999) to test our hypothesis that samples found outside the park would have more variable levels in thyroid hormone than samples found inside the park.

Collected scat samples had been in the field for varying amounts of time, which can affect sample quality (Millspaugh and Washburn 2004; Wasser et al. 2004) and thus needs to be accounted for in statistical models (Vynne et al. 2011c). Because we were collecting samples of unknown age, we used sample condition (amount of odor, moisture, presence of mold, invertebrates), as well as whether or not the sample was of adequate quality to amplify DNA, as proxies for age at the time of collection. We applied the same assessments as described by Wasser et al. (2004), as these were shown to be more accurate for determining sample quality than attempting to assign age categories to field-collected samples. We included sample condition variables and water weight (amount lost in drying), as well as collection year, in all candidate statistical models.

In addition to the above, we included each of the following variables in our fully parameterized candidate model for each hormone: amount of fruit in sample (FRUIT), whether or not the sample was found inside the park (PARK), distance to natural habitat patch (PATCHDIST), amount of cropland within 3 km^2 of sample (AMTCROP) or if <30% of area within 3 km^2 of sample is natural (NAT < 30), height of crops for samples found in cropland (HEIGHT, measured as above shoulders, below shoulders, below waist, or below knees), distance to pasture (PASTDIST), and distance to busy roads (ROADS) (Table 12.1).

SEASON (wet/dry), months of breeding (BREED), and months of harvest activity (January to February, June to July; HARVEST) were tested separately in candidate models for glucocorticoids. We used month, rather than season, as a covariate in the model for thyroid because thyroid hormone was expected to be more responsive to monthly pulses in food availability and to intraseasonal fluctuations in rain, crop fruiting phase, and prey availability. Because baseline progestin values in maned wolves were previously shown only to be significantly higher during breeding (March to May) and gestation (April to August) (Velloso et al. 1998; Wasser et al. 1995), we limited our analyses of landscape use response to these months only. AMTCROP was more influential in explaining hormone levels than NAT < 30 and so this variable was used in all final models.

Mean levels of glucocorticoids and thyroid hormone did not vary with gender ($F_{1,300} = 0.079$, $p = .780$ and $F_{1,202} = 2.5090$, $p = .115$, respectively), allowing samples to be combined by gender in subsequent analyses. Samples that did not amplify nDNA for gender were removed from thyroid hormone analyses because the mean difference in samples that failed to amplify nDNA for gender was significantly lower than that in those that amplified ($F_{1,382} = 19.2292$, $p < .001$), and thyroid hormone degrades more quickly than steroid hormones (SW, unpublished data). Progesterone analyses were restricted to samples obtained from females.

All hormone data were $\log_{10}$ transformed before statistical analysis to normalize the data. Other data were visually inspected to ensure linear fit was appropriate. We conducted model selection by starting with a fully parameterized model and using Akaike information criterion (AICc) to select the most parsimonious model. Model parameters were estimated through maximum likelihood estimation; nonsignificant factors that reduced the AICc for the model when removed (done iteratively) were eliminated from the final models. Likelihood ratio chi-square values of influential variables in models based on subsets of the data are reported directly in the text. Statistical analyses were conducted using JMP 8.0.2 (SAS Institute Inc., Cary, NC).

TABLE 12.1
Covariates Included in Generalized Linear Models Explaining Variation of Hormone Levels of Maned Wolf Scats and Their Significance within Whole Model Results, by Hormone

Covariate	Description	Glucocorticoid	Thyroid	Progesterone
Year	Survey year	NS	NS	NS
Water weight	% change in freeze dry	*NS*	*NS*	NS
Odor	Amount of odor in sample	S ↑ (*p* = .018)	S ↑ (*p* = .006)	S ↑ (*p* = .037)
Moisture	Sample was fresh, dry, or rained on at time of collection	NS	NS	NS
nDNA	Successful amplification for gender in sample	S ↑ (*p* = .031)	Controlled, excluded, failed	Controlled for females
FRUIT	Amount of fruit in a sample (none, some, all)	NS	S ↓(*p* = .001)	S ↓(*p* = .024)
SEASON	Wet/dry	*NS*	NS (month)	Controlled
BREED	Months of breeding (April to June) and/or gestation (May to August)	NS	*NS*	Controlled
HARVEST	Months of high human activity related to crop harvest	S ↑ (*p* = .001)	*NS*	*NS*
PARK	Sample located in or outside park	NS	NS	S ↑ (*p* = .046)
PATCHDIST	Distance from sample to natural habitat patch	S ↑ (interaction with HEIGHT, *p* = .001)	NS	*NS*
AMTCROP	Amount of cropland within 3 km² of sample	NS	S ↑ (*p* = .024)	S ↑ (*p* = .012)
HEIGHT	Crops above or below waist height	S ↓ as interaction with PATCHDIST	*NS*	NS
NAT<30	<30% of area within 3 km² of sample is natural	*NS*	*NS*	*NS*
PASTDIST	Distance to pasture	*NS*	S ↑ (*p* = .046)	*NS*
ROAD	Distance from sample to busy road	*NS*	NS	*NS*

NS indicates covariate was not significant for given hormone in full model; S↑ indicates there was significant, positive correlation between the covariate and the hormone; and S↓ indicates there was significant, negative correlation between the covariate and the hormone. Values in italics were dropped from most parsimonious models.

RESULTS

Distribution across the Landscape Matrix

We collected and DNA-confirmed 974 maned wolf scats in 3170 km of surveys. Maned wolves were detected in 63 of 65 survey quadrats, which were placed as far as 38 km from the park boundary. Whereas 38% of our survey effort was off roads and trails, about half the samples were found off roads and trails ($n = 469$ on neither, 429 on unpaved road, 45 on trail). While males and females occurred in equal proportions in our study area (Vynne 2010) and co-occurred throughout the landscape during all months of the year, males were significantly more likely to defecate on roads than females ($\chi^2_1 = 4.65$, $n_M = 213$, $n_F = 192$, $p = <.05$).

Contrary to our prediction, we detected maned wolf scats in highly converted areas more often than expected. Our sample recovery for confirmed maned wolf scats was 0.32 scats/km walked outside the park and 0.28 scats/km walked inside the park boundary. For surveys that occurred distant (>15 km) from the park boundary, the sample recovery was 0.38 scats/km; for those conducted in or within 15 km of the park border, it was 0.30 scats/km. Whereas 24% of the 20,000 random points generated from within our search tracks were in areas with <30% preserved, 31% of maned wolf scat samples were found in these more degraded areas.

Glucocorticoids

Glucocorticoid concentrations were highest during the months of active harvesting in the croplands (HARVEST, Table 12.1). There was a significant interaction between HEIGHT and PATCHDIST; glucocorticoid levels increased with distance from natural habitat patches when crop height was low but were unaffected by this distance when crop height was high (Figure 12.1). The significance of these relationships further increased when we limited our model only to samples located outside the park. When we limited the model to samples found inside the park ($n = 154$), the influence of harvest activity was nonsignificant, whereas season became significant, with glucocorticoid levels being highest during the dry season ($\chi_1^2 = 4.060$, $p < .05$). For samples at the park edge ($n = 73$), the amount of cropland within 3 km^2 of a sample significantly and positively predicted mean glucocorticoid levels ($\chi_1^2 = 14.615$, $p < .001$; Figure 12.2).

Thyroid Hormone

Contrary to our predictions, mean thyroid hormone levels of samples found outside the park did not vary more than those of samples found inside the park ($F_{1,564} = 0.618$, $p = .432$). Significant contributing variables to the GLM for thyroid hormone included amount of fruit in sample, amount of available cropland, and distance to pasture (Table 12.1). Thyroid hormone increased with proportion of available cropland within 3 km^2 of a sample. Samples found on or close to pastures, however, had lower levels of thyroid hormone and levels increased with distance from cattle pasture. Neither PARK nor PATCHDIST influenced whole model results, and removing these variables resulted in a more parsimonious final model.

Progesterone

Mean fecal progestin concentrations of female maned wolf samples were higher during gestation months than during the rest of the year ($F_{1,492} = 10.558$, $p = .001$). Levels peaked in April and showed

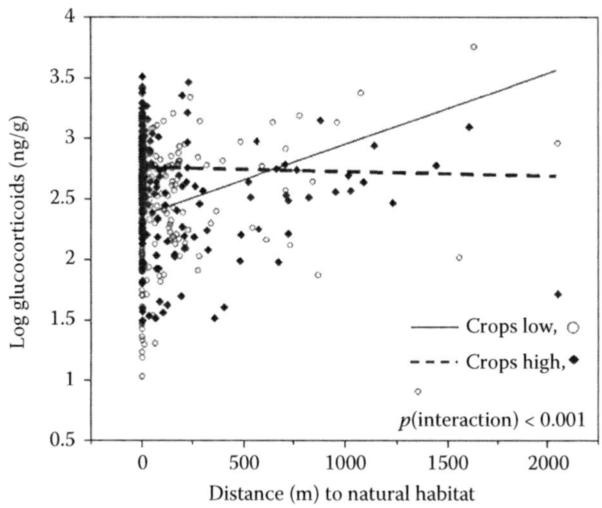

FIGURE 12.1 Levels of glucocorticoids found at various distances from natural habitat patches when crops were low and high.

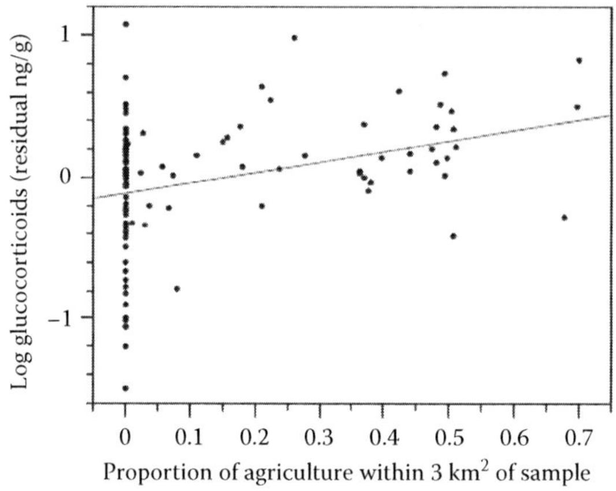

FIGURE 12.2 Levels of glucocorticoids of samples found inside the park relative to proportion of agriculture within 3 km^2 of the sample.

a clear drop after August, which is likely indicative that most pregnancies were initiated between April and June with parturition occurring between June and August. Restricting analyses to months of breeding and parturition, we found that samples found inside the park had higher mean progestin levels than those found outside the park (Figure 12.3), but for samples found outside the park, distance to the park was not significant. The mean level of progestins was also positively correlated with the amount of available cropland (AMTCROP, Table 12.1), such that females residing in the park but at the edge where cropland is available for foraging had the highest mean levels of fecal progestins. The most parsimonious GLM for female fecal progestins included FRUIT, PARK, and AMTCROP.

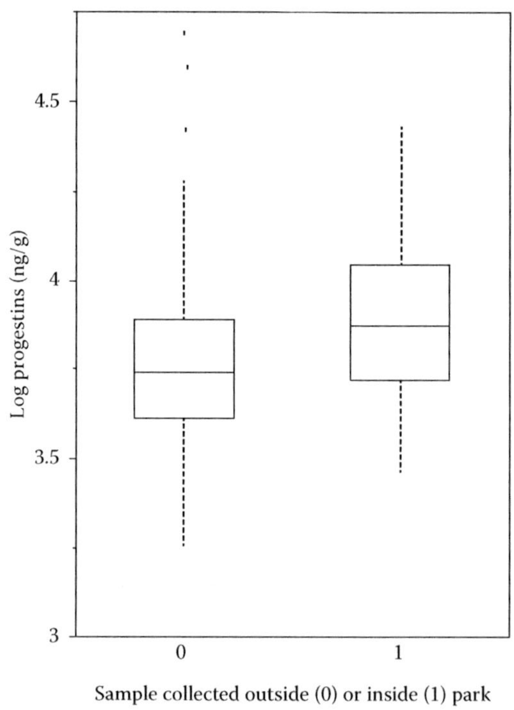

FIGURE 12.3 Levels of progestins in female maned wolf scat samples found inside and outside the park.

DISCUSSION

The results of this study suggest that maned wolves' extensive use of croplands (Vynne et al. 2011a,b) is likely driven by food availability. Scats collected from croplands had high levels of thyroid hormone, which indicates a relatively good nutritional status of maned wolves using these fields. Endocrine evidence that the maned wolves may be attracted to croplands for foraging on rodents is substantiated by the fact that maned wolf scats obtained from croplands had the highest prevalence of rodents as food items (Vynne, unpublished data). The maned wolf appears to be at the body size limit of the ability to subsist on smaller prey since most carnivores over 21 kg prey upon species that are as large as, or larger than, themselves (Carbone et al. 1999; Van Valkenburgh et al. 2004). Maned wolves may thus be protein limited and hence prefer to forage in areas that offer a high concentration of protein. The benefits of the high-protein food items available in croplands may outweigh negative influences of disturbance and psychological stress to which animals are subjected when ranging in the agricultural matrix. While we found a high ratio of protein to other items for scats found in cropland areas, this is not evidence that more protein for maned wolves is available in croplands than in nondisturbed areas. A follow-up study to understand food availability would be important before conclusions about the use of these areas can be confirmed.

Pastures tended to have very little standing vegetation at any time of the year and thus likely did not support significant rodent populations, thus explaining the result of lower thyroid hormone levels (indicating poorer nutritional status) in and around pastures. May-Júnior et al. (2009) conducted serum screening of maned wolves across a landscape and found lower mean corpuscular volume in maned wolves living near pastures at the park border than in wolves found inside the park or on the surrounding croplands. They suggested that these lower levels could be due to a nutritional deficiency in animals living near pastures. Maned wolves ranging in and around pastures may also be expected to have higher glucocorticoid levels since they may be susceptible to disease transmission from domestic animals (Deem and Emmons 2005; May-Júnior et al. 2009). Again, while our results point toward food resource availability driving both habitat selection (Vynne et al. 2011a) and physiological response, follow-up work on food resource availability is needed to confirm our conclusions.

Our finding that both female and male maned wolves received nutritional gains from their use of croplands provides optimism for the conservation of this species. However, ranging in croplands exposes individuals to increased susceptibility to collisions with vehicles, ingestion of toxins from chemical applications, physical and noise disturbance from machinery and clearing of land during harvest and planting periods, and reduced availability of habitat for den sites and resting. These costs to individuals ranging in the matrix were reflected by the physiological measures associated with increased disturbance and reduced amounts of native vegetation. Higher glucocorticoid levels were measured in samples that were deposited in agricultural fields and during times of year when there is increased human activity associated with crop harvest cycles. For individuals that used the park, increased glucocorticoids levels were associated with park edges near cropland. Similarly, pampas deer (*Ozotoceros bezoarticus bezoarticus*) living at the edge of the park had previously been shown to have higher levels of glucocorticoids than individuals living in the core of the park (Pereira et al. 2006).

Progesterone metabolite hormone concentrations further corroborated the results that suggested selection of croplands for foraging may have fitness trade-offs for maned wolves. High fecal progestin levels among female maned wolves are indicative of reproductive health during the mating season (Songsasen et al. 2006) and, in this study, progestin levels during the mating season were higher in female maned wolves with home ranges that overlapped with the park. Availability of habitat for undisturbed den sites likely contributes to a higher likelihood of reproductive success for individuals inside the park. Further work is needed to determine thresholds of natural habitat needed to maintain breeding in maned wolves, and at what levels conversion is likely to directly influence fitness.

It is important to consider that, in addition to natural variation due to the individual, breeding phase, life stage, or season, hormonal signals indicative of disturbance may be dampened by old or degraded samples (Vynne et al. 2011c). Although we were able to control for much of this variance, our measures are conservative since we were unlikely to have measured or controlled for all the variation due to sample degradation. Since fecal hormone measures are often metabolites of the parent hormone typically measured in blood, these metabolites may not entirely reflect biological activity of the parent hormone. While lack of significance of some covariates may have been due to study design, sample quality, or measurement issues, our results in general do appear to reflect expected trends when compared with other studies (Baker et al. 2013). Progestin levels reflected the expected seasonal variation previously reported for captive maned wolves (Velloso et al. 1998; Wasser et al. 1995) and glucocorticoid measures for free-ranging pampas deer in Emas National Park (Pereira et al. 2006).

Our results suggest that maned wolves may be fairly tolerant of high levels of habitat conversion as long as the matrix provides remnant vegetation for resting, den sites, and temperature buffering (Mourão et al. 2007), as well as adequate food and water supply (Emmons 2012). The results of this study provide cautious optimism for the future of the species in the face of a continuously changing landscape. At the same time, there is reason for concern. Maned wolves appear sensitive to disturbance and females have lower levels of hormones linked to reproductive success outside the legal reserve (of which there are relatively few throughout the maned wolves' range). Moreover, this cropland-dominated landscape is dynamic. Land clearing in the cerrado is continuing at a rapid pace and current trends toward sugar cane production for biofuels are bringing increased anthropogenic pressures to the region in the form of additional workers and factories as well as a change in the character of the landscape that was until now dominated by low-growth plantations such as soybean and cotton. Maned wolves may be able to habituate to land transformation for low-growing agriculture, but increased intensification, human pressures, and likelihood of disease introduction from domestic carnivores will make altered landscapes more hostile environments and threaten to isolate populations. Indeed, a recent study examining maned wolf survival rates over 16 years in this landscape showed a decline in the adult maned wolf survival rates corresponding to years of increased sugar cane expansion (Sollmann et al. 2010). Since this study was conducted at the onset of the sugar cane expansion, additional work is needed to understand the potential costs of not only conversion generally, but also conversion from a soy-dominated to a sugar cane-dominated landscape.

The maned wolf may indeed be an excellent flagship for the cerrado since, as a wide-ranging, large carnivore that requires open landscapes on which to establish home ranges, it appears it can also coexist where native cerrado vegetation and agriculture intermix. Like the cerrado, however, the maned wolves' status needs to be carefully monitored. With agricultural expansion continuing and likely to dominate throughout the distribution of the species, schemes to incentivize conservation on private lands will likely be necessary to ensure its long-term conservation and to ensure that the species' future does not depend on isolated pockets of protected lands, which are unlikely to alone ensure the persistence of the species.

ACKNOWLEDGMENTS

Sam Wasser provided laboratory space and overall project guidance. Rebecca Booth provided invaluable technical guidance in conducting the hormone assays and analysis. The Brazil Program of Conservation International, the Jaguar Conservation Fund, and the University of Brasilia assisted with many critical aspects of project inception and coordination. We acknowledge in particular the assistance of Leandro Silveira, Anah T. A. Jácomo, Ricardo Machado, and Jader Marinho-Filho. L. Erckman provided guidance on hormone assays and numerous field and laboratory staff and volunteers assisted with data collection and laboratory analyses. Thanks to the staff of Emas National Park, private landowners in the region, and the Brazilian Institute of Renewable Natural Resources and CNPq for project licensing. Funding was provided by the Morris Animal Foundation, TEAM Network of Conservation International, funded by the Gordon and Betty Moore Foundation, and a National Science Foundation Graduate Fellowship to CV.

REFERENCES

Baker, M. R., K. S. Gobush, and C. Vynne, C. 2013 . Review of factors influencing stress hormones in fish and wildlife. *Journal for Nature Conservation* 6.

Brazil. 1965. The Brazilian Forest Code. Law 4.771—September 15, 1965.

Brooks, T. M., S. L. Pimm, and J. O. Oyugi. 1999. Time lag between deforestation and bird extinction in tropical forest fragments. *Conservation Biology* 13: 1140–1150.

Brown, J. L., S. K. Wasser, D. E. Wildt, and L. H. Graham. 1994. Comparative aspects of steroid hormone metabolism and ovarian activity in felids, measured non-invasively in feces. *Biology of Reproduction* 51: 776–786.

Brown, J. L., S. K. Wasser, D. E. Wildt, L. H. Graham, and S. L. Monfort. 1997. Faecal steroid analysis for monitoring ovarian and testicular function in diverse wild carnivore, primate and ungulate species. *Zeitschrift fur Saugetierkunde* 62(Suppl 2): 27–31.

Cabezas, S., J. Blas, T. A. Marchant, and S. Moreno. 2007. Physiological stress levels predict survival probabilities in wild rabbits. *Hormones and Behavior* 51: 313–320.

Carbone, C., G. M. Mace, S. C. Roberts, and D. W. Macdonald. 1999. Energetic constraints on the diet of terrestrial carnivores. *Nature* 402: 86–288.

Carroll, C., R. F. Noss, P. C. Paquet, and N. H. Schumaker. 2004. Extinction debt of protected areas in developing landscapes. *Conservation Biology* 18: 1110–1120.

Creel, S., J. E. Fox, A. R. Hardy, J. Sands, B. Garrot, and R. O. Peterson. 2002. Snowmobile activity and glucocorticoid stress responses in wolves and elk. *Conservation Biology* 16: 809–814.

Cummings, D., J. L. Brown, M. D. Rodden, and N. Songsasen. 2007. Behavioral and physiologic responses to environmental enrichment in the maned wolf (*Chrysocyon brachyurus*). *Zoo Biology* 26: 331–343.

Deem, S. L. and L. H. Emmons. 2005. Exposure of free-ranging maned wolves (*Chrysocyon brachyurus*) to infectious and parasitic disease agents in the Noel Kempff Mercado National Park, Bolivia. *Journal of Zoo and Wildlife Medicine* 36: 192–197.

Dietz, J. M. 1984. Ecology and social organization of the maned wolf (*Chrysocyon brachyurus*). *Smithsonian Contributions to Zoology* 392: 1–51.

Eales, J. G. 1988. The influence of nutritional state on thyroid function in various vertebrates. *American Zoology* 28: 351–362.

Emmons, L. (ed.). 2012. The maned wolves of Noell Kempff Mercado National Park. *Smithsonian Contributions to Zoology* no. 639.

Goymann, W., E. Mostl, T. Vant Hof, M. L. East, and H. Hofer. 1999. Noninvasive fecal monitoring of glucocorticoids in spotted hyenas, *Crocuta crocuta*. *General and Comparative Endocrinology* 114: 340–348.

IUCN. 2009. *IUCN Red List of Threatened Species*. Version 2009.1. http://www.iucnredlist.org (accessed March 2010).

Jácomo, A. T. A., C. K. Kashivakura, C. Ferro, M. M. Furtado, S. P. Astete, N. M. Tôrres, R. Sollmann, and L. Silveira. 2009. Home range and spatial organization of maned wolves in the Brazilian grasslands. *Journal of Mammalogy* 90: 150–157.

Klink, C. and R. B. Machado. 2005. Conservation of the Brazilian Cerrado. *Conservation Biology* 19: 707–713.

May-Júnior, J. A., N. Songsasen, F. Cavalcanti, J. P. Santos, R. C. Paula, F. H. G. Rodrigues, M. D. Rodden, D. E. Wildt, and R. G. Morato. 2009. Hematology and blood chemistry parameters differ in free-ranging maned wolves (*Chrysocyon brachyurus*) living in the Serra da Canastra National Park versus adjacent farmlands, Brazil. *Journal of Wildlife Diseases* 45: 81–90.

Millspaugh, J. J. and B. E. Washburn. 2004. Use of fecal glucocorticoid metabolite measures in conservation biology research: Considerations for application and interpretation. *General and Comparative Endocrinology* 138: 189–199.

Mourão, G. and M. Í. Medri. 2007. Activity of a specialized insectivorous mammal (*Myrmecophaga tridactyla*) in the Pantanal of Brazil. *Journal of Zoology* 271: 187–192.

Oppenheimer, J. H. 1999. Evolving concepts of thyroid hormone action. *Biochimie* 81: 539–543.

Pereira, R. J. G., J. M. B. Duarte, and J. A. Negrão. 2006. Effects of environmental conditions, human activity, reproduction, antler cycle and grouping on fecal glucocorticoids of free-ranging Pampas deer stags (*Ozotoceros bezoarticus bezoarticus*). *Hormones and Behavior* 49: 114–122.

Pride, R. E. 2005. High faecal glucocorticoid levels predict mortality in ring-tailed lemurs (*Lemur catta*). *Biology Letters* 1: 60–63.

Rodden, M., F. Rodrigues, and S. Bestelmeyer. 2004. Maned wolf *Chrysocyon bracyurus*. In C. Sillero-Zubiri, M. Hoffman, and D. W. Macdonald (eds), *Canids: Foxes, Wolves, Jackals and Dogs, Status Survey and Conservation Action Plan*, pp. 38–43. IUCN/SSC Canid Specialist Group, Gland, Switzerland and Cambridge, UK.

Sands, J. and S. Creel. 2004. Social dominance, aggression and fecal glucocorticoid levels in a wild population of wolves, *Canis lupus*. *Animal Behaviour* 67: 387–396.

Silveira, L., M. M. Furtado, N. M. Tôrres, R. Sollmann, G. Uhl, and A. T. A. Jácomo. 2010. Maned wolf density in a central Brazilian grassland reserve. *Journal of Wildlife Management* 73: 68–71.

Sollmann, R. A., M. M. Furtado, T. A. Jácomo, N. M. Tôrres, and L. Silveira. 2010. Maned wolf survival rate in central Brazil. *Journal of Zoology* 282: 207–213.

Songsasen, N., M. Rodden, J. L. Brown, and D. E. Wildt. 2006. Patterns of fecal gonadal hormone metabolites in the maned wolf (*Chrysocyon brachyurus*). *Theriogenology* 66: 1743–1750.

Trolle, M., A. J. Noss, E. S. de Lima, and J. C. Dalponte. 2006. Camera-trap studies of maned wolf density in the Cerrado and the Pantanal of Brazil. *Biodiversity and Conservation* 16: 1197–1204.

UNEP-WCMC. 2010. *The World Database on Protected Areas (WDPA): Annual Release* [On-line]. Cambridge, UK: UNEP-WCMC. Available at http://www.wdpa.org (accessed May 2010).

Van Valkenburgh, B., X. Wang, and J. Damuth. 2004. Cope's rule, hypercarnivory, and extinction in North American canids. *Science* 306: 101–104.

Velloso, A. L., S. K. Wasser, S. L. Monfort, and J. M. Dietz. 1998. Longitudinal fecal steroid excretion in maned wolves (*Chrysocyon brachyurus*). *General and Comparative Endocrinology* 112: 96–107.

Vynne, C. 2010. Landscape use by wide-ranging mammals of the Brazilian Cerrado. PhD Dissertation, University of Washington, Seattle.

Vynne, C., J. L. Keim, R. B. Machado, J. Marinho-Filho, L. Silveira, M. J. Groom, and S. K. Wasser. 2011a. Resource selection and its implications for the wide-ranging mammals of the Brazilian Cerrado. *PLoS ONE* 6(12): e28939.

Vynne, C., J. L. Keim, R. B. Machado, J. Marinho-Filho, L. Silveira, M. J. Groom, and S. K. Wasser. 2011b. Effectiveness of scat-detection dogs in determining species presence in a tropical savanna landscape. *Conservation Biology* 25: 154–162.

Vynne, C., M. R. Baker, Z. K. Breuer, and S. K. Wasser. 2011c. Factors influencing degradation of DNA and hormones in maned wolf scat. *Animal Conservation* 15: 184–194.

Wasser, S. K., R. Thomas, P. P. Nair, C. Guidry, J. Southers, J. Lucas, D. E. Wildt, and S. L. Monfort. 1993. Effects of dietary fibre on faecal steroid measurements in baboons (*Papio cynocephalus cynocephalus*). *Journal of Reproduction and Fertility* 97: 569–574.

Wasser, S. K., A. L. Velloso, and M. D. Rodden. 1995. Using fecal steroids to evaluate reproductive function in female maned wolves. *Journal of Wildlife Management* 59: 889–894.

Wasser, S. K., S. Papageorge, C. Foley, and J. L. Brown. 1996. Excretory fate of estradiol and progesterone in the African elephant (*Loxodonta africana*) and patterns of fecal steroid concentrations throughout the estrous cycle. *General and Comparative Endocrinology* 102: 255–262.

Wasser, S. K., K. E. Hunt, J. L. Brown, K. Cooper, C. M. Crockett, U. Bechert, J. J. Millspaugh, S. Larson, and S. L. Monfort. 2000. A generalized fecal glucocorticoid assay for use in a diverse array of nondomestic mammalian and avian species. *General and Comparative Endocrinology* 120: 260–275.

Wasser, S. K., B. Davenport, E. R. Ramage, K. E. Hunt, M. Parker, C. Clarke, and G. Stenhouse. 2004. Scat detection dogs in wildlife research and management: Applications to grizzly and black bears in the yellowhead ecosystem, Alberta, Canada. *Canadian Journal of Zoology* 82: 475–492.

Wasser, S. K., J. C. Azkarate, R. K. Booth, L. Hayward, K. Hunt, K. Ayres, C. Vynne, K. Gobush, D. Canales-Espinosa, and E. Rodríguez-Luna. 2010. Non-invasive measurement of thyroid hormone in feces of a diverse array of avian and mammalian species. *General and Comparative Endocrinology* 168: 1–7.

Wingfield, J. C. and M. Ramenofsky. 1999. Hormones and the behavioral ecology of stress. In: P. H. M. Balm (ed.), *Stress Physiology in Animals*, pp. 1–51. Sheffield Academic Press, Sheffield.

Zar, J. H. 1999. *Biostatistical Analysis*, 4th edn. Prentice Hall, Upper Saddle River, NJ.

13 The Maned Wolf Conservation Project

Serra da Canastra, Minas Gerais, Brazil

Rogério Cunha de Paula, Flávio H.G. Rodrigues, Marcelo Ximenes Bizerril, Nucharin Songsasen, Eduardo Eizirik, Ronaldo Gonçalves Morato, Jean Pierre Santos, Fernanda Cavalcanti Azevedo, Joares Adenilson May Jr., Carla Cruz Soares, Fabiana Lopes Rocha, and Ricardo Corassa Arrais

CONTENTS

Introduction 177
Study Site 178
The Maned Wolf Conservation Project 179
Research 179
Education 180
Conservation 180
History of the Project 181
Phase I (2000–2002): Tourism and Altered Behavior 182
Phase II (2004–2008): Research, Education, and Conservation 182
Phase III (2008–2010): Healthy Wolves, Healthy System 186
Phase IV (2010 to present): Gathering, Informing, Expanding 187
Participating Institutions 189
Overview of Results 189
References 191

INTRODUCTION

Conservation programs that focus on a charismatic flagship species have proven particularly successful in promoting the preservation of critical habitats and ecosystems (Leader-Williams and Dublin 2000). Top predators as "umbrella" species serve an additional important role because their survival entails that of a host of other species (Ray 2005). Large carnivores have been the center of conservation programs worldwide for several reasons: they are key species for ecosystem protection and restoration, site-based conservation planning, and the monitoring of biodiversity status (Ray 2005). Among several carnivore species in South America, the maned wolf is one of the most efficient representative species for the success of a conservation program, due to its wide distribution range, particular ecology and behavior, and lack of threat to human life (despite nominal damage to the local economy). Throughout the world, endangered species have been used as ambassadors to link people to the value of wildlife and the ecosystems they inhabit (Leader-Williams and Dublin 2000). By protecting endangered species, one may well provide an incentive for stakeholders for habitat maintenance or restoration, which in turn results in improvement of their lives (Entwistle and Dunstone 2000). Conservation programs involve several stakeholders in the planning of conservation practices associated to education, scientific research, and public policy application (Reyers et al. 2010). This "transdisciplinary" approach has been presented as one of the solutions for balancing wilderness maintenance, socioeconomics, and culture (Jerusalinsky et al. 2010; Reyers et al. 2010). Following this approach, protecting the maned wolf helps to preserve species' natural habitat, the cerrado biome in Brazil, which is considered one of the world's biodiversity hotspots (Myers et al. 2000). The Maned Wolf Conservation Project is an example of how to use a flagship species to raise people's awareness of wilderness protection, together with utilizing a transdisciplinary approach to promote conservation of a threatened species and improve the relationship of a local community with nature. The project was designed to generate new information on species biology in a way that would serve as a model for developing other research/outreach-based initiatives for conserving species and habitats within the cerrado or other threatened ecosystems throughout the country or the continent.

Study Site

The project took place in the Serra da Canastra region, Minas Gerais State, southeastern Brazil. The region is under the dominium of the cerrado biome. The cerrado is listed as one of 25 biodiversity hotspots (Myers et al. 2000) and is home to 935 species of birds and nearly 200 mammals. During the past 50 years, the cerrado has been extensively developed for agriculture; more than 80% of the ecosystem is directly used by humans. Clearly, ongoing development of this critical ecosystem and intense human occupation (which are unlikely to stop) pose significant threats to wildlife, including the maned wolf.

The project site ranges throughout 400,000 ha of altered and protected areas. The Serra da Canastra National Park (SCNP), created 40 years ago, encompasses a total of 200,000 ha of which 71,525 ha have been regulated and managed since 1978. Mammal diversity is high, with approximately 40 species (excluding small mammals and bats), and carnivores represent nearly half of the terrestrial mammal species. The region, mainly the protected areas, constitutes an important sanctuary for threatened species such as the maned wolf, the giant anteater, and the giant armadillo. Besides the SCNP, the region is made up of small ranches (landholdings of 100 ha or less), in all of which planting of forage grass is evident. The primary economic activity in the region is cattle ranching; agricultural activities are mostly confined to subsistence plantings, except for large areas prepared for coffee and corn plantation (Mota and Paula 2006; Paula and Mota 2010).

Local people have been coexisting with wildlife for hundreds of years in the region. Although everyday encounters are often ignored by both sides, occasionally losses for both human and wildlife can be observed as a consequence of livestock predation and direct persecution to carnivore species, mainly pumas and maned wolves. Despite severe habitat disturbance in the surrounding areas, population growth of several wildlife species, including the maned wolf, has been observed within the SCNP during the past years. This demonstrates the value of the park via its physical (water, soil) and biological (richness in biodiversity) aspects. Due to the high density of the maned wolf in this region and the special relationship between people and nature, the region represents the perfect site for long-term population monitoring and practices for maned wolf conservation. Furthermore, by improving the relationship between local people and wildlife by emphasizing a healthy coexistence, it is possible that success stories generated in Serra da Canastra region can be applied throughout the entire range of the maned wolf.

THE MANED WOLF CONSERVATION PROJECT

The Maned Wolf Conservation Project arose from the personal initiative and dedication of two researchers investigating the maned wolf in different areas. Rogério Cunha de Paula started to work with the species in the Serra da Canastra region while Flávio H. G. Rodrigues initiated his investigations in central Brazil. The project was initiated based on the desire to build a conservation model for the maned wolf. The aim is conservation of the maned wolf through extensive research, public awareness, and applied conservation actions of a specific population, and further throughout its distribution range.

The Maned Wolf Conservation Project was planned to provide long-term population monitoring and constant evaluations of the threats to the species in the Serra da Canastra region while management actions and strategies were being implemented.

We believe that by understanding the biology of this flagship species, assessing the impact of human disturbance on maned wolf health and survivorship, and using our experiences and stories to educate the local community, we would contribute significantly to conserving both the species and the ecosystem, as well as improve the relationships between humans and nature.

To achieve our mission and objectives, the idea has always been to build a team of researchers with different expertise and involve several institutions in network-based research and conservation actions. Although the maned wolf has always been the target element of the project, relationships with other species and especially the role of the local community in conservation issues are some of the important facets of the complex polygon that emerged when building the project.

In summary, the aims of the project have resulted in it becoming transdisciplinary, combining the following elements:

Research: through ecology, behavior, health and epidemiology, genetics, reproduction
Education: through discussion of environmental problems, communication, and cultural approaches
Conservation actions: through management of observed impacts, implemented by the fusion of scientific data generated by the research and the strategies proposed by the educational program

The connection of these interfaces has resulted in a better plan for the species' survival. We thus used these three pillars as the foundation of our project structure.

Research

With the exception of feeding ecology, little is known about the maned wolf in nature, and there are insufficient data to predict future conservation or sustainability status (Paula et al. 2008). The habitat of this species is undergoing rapid conversion, and it is essential to determine the influence of human development on the species' survival. Field initiatives aiming to improve our fundamental understanding of the maned wolf's biology, ecology, behavior, health, reproduction, and genetics, while determining the risks imposed by habitat alterations and increasing human disturbance, are necessary to achieve our main objectives of evaluating the species' long-term survival.

The research component of this project includes assessing social behaviors and characterizing individual territories, food resources, and habitat integrity. In order to apply management actions, it is necessary to analyze specifically the habitat use and dispersal rates of animals that inhabit areas with close contact with humans or areas under anthropogenic activities (such as farming, ranching, mining, and tourism). Considered as one of the main emerging threats to carnivore populations, the spread of diseases from domestic animals to wildlife can be a serious threat to the wolves. Thus, it is extremely important to investigate epidemiological risks to maned wolves from direct and indirect contact with domestic animals. Genetics can not only provide a background to understand how gene diversity influences species' survival but also show how resistant wolves can be to natural and artificial adversities. Through hormone analysis, we can also understand and perceive the influence of several variables on the species' physiology. Finally, by verifying the main causes of wolf mortality or reduced reproductive success and attempting to control the unnatural causes, we might make the conservation plan for the species more successful.

A detailed study, involving capturing and monitoring wolves using VHF, GPS/VHF, and satellite transmitters (in different phases of the study), has provided high-quality information. The intensive monitoring of collared animals allows us to build models for improved conservation through complete data collection not only for ecological or behavioral needs, but also for all the thematic areas. By following animal tracks, we can provide a database for answering other questions on health, genetics, and reproduction. The data obtained from monitoring of the wolves throughout the project years, combined with various other information generated, provide plenty of possibilities of use for scientific matters, species and habitat management, and public policies. Thus, this will certainly aid habitat restoration, corridor creation, and local population management. Thereafter it can be used as a model for the conservation of the species in general.

Education

We strongly believe that the maned wolf is an ambassador for making people aware of the value of wildlife and the preservation of the cerrado. In this context, we have initiated a community outreach program to raise public awareness in the Serra da Canastra region. When educating local communities about the species and its conservation, the long-term survival of this species can be improved. Protecting the maned wolf helps to preserve the cerrado and may well provide an incentive for at least some habitat restoration.

The conservation impact of our project is related to the stories—scientific or based on other matters—that we can generate and our experiences that can be passed on to local communities through our growing outreach and awareness efforts.

Clearly, some of the best ways to preserve local habitats and their biodiversity are (1) to communicate the interesting biology and role of species in local ecosystems and communities, (2) to manage conflicts between local people and wildlife, (3) to make locals perceive and understand the need for coexistence with wild species and for preservation of wild habitats, (4) to increase pride within the community, helping them to recognize the positive values of being inhabitants of that place, and (5) to integrate the community and initiate discussion of topics of major concern, including those related to nature. In this way, we tried to bring people together and initiate discussion and debate on issues important to them and what they think is the best way to become integrated in their natural, rural, and urban environments. Our objectives were to understand the community's needs and to encourage shy local people—from children to elders—to discuss anything, thereby learning how to express themselves. The next step was to debate how to live in harmony with wildlife, especially with the wolves.

Conservation

In general, people in Serra da Canastra have mixed feelings toward the maned wolf. The species has been used as a central figure to explain the complexity and fragility of the local ecosystem. We try to raise awareness through the presentation of the beauty of the region, contrasting with negative impacts that lead to the destruction of nature (fire, hunting, deforestation, etc.). Presenting the two sides, we have been trying to engage locals in several self-sustained actions that improve the environment's status in the region, including adapting harvest practices, applying management actions on livestock depredation by predators, and providing information on the ecological functions of the wild species.

Positive conservation impacts will be related to the stories that we can produce and our experiences that can be told to local communities through our growing outreach and awareness efforts. Clearly, one of the best ways to preserve the cerrado and its biodiversity is to communicate the interesting biology and role of species in local ecosystems and communities. We believe that by involving the local community, the research can be more effectively and widely transmitted to local people, who ultimately will be responsible for preserving this habitat and its wildlife in perpetuity. One of the main tasks is to convince farmers that people and wolves can surely coexist.

Conflicts between maned wolves and local populations have been suggested as one of the main threats to the species' maintenance. Although official records of direct persecution resulting in the death of maned wolves are unusual in the Serra da Canastra region, local farmers tend to always blame the wolves for any livestock losses. This attitude is concerning, since awareness and tolerance of the species is low, sometimes leading to persecution and killings.

As farmers raise their avian livestock in an overexposed environment, simple preventive methods would improve their breeding and, consequently, production. However, on top of their losses, farmers cannot afford to build facilities to raise their chickens in enclosed spaces. One option was to test prototypes of chicken coops to prove to local people how effective they were at protecting from predators and improving production.

Since 2005, we have visited schools in the entire regions (mostly rural schools) to speak to teachers and students about conservation of the cerrado and its flagship species, the maned wolf. This has resulted in an improvement in local people's attitudes toward the wolves. Because human and wolf conflict is one of the critical conservation threats to this species, we have been visiting farmers and helping them to solve problems, as well as educating them about the importance of maned wolves to cerrado habitats.

Besides the direct relationship with humans resulting in conflicts, infectious diseases caused by pathogens spread from domestic animals can represent a serious threat to wildlife, in this case to the maned wolf. Maned wolves living in rural areas have been exposed to several domestic animal diseases. We found several types of pathogens in the wild population but the canine parvovirus (CPV), a disease that causes high mortality rates in young individuals, is of major concern. Our previous study demonstrated that anthropogenic factors (i.e., farming, ranching, and tourism) significantly affect the general health status of maned wolves and the population might be at risk due to infectious diseases. Undoubtedly, not only CPV but also several other diseases from domestic dogs must be contained to avoid drastic reductions in the wildlife population due to this factor. Thus, campaigns to vaccinate domestic dogs and reduce their density were proposed as necessary in the surroundings of SCNP.

The information generated in this project has been shared with governmental agencies at state and federal level, and specifically with CENAP (National Research Center for Carnivore Conservation), the research center within the Environmental Ministry responsible for developing leading research, policies, and conservation actions for carnivores in Brazil. This is a way to ensure the application of all the scientific information collected in the field. By collecting and sharing scientific information and through direct actions involving the local people we have experienced better conservation actions recognized by the local authorities and general public.

History of the Project

The prequel of the project was an investigation of the impact of human activities on mammals in the Serra da Canastra region in 1998. The results showed that farming and ranching had a negative influence on species' diversity and abundance in areas surrounding the national park. However, persistence of the maned wolf in altered areas suggested that this species tolerates human activities (depending on the type of activity and degree of severity), especially when levels of habitat disturbance are low. From this point, we decided to perform a more complex evaluation of several aspects of maned wolf conservation, including the relationship with the local people. Thus, the specific investigation of the maned wolf was started in 2000.

Phase I (2000–2002): Tourism and Altered Behavior

The Maned Wolf Conservation Project arose from concern about the effects of the constant feeding of wild wolves by park rangers, tourist guides, and visitors around the areas near the park's field base. A short-term survey from 2000 to 2002 was conducted through direct behavioral observations and analysis of fecal samples. Several negative results could be clearly seen, and alterations in the animals' ecology and behavior were found.

At that time, seven adult animals (four females and three males) were identified (by individual marks on the mane, throat, tail, and legs) through visual observations and camera trapping. Despite the maned wolf's solitary habit, sometimes several of them (at most five individuals) were observed no more than 10 m apart, waiting in the middle of the day (against their natural crepuscular/nocturnal activity pattern) for the scheduled food provided by people (leftovers from rangers' meals and snacks from guides and tourists). Trash containers were visited by wolves during the day and night, and food, bottles, and cans were taken by the seven individuals around the field base. Pieces of toilet paper, plastic (from bags and caps), aluminum foil, chicken, and pork bone, among many other things, were found in their scats. We analyzed 93 fecal samples and 69 of them had strange elements. Thus, only 26% of the samples had only natural items defecated (hair, bones, scales, seeds, fruit parts). Considering that 46 samples (almost half of the total scats observed) did not have any natural items, we assumed that the artificial feeding might be harmful for the animals. Differences in behavior, feeding ecology, activity patterns, and perhaps health (although not physiologically evaluated), besides an incredibly high tolerance to humans, motivated us to initiate a campaign not only on the site but also in the entire region to stop feeding the wolves. We observed three wolves with broken teeth, secretion in the eyes, and poor coat condition; many wolves had chronic diarrhea.

The desire to stop the artificial feeding was the first attempt to involve the local community in a positive action toward maned wolf conservation. Posters and flyers were distributed in local shops, restaurants, and hotels, and at the park gates and visitor center, among other places. Additionally, talks to tourist guides and park rangers were made and in 4 months, the feeding practice was greatly reduced. Although the campaign was well assimilated in local towns and among tourists, several park employees continued to feed the animals (suggesting that the individuals were depending on their feeding). The garbage cans were replaced in the entire protected area (not only at the feeding site) to prevent access by the animals. Over a period of 9 months, only three animals were regularly using the specific site. One of them, a female, which was the only one left after 11 months, had made her den to deliver five puppies at the site. The litter (and the high tolerance of mother and puppies to humans) was the strong reason for punishments from the park's administration for employees that persisted in feeding the wolves. In March 2001, the artificial feeding was totally abolished from SCNP and the surroundings. Two young wolves were captured, ear-marked, and observed together with the three brothers and mother for 11 months. After this period, they all dispersed, and we lost visual contact with all of them.

All this experience of observations of altered behavior, ecology, and perhaps health raised several questions to be answered. That was the first step of the Maned Wolf Conservation Project.

Phase II (2004–2008): Research, Education, and Conservation

From these earlier studies, a desire to observe the differences between Serra da Canastra's maned wolf population and that of the late 1970s research (Dietz 1984), and begin a detailed transdisciplinary study on the species, also considering the effects of human activities on its life, attracted the expertise of two researchers. From a compilation of what was available on the species and the gaps in the information, we proposed to conduct a long-term study by comparing several aspects of the maned wolf's life in protected versus disturbed areas.

The early components of this complex project began in December 2003 and the first captures began in early 2004. The activities of this second phase committed eleven researchers, including biologists and veterinarians, from nine different institutions.

In this second phase, the project had three primary elements: (1) ecology and behavior, (2) health and reproduction, and (3) capacity building and education. Addressing all of these components in parallel was essential to successfully achieve our goal of establishing a multi-institutional and trans-disciplinary long-term project. Our specific objectives were to determine (1) territorial differences among individual wolves and their social behavior as well as food availability, genetics, and specific responses to habitat quality and human activities; (2) dispersion rates of animals that inhabit protected areas compared to those in close contact with humans or areas under anthropogenic activities; (3) the health and reproduction of the local wolf population; (4) epidemiological risks of direct and indirect contact with domestic animals; and (5) potential risks of the presence of maned wolves to avian livestock. We also aimed to establish an environmental education program with the local communities focusing on the wolves' conservation.

Capturing and Monitoring: Home Range, Habitat Use, Social Behavior, Genetics

Although the following procedures are presented here as part of the second phase of the project, they are still being used now. Several potential areas inside and outside the protected area have been continually used for the capture of maned wolves. The animals are captured in live traps baited with cooked chicken and sardine. Once captured, the wolves are anesthetized and weighed. Body measurements are taken and biological samples (i.e., blood, urine, feces, semen, tissue, hair, etc.) are collected for epidemiological analysis and further evaluations on reproduction status and genetics (these two last analyses were specific to this phase). Each captured animal is marked with an ear-tag and then fitted with a radio collar (VHF, VHF/GPS) in order to monitor its activity patterns, home range, living areas, habitat preference, dispersal movements, behavioral patterns, and individual discrepancies. Animals were fitted with VHF collars (Wildlife Materials in 2004, Telonics in 2004, and ATS from 2004 through 2009) and since 2007 have been fitted with VHF/GPS collars (Lotek).

The captures from 2004 through 2008 were concentrated in the eastern sector of the national park, borders and surrounding areas, within a maximum of 10 km from the border. Until 2007, the collared animals were monitored on a regular basis. A protocol with at least two scans per day (day and night) of the entire covered area used to be conducted by the field team. In addition, the team spent time searching for collared animals to perform ethograms for behavior analysis. Records were kept of all encounters between individuals or pairs as well as all territorial markings (feces, urine); this information was compiled with data on living areas with the purpose of determining social aspects of territoriality.

The home ranges were calculated and analyzed with the Kernell and minimum convex polygon methods (in order to compare with Dietz's analysis) using the software LOAS and ArcView (using the extension Animal Movement). It was intended to monitor all the young animals captured (mainly animals occupying disturbed areas) more intensively in order to access information on parental behavior and dispersal patterns during the period that the offspring separate from the parents. We tried to intensify the monitoring until a young wolf had established its own home range. The dispersion distance was then measured.

From the captures and blood sampling, we performed a genetic evaluation, with the following goals: (1) We found the individual gene mark, which includes the "family" group compositions from studies on molecular ecology through the findings on microsatellite loci. From the genealogic tree, we could make inferences on social structure, territoriality, tolerance on home range overlapping, and dispersal movements. (2) Furthermore, we analyzed the genetic variability of the population compared with genetic information already known from other populations throughout the country.

Feeding Ecology and Habitat Use

Information on resource availability was collected and analyzed in specific surveys based on diet (through feces analysis). Data on fruit seasonality and small mammal dynamics were also obtained and analyzed together with data on habitat use and living areas. Sampling sites for the assessments were randomly chosen, and each site was sampled during wet and dry seasons. A comparison of all the sites assessed was made based on species' relative abundance from all the data acquired on all the taxonomic groups (vegetable and animal). The precise data on food availability that were assessed were related to the home range extensions, and to the size and exact location of each territory within the home ranges. Using this analysis, it was possible to indicate more suitable areas for maned wolves' conservation, based on food availability.

Health, Reproduction, and Epidemiology

An assessment of biomedical status of individual wolves was constantly performed over the period (with a minimum recapture interval of 2 months) to minimize or eliminate any seasonal effects. Upon capture, individual wolves were anesthetized and assessed for general status and body condition. All the clinical measures, besides notes on body condition (including weight and teeth), were taken as important observations for further comparisons. Besides the clinical monitoring of the entire population, as mentioned before, we aimed to assess the level of threat of domestic animals (mainly dogs) to the wolf population. Thus, for this specific objective, captures outside the park, where the animals live in close contact with domestic animals, were essential.

Biological samples (blood, feces, and urine) of wolves were collected and blood from the animals captured from 2004 through 2007 was analyzed. Both biochemical and epidemiological evaluations were performed. For the epidemiological analysis, we tested maned wolves for canine distemper virus (CDV), CPV, brucellosis, rabies, leptospirosis, toxoplasmosis, neosporosis, dirofilariosis, leishmaniosis, and canine ehrlichiosis. In addition, biological samples were collected from domestic dogs on the farms in the surrounding areas. Urine and feces were collected respectively by direct punction into the bladder and mechanical extraction from the rectum for parasitological tests. Moreover, genital scrubs on females were made with cotton swabs to check for other parasites. Finally, external parasites (mostly ticks) were collected using specific protocols to investigate infestation in protected versus nonprotected areas and the potential diseases carried by the wolves as hosts.

All captured individuals were evaluated for their "reproductive success." This was defined as adult pairs producing offspring during the 1-year assessment interval. Pup survival to weaning and pattern of pup dispersal were also monitored. All these observations were performed by radiotelemetry information and direct observation. Furthermore, semen samples were collected from males during the breeding season and assessed for seminal characteristics (i.e., volume, motility, and morphology). Comparison of seminal characteristics of wolves living in different areas was performed to determine the impact of human disturbance on male reproductive fitness.

Education

Concerning the environmental education, we worked from 2004 through 2010 based on three approaches. The starting point was the beginning of a close interaction with the community in order to learn their culture and understand their perceptions about nature and feelings toward the national park. In order to gather information, chats and interviews were conducted throughout the region with distinct stakeholders, including all generations. Secondly, we focused on teaching environmental subjects in schools, both in local towns and in rural areas. The main objective was to introduce topics related to the space where the children live, the cerrado, through seminars with teachers and talks to the children. This activity promoted discussion among teachers and even at a political level (in the districts' jurisdictions) on an education plan using environmental themes as one of the priorities in local education. The seminars involved close to 100 teachers from two municipalities in the region.

The final objective was to open up communication regarding the activities and purposes of the Maned Wolf Project and the needs and doubts of the community, especially with the farmers. It was hoped that the local community speaking out would result in changes in the way they manage the land, generating positive changes in natural resource use and even their local economy.

After a 1-year period of observation and interactions with the local community, we identified three major aspects to be considered: (a) the negative attitude of the previous land owners that was caused 35 years before when the national park was established by the federal government; (b) the lack of enthusiasm for the culture and environment of the region; and (c) the shortage of opportunities and meetings where discussion about local issues takes place.

Considering these facts, in order to have a successful environmental education initiative that eventually would evolve into a program, we aimed for the following:

a. To promote the region's cultural and environmental potential
b. To improve the setting for meetings and discussion about local issues
c. To encourage the local community to live in a sustainable way, respecting the natural resources, and in harmony with wildlife

Nevertheless, we did not consider only environmental problems related to this species. We chose methodologies that would increase community participation and discussions on the local problems such as tourism, local economy, and public health, among others. Thus, to achieve our goals, two activities were carried out: a collective production of a book about the region and the creation of rural movie sessions.

- Community book. Over 5 months, a group of 30 members of the community—such as teachers, students, farmers, tourism entrepreneurs, and public employees—worked on discussing several topics that resulted in the writing of a book about cultural, historical, environmental, and economic aspects of the Serra da Canastra region.
- Rural movie sessions. The "Cine Lobo" consisted in the exhibition of a 15-minute film produced by the project team. The project generated three short movies showing information about the research project, local natural qualities, conflicts between farmers and maned wolves, and public health. The short movies were presented at schools, farms, and small villages and were followed by commercial movies of interest to the community and later by a discussion of the issues involved, including social, environmental, economic, and others.

Human–Maned Wolf Conflicts

Livestock depredation is one of the main causes of conflicts that frequently result in the persecution of maned wolves due to economic losses. The information collected from 2006 to 2007 showed that 83% of the farms in the Serra da Canastra region (n = 120) had conflicts with carnivore species and the maned wolf was indicated as a species responsible for livestock losses. Considering this, we ran quantitative and qualitative analyses on avian depredation and the real impact of the maned wolf on local economies. Further, we implemented management and preventive methods to reduce the depredation. Twenty farms (of the 120 visited) were chosen (based on depredation rates and willingness of landowners to commit to the project's suggestion of sustainable land use) to systematically evaluate the impact of maned wolves on livestock losses. Much information on these farms has been recorded, including the existing facilities (i.e., chicken coops or similar structures to keep domestic animals) and other factors that might keep predators away (e.g., domestic dogs, scarecrows). We visited them on a weekly basis, recording all the attacks as well as the perceptions of landowners toward related species that caused the conflicts. Additionally, notes on the farmers' attitudes toward isolated attacks and the conflicts in general are registered.

The conflict management was evaluated through intensive monitoring before and after the use of preventive methods (domestic dogs, scarecrows, and chicken coops). The preventive options were analyzed separately in order to highlight the most effective method. The presence and absence of other species of predators involved in conflicts were recorded separately by the active search of indirect signs and camera trapping. The quantification of the attacks was performed. The cost of animals preyed upon by wolves and other predators was based on the local price of chicken in addition to the raising costs to evaluate the real economic impact to a single farmer and from a general regional point of view.

Phase III (2008–2010): Healthy Wolves, Healthy System

During the previous 5 years of the project, we studied the ecology, health, genetics, and reproduction of wolves living in the Serra da Canastra region. We observed that human activities significantly affect wolf ecology and health and perhaps affect reproductive success. However, how the species perceives and uses the altered landscapes was not observed extensively, resulting in the generation of little information. We also found that there is a potential risk of disease spillover from domestic dogs to maned wolves. Considering this situation, we entered the third phase of our project when we started meeting previous objectives and defining new topics to investigate. Using the information generated, and considering the difficulties in managing the scarce financial resources available for long-term projects, we decided what research and actions to prioritize in order to succeed in conservation of the maned wolf. Thus, for this part of the project, our goals were as follows:

- To evaluate the maned wolf habitat's availability and use
- To investigate the impact of domestic animals' diseases on the maned wolf population
- To expand serological assessment to a larger number of wolves and dogs, especially in areas where there is evidence of disease exposure
- To identify various factors contributing to disease risk, including size and location of domestic and wild canids, type of habitat, and anthropogenic factors
- To evaluate the contact rates between maned wolves and domestic dogs in the surrounding areas of SCNP
- To improve and expand interaction with the local community through the educational methods used in Phase II regarding topics of public health and their interactions on the welfare of domestic animals
- To expand the conflict assessment and management through the implementation of preventive methods in other areas of the region

Animals continued to be captured in this phase, but at lower frequencies than before. Other sites were identified for capturing maned wolves to achieve the above objectives. The objectives were more specific and consequently the sampled area was in nonprotected sites. We concentrated our capture activities outside the park in this period. Captures within the protected area were for the purpose of uncollaring the wolves and monitoring in Phase II. The captured animals in this period were fitted exclusively with GPS/VHF collars (Lotek) in order to monitor activity patterns, home ranges in disturbed landscape, living areas, habitat preferences, dispersal movements, behavioral patterns, and responses to human activities, besides the contact rates with domestic animals and epidemiological analysis. The collars were programmed to register locations at 4-h intervals (dropping to 2-h intervals during the breeding season from May through August).

Natural habitats and disturbed areas were assessed in the entire region, inside and outside the protected area. This assessment was correlated with information collected from the animals monitored in the same areas in order to analyze habitat use based on habitat spatial quality. With the aim of mapping potential areas and paths for dispersion, all the environmental variables related to habitat suitability for maintenance of the species were considered. The analysis followed three processes, in a way of overlapping layers of information: (1) Species–environment relationship analysis: Assessment of species' characteristics such as habitat preference, sensitivity to impacts, and general ecological requirements, among other factors. In this procedure, information that was already known about the species in general and in the study area was used. (2) The second layer was built from monitoring by radio telemetry, observations, and mapping concerning individual animals. (3) Analysis of maps and satellite images composed the third layer, where all the possible spatial, geomorphologic, and topographic information was correlated to vegetation maps. Together these formed the background information used to define potential areas for dispersion and establishment in order to preserve or recover ecological integrity.

Furthermore, the proposed objectives were designed to systematically determine the spread of several diseases in the maned wolf population. Although we had already obtained serum samples from several individuals in Phase II (maned wolves and domestic dogs), increasing the sample size and obtaining samples in multiple years allowed us to accurately determine the spatial transmission of these infectious diseases. By using remote sensing technology, landscape epidemiology provides a tool for mapping the spatial distribution of conditions favorable to infection, based on parameters relevant to pathogen presence and abundance. Therefore, we used remote sensing technology to identify the spatial dynamic of several diseases and the contact rates to domestic dogs in the Serra da Canastra region. Although there was no direct evidence of dogs interacting with wolves, a survey and mapping of domestic dogs and cats and their serology were performed in order to overlap the information of rural areas to wolves' home range.

During this period, we visited schools and spoke to teachers and students about conservation of the cerrado and its flagship species, the maned wolf. This resulted in an improvement in local people's attitudes toward the wolves. Because human and wolf conflict is one of the critical conservation threats to this species, our field researchers visited farmers and helped them to solve problems of wolves' predation of chickens. For example, they demonstrated an alternative way to raise chickens (by building chicken coops), as well as educating farmers about the importance of maned wolves to cerrado habitats. We also conducted a preliminary study to determine the proportion of chicken predation by maned wolves and have found that the frequency of chicken depredation by maned wolves varies among regions. Using this perspective, we produced and locally distributed a brochure that explains, step by step, why and how to build a chicken coop. Then, to assist the interested farmers, we were simply advising and stopped providing materials and labor to build the preventive units. Ten units were built by the project some years ago, and since then we have observed 16 new units throughout the range, built by the landowners. Thus, in this phase, we continued to assist the farmers in conflict management not only directly but also through the "Cine Lobo." We produced a short narrative movie that discusses the maned wolf–farmer conflict and how people and wildlife can coexist in the region (available at http://vimeo.com/11825505).

The education actions were based on not only conflict management but also the needs observed during the previous years of contact with the local population. During this phase, we developed a booklet to raise awareness in the community of public health involving healthy everyday practices for people in towns and rural areas. The domestic animals' health and welfare were important topics to be discussed, not only due to the large consumption of animals and animal products (pigs, chicken, and cattle meat, milk, cheese), but also due to the bad disposal or mistreatment of domestic animals in close contact with the local population and wildlife. Besides the printed material, once more we produced a short video to illustrate the problems and present solutions; this was presented at rural movie sessions, local schools, festivals, and other gatherings (available on http://vimeo.com/6391391). Additionally in this phase, the project, in association with Brasilia University, conducted a community communication course aiming to teach environmental education practices. The course was open to the local people, who learned and produced several materials: a radio campaign, a newspaper, and a short video, among others, based on environmental topics.

Phase IV (2010 to Present): Gathering, Informing, Expanding

Most of the data obtained in Phase III are still under analysis, especially those on spatial ecology and ecological epidemiology. However, while evaluating the data from the third phase, especially those on landscape ecology and epidemiology, more questions were generated and the need to expand not only the scientific database but also the educational actions and the conflict assessments and management to other areas within the region was recognized. To fulfill these needs, the project initiated its fourth phase.

The Maned Wolf Conservation Project has always pursued the spread of information as one of its main principles. Scientific and general information has been disseminated only locally, through educational activities and strategies that we have implemented since 2000. From 2004, the project gained the attention of TV, magazines, newspapers, and radio reports regionally and nationally. In many cases, we engaged the local people in the reports (especially those for TV) in order to raise their self-esteem showing the community that several of their own people were involved in good environmental practices.

Environmental education, outreach, and building capacity within local communities are essential components of our maned wolf conservation initiative. Since the project's onset, we have made progress toward these objectives. Our efforts to visit farms and schools also seem to be improving the attitudes of local people toward not only the maned wolf, but also the value of wildlife in general and the cerrado ecosystem. Some of our early success has been due to the videos shared in movie sessions, the community book (available at http://www.procarnivoros.org.br/pdfs/Um_lugar_chamado_canastra.pdf) that is nowadays being used in local schools, the public health campaign, and the conflict management. In all cases, we had deep involvement of the local community. Drawing on the belief that there is much more to be accomplished in terms of community outreach and building relationships with local people, we structured a new phase of the project based on environmental education and public awareness through several communication practices.

New areas were chosen in order to continue to assist the local community in conflict management. The same approach used in Phases II and III will be used in other areas that have not yet been visited. In the coming years, we intend to replicate the methods that we have used in Serra da Canastra in other regions. Thus, the objectives for this phase, besides continuing to evaluate and manage the conflicts, are to focus on public awareness not only for the Serra da Canastra region but also throughout the country, with the collaboration of several partners.

A massive maned wolf communication campaign was initiated in 2012. The two main steps are as follows: (1) the development of a wolf character to be incorporated in written materials concerning environmental education (flyers and pamphlets that emphasize the importance of maned wolf conservation and cerrado protection) and these materials to be distributed to local people, children in schools, and tourists and in strategic locations in large cities. Additionally, we have started to work with local, regional, and national media (television, newspaper, magazines, and radio) to inform the public about conservation issues critical to the sustainability of the cerrado and wildlife status and importance. Besides reaching landowners and people in conflict with the species, another goal is to (2) reduce the impact of human activities on wildlife, mainly on maned wolves. Furthermore, the maned wolf character will be the center of an educational campaign, "Sou Amigo do Lobo" ("I'm a Friend of the Maned Wolf"), to be started in 2013, where several products and benefits will be arranged with local schools, businesses, hotels, shops, tourist centers, and so forth. In addition to visiting schools, our team will lead field trips for students of local schools and universities to enhance their appreciation of wildlife and cerrado habitats and increase efforts toward protecting them.

As the final products of this phase, we intend to achieve the following:

- To have the maned wolf nationally recognized as the cerrado's ambassador and as the animal symbol for the biome. Consequently, we will try to define the species as the totem of the Serra da Canastra region.
- To engage Serra da Canastra's community in discussions and meetings to define new policies for environmental matters.
- To have the natural attributes of the Serra da Canastra region respected and used in an environmental rational, lowering the negative effects of human activities on wildlife and the landscape in general.

Thus, the actions and results produced by this phase should benefit the local community in two main ways: (1) the valuation of nature and culture; and (2) the improvement of local articulation in order to discuss their problems. As part of this context, we plan to invest more time in and use different approaches to reach out to kids and teenagers, providing favorable conditions for communication within their educational processes. Indirectly, a society that is better prepared to discuss different topics, and that has been protecting the wildlife and preserving their suitable habitats, might increase tourism in the region.

As this has been successful in Serra da Canastra, we aim to disseminate all of our methods and strategies and adapt them in other regions where maned wolf conservation is already compromised.

Participating Institutions

The following institutions participated in the project by coordinating or providing support on data analysis, personnel resources, infrastructure, and so forth: Instituto Pró-Carnívoros; Centro Nacional de Pesquisa e Conservação de Mamíferos Carnívoros, CENAP; Universidade Federal de Minas Gerais, UFMG; Universidade de Brasília, UnB; Centro Universitário de Brasília, UNICEUB; Smithsonian Conservation Biology Institute (formerly National Zoo's Conservation and Research Center); Faculdade de Medicina Veterinária e Zootecnia, FMVZ/USP; Pontifícia Universidade Católica do Rio Grande do Sul, PUCRS; Universidade Federal do Paraná, UFPR; Fundação Oswaldo Cruz, FIOCRUZ; Pontífice Universidade Católica, PUC/Arcos; and Universidade Estadual Norte Fluminense, UENF.

OVERVIEW OF RESULTS

The impressive amount of data generated by the research over 8 years of investigations, in addition to the feedback from the educational approaches and conservation practices, resulted in great contributions and good examples of how a transdisciplinary and multi-institutional project can grow into a program and succeed in linking science, education, politics, and economics.

Not all the data have been analyzed and processed yet, especially those related to Phases III and IV. New information not only on the species, but also on the methods we are using has already been published in peer-reviewed articles and books (May-Júnior et al. 2009; Rocha and Bizerril 2009; Rocha et al. 2013; Bizerril et al. 2011; Paulino et al. 2008; Spercoski et al. 2012). Most of the information acquired in Phase II is available in academic theses and monographs (Amboni 2007; Corral 2007; Leão 2009; Santos 2007; Azevedo 2008; Spercoski 2007; Rodrigues 2009), abstracts of both national and international conferences, and reports for sponsors. Additionally, educational folders, flyers, and booklets were produced to support educational activities. All the scientific and educational material published from the project can be downloaded in the publications section of the Pró-Carnívoros Institute website (http://www.procarnivoros.org.br). The short videos produced in the project are available at http://vimeo.com/comcomunitaria.

In Table 13.1 we present some information related to the research activities, and the educational and conservation practices.

From 2004 to 2007, most of the project activities were based on capture and monitoring. Thus, the year 2006 presented the largest number of animals trapped (Figure 13.1).

Following the proposed objectives, the capturing was intense from 2004 through 2007. In Phase III they were lower and in Phase IV they were reduced to two capture expeditions per year. Figure 13.2 illustrates the sequence of capture events on a yearly basis. More information on the capturing events and results are available in Table 13.1.

One major finding in the current study, which differs greatly from that of Dietz, who studied maned wolves within the same area three decades earlier, is that wolf density in this study seems to be much higher than that reported by Dietz (1984). In the current study, several adult wolves occupied an overlapped home range, while Dietz reported that only one male and one female shared the same area. The differences in the management of the area (from the late 1970s compared to this

TABLE 13.1
Project Remarks from 2004 to 2011

Captures and monitoring	303 capture events (7423 trap-nights) 58 maned wolves captured (28 females and 30 males): 6 young (3–12 months old), 22 subadults (1–3 years old), 30 adults (3–10 years old) High recapture rate: 28 animals captured 2–68 times; 30 wolves captured only once 43 animals monitored by VHF telemetry (1.278 locations) 9 animals monitored by GPS/VHF telemetry (23.384 locations) 25 individuals living exclusively within protected areas; 19 wolves on park's border (within 1 km and off limits); 20 animals specifically in farmland
Ecology	Home range varying 15–114 km^2 (average 51 km^2) Home ranges within the SCNP larger than those outside No differences between males' and females' home ranges Average for females: 63 km^2 (usually dropped to half size during breeding season) Average for males: 42 km^2
Social behavior	Overlapping of home ranges reached 92% for a defined pair Tolerance among wolves, from higher to lower (observed through overlapping home ranges): male–female > female–female > male–male
Demography	Serra da Canastra has been identified as the site with the highest maned wolf population density: 0.08 individuals/km^2
Reproduction	First maned wolf sperm collection in the wild Variation of levels of cortisol metabolites in feces; possible to conclude that wolves living in area surrounding the park may have been exposed to events that lead to severe stress processes, opposed to those living within the protected areas
Genetics	Development of maned wolf's genetic markers Definition of familiar groups among the captured animals and observation of immigration of individuals in the protected area, adding new genetic material in local gene pool
Health and epidemiology	Pathogens detected in maned wolves: canine distemper, CPV, adenovirus, coronavirus, rabies, Chagas' disease, leishmaniosis, leptospirosis, and toxoplasmosis Vaccination of domestic dogs in the surroundings of the SCNP: 2500 vaccines (in 2004, 2005, 2006, 2009, 2010)
Training and capacity building	Hosted 3 field courses (Concepts and Methodologies on Carnivore Conservation), with a total of 94 Brazilian and foreign students trained 32 students and professionals trained in internships of a minimum of 1 month 2 PhD theses 7 master's theses 3 graduation monographs 14 presentations at international and national conferences
Conflict management	Building of 10 chicken coops in local farms to prevent aviary depredation Significant reduction of maned wolf killings Significant increase of tolerance toward maned wolf (reduction of hunting) Conflict zones for maned wolf depredation defined through ecological modeling to direct efforts on conflict management
Education	86 lectures in local and rural schools about the importance of the cerrado ecosystem and presentation of the project Distribution of the project DVD (containing the short educational movies produced by the project team) to local institutions, schools, and local people Publication of the community book *Um Lugar Chamado Canastra* (*A Place Called Canastra*) and distribution of 4000 copies in the region Publication and distribution of 1000 copies of the children's book *Lá no Coração* (*Into the Heart*)
Funding	Around US$600,000 from Brazilian and international funding agencies, direct sponsors, and private companies

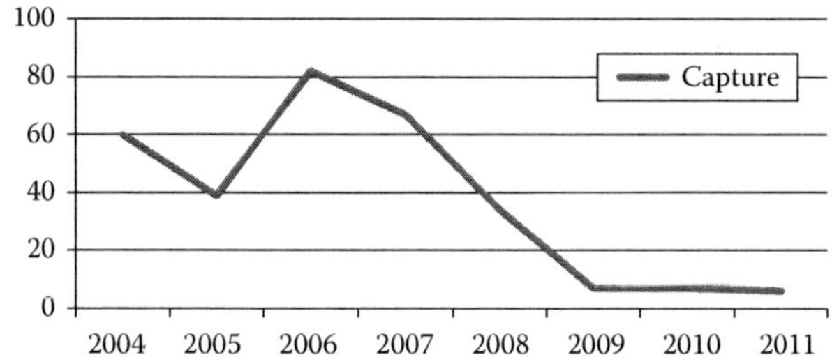

FIGURE 13.1 Maned wolf capture events from 2004 to 2011 in Serra da Canastra region.

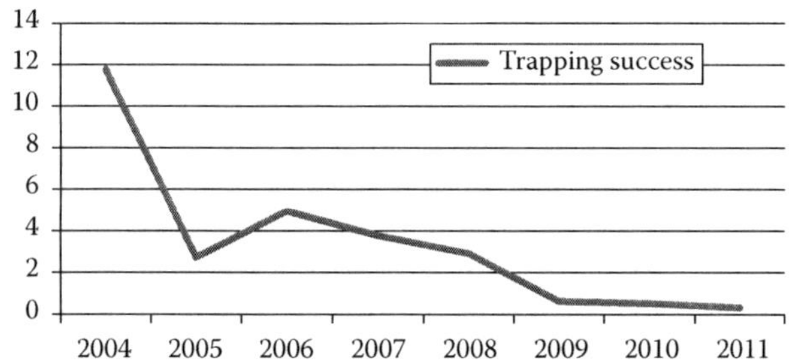

FIGURE 13.2 Maned wolf trapping success from 2004 to 2011 in Serra da Canastra region.

period) surely contributed to the differences noticed in the maned wolf population and the use of Serra da Canastra and its surroundings. The area began to be managed as a fully protected area only in 1979, making a large difference in the quality of habitats within the national park.

Because CENAP/ICMBio (a government research center within the Environmental Ministry) is responsible for developing and implementing environmental policies related to carnivore species, its participation in the project organization ensures that knowledge gained from our investigations has an impact on decision-making processes and on the implementation of the conservation action plan for the maned wolf.

In a general analysis of the achievements of the Maned Wolf Conservation Project in the last 12 years, we are confident that the presence of institutions with activities based in research, education, and public policies, among other topics of conservation, has been important in generating a high quantity and quality of information. This was essential for the use and spread of applied actions both locally and nationally. However, the involvement of the local community, which has been present in almost all the activities, has been crucial to the success of achieving our objectives. With that, we have indications that maned wolf hunting has decreased significantly in the region and local acceptance of the maned wolf has improved greatly, as well as people's awareness of nature in general. This has been possible mainly due to the intense involvement of local institutions and people from different backgrounds in planning and executing several activities in a way that everyone learns and evolves toward a common objective: a better, sustainable life depending on natural resources and the preservation of wildlife.

REFERENCES

Amboni, M.P.M. 2007. Dieta, disponibilidade alimentar e padrão de movimentação do lobo-guará, *Chrysocyon brachyurus*, no Parque Nacional da Serra da Canastra, MG. Dissertação de Mestrado, Universidade Federal de Minas Gerais, Belo Horizonte.

Azevedo, F.C. 2008. Área de vida e organização espacial de lobos-guará (*Chrysocyon brachyurus*) na região do Parque Nacional da Serra da Canastra, Minas Gerais, Brasil. Dissertação de Mestrado, Universidade Federal de Minas Gerais, Belo Horizonte.

Bizerril, M.X., Soares, C.C., Santos, J.P. 2011. Linking community communication to conservation of the maned wolf in central Brazil. *Environmental Education Research*, 17:815–817.

Corral, L. 2007. Avaliação da predação de criações domésticas por lobo-guará (*Chrysocyon brachyurus*) no entorno do Parque Nacional da Serra da Canastra, MG, Brasil. Dissertação de mestrado, Universidade Federal de Minas Gerais, Belo Horizonte.

Dietz, J.M. 1984. Ecology and social organization of the maned wolf (*Chrysocyon brachyurus*). *Smithsonian Contributions to Zoology*, 392:1–51.

Entwistle, A., Dunstone, N. 2000. Future priorities for mammalian conservation. In: A. Entwistle, N. Dunstone (eds), *Priorities for the Conservation of Mammalian Diversity—Has the Panda Had Its Day?* Cambridge University Press, Cambridge, UK.

Jerusalinsky, L., Teixeira, F.Z., Lokschin, L., Alonso, A., Jardim, M.M.A., Cabral, J.N.H., Printes, R.C., Buss, G. 2010. Primatology in south Brazil: A transdisciplinary approach to the conservation of the brown-howler-monkey *Alouatta guariba clamitans* (Primates, Atelidae). *Iheringia Série Zoologia*, 100:403–412.

Leader-Williams, N., Dublin, H.T. 2000. Charismatic megafauna as "flagship species." In: A. Entwistle, N. Dunstone (eds), *Priorities for the Conservation of Mammalian Diversity—Has the Panda Had Its Day?* Cambridge University Press, Cambridge, UK.

Leão, R.S. 2009. Ocorrência de mercúrio em lobo-guará (*Chrysocyon brachyurus*) e components de dieta do Parque Nacional da Serra da Canastra e a'reas do entorno, Minas Gerais, Brasil. Dissertação de Mestrado, Universidade Federal do Mato Grosso do Sul, Campo Grande.

May-Júnior, J.A., Songsasen, N., Azevedo, F.C., Santos, J.P., Paula, R.C., Rodrigues, F.H.G., Rodden, M.D., Wildt, D.E., Morato, R.G. 2009. Hematology and blood chemistry parameter differs in free-ranging maned wolves (*Chrysocyon brachyurus*) living in the Serra da Canastra National Park (Brazil) versus adjacent farmlands. *Journal of Wildlife Diseases*, 45(1):81–90.

Mota, L.D., Paula, R.C. 2006. *Serra da Canastra.* São Paulo, Aori.

Myers, N., Mittermeier, R.A., Mittermeier, C.G., Fonseca, G.A.B., Kent, J. 2000. Biodiversity hotspots for conservation priorities. *Nature*, 403:853–858.

Paula, R.C., Médici, P., Morato, R.G. (eds) 2008. *Plano de ação para a conservação do lobo-guará: Análise de viabilidade populacional e de habitat.* Edições IBAMA, Brasília.

Paula, R.C., Mota, L.D. 2010. *Serra da Canastra: Diversidade infinita.* Gamba Productions, Sao Paulo.

Paulino, F.O., Bizerril, M.X.A., Coelho, J.F.G., Mendes, J.S., Pedrosa, L.L. 2008. Fala canastra! Ações de comunicação comunitária para a educação ambiental na Serra da Canastra. *Participação.* Revista do Decanato de Extensão da Universidade de Brasília, v. 8, pp. 58–66.

Ray, J.C. 2005. Large carnivorous animals as tools for conserving biodiveristy: Assumptions and uncertanties. In: J.C. Ray, K.H. Redford, R.S. Steneck, J. Berger (eds), *Large Carnivores and the Conservation of Biodiversity.* Island Press, Washington, DC.

Reyers, B., Roux, D.J., Cowling, R.M., Ginsburg, A.E., Nel, J.L., O'Farrell, P. 2010. Conservation planning as a transdisciplinary process. *Conservation Biology*, 24(4):957–965.

Rocha, F.L, Bizerril, M.X.A. 2009. Promoção da saúde na Serra da Canastra: Aliando preservação ambiental e saúde pública. In: F.O. Paulino. *Comunicação e Saúde.* Casa das Musas, Brasília.

Rocha, F.L., Roque, A.L.R., Arrais, R.C., Santos, J.P., Lima, V.S., Xavier, S.C.C., Cordeir-Estrela, P., D'Andrea, P.S., Jansen, A.M. 2013. *Trypanosoma cruzi* TcI and TcII transmission among wild carnivores, small mammals and dogs in a conservation unit and surrounding areas, Brazil. *Parasitology*, 140(2):160–170.

Rodrigues, M.L.F. 2009. Diversidade genetica e estrutura populacional do lobo-guará (*Chrysocyon brachyurus*). Dissertação de Mestrado. Pontifícia Universidade Católica do Rio Grande do Sul, Porto Alegre.

Santos, J.P. 2007. Análise quantitativa e métodos preventivos de predação de animais domésticos por canídeos selvagens no entorno do Parque Nacional da Serra da Canastra. Monografia de Graduação. Instituto Superior de Educação—UNIFOR-MG, Formiga.

Spercoski, K.M. 2007. Monitoramento a longo prazo da excreção de corticóides fecais em lobos-guará (*Chrysocyon brachyurus*) de cativeiro e vida livre: Uma contribuição para o manejo e a conservação da espécie. Dissertação de Mestrado, Universidade Federal do Paraná, Curitiba.

Spercoski, K.M., Morais, R., Morato, R., Paula, R.C., Azevedo, F.C., May-Júnior, J.A, Santos, J. P., Reghelin, A.L., Wildt, D., Songsasen, N. 2012. Adrenal activity in maned wolves is higher on farmlands and park boundaries than within protected areas. *General and Comparative Endocrinology*, 179(2):232–240.

14 Maned Wolf Conservation in the South of Brazil and Uruguay

Diego Queirolo, Cibele Indrusiak, Lorena Coelho, Mariana Cosse, and Graziela Dotta

CONTENTS

Introduction 193
Historic and Current Knowledge 194
Threats and Conservation 197
Future Conservation Research 198
References 199

INTRODUCTION

The maned wolf (*Chrysocyon brachyurus*) (Illiger, 1815) is considered a threatened species in all states of the southern regions of Brazil, and is listed as Vulnerable (VU) in Paraná (GEP 2010), Critically Endangered (CR) (C2ai, D2) in Santa Catarina (CONSEMA 2011), and Critically Endangered (CR) (A3, B2, C2, D3, E3) in Rio Grande do Sul (GERS 2002). In the Brazilian assessment, it is classified as Vulnerable (VU) (A3c) (MMA 2003) and recently as Vulnerable (VU) (A3ce) (Paula et al. forthcoming). In Uruguay, there is no list of threatened species following the IUCN standard (IUCN 2011), but on the National Protected Areas System program, the maned wolf has been classified as a "priority species for conservation" (Soutullo et al. 2009). Finally, it is a Near Threatened species (NT) (3.1) at global level (Rodden et al. 2008), and it is also on Appendix II of the Convention on International Trade in Endangered Species of Wild Fauna and Flora (CITES 2012).

It is a canid species associated with open areas, featuring herbaceous vegetation and shrubland, such as the cerrado, the Chaco, swamp and wetland areas, "palmares" (areas dominated by palm species), and grasslands (Dietz 1985; Rodden et al. 2008; Queirolo et al. 2011). In the cerrado of Goiás State, central Brazil, Rodrigues (2002) reports that the maned wolf forages in "campo sujo" (grassland habitat with sparse shrubs) and rests in areas with tall grasses (up to 2 m high). Queirolo and Motta-Júnior (2007) point out that at Serra da Canastra National Park more than 80% of the prey species consumed by the maned wolf are inhabitants of open physiognomy areas.

Dietz (1984) suggests that *C. brachyurus* is a solitary species, and is more active at dusk and night. However, recent studies suggest a more social behavior, with the species showing intense interaction among individuals and even between males, females, and their offspring (de Melo et al. 2007; Jácomo et al. 2009). Maned wolves weigh about 24 kg and are about 1 m high. Despite its large body size, the species feeds mainly on small to medium vertebrates, such as rodents, birds, and reptiles, as well as on insects, eggs, and fruits, this last item making up a large proportion of total consumption (Dietz 1984; Motta-Júnior et al. 1996, 2002; Queirolo & Motta-Júnior 2007; Rodrigues et al. 2007; Bueno and Motta-Júnior 2009). *C. brachyurus* has a complex vocal range

(Peters & Wozencraft 1989) that is used, among other things, to mark out territory, and its various calls have resulted in many myths and folktales surrounding the species.

HISTORIC AND CURRENT KNOWLEDGE

The maned wolf is a species with few records, leading to an assumption that it has always been rare in its southern distribution region. In Uruguay it was not included in any list of mammal species until two illegal hunting incidents involving adult male individuals: the first in 1989 in the Department of Río Negro (32°35′45″S; 58°08′46″W) and the second in 2006 in the Department of Cerro Largo (32°14′43″S; 54°03′30″W). Both specimens are now part of the Collection of Mammals in the National Museum of Natural History in Montevideo, Uruguay. Moreover, in the area in Cerro Largo where one individual was found, a fecal sample from another specimen was obtained.

Mones and Olazarri (1990) cite several authors that have mentioned a historic presence of the maned wolf in Uruguay. However, there is no evidence to prove species records, as these are based only on peoples' reports or personal observations. According to Mones and Olazarri (1990), the first proven record of maned wolves in Uruguay was in 1923, by Larrañaga, in a manuscript written almost 100 years before its publication. Azara (1802) mentioned the species' presence in Uruguay extending to the Río de la Plata (La Plata River), and Burmeinster (1861) suggests that the maned wolf could be found particularly in the west of the Río de la Plata. These two reports refer to the territory where Uruguay and south Brazil are today.

Figueira (1894) mentions the rare presence of *C. brachyurus* at the end of the nineteenth century and states that the species could only be seen in the departments of the north and east of Uruguay. Moreover, Acevedo Díaz (1890) considers that "the maned wolf" is "on his way to complete extinction" and that in those times finding individuals of the species to the north of Negro River was sporadic. Furthermore, Devincenzi (1935) has no doubt concerning the presence of the maned wolf in the nineteenth century. However, Arredondo (1959) distrusts the previous information, stating that "in Cuareim and Yaguarón I have heard that some people have seen maned wolves, occasionally." Finally, Ximénez et al. (1972) considered that *C. brachyurus* was a species not confirmed in Uruguay. From this point in time, *C. brachyurus* was not mentioned in any of the following lists of mammal species of Uruguay (Langguth 1976; Achaval 1989). Nevertheless, people living in the countryside have made reports and declarations about the presence of the maned wolf over time, including recently (Mones and Olazarri 1990; Prigioni and Sappa 2003).

It is hard to find documents showing the historic presence of the maned wolf in the Rio Grande do Sul (RS) region. Ihering (1927) considers *C. brachyurus* as a rare species, mentioning that it could be found on "the large wetlands in the south of Rio Grande do Sul." Comparing this affirmation with Figueira (1894), who states the occurrence of maned wolves in east Uruguay, we remark that this region is a continuum of the wetlands mentioned by Ihering. Breyer (1979) records the presence of *C. brachyurus* in all municipalities of Rio Grande do Sul using interviews. This author verifies the regular presence of the maned wolf in central, western, and southwestern areas of the state, besides a small patch in the northeast region and intermittent occurrence in the extreme north, south, southeast, coastal plain, and in the capital of Rio Grande do Sul State, Porto Alegre. The highest frequency of reports was in Serra do Caverá, municipality of Alegrete, and in the Campos de Cima da Serra (upland grasslands), in the north. Indrusiak and Eizirik (2003) describe the presence of *C. brachyurus* in Rio Grande do Sul, mainly in the northeastern municipalities and in the southern part of the state, including municipalities on the boundary with Uruguay, such as Dom Pedrito. More recent records on maned wolf occurrence in the state are from São Gabriel (2009; camera trap photograph taken by Leandro Chisté Pinto, *Zero Hora Newspaper* 27/08/09) and Santiago (2007; T. Silva, personal communication) municipalities, located in the central–southwestern and central–western regions of Rio Grande do Sul, respectively. Moreover, one individual has been photographed in the National Forest of São Francisco de Paula, in the east–northeast (R. Marques, personal communication). Once again, it is possible to establish an association between the species' presence in northeastern Uruguay and in west–southwestern Rio Grande do Sul.

Cimardi (1996) considers the species as probably extinct in the state of Santa Catarina, but Cherem et al. (2004) mention that the last reports of the maned wolf are from 1996, in the grasslands of Lages, near the boundary with Rio Grande do Sul. Historic records have mentioned the species' presence in Campos Novos (Azevedo et al. 1982), also close to Rio Grande do Sul. Recently, Bazilio et al. (2009) record the presence of *C. brachyurus* in the north of Santa Catarina (municipality of Três Barras), near the state of Paraná. The species is presumably rare in Paraná, since historic records are almost inexistent. Currently, the maned wolf can be found in central and north Paraná (Paula et al. 2008), following open vegetation habitats (cerrado and Campos Gerais). The presence of the maned wolf in Paraná might be increasing because the species could be using areas previously dominated by Atlantic Forest vegetation but that have been recently converted to open habitats, as suggested by Queirolo et al. (2011).

Chebez (2008), reporting the occurrence of *C. brachyurus* in Argentina, describes the appearance of the species in Misiones, the province neighboring Rio Grande do Sul. The author refers to a capture event of a maned wolf in 1989, in the municipality of Garruchos (RS), nearest to the Uruguay River. Chebez (2008) also describes the species' distribution in Argentina, mentioning a series of localities in the provinces of Corrientes and Entre Ríos, both along the Uruguay River, fronting Rio Grande do Sul and Uruguay. One of the most interesting aspects of this report relies on the following statement: "people talk about some erratic animal south of the Gualeguaychú river mouth," this locality being almost in front of the place where one specimen was captured in 1989 in the Uruguayan department of Río Negro.

The lack of information on current and historic distribution of *C. brachyurus* was highlighted by specialists at the Maned Wolf Action Plan: Population and Habitat Viability and Assessment workshop, held in Brazil in 2005 (Paula et al. 2008). South Brazil and Uruguay, together with the Pampean region in Argentina, are considered one of the most modified and threatened habitats of South America (Bilenca and Miñaro 2004) due to strong anthropogenic pressures, such as the conversion of natural grasslands in agricultural areas and for livestock farming. Because of those habitat modifications, recent geographic distribution of *C. brachyurus* has been facing changes, showing a large decrease in its austral range (Queirolo et al. 2011).

When conservation of biodiversity is advocated, it is essential to understand the recent distribution of species in detail (Peterson et al. 2002). Aiming to achieve this understanding, in recent years species distribution models have been developed in an attempt to answer questions about ecology, biogeography, and conservation, as well as looking at climatic change and aspects of species management (Guisan and Zimmermann 2000; Franklin 2010). Distribution models connect a set of environmental variables to data on species occurrence in a particular region (Pearson 2007), and they can be used to predict suitable habitats for a specific species (Phillips et al. 2006; Franklin 2010). Here, we used *C. brachyurus* records obtained from Queirolo et al. (2011) and environmental variables that were analyzed together in order to identify the environmental conditions associated with the species distribution and create a potential distribution map (Figure 14.1). It is clear from the map that north Uruguay and Rio Grande do Sul State (darker portions on the map) are areas extremely relevant for the occurrence of the maned wolf, suggesting that in terms of climate the region continues to be suitable for the species.

Future potential distribution of *C. brachyurus*, using scenarios that match three of the four scenarios developed by the Intergovernmental Panel on Climate Change (IPCC), is shown in Figure 14.2. The scenarios represent possible evolution of future greenhouse gas (GHG) emissions based on hypotheses about socioeconomic development and its connections (Montes 2010). They represent the future from two approaches: (1) development focused on environmental and economic aspects, and (2) a world with global or regional growing patterns (Cuesta et al. 2009). Scenario A1B implies mid–high emission, which means fast regional economic development driven by the use of new and more efficient technologies, with a balance between the use of fossil and nonfossil fuels. On the other hand, scenario A2A implies high GHG emissions, with stable population growth,

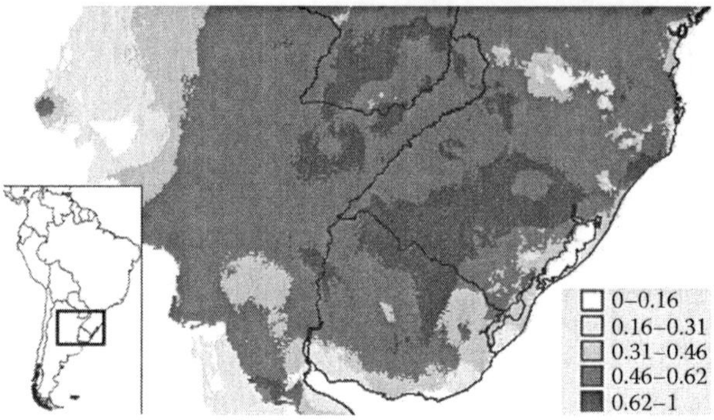

FIGURE 14.1 Predictive models of potential geographic distribution for the maned wolf in south Brazil and Uruguay. MaxEnt 3.3.3 software (From Phillips, S.J., et al., *Ecological Modelling*, 190, 231–259, 2006) was used to combine current occurrence records of *C. brachyurus* (From Queirolo, D., et al., *Oryx*, 45(2), 296–303, 2011) and 10 climatic variables obtained from WorldClim database (http://www.worldclim.org) (From Hijmans, R.J., et al., *International Journal of Climatology*, 25, 1965–1978, 2005). Label values closer to 1 (darker areas) indicate a higher probability of species occurrence.

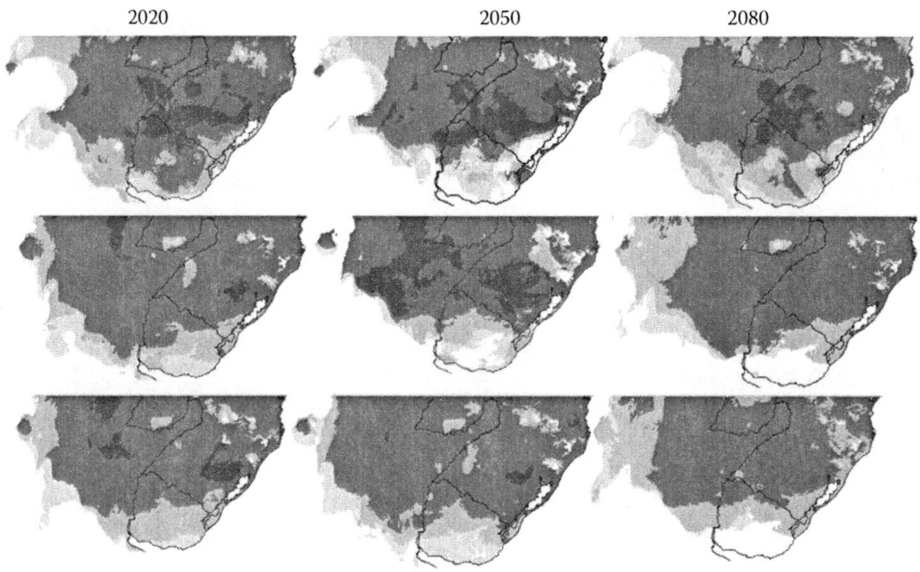

FIGURE 14.2 Predictive models of potential geographic distribution for the maned wolf in south Brazil and Uruguay. MaxEnt 3.3.3 software (From Phillips, S.J., et al., *Ecological Modelling*, 190, 231–259, 2006) was used to combine current records of occurrence of *C. brachyurus* (From Queirolo, D., et al., *Oryx*, 45(2), 296–303, 2011) and future climatic scenarios (A1B, A2A, and B2A) projected by the HadCM3 model developed by the Hadley Centre Coupled Model to 2020, 2050, and 2080, using a spatial resolution of 2.5 arc minutes. Scenarios obtained from WorldClim database (http://www.worldclim.org) (From Hijmans, R.J., et al., *International Journal of Climatology*, 25, 1965–1978, 2005). As in Figure 14.1, darker areas have a higher probability of occurrence of maned wolves.

regionally adjusted economic development, and a disconnected and slow improvement in the use of new technologies. Finally, scenario B2A describes a development model in which local solutions to economic, social, and environmental sustainability issues prevail. This scenario implies low GHG emissions, and it is aligned to environmental protection and social equality, focusing on local and rational levels (IPCC 2007; Cuesta et al. 2009).

In Figure 14.2, we can see that in scenarios A2A and B2A, areas with excellent probabilities of occurrence of *C. brachyurus* (darker portions) are fragmented and reduced and by 2080 none of these areas are left. On the other hand, in scenario A1B, areas with excellent probabilities of occurrence are reduced, but do not disappear in 2080. Prevosti et al. (2004) mention that maned wolf occurrence could be connected to a limited range in temperature that would be the major component affecting species distribution if it was driven only by climatic factors. These authors suggest that during the Holocene, *C. brachyurus* expanded its distribution area due to an increase in temperature. Scenario A1B implies mid–high GHG emissions (global warming) and an increase in temperature, although to a lesser degree than scenario A2A.

This argument means that maned wolf distribution could be benefited by scenario A1B, as has occurred in the past, but it implies not only a fast global economic growth but also a rapid introduction of newer and more efficient technologies, which could end up restricting the maned wolf's geographic distribution, for instance, by habitat reduction. On the other hand, taking into account only climatic factors, it is expected that the potential distribution of *C. brachyurus* could be maintained in scenario A2A, the high emissions scenario, since this model does not consider several other factors (ecological and anthropic) that can also affect species distribution. These results could attest that more knowledge is needed about the main factors that affect maned wolf distribution if we intend to improve the conservation efforts of the species.

Recently, some molecular ecology techniques have been used on the three specimens found in Uruguay (two captured individuals and one fecal sample). It is valuable to highlight here the importance of the use of noninvasive techniques in the study of species that are elusive and/or have low population density, such as the maned wolf in Uruguay and south Brazil. Molecular characterization of the specimens from Uruguay allowed us to establish high allotypic diversity to the mitochondrial control region, since three distinct haplotypes were obtained from the three samples (González et al. 2009). The three individual samples were determined as males by analyzing a fragment of gene ZFX-ZFY (Mannise et al. 2012).

On the other hand, preliminary analysis of microsatellite loci allows us to hypothesize that the samples from Centurión locality, Department of Cerro Largo, are of a male and its offspring. We used nine molecular markers and the samples shared at least one allele, an expected pattern of Mendelian inheritance (Mannise et al. 2011). Based on these results, we can state that there are reproductively active specimens inhabiting the region, since a familial group has been found in Cerro Largo. Nevertheless, this might be a region with individuals in transit.

THREATS AND CONSERVATION

Because the maned wolf is a typically nonforest species, we could suppose that the sharp reduction of its area of occurrence was, at least in part, a consequence of the agriculture and livestock expansion over grasslands and savannas in south Brazil and Uruguay (Queirolo et al. 2011). Increase in afforestation and climatic changes, particularly since the Holocene (12,000 years BP), can also be considered as important factors driving the species' geographic distribution (Prates Jr 2008). In addition, removal of large patches of Atlantic Forest, mainly in Paraná and Santa Catarina States, could be permitting an expansion of *C. brachyurus* distribution with individuals coming from the cerrado of central São Paulo and southwest Minas Gerais States and moving from north to south. In this case, we would have individuals from central Brazil being in contact with those from Rio Grande do Sul State, something that has never occurred in the past. Moreover, hunting for several reasons, such as fear and livestock (particularly sheep) and domestic bird protection, as well as to obtain amulets, is also an important agent in the species' decrease in the region (Indrusiak and Eizirik 2003). Further, in these southern grasslands it is common practice to use poison to eliminate predator species, and this practice could cause the death of some maned wolves. Finally, the use of nonselective traps could occasionally capture maned wolves and end up killing them.

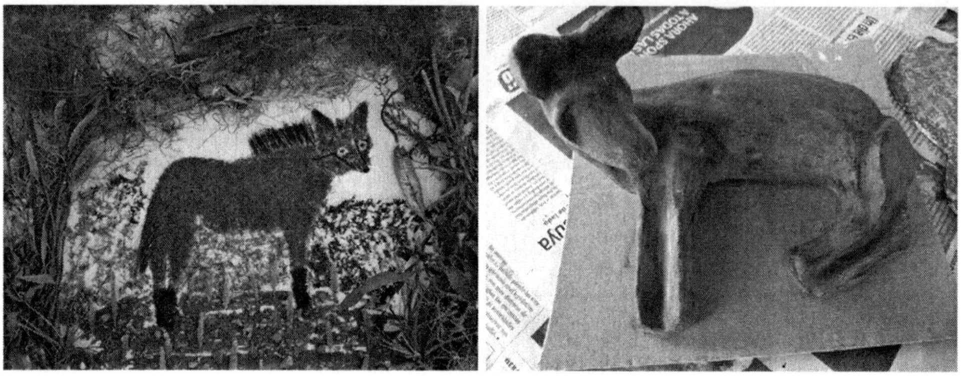

FIGURE 14.3 Maned wolf representations made with local materials, such as plant species and horsehair (*left*) and clay (*right*). Both were obtained during workshops about biodiversity and plastic expression with rural women in Centurión (Cerro Largo Department), Uruguay.

Chebez (2008) considered that road hit is an important element of *C. brachyurus* mortality in Argentina. In Rio Grande do Sul and in Uruguay, it cannot be considered as an important threat, but in Paraná there have been some hit events recently (Paula et al. 2008), leading us to assume that in this region, as maned wolves are more frequent, road collisions can represent a threat for the species' survival. Finally, Chebez (2008) also mentions several diseases of *C. brachyurus* individuals, such as cystinuria, the renal parasite *Dioctophima renale*, and parvovirus, which increase the species' vulnerability, a fact also highlighted by May-Junior et al. (2009).

It is important to highlight the role of environmental education in the region, increasing community knowledge about native species as well as amplifying their participation in conservation actions. These actions are essential to minimize socioeconomic conflict, not only by modifying people's beliefs about the maned wolf being a competitor (e.g., by using management methods to minimize predator conflicts with domestic animals; considering the reduction or cessation of trapping/killing canid species, which can affect other species as well), but also by increasing community involvement in conservation and biology aspects of *C. brachyurus*.

A project aiming to instigate awareness of natural diversity and cultural richness has been developed with the local community of Centurión in the Department of Cerro Largo, Uruguay (sponsored by Global Green Grant [PR52-606]; Eco-Compromiso, 2009, World Bank; and Agencia Nacional para La Investigación y La Innovación, Uruguay, 2008–2010 [PR-PCTI200940]). Working with children and rural women in Plastic Expression Workshops (Figure 14.3) allowed the community to become more familiar with several aspects of the local fauna from a creative and aesthetic point of view, and it also helped to collate bibliographic references on the species, material that was used during the development of the workshop activities. Through the different activities developed in this area, we have noticed a new perception of the species and the need to protect biodiversity. Moreover, the activities have helped in multiplying the potential for sustainable development in the locality. Small landowners are coordinating different tasks aimed at ecotourism entrepreneurship, to speed up the process of inclusion of the area in the National System of Protected Areas (Cosse and Mannise 2008). In 2011, injunction reliefs were taken to stop afforestation, after demands made by the local community. We consider this process can be used to demonstrate the importance of projects involving local communities when one aims to protect biodiversity in an area.

FUTURE CONSERVATION RESEARCH

In 2005, the elaboration and implementation of the Action Plan for Maned Wolf Conservation were set out during a workshop in which researchers from all countries where the species is found took part. The workshop resulted in a publication (Paula et al. 2008) stating some objectives to assist the development of regional conservation programs for the maned wolf. Some questions arise from this paper that we consider essential when suggesting the next steps for maned wolf conservation in south Brazil and Uruguay. One of the principal aims is to take into account the differences in policies and legislation between both countries and, more than this, even among different states in Brazil; this will highlight the deficiencies in and lack of proper policies concerning threatened species. As such, it is important to promote links between research institutions, governmental agencies, sponsorship agencies, and nongovernmental organizations (Paula et al. 2008).

Actions that focus on maned wolf populations (i.e., mortality and birth rates) are key to take into account when putting forward conservation projects for both the species and its habitat; for instance, encouraging the creation and maintenance of habitat corridors for the species. As the regions we are focusing on in this chapter are similar, new projects and conservation activities have to be adapted accordingly. The currently low (or maybe naturally rare) abundance of the maned wolf can be used as a sign to start working on habitat conservation instead of targeting only at the species level. Moreover, conservation programs that target more than one species could be more cost-effective for the species' conservation. Including other medium to large mammal species that are threatened or have low abundance, such as the puma (*Puma concolor*), as suggested by Martínez et al. (2010), and the Pampas cat (*Leopardus colocolo*), among others, could be a good way to improve conservation programs in the region.

Following this, continuation and extension of the project at Centurión in Uruguay, previously mentioned, should be the first step for future conservation initiatives in the region. Also, similar projects should be carried out in different areas, for instance Farrapos, western Uruguay; the region between Serra do Caverá, Alegrete Municipality, and the city of São Gabriel (both in southwestern Rio Grande do Sul); as well as the grasslands near Cambará do Sul, Lajes, and Campos Novos Municipalities, northeast of Rio Grande do Sul and southern Santa Catarina. Finally, Campos Gerais region, in Paraná, should also be considered, as recommended by Pontes Filho et al. (1997).

On the other hand, it is essential to bring environmental changes caused by monocultures to people's attention. Afforestation (*Pinus* spp. and *Eucalyptus* spp., for example) and agriculture monocultures (e.g., soya and rice) are the main threats to habitat maintenance and, as such, for maned wolf survival. Moreover, the requirement to propose, execute, and comply with the ecologic zoning plans demanded by law in all Brazilian states is a first step to follow and enforce valid legislation.

Finally, improving the effectiveness of environmental education programs for conservation of the maned wolf can help to reduce human–predator conflicts in the rural areas. These programs are essential, particularly in those regions where the maned wolf is little known or where its presence is linked to domestic animal predation.

REFERENCES

Acevedo Díaz, E. 1890. *Nativa*. La Obrera Nacional, Montevideo.

Achaval, F. 1989. Lista de especies de vertebrados del Uruguay. Parte 2: Anfíbios, reptiles, aves y mamíferos. Facultad de Humanidades y Ciencias. Universidad de la República, Uruguay.

Arredondo, H. 1959. *Notas zoológicas uruguayas (de mis memorias)*. Imprenta LIGU, Montevideo.

Azara, F. 1802. *Apuntamientos para la historia natural de los cuadrúpedos del Paraguay y Río de la Plata*. Imprenta de la Viuda de Ibarra, Madrid.

Azevedo, T.R., El Achkar, D., Martins, M. de F., and Ximénez, A. 1982. Lista sistemática dos mamíferos de Santa Catarina conservados nos principais museus do estado. *Revista Nordestina de Biologia*, 5(1): 93–104.

Bazilio, S., Schemczssen, Z., and Marques, A.C. 2009. Registro visual do lobo-guará, *Chrysocyon brachyurus* (Illiger, 1815) (Mammalia: Carnivora: Canidae) na Floresta Nacional de Três Barras, SC. *Biotemas*, 22(4): 133–137.

Bilenca, D. and Miñaro, F. 2004. *Identificación de áreas valiosas de Pastizal (AVPs) en las pampas y campos de Argentina, Uruguay y sur de Brasil*. Fundación Vida Silvestre Argentina, Buenos Aires.

Breyer, F.R.S. 1979. O guará no Rio Grande do Sul. Report IBDF, Brasilia.

Bueno, A.A. and Motta-Júnior, J.C. 2009. Feeding habits of the maned wolf, *Chrysocyon brachyurus* (Carnivora: Canidae), in southeast Brazil. *Studies on Neotropical Fauna and Environment*, 44(2): 67–75.

Burmeinster, H. 1861. *Reise durch die La Plata-Staaten, MIT besonderer Rücksicht auf die physische Beschaffenheit und den Culturzustand der Argentinischen Republik. Ausgeführt in den Jahren 1857, 1858, 1859 und 1860*. H. W. Schmit, Halle.

Chebez, J.C. 2008. *Los que se van. Fauna Argentina amenazada*. Vol. 3. Editorial Albatros, Buenos Aires.

Cherem, J.J., Simões-Lopes, P.C., Althoff, S., and Graipel, M.E. 2004. Lista dos mamíferos do Estado de Santa Catarina, sul do Brasil. *Mastozoologia Neotropical*, 11(2): 151–184.

Cimardi, A.V. 1996. *Mamíferos de Santa Catarina*. FATMA, Florianópolis.

CITES. 2012. Apéndices I, II y III. Convención sobre el Comercio Internacional de Especies Amenazadas de Fauna y Flora Silvestres. CITES, Geneva. Available at: http://www.cites.org/esp/app/2012/S-2012Apr03.pdf. Accessed 3 April 2012.

CONSEMA (Conselho Estadual do Meio Ambiente do Estado de Santa Catarina). 2011. Resolução Consema n° 002, de 6 de dezembro de 2011. Lista oficial de espécies da fauna ameaçadas de extinção no Estado de Santa Catarina. *Diário Oficial Estado de Santa Catarina*, 19.237: 02–08.

Cosse, M. and Mannise, N. 2008. Participación comunitaria y conservación de "aguará guazú". IX Jornadas de Zoología del Uruguay. 10–13 December, Montevideo.

Cuesta, F., Peralvo, M., and Valarezo, N. 2009. Los bosques montanos de los Andes tropicales. Una evaluación regional de su estado de conservación y de su vulnerabilidad a efectos del cambio climático. Serie Investigación y Sistematización #5. Programa Regional ECOBONA-INTERCOOPERATION, Quito.

Devincenzi, G.J. 1935. Mamíferos del Uruguay. *Anales del Museo de Historia Natural de Montevideo*, (2)4(10): 1–96.

Dietz, J.M. 1984. Ecology and social organization of the maned wolf (*Chrysocyon brachyurus*). *Smithsonian Contribution to Zoology*, 392: 1–51.

Dietz, J.M. 1985. *Chrysocyon brachyurus*. *Mammalian Species*, 234: 1–4.

Figueira, J.H. 1894. Contribución á la fauna uruguaya. Enumeración de mamíferos. *Anales del Museo Nacional de Montevideo*, 1: 187–217.

Franklin, J. 2010. *Mapping Species Distributions. Spatial Inference and Prediction*. Cambridge University Press, Cambridge, UK.

GEP (Governo do Estado do Paraná). 2010. Decreto Estadual n° 7.264, de 1° de junho de 2010. Lista Vermelha dos Mamíferos Ameaçados de Extinção no Estado do Paraná. *Diário Oficial do Paraná*, 8233: 83.

GERS (Governo do Estado do Rio Grande do Sul). 2002. Decreto Estadual n° 41.672, de 11 de junho de 2002. Lista de Espécies da Fauna Ameaçada de Extinção no Rio Grande do Sul. *Diário Oficial do Rio Grande do Sul*, 110: 1–4.

González, S., Franco, R., Cosse, M., Emmons, L., Vynne, C., Duarte, J., and Maldonado, J.E. 2009. Phylogeography and Pleistocene demographic history of the endangered maned wolf (*Chrysocyon brachyurus*). 10th International Mammalogical Congress, Mendoza, Argentina.

Guisan, A. and Zimmermann, N.E. 2000. Predictive habitat distribution models in ecology. *Ecological Modelling*, 135: 147–186.

Hijmans, R.J., Cameron, S.E., Parra, J.L., Jones, P.G., and Jarvis, A. 2005. Very high resolution interpolated climate surfaces for global land areas. *International Journal of Climatology*, 25: 1965–1978.

Ihering, H. 1927. Os mammiferos do RS. *Revista do Museu e Archivo Público do RS*, 19: 41–77.

Indrusiak, C. and Eizirik, E. 2003. Carnívoros. In: Fontana, C.S., Bencke, G.A., and Reis, R.E. (Orgs.) (eds), *Livro Vermelho da Fauna Ameaçada de Extinção no Rio Grande do Sul*. EDIPUCRS, Rio Grande do Sul, Brazil, pp. 507–533.

IPCC (Intergovernmental Panel on Climate Change). 2007. Cambio climático 2007, Informe de síntesis. Contribución de los Grupos de trabajo I, II y III al Cuarto Informe de evaluación del Grupo Intergubernamental de Expertos sobre el Cambio Climático IPCC. Geneva, Switzerland.

IUCN Standards and Petitions Subcommittee. 2011. Guidelines for using the IUCN Red List categories and criteria. Version 9.0. Prepared by the Standards and Petitions Subcommittee. http://www.iucnredlist.org/documents/RedListGuidelines.pdf.

Jácomo, A.T.A., Kashivakura, C.K., Ferro, C., Furtado, M.M., Astete, S.P., Tôrres, N.M., Sollmann, R., and Silveira, L. 2009. Home range and spatial organization of maned wolves in the Brazilian grasslands. *Journal of Mammalogy*, 90(1): 150–157.

Langguth, A. 1976. Mamíferos. In: Langguth, A. (ed.), *Lista de las especies de vertebrados del Uruguay*. Museo Nacional de Historia Natural, Facultad de Humanidades y Ciencias, Montevideo, pp. 1–6.

Larrañaga, D.A. 1923. Zoología. In: *Escritos de Don Dámaso Antonio Larrañaga*, part 2. Instituto Histórico y Geográfico del Uruguay, Montevideo, pp. 297–471.

Mannise, N., Cosse, M., Repetto, L., Franco, R., Maldonado, J., and González, S. 2011. Caso de estudio: Caracterización y presencia de aguara guazú (*Chrysocyon brachyurus*) en Uruguay empleando herramientas de ecología molecular. Segundas Jornadas de Genética del Uruguay, 21–23 July, Montevideo.

Mannise, N., Cosse, M., Repetto, L., Franco, R., Maldonado, J., and González, S. 2012. Análisis genético sobre la presencia de aguará guazú (Carnivora, Mammalia) en Uruguay. *Boletín de la Sociedad Zoológica del Uruguay*, 21(1–2): 30–38.

Martínez, J.A., Rudolf, J.C., and Queirolo, D. 2010. *Puma concolor* (Carnívora, Felidae) en Uruguay: Situación local y contexto regional. *Mastozoología Neotropical*, 17(1): 153–159.

May-Junior, J.A., Songsasen, N., Azevedo, F.C., Santos, J.P., Paula, R.C., Rodrigues, F.H.G., Rodden, M.D., Wildt, D.E., and Morato, R.G. 2009. Hematology and blood chemistry parameters differ in free-ranging maned wolves (*Chrysocyon brachyurus*) living in the Serra da Canastra National Park versus adjacent farmlands, Brazil. *Journal of Wildlife Diseases*, 45(1): 81–90.

de Melo, L.F.B., Sábato, M.A.L., Magni, E.M.V., Young, R.J., and Coelho, C.M. 2007. Secret lives of maned wolves (*Chrysocyon brachyurus* Illiger, 1815): As revealed by GPS tracking collars. *Journal of Zoology*, 271: 27–36.

MMA (Ministério do Meio Ambiente do Brasil). 2003. Instrução Normativa nº3 de 27 de maio de 2003. Diário Oficial da União, Seção 1, nº101, 28/05/2003: 88–97.

Mones, A. and Olazarri, J. 1990. Confirmación de La existência de *Chrysocyon brachyurus* (Illiger) en el Uruguay (Mammalia: Carnivora: Canidae). *Comunicaciones Zoológicas del Museo de Historia Natural de Montevideo*, XII(174): 1–6.

Montes, V.C. 2010. Posibles impactos del cambio climático e la agricultura chilena. INIA Tierra Adentro, 44–47.

Motta-Júnior, J.C., Talamoni, S.A., Lombardi, J.A., and Simokomaki, K. 1996. Diet of the maned wolf, *Chrysocyon brachyurus*, in central Brazil. *Journal of Zoology*, 240: 277–284.

Motta-Júnior, J.C., Queirolo, D., Bueno, A.A., and Belentani, S.C.S. 2002. Fama injusta. *Ciência Hoje*, 31(185): 71–73.

Paula, R.C. de, Médici, P., and Morato, R.G. (eds). 2008. Plan de acción para la conservación del aguará guazú. Análise de viabilidad poblacional y de hábitat. [Maned wolf action plan. Population and habitat viability assessment.] Instituto Chico Mendes de Conservação da Biodiversidade, Brasilia.

Pearson, R.G. 2007. Species distribution modeling for conservation educators and practitioners. Synthesis. American Museum of Natural History. Available at http://ncep.amnh.org.

Peters, G. and Wozencraft, W.C. 1989. Acoustic communication by fissiped carnivores. In: Gittleman, J.L. (ed.), *Carnivore Behavior, Ecology, and Evolution*. Cornell University, New York, pp. 14–56.

Peterson, A.T., Ball, L.G., and Cohoon, K.P. 2002. Predicting distributions of Mexican birds using ecological niche modelling methods. *International Journal of Interoperability in Business Information Systems*, 144: 27–32.

Phillips, S.J., Anderson, R.P., and Schapire, R.E. 2006. Maximum entropy modeling of species geographic distributions. *Ecological Modelling*, 190: 231–259.

Pontes Filho, A., Silva, C.B.X., Lange, R.R., and Cavalcanti, R.K. 1997. Projeto lobo-guará—Contribuição à conservação ambiental dos campos gerais do Paraná - Brasil. *Proceedings of the First Brasilian Congress of Conservation Units*, 2: 848–860.

Prates Jr., P.H. de S. 2008. Diversidade genética e história evolutiva do lobo-guará (*Chrysocyon brachyururs*). Tese de Doutorado. Pontifícia Universidade Católica do Rio Grande do Sul, Porto Alegre.

Prevosti, F.J., Bonomo, M., and Tonni, E.P. 2004. La distribución de *Chrysocyon brachyurus* (Illiger, 1811) (Mammalia: Carnivora: Canidae) durante el Holoceno en la Argentina: Implicancias paleoambientales. *Mastozoología Neotropical*, 11(1): 27–43.

Prigioni, C. and Sappa, A. 2003. El aguará guazú. Ficha Técnica, *Bañados del Este*, 9(21). Available at http://www.probides.org.uy/.

Queirolo, D. and Motta-Júnior, J.C. 2007. Prey availability and diet of maned wolf in Serra da Canastra National Park, southeastern Brazil. *Acta Theriologica*, 52(4): 391–402.

Queirolo, D., Moreira, J.R., Soler, L., Emmons, L.H., Rodrigues, F.H.G., Pautasso, A., Cartes, J.L., and Salvatori, V. 2011. Historical and current range of the near threatened maned wolf *Chrysocyon brachyurus* in South America. *Oryx*, 45(2): 296–303.

Rodden, M., Rodrigues, F., and Bestelmeyer, S. 2008. *Chrysocyon brachyurus*. In: IUCN 2010. *IUCN Red List of Threatened Species*. Version 2010.4. http://www.iucnredlist.org. Accessed 17 December 2010.

Rodrigues, F.H.G. 2002. Biologia e conservação do lobo-guará na Estação Ecológica de Águas Emendadas, DF. Tese de Doutorado, Universidade Estadual de Campinas, Brasil, 96p.

Rodrigues, F.H.G., Hass, A., Lacerda, A.C.R., Grando, R.L.S.C., Bagno, M.A., Bezerra, A.M.R., and Silva, W.R. 2007. Feeding habits of the maned wolf (*Chrysocyon brachyurus*) in the Brazilian Cerrado. *Mastozoología Neotropical*, 14(1): 37–51.

Soutullo, A., Alonso, E., Arrieta, D., Beyhaut, R., Carreira, S., Clavijo, C., Cravino, J., et al. 2009. Especies prioritarias para la conservación en Uruguay 2009. Serie de Informes nº 16, Proyecto Fortalecimiento del Proceso de Implementación del Sistema de Áreas Protegidas, Sistema Nacional de Areas Protegidas de Uruguay.

Ximénez, A., Langguth, A., and Praderi, R. 1972. Lista sistemática de los mamíferos del Uruguay. *Anales del Museo de Historia Natural de Montevideo*, (2)7(5): 1–49.

15 Maned Wolf in Argentina

Lucía Soler

CONTENTS

Introduction 203
Threats 204
Domestic Dogs 204
Parasites and Infectious Agents 204
Beliefs and Legends 205
Conflicts 205
Stochastic Factors and Road Kills 206
Conservation Projects 207
Northeastern and Northwestern Regions 207
Conservation of Wild Carnivores in Northeastern Argentina 207
Central Region 209
Maned Wolf in the Bajos Submeridionales 209
Ñangarekó Project 210
Maned Wolf Conservation Project: Buenos Aires Zoo 211
Central–Northeastern Region 211
Genetics Applied to Conservation of Endangered Species and Their Habitats 211
Rescue and Rehabilitation Program of Maned Wolf in Argentina: Fundación Temaikén (Buenos Aires) 212
Actions in Argentina: Overview and Future Perspectives 212
References 216
Appendix: Projects and Leaders 219

INTRODUCTION

At present, the maned wolf has a discontinuous distribution in Argentina: eastern Formosa, Chaco, and Corrientes; central and northern Santa Fe; northern Entre Rios; southern Misiones; northeastern Córdoba; and southeastern Santiago del Estero (Becacceci 1992a; Galliari et al. 2004; Haro et al. 2001; Queirolo et al. 2011; Pautasso & Fandiño 2008). *Chrysocyon brachyurus* occurs spontaneously in open areas covered with natural grasslands and in matrices of native forest and caranday and pindó palms (Chebez 2008). It occurs naturally in natural grasslands, marshes, swamps, and palm and in closed forests. However, it also inhabits modified landscapes such as deforested areas (Orozco et al. 2009) and agricultural mosaics with livestock farming and exotic forest (*Eucalyptus* spp. and *Pinus* spp.) (Michelson 2005; Soler 2010a; Soler et al. 2011); individuals have also been recorded in the vicinity of urban and semiurban areas (Orozco et al. 2009). In other countries it may reach up to 1500 m above sea level and rainforests (e.g., Aragona & Setz 2001); however, in Argentina it only occurs in low areas. In the past, it was present in the lowlands of southern Santa Fe and Entre Rios and in the dry areas of Chaco, but this information is not well documented. In southern Misiones, there are some isolated records in zones where it has not been seen before (Chebez 2008; Chebez & Massoia 1996).

Based on archeological and paleontological records, the presence of the maned wolf was confirmed in Buenos Aires (Garcia Esponda et al. 2001; Prevosti et al. 2004). The current distribution is limited by temperatures not below 19°C and precipitation between 1500 and 550 mm (Prevosti et al. 2004). However, from the 1980s the upper limit of the rainfall range began to increase as its presence was recorded in zones where it had never occurred before or where it had been absent for many years (Pautasso 2009). Because the information is limited, in particular for some unexplored areas, we cannot be sure that we are experiencing a geographic expansion (Pautasso 2009; Queirolo et al. 2011). It has been proposed on several occasions that the populations of the maned wolf in Argentina are not on a spatial continuum (de Paula Cunha et al. 2008). Small fragmented populations have been documented in southern Santiago del Estero (Bañados del Río Dulce) and northeastern Córdoba (Laguna de Mar Chiquita). New locations were confirmed on the edge of its distribution, particularly in Chaco (Salvatori et al. 2004), Córdoba, Santa Fe, and Santiago del Estero (Miatello & Cobos 2008; Richard et al. 1999).

C. brachyurus was listed as Vulnerable (Reca et al. 1996), following national legislation criteria (Decree No. 691/81, Law No. 22.421/81). In 1997, it was again categorized as having the same conservation status, following the criteria A1acde, B1+2abcd; C2a (Ojeda & Díaz 1997). In 2000 and 2012, it was classified as Endangered in accordance with the criteria A1acde+2ce; C2a (Díaz & Ojeda 2000; Ojeda, et al. 2012).

Some attempts have been made to increase regional and national protection. Santa Fe, Corrientes, and Chaco declared the species as a Provincial Natural Monument under Law No. 12.182, Decree No. 1.555/92, and Law No. 4.306, respectively, in order to ensure the maned wolf's protection and to recover the status of wild populations. These legislative statements imply (a) the total and permanent prohibition of poaching; (b) the absolute prohibition on capture, harassment, persecution, possession, traffic, and sale of individuals or elements made with parts of the maned wolf; and (c) the confiscation of captured individuals, animal parts, products made with maned wolf parts, and all elements used to capture and kill maned wolves.

There is a dearth of information about populations in conservation units. Within the maned wolf's current distribution there are 174 protected areas, including Ramsar sites. The canid was recorded in 21 (12%) of them and 4 are national parks: Copo National Park (NP), Chaco NP, Río Pilcomayo NP, and Mburucuyá NP (Orozco et al. 2009).

THREATS

Domestic Dogs

Attacks by dogs were observed with high frequency in several regions of Argentina (Orozco et al. 2009; Soler et al. 2006b). There are many cases of maned wolf death by dogs accompanying rural residents in the field, who often encourage their dogs to attack and kill carnivores (felids, raccoons, foxes, etc.) (Soler et al. 2006b).

Parasites and Infectious Agents

A high prevalence of antibodies to *Neospora caninum*, canine coronavirus, canine adenovirus, canine distemper, and *Toxoplasma gondii* was detected, suggesting that dogs represent a high risk to canid conservation (Cáceres et al. 2009; Pautasso 2009; Sisa et al. 2006). Clinical signs and pathological findings were compatible with canine distemper and rabies in maned wolves in Corrientes (Lertora et al. 2008). Other parasites recorded are *Capillaria* sp. (Beldoménico et al. 2002; Mussart et al. 2003), *Alaria alata*, described for the first time in captivity (Martínez et al. 2005; Rigonutto et al. 2000), *Cryptosporidium* sp. (Santa Cruz et al. 2003), and *Dioctophyma renale* (Beccaceci 1990; Estévez et al. 1996; Mussart et al. 2003; Sisa et al. 2006; Soler et al. 2005b, 2009b). Cases of cystinuria were found in both wild and captive individuals (Mussart & Coppo 1999; Mussart et al. 2003), but the impact of that disorder on wild populations remains unknown. Evidence of just one case of tuberculosis (Becacceci 1992b) and another of histiocytic lymphoma (Soler et al. 2006c) was found.

Beliefs and Legends

These would be silent threats, as rural people are reluctant to recognize and talk about their ancestral beliefs. The legend of lobizón and folkloric uses of body parts in rituals are common in rural communities of Corrientes and Chaco provinces (Soler 2008b). In southern Santiago del Estero and northern Santa Fe, the legend does not seem to have influence; in southern–central Santa Fe, people use maned wolf leather to cure hemorrhoids (Orozco et al. 2009).

Conflicts

This threat and road kill are widespread in Argentina, but in some regions conflicts occur with high frequency. Although a high percentage of rural people do not consider the species detrimental, the maned wolf is often captured and/or killed (Figure 15.1). The causes are based on cultural issues (fear, curiosity, and popular beliefs about "powers" conferred upon it) that have been passed on from generation to generation (Salvatori et al. 2004; Soler 2008b). In Santiago del Estero, Santa Fe, Chaco, and Corrientes the mortalities are possibly related to the conspicuousness of this canid and the ignorance about the species (Orozco et al. 2009; Soler & Carenton 2008). Nevertheless, the maned wolf is accepted in rural areas of Corrientes and Chaco, and many rural residents consider it a positive species and necessary for the environment (Soler et al. 2006a; Soler 2008b), using it as an emblematic carnivore to promote nature conservation and the habitat (Soler & Carenton 2008; Soler et al. 2006a).

In a few cases, ranchers consider the maned wolf to be a predator of small livestock (goats and sheep). Predation on calves and foals was suggested rarely and only in areas where the canid is rare and attacks may be confused with those of pumas or feral dogs. A survey of conflicts carried out in Chaco and Corrientes between 2002 and 2004 (N = 600 interviews) showed that the maned wolf rarely attacks domestic mammals; predation cases were mainly on poultry (Salvatori et al. 2004; Soler 2008b).

The negative perception of the maned wolf in different regions of Argentina is associated with the local or regional culture and the relationship between man and nature (Orozco et al. 2009; Soler et al. 2006b). In the north of the Bajos Submeridionales (Santa Fe), hunters have a negative view

FIGURE 15.1 Individuals shot by rural residents. Dogs and firearms are a widespread practice throughout the maned wolf's range. (Photo: Juan Martín Mastropaolo.)

because maned wolves often take captured prey (Pautasso 2003). In spite of this, a positive perception persists among rural residents, probably because they are not directly related to the commercial harvesting of wildlife (Orozco et al. 2009). Between 1997 and 2008, about 36 individuals were recorded as captured (including some historical cases) and 57 were recorded as killed for different reasons such as hunting, road kill, and droughts (Pautasso et al. 2010). Hunting and trapping represented 62% of deaths and the causes for direct persecution were varied. In Santiago del Estero during the last decade, there was a significant growth in trophy hunting; demand for several species, including the maned wolf, which are sold by rural residents, was detected (Orozco et al. 2009); 33% of respondents had at least one maned wolf in captivity (Orozco et al. 2005).

Illegal trade is not a common activity in Argentina but is widely distributed and includes primarily the sale of puppies or young individuals to clandestine zoos or private collections (Orozco et al. 2009).

Stochastic Factors and Road Kills

Floods and droughts are natural stochastic factors that affect the species (Beccaceci 1992a, 1993). During recent decades (from 1970), beneficial initially wet periods followed by prolonged dry periods became more frequent. This situation has been aggravated by the creation of dams and pipeline wetlands, by the increase in the stocking in marginal areas, and by a growing trend of replacement habitats for crops (Orozco et al. 2009).

Both in Santa Fe and Corrientes, rural residents commented that the intensified precipitation favored the presence of the species (Soler 2008b), while in Santiago del Estero the increase of sightings would be associated with dry seasons (Marcela Orozco, personal communication). Droughts cause the movement of individuals toward artificial water supplies for livestock, increasing the exposure of maned wolves to be captured or killed (Pautasso & Fandiño 2008; Pautasso et al. 2010). Between 2006 and 2008, at least 21 individuals were recorded as deceased from several causes in Santa Fe alone (Orozco et al. 2009).

Road kills on high-speed roads are historical in Argentina. There is a concentration of records in central–western Santa Fe and southeastern Santiago del Estero (Pautasso 2009; Pautasso et al. 2005; Richard et al. 1999), as well as in northern and northeastern Corrientes, where national routes are distributed (Figure 15.2).

The loss of natural habitats in combination with long periods of drought forces maned wolves to move in search of suitable environments, following watercourses; during these trips, many die. This threat could increase in the near future (Figure 15.3; Pautasso et al. 2005). In that sense, Torres et al.

FIGURE 15.2 Collisions on high-speed paved roads are a constant and growing threat in Argentina. (Photo: Juan Martín Mastropaolo.)

FIGURE 15.3 The occurrence of individuals wandering in cities is growing in Argentina. A minority are rescued and released. (Photo: Juan Martín Mastropaolo.)

(2013) conducted a mapping of more stable environments for the species and claim that the climatic change expected for the next 40 years will be an important factor to be considered in conservation strategies of the maned wolf in Argentina.

CONSERVATION PROJECTS

Northeastern and Northwestern Regions

Conservation of Wild Carnivores in Northeastern Argentina

Since early 2002, the nongovernmental organization (NGO) Huellas has been surveying in northeastern Argentina. The proposal comprises ecological research and conservation action on 12 carnivore species (felids: *Leopardus geoffroyi*, *L. colocolo*, *Puma yagouaroundi*, *P. concolor*; canids: *Cerdocyon thous*, *Lycalopex gymnocercus*, *C. brachyurus*; mustelids: *Galictis cuja*, *Conepatus chinga*, *Lontra longicaudis*; procionids: *Procyon cancrivorous*, *Nasua narica*) (Tables 15.1 and 15.2).

Numerous strategies were implemented with different levels of scope (Alegre 2007a; Soler 2008a; Soler et al. 2008a), reaching approximately 5000 students in Corrientes and Chaco. The methodologies that generated the most positive connections between children and the species included free

TABLE 15.1
***In Situ/Ex Situ* Projects, 2000–2012**

Region of Argentina	Province	Total Distribution (%)	No. of Projects (1990–2000)	No. of Projects (2001–2011)	No. of Projects to 2012
Northwest	Formosa	11	1	0	0
	Chaco	18	1	2	1
Centre	Santa Fe	14	2	2	2
	Santiago del Estero	23	1	1	1
	Córdoba	2	0	0	0
Northeast	Corrientes	2	0	0	0
	Misiones	2	1	1	0
	Entre Ríos	2	0	1	1

TABLE 15.2
Number of Projects per Province, in Argentina, between 2000 and 2012

	Provinces							
Project	**Córdoba**	**Entre Ríos**	**Santa Fe**	**Chaco**	**Santiago del Estero**	**Corrientes**	**Formosa**	**Misiones**
Research	1	–	1	1	1	1	–	–
Conservation education	1	–	1	2	1	1	–	–
Conservation actions	1	–	1	–	–	1	–	–

drawing, painting silhouettes, and puppets telling stories about the life history of *C. brachyurus* (Soler et al. 2008b).

Between 2004 and 2009, the project organized five rural community encounters, which constituted a powerful social tool to spread the information on the plight of the species (Alegre 2007b; Soler et al. 2008b). The most successful were those where images and videos of the canid were used, and local and regional artisans, folkloric groups, and the Catholic Church took part. The maned wolf was a good educational resource to address themes such as biodiversity loss and environmental changes (Figures 15.4 through 15.6; Soler 2010b).

The project ran mainly in private areas with agriculture and livestock and in protected areas of Chaco and Corrientes. Indirect and direct evidence and interviews with rural residents are the methods commonly applied to establish the presence/absence of the maned wolf. Biological samples were also used to study the health status of the maned wolf and other wild carnivores.

During 2004 and 2005, two adults (one male and one female) were equipped with VHF radio collars. The study area included private fields under intense anthropic use. The female's home range was partially contained in the male's range. They showed peaks of activity between 7:00 p.m. and 9:00 a.m., and most of the fixes of inactivity were located in open habitats (Soler 2008b; Soler et al. 2011). The most important annual food items were rodents, native palm fruits, fish, and shellfish (Soler 2008b). Information collected by Huellas in Chaco and Corrientes indicated flexibility in the spatial, temporal, and trophic ecology, particularly in disturbed habitats (Soler & Cáceres 2009;

FIGURE 15.4 Workshop for children in rural schools of Corrientes province (2009). A girl paints the silhouette of a maned wolf as part of an educational activity. (Photo: Lucía Soler.)

FIGURE 15.5 Children's workshop in a town in northern Argentina (2008). Kids doing free drawing based on a story of a young maned wolf that must begin to establish its own territory. (Photo: Lucía Soler.)

Soler et al. 2011), features that could be key to their preservation, considering that the largest populations in Argentina are in private fields and not in protected areas. The maned wolf appears to be a species adaptable to human activities and changes in a landscape made up of a matrix of natural wetlands, grasslands under livestock use, native forests, exotic forests, and small towns with a permanent human presence.

The project obtained the WAZA Branding Project No. 06031 in 2006. It has had financial support from the Zoo Amnéville (France), Zoo Doué la Fontaine (France), Cerza Conservation (France), Peaugres Safari Zoo (France), Societé Zoologique de Paris (France), John Ball Conservation Society (U.S.), Idea Wild (U.S.), Rufford Foundation (U.K.), Survival Plan Committee of the Maned Wolf (MW/SSP) through Friends of the Dickerson Park Zoo (U.S.), Ministry of Culture: Socio-Community Projects (Argentina), Brookfield Zoo (Chicago Zoological Society, U.S.), Abilene Zoo (U.S.), and World Association of Zoos and Aquariums (Switzerland). In collaboration with the Instituto de Ecologia Aplicada di Rome (Dr. Valeria Salvatori), financial support was obtained from Rufford Foundation (U.K.), Sophie Dandford (U.S.), and Metrozoo Park (U.S.).

Central Region

Maned Wolf in the Bajos Submeridionales

Since 2000, three independent projects have been developed in the Bajos Submeridionales of Santa Fe province. This region is characterized by a depressed and flood plain that extends in a swampy coast on the Río Paraná and is one of the most important regions for the species in the province. The education activities were organized in two areas, Los Amores and Fortin Olmos, and at the Museo Provincial de Ciencias Naturales "Florentino Ameghino," in Santa Fe City. All the projects

FIGURE 15.6 Conservation education campaigns (2007). Organized by Huellas NGO project. (Photo: Lucía Soler.)

included interviews with rural people and workshops in rural schools. On several occasions they were assisted by the Ministerio de Ambiente of Santa Fe.

Previously, in 1996, Laura Steiman also carried out community encounters and surveyed rural residents, in northern Rio Salado, to define conflicts and their images of the maned wolf. The proposal aimed to study the *ex situ* social and reproductive behavior of the canid. She also developed methodologies including mainly sightings of animals, registration of tracks, and so forth. The results of this project were not published, but a website was developed: http://www.aguaraguazu.org.ar.

In addition, presence/absence information was mapped and dead individuals were deposited in the Museo "Florentino Ameghino" (Pautasso & Fandiño 2008), one of the most complete collections of maned wolf skulls and postcranial skeletons in Argentina. A total of 15 individuals made this public collection one of the largest in the country.

In 2009, the experience and information about the species was compiled and published, resulting in the *Conservation and Management Plan for the Maned Wolf in Santa Fe* (Pautasso 2009), which became the first complete document in Argentina to consider a protocol for the rescue, rehabilitation, and release of the species.

Ñangarekó Project

This project was carried out by the Laboratorio of Eco-Epidemiología, Facultad de Ciencias Exactas y Naturales of the Universidad de Buenos Aires in southeastern Santiago del Estero between 2004 and 2006. It was supported by the Veterinary Field Program (Wildlife Conservation Society, U.S.), Oregon Zoo (U.S.), Foundation's Future for Wildlife Conservation Fund, and Universidad de Buenos Aires (Orozco et al. 2009). The study was focused on confirming the presence of maned wolf populations and establishing critical areas. Local threats were evaluated and an investigation on diseases prevalent in domestic dogs was carried out. Educational strategies were conducted in rural communities including awareness and health prevention campaigns. Illegal trade and road kill of maned wolves were the two most serious threats recorded in this region and to address this, an education campaign to mitigate kills on high-speed roads was implemented during 2005. The demand for individuals decreased during 2004 and 2006, but has increased again since 2007, when the project ended (Orozco et al. 2009). During 2005, a survey was conducted in six villages of the Bañados del Rio Dulce System to identify areas of presence and frequency of occurrence. Forty-two people were interviewed and during the visits blood samples were obtained from 130 dogs. There was a high prevalence of antibodies to *N. caninum* and canine coronavirus (67%), canine adenovirus (59%), *Dirofilaria immitis* (58%), canine distemper (57%), and *T. gondii* (33%). These results suggest that dogs are a threat to wild carnivore conservation in the surveyed area. A rabies vaccination plan for pet dogs and cats was implemented (Pautasso 2009). All of the rural residents responded that they knew the maned wolf and 95% had observed one at least once in the wild. They described the presence of the species in the area from the 1990s, and most of the records had occurred since 1995. The maned wolf would have colonized the area in recent times and it is possible that it had been in closer contact with human settlements because of the reduction and fragmentation of habitat due to agricultural expansion; 33% of the respondents had captured maned wolves at least once (Orozco et al. 2005).

Maned Wolf Conservation Project: Buenos Aires Zoo

Since 1999, this project has focused mainly on *ex situ* management; with proposals to adapt the enclosures to the requirements proposed by the manual of the Maned Wolf Species Survival Plan of the International Union for Conservation of Nature (Infantino et al. 2009). Modifications in health and nutritional management have increased reproductive success, with a total of nine pups from pregnancies in 2000, 2002, and 2008 raised favorably. They also studied several aspects of the development and growth of the offspring in captivity (Pino et al. 2001; Orozco et al. 2001a,b) and the influence of environmental enrichment on the behavior of the species, demonstrating the effectiveness of enrichment in reducing stereotyped behavior of animals kept under controlled conditions (Infantino et al. 2009).

The Maned Wolf *Ex Situ* National Registry (Studbook) was coordinated by this zoo and during 2001 a database on individuals kept in captivity was created with the objectives of enhancing the exchange between Argentinean zoos and guiding inbreeding for conservation purposes. In 2004, the results were disseminated in a digital version, but despite these efforts, the information has not been implemented. Nonetheless, since 2011 the National Registry has been reviewed within the framework of the National Action Plan in an investigation considering genetic information conducted by Vanina Raimondi (Museo Argentino de Ciencias Naturales "Bernardino Rivadavia" [MACN], Consejo Nacional de Investigaciones Científicas y Técnicas [CONICET]).

The Buenos Aires Zoo project considered an *in situ* proposal in Chaco province during 2008 and 2009. Awareness campaigns, broadcast media (TV and local radio), and training of teachers in rural areas were used to carry out *in situ* conservation objectives. The *ex situ* component aimed to strengthen and deepen zoo visitors' knowledge, raise awareness of the care and conservation of the natural habitat, and promote active participation in solving environmental problems. The zoo's education department implemented participation strategies, mainly with children, who make up about 70% of visitors: "young carers" program (6–12 years old), winter camps (3–12 years old), a course for guides (university students), and training courses (teachers).

Central–Northeastern Region

Genetics Applied to Conservation of Endangered Species and Their Habitats

This study by the Laboratorio de Herramientas Moleculares (MACN, CONICET, Buenos Aires) includes population genetic variability analysis and environmental diagnosis of the maned wolf and pampas deer (*Ozotoceros bezoarticus*) in wetland areas of Santa Fe and Corrientes provinces. DNA is obtained from scats, blood samples, museum collections, and tissue samples from dead animals. The research aims to establish levels of inbreeding, effective population numbers, migration information, and population structure. The comparison of populations separated by geographical and anthropogenic barriers will allow phylogeographic relationships to be established and management units and evolutionarily significant units (ESU) to be identified. Thus, core populations with singular genetic variability can be preserved (Pautasso 2009).

Rescue and Rehabilitation Program of Maned Wolf in Argentina: Fundación Temaikén (Buenos Aires)

The objectives of this project are to restore injured individuals affected by human activities and natural phenomena, to reallocate the rehabilitated individuals in their habitat, and to promote the decrease of the detrimental effects of human activities. In cases in which return to the wild is not possible, the program identified an alternative, to reach a consensus with the government agencies to include the individuals in *ex situ* conservation plans (González Ciccia 2010).

At the end of 2009, an adult male was rescued in a town located in Córdoba province (Colonia Vignaud). The animal was transferred to the Foundación Temaikén rescue centre (CRET), where it remained almost completely isolated for several months until release. At the same time, in the area of the rescue an awareness, education, and outreach plan was implemented, which was organized in four stages that included diagnosis, integration with the local people, and implementation of an educational program. The diagnosis was carried out through oral interviews with about 300 residents; 73% of them did not know the species, however, 76% were interested in the conservation of the maned wolf.

The program reached about 1500 residents of eight villages (Cologne Vignaud, Marull, Balnearia, Miramar, Altos del Chipión, Paquita, La Posta, Villa Rosario Saladillo). In November 2010, the maned wolf was released, equipped with a VHF radio collar VHF, and nowadays the individual is being monitored by park rangers of the Mar Chiquita Reserve, volunteers, and Temaiken researchers.

ACTIONS IN ARGENTINA: OVERVIEW AND FUTURE PERSPECTIVES

Between 1990 and 2009, 31 documents were written: 31% were reports from workshops, conferences, and popular journals; 30% were abstracts presented at scientific conferences; and 29% were published in national and international journals; 24% of the documents considered captive individuals. Maned wolf distribution was the topic most frequently published (32%), followed by reports concerning the health status of wild (22.5%) and captive (22.5%) individuals. Conflicts and environmental issues were rarely addressed. Regarding the scope, 48% were regional publications, 32% local, and the remainder had international impact.

Since 2002, when the first national conservation workshop took place knowledge of the species has advanced; however, most of the recognized conservation problems remain unresolved and in many cases have worsened, probably due to the magnitude of the objectives proposed in 2002 and the scarce or absent support from the wildlife agencies in each province. Local actions and individual contributions by the projects described were appropriate conservation initiatives on a local scale, mainly to reverse the negative perception toward *C. brachyurus* in some areas. However, they tend to be ineffective and insufficient in providing solutions to conservation problems of high impact, such as road kills on high-speed routes, direct persecution, and loss of habitat due to fragmentation or artificial habitat replacement (Orozco et al. 2009).

The workshops were organized to define guidelines for effective conservation. However, it was not possible to maximize the efforts of the workshops until 2005, when the Grupo Argentino Aguará Guazú (GAAG; Argentinean Maned Wolf Group) was organized, comprising 16-member institutions (zoos, NGOs, research groups of universities and museums) (González Ciccia 2009).

The GAAG emerged with a national strategy to develop, manage, execute, and monitor the action plan. The group also proposes to increase knowledge about this carnivore and to promote projects whose actions minimize threats.

Between 2002 and 2011, nine regional and national workshops (Table 15.3) were carried out in order to (1) list and map threats, (2) recognize and prioritize conflicts in the wild and problems in captivity, (3) identify and prioritize strategies and actions for *ex situ* and *in situ* conservation, and (4) develop recommendations for conservation in the wild and management in captivity. The results achieved by the GAAG are as follows: (1) development of a management, rehabilitation, and release protocol for confiscated individuals; (2) characterization of educational strategies and methodologies adapted to each region and evaluation to bring nominal support to *in situ* and *ex situ* educational and research; (3) identification and prioritization of threats to the eight provinces where the species occurs, and (4) mapping of the areas critical to conservation efforts (González Ciccia et al. 2012).

The National Action Plan workshops took place under the coorganization of the Dirección de Fauna Silvestre de Nación, Departamento de Ambiente y Desarrollo Sostenible. This participation encouraged the provinces to become more aware of the necessity and urgency of a national consensus (Soler & González Ciccia 2010; González Ciccia et al. 2012). Most of the provinces pledged to take specific actions to make the National Action Plan a reality.

In 2005, members of the GAAG took part in the First International Workshop on Population and Habitat Viability of the Maned Wolf, held in Mina Gerais, Brazil (de Paula Cunha et al. 2008). Some of the results have been partially applied in Argentina to promote maned wolf conservation, but actions and activities proposed in 2005 to be developed and implemented together have not yet materialized.

Another action developed by the GAAG during 2009 was the coproduction of the video in Spanish called *Maned Wolf: Face and Cross of an Endangered Species* (*Aguará guazú: Cara y Cruz de una especie en extinción*). The images were obtained and edited by Canal Plus, Sogecable (Spain). The film is an educational resource that updates knowledge of the species in Argentina and is intended for a general audience. Institutional participants were the Museo Argentino de Ciencias Naturales "Bernardino Rivadavia" (Dr. David Flores, Sociedad Argentina para el Estudio

TABLE 15.3
Workshops 2002–2011

- Conservation of Maned Wolf in Argentina and Surrounding Countries I Workshop (2002, Corrientes. Organized by Huellas and Oikoveva NGOs)
- Conservation of Maned Wolf in Argentina and Surrounding Countries II Workshop (2003, Resistencia, Chaco. Organized by Buenos Aires Zoo and Dirección de Fauna, Parques y Ecología de Chaco)
- Conservation Education of the Maned Wolf in Argentina (2003, Corrientes city. Organized by Huellas NGO and Dirección de Recursos Naturales de Corrientes)
- Maned Wolf Conservation Priorities in Argentina (2005, Escobar, Buenos Aires. Organized by Fundación Temaikén)
- Health Status of *Chrysocyon brachyurus* in Argentina. In the V Argentine Congress of Parasitology (2009, La Plata, Buenos Aires. Organized by Huellas NGO and Ñangarekó Project)
- National Action Plan for the Conservation of *Chrysocyon brachyurus* in Argentina I Workshop (2009, Roque Sáenz Peña, Chaco. Organized by Fundación Temaikén and Complejo Ecológico Municipal Roque Sáenz Peña)
- National Action Plan for the Conservation of *Chrysocyon brachyurus* in Argentine II Workshop (2010, Córdoba. Organized by Foundación Temaikén, Secretaria de Medio Ambiente de Córdoba, Córdoba Zoo, Dirección de Fauna Silvestre de Nación–Departamento de Ambiente y Desarrollo Sostenible)
- National Action Plan for the Conservation of *Chrysocyon brachyurus* in Argentina. Education strategies III Workshop (2010, Mburucuyá, Corrientes. Organized by Huellas NGO and Fundación Temaikén)
- National Action Plan for the Conservation of *Chrysocyon brachyurus* in Argentina. Management and *Ex Situ* Conservation IV Workshop (2011, Buenos Aires. Organized by Buenos Aires Zoo and Fundación Temaikén)

de los Mamíferos); Proyecto Genética Poblacional de *C. brachyurus* en Argentina, CONICET (Lic. Vanina Raimondi); Universidad de Buenos Aires, Ñangarekó Project, Universidad de Buenos Aires (Dr. Marcela Orozco); Fundación Temaikén (Lic. Paula González Ciccia); Complejo Ecológico Roque Sáenz Peña (Chaco); Parque Nacional Mburucuyá (Administración de Parques Nacionales [APN], Corrientes); Secretaria de Medio Ambiente y Desarrollo de la Nación; and Asociación Huellas project. This educational material is the first national collation of information about the advances and critical status of *C. brachyurus* in Argentina and it includes recommendations on actions needed for its long-term conservation.

The experiences of the projects provided basic information on successful activities that could be implemented, paying particular attention to new areas. Nevertheless, isolated efforts that do not benefit the species were identified as threats in the workshops by SWOT analysis (Soler et al. 2008b).

The workshops favored the planning of conservation strategies; however, we lack information about the population status in areas where maned wolf populations have been eradicated or severely reduced. It is imperative that the provincial governments take an active role in formulating future action plans, contributing to the development of the projects with human resources and financial support.

The lack of continuity in governmental decisions and the concomitant changes in conservation policies are a permanent threat in Argentina, so it is necessary to establish mechanisms to ensure long-term implementation of the action plan. A clear indication of the lack of agreed conservation policies relates to the protocol to release seized animals: at present only three provincial wildlife agencies (Córdoba, Chaco and Santa Fe) have implemented and participated in rescue, rehabilitation, and release of several maned wolves and in only one case (Córdoba) is the maned wolf being monitored postliberation (Figures 15.7 through 15.9).

A positive outlook for achieving the conservation of the maned wolf will depend on addressing the following priority issues in the short (5 years) and medium term (10 years): (1) Collect more accurate information on the species' distribution, particularly in border areas where the canid may be in increased conflict and important populations are decreasing. (2) Monitor conflicts with the maned wolf, considering regional differences. (3) Increase knowledge of the ecology and behavior in disturbed areas, particularly considering that one of the threats to the species is environmental transformations that expose it to greater proximity to humans.

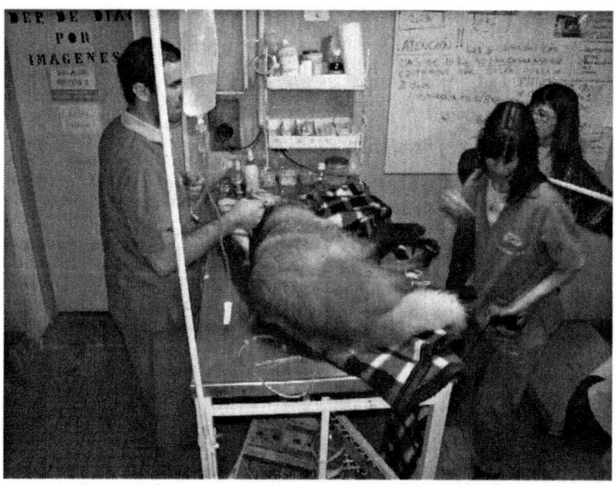

FIGURE 15.7 A maned wolf rescued and transported to an academic centre where a team of veterinarians stabilized the individual. (Photo: Juan Martín Mastropaolo.)

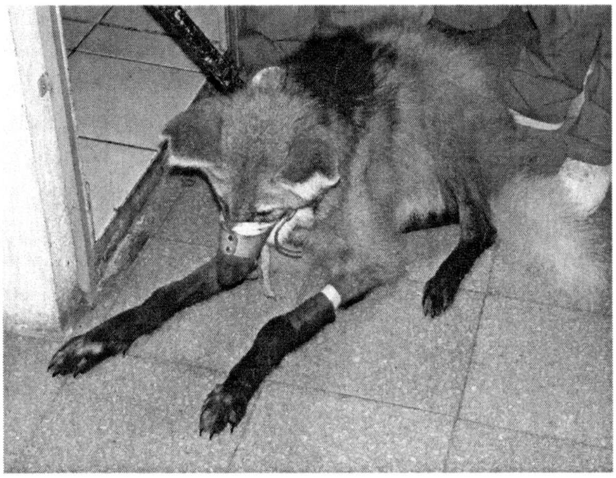

FIGURE 15.8 The maned wolf in recovery after surgery. (Photo: Juan Martín Mastropaolo.)

FIGURE 15.9 Most of the rescued animals do not return to nature. This individual remains at a zoo in a precarious condition and does not receive specific care for rehabilitation. (Photo: Juan Martín Mastropaolo.)

(4) Analyze the genetics of populations, in particular those that are isolated. (5) Establish the health status of the wild population. (6) Develop more educational programs and communication (Soler et al. 2009).

Over 10 years of interdisciplinary cooperation, we recommended an efficient social network and conservation education campaigns as essential and indispensable instruments for the survival of wild populations. The educational process must be accompanied by research based on priority issues (González Ciccia et al. 2012) that have been listed and detailed in several national meetings (Soler et al. 2005a). Achieving maned wolf conservation goals will require national consistency. If wild populations are to survive, long-term human and economic resources are required, but coordination between provinces is also needed to capitalize on all the efforts being made.

C. brachyurus does not yet have very critical conservation status in Argentina, which gives us the possibility and responsibility to continue working proactively to prevent worsening of the situation of native populations in the wild.

REFERENCES

Alegre S. 2007a. Análisis del programa de educación del proyecto "Conservación del aguará guazú en Corrientes," desde la óptica de la complejidad de la Educación Ambiental. [Education program analysis of the project "Conservation of maned wolf in Corrientes," from the complexity of the environmental education.] Carrera de especialización Educación en ambiente para el desarrollo sustentable. Convenio Ctera, Universidad del Comahue.

Alegre S. 2007b. Talleres comunitarios de educación ambiental para la introducción del pensamiento ambiental a nivel local. [Community environmental education workshops: An introduction to local environmental thought]. *DELOS: Desarrollo Local Sostenible*, 1(1):1–13.

Aragona M. & Setz E.Z.F. 2001. Diet of the maned wolf, *Chrysocyon brachyurus* (Mammalia: Canidae), during wet and dry seasons at Ibitipoca State Park, Brazil. *Journal of Zoology of London*, 254:131–136.

Beccaceci M.D. 1990. First finding in Argentina of the parasite *Dioctophyma renale* (Goeze, 1782) in a maned wolf, *Chrysocyon brachyurus*, in the wild. *IUCN Veterinary Group Newsletter*, 5:8–10.

Becacceci M.D. 1992a. The maned wolf, *Chrysocyon brachyurus*, in Argentina. In: *International Studbook for the Maned Wolf* Chrysocyon brachyurus *(Illiger, 1811)* (Matern, B., ed.). Frankfurt Zoological Garden, Frankfurt, pp. 50–56.

Becacceci M.D. 1992b. Tubercolosis in maned wolf. *IUCN/SSC Veterinary Group Newsletter*, 6:12.

Beccaceci M.D. 1993. El aguará guazú (*Chrysocyon brachyurus*) en la provincia de Corrientes. [The maned wolf in Corrientes province.] *Facena*, 10:19–31.

Beldoménico P.M., Hunziker D., López Taverna L. & Reif P.K. 2002. Capillaridae eggs found in the urine of free ranging maned wolf from Argentina. *Memorias do Instituto Oswaldo Cruz*, 97(4):509–510.

Cáceres F., Soler L. & Sisa A. 2009. Registro de enfermedades infecciosas y parasitarias en aguará guazú y otros cánidos silvestres simpátricos, en el nordeste de Argentina. [Records of infectious and parasitic diseases in maned wolf and other sympatric canids in northeastern Argentina.] In: *Estado de conocimiento y conservación del aguará guazú (*Chrysocyon brachyurus*) en la Provincia de Santa Fe, Argentina. Plan de acción para la conservación del aguará guazú en Santa Fe*. Versión 01, pp. 49–52.

Chebez J.C. 2008. *Los que se van. Fauna argentina amenazada.* [*Those Who Leave. Threatened Wildlife of Argentina*]. Vol. 3. Editorial Albatros, Buenos Aires.

Chebez J.C. & Massoia E. 1996. Mamíferos de la provincia de Misiones. In: *Fauna Misionera. Catálogo sistemático y zoogeográfico de los vertebrados de la provincia de Misiones (Argentina).* (Chebez, J.C., ed.). L.O.L.A., Buenos Aires, pp. 180–206.

Díaz M. & Ojeda R. (eds). 2000. *Libro rojo para especies de mamíferos amenazadas de Argentina.* [*Red Book for Threatened Mammal Species in Argentina.*] Sociedad Argentina para el Estudio de los Mamíferos (SAREM), Argentina.

Estévez J.O., Maubecín E.G. & Mentzel R.E. 1996. Hallazgo de *Dioctophyma renale* (Goeze, 1782) en un aguará guazú (*Chrysocyon brachyurus*) y su tratamiento por nefrectomía. [Finding of *Dioctophyma renale* (Goeze, 1782) in a maned wolf and its treatment by nephrectomy.] *Correo Veterinario*, 160:11–12.

Galliari C., Cirignoli S. & Horlent N. 2004. La expansión de la geonomía del aguará guazú (*Chrysocyon brachyurus*) y su paradójica probabilidad de extinción local a corto plazo. [The expansion of the maned wolf distribution and its paradoxical probability of local extinction in a short term.] I Congreso Nacional de Biodiversidad, Escobar, Argentina.

Garcia Esponda C.M., Acosta A.A., Loponte D.N. & De Santis L. 2001. Registro de *Chrysocyon brachyurus* (Carnivora: Canidae) en contextos arqueológicos en el noreste de la provincia de Buenos Aires. [Registration of *Chrysocyon brachyurus* in archaeological contexts in the northeastern Buenos Aires province.] *Mastozoología Neotropical*, 8(2):159–163.

González Ciccia P. 2009. Present and future of maned wolf. Recent update. *WAZA News*, 4:18–19.

González Ciccia P.M. 2010. Rescate y reintroducción del aguará guazú (*Chrysocyon brachyurus*) en los Bañados de Mar Chiquita. [Rescue and reintroduction of the maned wolf in Bañados de Mar Chiquita (Córdoba).] XXIII Jornadas Argentinas de Mastozoología, Bahía Blanca, Buenos Aires, Argentina.

González Ciccia P., Soler L. & Orozco M. 2012. Linking institutional cooperation for the conservation of *Chrysocyon brachyurus* (Order Carnivora) in Argentina. IV International Wildlife Management Congress: Cooperative Wildlife Management Across Borders: Learning in the Face of Change. Durban, South Africa.

Haro J.G., Torres R.M. & Bistoni M.A. 2001. Presencia del aguará guazú *Chrysocyon brachyurus* en el sistema de humedales Laguna Mar Chiquita—Bañados del Río Dulce. [Presence of maned wolf in the wetlands system Laguna Mar Chiquita-Bañados del Río Dulce (Córdoba).] *FACENA*, 17:95–98.

Infantino E., Rago M.V., Barrios L. & Ricci S. 2009. Enriquecimiento ambiental en aguará guazú (*Chrysocyon brachyurus*). [Environmental enrichment in maned wolf (*Chrysocyon brachyurus*).] In: *Estado de conocimiento y conservación del aguará guazú (*Chrysocyon brachyurus*) en la Provincia de Santa Fe, Argentina. Plan de acción para la conservación del aguará guazú en Santa Fe*, Versión 01, pp. 79–80.

Lertora J., Mosqueda M., Mohr B. & Frias M. 2008. Reporte de distemper canino en aguará guazú (*Chrysocyon brachyurus*). [Report of canine distemper in maned wolf.] Sexta reunión Argentina de Patología Veterinaria. Facultad de Ciencias Veterinarias. Universidad Nacional del Nordeste, Corrientes, Argentina.

Martínez F.A., Binda J.L. & Maza Y. 2005. Determinación de platelmintos por coprología en carnívoros silvestres. [Coprological survey on wild carnivores.] Comunicaciones Científicas y Tecnológicas. Secretaria General de Ciencia y Técnica. Universidad Nacional del Nordeste, Corrientes, Argentina.

Miatello R. & Cobos V. 2008. Nuevos aportes a la distribución del aguará guazú (*Chrysocyon brachyurus*, Mammalia, Canidae) en las Provincias de Córdoba y Santiago del Estero, Argentina. [New records on maned wolf distribution in Cordoba and Santiago del Estero provinces, Argentina.] *Mastozoología Neotropical*, 15(2):209–213.

Michelson A. 2005. Predicción de aptitud de hábitat para aguará guazú (*Chrysocyon brachyurus*) en un paisaje antropizado del NE de Corrientes, Argentina. [Prediction of habitat quality for maned wolf in anthropized landscapes in the northeast of Argentina.] Degree Thesis in Biology. Facultad de Ciencias Exactas y Naturales, Universidad de Buenos Aires.

Mussart N.B. & Coppo J.A. 1999. Cystine nephrolithiasis in an endangered canid, *Chrysocyon brachyurus* (Carnivora: Canidae). *Revista de Biología Tropical*, 47(3):623–626.

Mussart N.B., Solís G.A., Arzuaga S.M. & Coppo, J.A. 2003. Determinaciones hematológicas y urinarias en aguará-guazú (*Chrysocyon brachyurus*) en cautiverio en el nordeste argentino. *Revista Veterinaria*, 14(2):79–84.

Ojeda R.A., Chillo V. & Díaz Isenrath G.B. 2012. *Libro rojo de mamíferos amenazados de la Argentina.* [*Red Book of Threatened Mammals of Argentina.*] SAREM, Buenos Aires.

Ojeda R.A. & Díaz G.B. 1997. *Libro rojo de mamíferos y aves amenazados de Argentina. Sección II: Mamíferos.* [*Red Book of Threatened Mammals and Birds of Argentina. Section II: Mammals.*] FUCEMA, SAREM, AOP, and APN, Buenos Aires.

Orozco M., Ceballos L., Pino M., Rosso S., Petrocco F. & Gürtler R.E. 2005. Presence of *Chrysocyon brachyurus* (maned wolf) in southeastern Santiago del Estero, Argentina. XX Jornadas Argentinas de Mastozoología, Buenos Aires, Argentina.

Orozco M., Pino M., Linares F., Bauzá A. & Díaz M. 2001a. The maned wolf conservation project (*Chrysocyon brachyurus*). *Zoológica Latinoamericana*, 1:30.

Orozco M., Pino M. & Díaz M. 2001b. Comparative report between deciduous dental chronology of domestic canine (*Canis familiaris*) and maned wolf (*Chrysocyon brachyurus*). X ALPZA Congress, Buenos Aires, Argentina.

Orozco M.M., Ceballos L.A., Pino M.C., Rosso S., Petrocco F. & Gürtler R.E. 2005. Presencia de *Chrysocyon brachyurus* (aguará guazú) en el sudeste de Santiago del Estero. [Presence of *Chrysocyon brachyurus* in the southeastern Santiago del Estero.] XX Jornadas Argentinas de Mastozoología, Buenos Aires, Argentina.

Orozco M., Pautasso A.A. & Rago V. 2009. (in press). Aguara guazú (*Chrysocyon brachyurus*) en Argentina. Situación actual y acciones de conservación. [Maned wolf (*Chrysocyon brachyurus*) in Argentina. Current status and conservation actions.] Manuscript submitted to the Dirección de Fauna Silvestre, Secretaria de Ambiente y Desarrollo Sustentable, Argentina.

Paula R.C. de, Medici P. & Goncalves Morato R. 2008. Population and viability assessment (PHVA)—Argentina. In: *Maned Wolf Action Plan. Population and Viability Assessment.* Instituto brasileiro do Meio ambiente e dos Recursos naturais Renováveis centro nacional de informação, Tecnologias ambientais e Editoração. Edições Ibama, Brasilia, pp. 103–114.

Pautasso A.A. 2003. Aprovechamiento de fauna silvestre por pobladores rurales en la fracción norte de los Bajos Submeridionales de la provincia de Santa Fe, Argentina. [Use of wildlife by rural people in the northern area of Bajos Submeridionales of Santa Fe province, Argentina.] *Comunicaciones del Museo Provincial de Ciencias Naturales "Florentino Ameghino"* (nueva serie), 8(2):1–66.

Pautasso A.A. & Fandiño B. 2008. Aguará guazú (*Chrysocyon brachyurus*). In: *Mamíferos de la provincia de Santa Fe, Argentina.* [*Mammals of Santa Fe province, Argentina.*] Comunicaciones del Museo Provincial de Ciencias Naturales "Florentino Ameghino" (nueva serie), 13(2):1–253.

Pautasso A.A., Chersich D., Peña M., Virasoro C., Soler L., Fandiño B., Raimondi V., Senn A. & Mastropaolo J.M. 2005. Asfaltado de rutas en los Bajos Submeridionales: Previsibles impactos sobre la fauna. [Asphalting of roads in the Bajos Submeridionales (Santa Fe): Predictable impacts on wildlife.] *Revista Ecológica*, 9:12–15.

Pautasso A.A. (ed.). 2009. Estado de conocimiento y conservación del aguará guazú (*Chrysocyon brachyurus*) en la Provincia de Santa Fe, Argentina. Plan de Acción para la Conservación del aguará guazú en Santa Fe, Versión 01. [Knowledge and conservation status of the maned wolf in Santa Fe province, Argentina. Action Plan for the Conservation of the maned wolf in Santa Fe, Version 01.] *Biológica- Naturaleza, Conservación y Sociedad*, 11:1–123.

Pautasso A.A., Raimondi V.B., Bierig P.L. & Leiva, L.A. 2010. Mortalidad de venado de las pampas (*Ozotoceros bezoarticus*) y aguará guazú (*Chrysocyon brachyurus*) en represas de almacenamiento de agua en los Bajos Submeridionales de Santa Fe, Argentina. [Mortality of pampas deer and maned wolf in water storage dams in the Bajos Submeridionales of Santa Fe, Argentina.] Nótulas Faunísticas—Segunda Serie, 52:1–6.

Pino M., Orozco M. & Díaz M. 2001. Report on growth parameters for maned wolf pups (*Chrysocyon brachyurus*). X ALPZA Congress, Buenos Aires, Argentina.

Prevosti F.J., Bonomo M. & Toni E.P. 2004. La distribución de *Chrysocyon brachyurus* (Illiger, 1811) (Mammalia: Carnivora: Canidae) durante el holoceno en la Argentina: implicancias paleoambientales. [The distribution of *Chrysocyon brachyurus* (Illiger, 1811) (Mammalia: Carnivora: Canidae) during the Holocene in Argentina: paleoenvironmental implications.] *Mastozoología Neotropical*, 11(1):27–43.

Queirolo D., Moreira J.R., Soler L., Emmons L., Rodrigues F.H.G., Pautasso A., Cartes J.L. & Salvatori V. 2011. Historical and current range of the near threatened maned wolf *Chrysocyon brachyurus* in South America. *Oryx*, 45(2):296–303.

Reca A.R., Ubeda C. & Griguera D.1996. Prioridades de conservación de los mamíferos de Argentina. [Conservation priorities of Argentinean mammals.] *Mastozoología Neotropical*, 3(1):87–117.

Richard E., Giraudo A. & Abdala C. 1999. Confirmación de la presencia de Aguará Guazú (*Chrysocyon brachyurus*) en Santiago del Estero, Argentina. [Confirmation of the presence maned wolf in Santiago del Estero, Argentina.] *Acta Zoológica Lilloana*, 45(1):155–156.

Rigonutto T., Martínez F., Nuñez S., Troyano J.C., Gauna Añasco L., Duchene A., Stancato M. & Sicardi S. 2000. Hallazgo de *Alaria* sp. (Trematoda, Strigeiidae) en carnívoros silvestres. [Finding of *Alaria* sp. (Trematoda, Strigeiidae) in wild carnivores.] Comunicaciones Científicas y Tecnológicas. Secretaria General de Ciencia y Técnica, Universidad Nacional del Nordeste, Corrientes, Argentina.

Salvatori V., Soler L., Cáceres C., Pérez P., Gori M., Fleita A. & Cuello P. 2004. Estimating presence of maned wolf in northern Argentina from local knowledge: Preliminary results. Memorias del VI Congreso Internacional para el manejo de fauna silvestre en la Amazonía y Latinoamérica: lecciones aprendidas, Iquitos, Perú.

Santa Cruz A.C., Gómez L.G., Comolli J.A., Torres Jiménez G., González J.A. & Ortiz J.C. 2003. Detección de *Cryptosporidium* sp. en aguará guazú (*Chrysocyon brachyurus* Illiger, 1815) del zoológico de Sáenz Peña, Chaco. [Detection of *Cryptosporidium* sp. in maned wolf (*Chrysocyon brachyurus* Illiger, 1815) Sáenz Peña Zoo, Chaco.] Comunicaciones Científicas y Tecnológicas. Secretaria General de Ciencia y Técnica, Universidad Nacional del Nordeste, Corrientes, Argentina.

Sisa A., Soler L., Caffe L. & Casanave E.B. 2006. Resultados de estudios sanitarios en *Chrysocyon brachyurus* y otros cánidos simpátricos en el nordeste de Argentina. [Results of health studies in *Chrysocyon brachyurus* and other sympatric canids in northeastern Argentina.] *XXI Jornadas Argentinas de Mastozoología*, Tafí del Valle, Tucumán, Argentina.

Soler G.L. 2008a. Wild carnivores: From conflict to tolerance. *Journal of the International Zoo Educators*, 44:47–50.

Soler G.L. 2008b. Aspectos ecológicos y perspectivas de conservación de *Chrysocyon brachyurus* en Argentina. [Ecological aspects and conservation perspectives of *Chrysocyon brachyurus* in Argentina.] Magister thesis dissertation in biological science. Departamento de Biología, Bioquímica y Farmacia. Universidad Nacional del Sur, Bahía Blanca.

Soler G.L. 2010a. Wild carnivores: Advances in knowledge and conservation strategies in Argentina. Recent update. *WAZA News*, 2:22–23.

Soler G.L. 2010b. Evaluación de estrategias educativas: Proyecto "Conservación de los carnívoros del nordeste argentino." [Evaluation of educational strategies: "Carnivore conservation project in northeastern Argentina."] *Biológica- Naturaleza, Conservación y Sociedad*, 12:37–47.

Soler G.L. & Cáceres F.S. 2009. Uso de hábitat por *Chrysocyon brachyurus*: un análisis de flexibilidad comportamental como herramienta para la conservación. [Habitat use by *Chrysocyon brachyurus*: An analysis of behavioral flexibility as a tool for the conservation.] In: *Estado de conocimiento y conservación del aguará guazú (*Chrysocyon brachyurus*) en la Provincia de Santa Fe, Argentina. Plan de acción para la conservación del aguará guazú en Santa Fe*, Versión 01, pp. 25–27.

Soler G.L. & Carenton J.M. 2008. The emblematic maned wolf: A carnivore for the conservation of wetlands in Argentina. *WAZA Magazine*, 10:4–7.

Soler G.L. & González Ciccia P. 2010. Acciones para la conservación del aguará guazú (*Chrysocyon brachyurus*) en Argentina: resultados preliminares. [Actions for the conservation of maned wolf in Argentina: Preliminary results.] XXIII Jornadas Argentinas de Mastozoología, Bahía Blanca, Buenos Aires.
Soler G.L., Carenton J.M., Birochio D., Salvatori V., Orozco M.M., Rosso M.S., Carpinetto M., et al. 2005a. Problems and recommendations for the conservation of maned wolf in Argentina. Results from the First Workshop of *Chrysocyon brachyurus* in Argentina and surrounding countries: Conservation *in situ* and *ex situ*. *Endangered Species UPDATE*, 22(1):1–9.
Soler G.L., Carenton J.M., Cowper Coles R., Pérez P., Fleita A., Cáceres C. & Acuña F. 2005b. Nuevos registros de *Dioctophyma renale* en *Chrysocyon brachyurus* en vida silvestre. [New records of *Dioctophyma renale* in *Chrysocyon brachyurus*.] Publicación del Grupo Fauna Casilda. Facultad de Ciencias Veterinarias de la Universidad Nacional de Rosario. *Nowet*, 3(3):22–27.
Soler G.L., Fleita A., Carenton J.M., Cuello P., Pérez P. & Palacios M.J. 2006a. El uso de una especie en extinción como herramienta en educación ambiental: Experiencias pedagógicas en la Provincia de Corrientes, Argentina. [The use of an endangered species as a tool in environmental education: Pedagogical experiences in Corrientes province, Argentina.] Publicación de la Asociación de Docentes en Ciencias Biológicas de la Argentina. *Revista de Educación en Biología*, 9(2):29–35.
Soler G.L., Salvatori V. & Fleita, A. 2006b. Relevamiento de conflictos con carnívoros silvestres en el nordeste de Argentina. [Survey of conflict with wild carnivores in northeastern Argentina.] Congreso Sudamericano de Mastozoología, Gramado, Brazil.
Soler G.L., Sisa A., Fleita A., Carenton J.M., Sellarés M.E., Pérez P., Gómez G., Caffe G. & Holovate R. 2006c. Registro de linfoma histiocítico en un individuo de lobo de crín (*Chrysocyon brachyurus*) en Argentina. [Maned wolf histiocytic lymphoma record in Argentina.] I Congreso Sudamericano de Mastozoología, Gramado, Brazil.
Soler G.L., Cáceres F.S., Temperoni R.V., Nigro N. & Ciuccio M. 2008a. Evaluando la relación entre los niños y la naturaleza a través de cuentos y dibujos. [Assessing the relationship between children and nature through stories and pictures.] VIII Jornadas Nacionales: III Congreso Internacional de Enseñanza de la Biología, Mar del Plata, Argentina.
Soler G.L., Rosso M.S. & Casanave E.B. 2008b. Talleres comunitarios de educación ambiental: Una herramienta para la conservación de especies conflictivas y en peligro de extinción. [Community workshops on environmental education: A tool for the conservation of conflict and endangered species.] Libro de Actas de las "II Jornadas Nacionales Interdisciplinarias de Investigación y Educación." Desafíos y Tensiones actuales, Bahía Blanca, Buenos Aires.
Soler L., Cáceres F., Palacios González M.J., Rodden M., Carenton J.M., Minini S. & Casanave E.B. 2009a. Conservation perspective on maned wolf *Chrysocyon brachyurus* in Argentina. 10th International Mammalogical Congress, Mendoza, Argentina.
Soler G.L., Sisa A., Casanave E.B., Cáceres F.S., Palacio González M.J., Sellares M.E., Eclessia O. & Zajarevich S. 2009b. *Dioctophyma renale* en *Chrysocyon brachyurus* (Mammalia, Carnivora): ¿Una amenaza latente?. [*Dioctophyma renale* in *Chrysocyon brachyurus* (Mammalia, Carnivora): A potential threat?] V Congreso Argentino de Parasitología, La Plata, Buenos Aires.
Soler L., Luengos Vidal E.M., Casanave E.B., Chtcherbine A., Peris Campodarbe A., Iaconis K. & Palacios González M.J. 2011.Uso de hábitat, home range, patrón de actividad y dieta del aguará guazú, *Chrysocyon brachyurus,* en un paisaje fragmentado del Nordeste argentino. [Habitat use, home range, activity pattern and diet of maned wolf in a fragmented landscape of northeast Argentina.] X Congreso de la SECEM (Sociedad Española para la Conservación y Estudio de los Mamíferos), Málaga, España.
Torres R., Jayat P. & Pacheco S. 2013. Modelling potential impacts of climate change on the bioclimatic envelope and conservation of the Maned Wolf (*Chrysocyon brachyurus*). *Mammalian Biology*, 78:41–49.

APPENDIX: PROJECTS AND LEADERS

Museo Municipal Florentino Ameghino de Santa Fe
Andres Pautasso
andrespautasso@yahoo.com.ar

Ñangarekó Project, Universidad de Buenos Aires (UBA)
D. V. M. Marcela Orozco
marcelaorozco.vet@gmail.com

Rescue, rehabilitation, and release of maned wolves: Fundación Temaikén
Biologist Paula González Ciccia
pgonzalez@temaiken.org.ar

Maned wolf *ex situ* conservation: Zoológico de la Ciudad de Buenos Aires
Stella Maris Velázquez
huanca@gmail.com; svelazquez@zoobuenosaires.com.ar

Maned wolf population genetics: Museo Argentino de Ciencias Naturales Bernardino Rivadavia (MACN)
Biologist Vanina Raimondi
licenbio2@gmail.com

Conservation of Wild Carnivores in Northeastern Argentina: Huellas NGO
MSc. Lucía Soler
lucia.soler@huellas.org.ar

16 Environmental Influences on Maned Wolf Ecology in Bolivia

Louise H. Emmons

CONTENTS

Introduction 221
Study Area and Methods 222
Summary of Previous Research 223
Resource Variability 224
Climate Change 224
Maned Wolves and Fire 225
Fruit Resources and Fire 226
Rodent Prey and Fires 227
Rodent Variability 228
Invasive Grasses 229
Disease Outbreaks 229
Discussion 230
Suggestions for Action 232
Acknowledgments 232
References 233

INTRODUCTION

The savannas that encompass the Bolivian maned wolf habitat are divided among three biogeographical regions contiguous with Brazil: (1) the Beni–Pampas del Heath seasonal wetlands, of Beni and La Paz Departments (with a tiny extension into Peru), bordering Rondônia; (2) the cerrado of northeastern Santa Cruz Department, bordering Rondônia and Mato Grosso; and (3) the Bolivian Pantanal on the southeastern border of Santa Cruz Department, contiguous with Mato Grosso do Sul and Paraguay (Queirolo et al. 2011; Muir and Emmons 2012). Most of these savannas are in geologic basins with poor drainage that experience seasonal flooding from late January to May. Along with fire, the flooding maintains the grasslands as a disclimax under climatic conditions that today would favor the existence of forests (Hirota et al. 2011; Mayle et al. 2007; Staver et al. 2011). Maned wolf habitat in natural savannas of Bolivia thus depends for its persistence on a fragile balance of tree suppression by the opposing forces of fire and water, which in turn are dependent on climate cycles and human activities. As well as influencing global phase changes between forest and savanna, short-term variations in fire and flooding can influence the food resource base of maned wolves and, consequently, the local carrying capacity of the habitat and resultant maned wolf population density.

Maned wolves have been little studied in Bolivia, where the absence of all-season roads hinders access to much of the maned wolf range. All published research and unpublished theses have been based in Noel Kempff Mercado National Park (NKP), in the department of Santa Cruz, which is likewise our study area (Lilienfeld 2000; Robbins and Deem 2002; Rumiz and Sainz 2002; Deem and Emmons 2005; Bronson et al. 2008; Deem et al. 2008; Castro 2010; Cuellar 2011; Emmons 2012).

We followed a small population of maned wolves in the Los Fierros savanna of NKP from 2001 to 2012, largely by GPS collar technology (2003–2011; Emmons 2012). During our study, we witnessed steep declines in the maned wolf numbers in the study area and in the rodent species richness and biomass in the savanna on which they live (Emmons 2009; Emmons 2012; Castro and Emmons 2012). Muir and Emmons (2012) treated local and global conservation and management issues based on our study and made suggestions for conservation action. My goal in this chapter is to present some additional observations on environmental factors that may influence maned wolf resource levels and, consequently, the territory and population size of breeding maned wolves in our study area, and perhaps throughout their geographic range. Some or all factors may have contributed to the recent decline in maned wolf numbers and are of relevance to future conservation of the species. As a frame of reference, I first briefly describe the study area and research methods and give a short summary of our findings on behavioral ecology.

STUDY AREA AND METHODS

We studied maned wolves in the 220 km^2 Los Fierros savanna (S 14°67′, W 60°91′; 190 m elev.). NKP is a 10,000 km^2 UNESCO World Heritage Site on the border of Brazil. Its geology and biota are described in Killeen and Schulenberg (1998), and the mammals are described in Emmons et al. (2006). About a third of NKP is fragmented into savannas ("pampas") embedded in humid to dry forest formations (Killeen and Schulenberg 1998). Maned wolves have been recorded in all of the savannas (Rumiz and Sainz 2002), but we do not know if all the savannas include breeding pairs. Adjacent maned wolf populations on NKP savannas of the Huanchaca Meseta to the north and cattle estancias to the south can interchange with the study population at Los Fierros.

There is now little human activity in the study area, but before establishment of NKP, Los Fierros savanna was used for raising cattle on natural grasslands, with annual burning (R. Choré, pers. comm.). The forests were logged for *Swietenia macrophylla* (mahogany) and *Amburana cearensis* ("roble"), and rubber was extracted from adjoining forests. Now, a few tourist vehicles cross the pampa in the dry season, in transit from the nearest community (38 km away through forest). Vehicles must raft across the river into the park, which limits unauthorized entry. There is no hunting in the study area, and there are no domestic animals on the pampa.

The Los Fierros savanna is the only area of NKP with road access and feasibility to study maned wolves by trapping. The southern half of this savanna is in the privately managed El Refugio Huanchaca Biological Station (El Refugio), where our research was based. The savanna is surrounded by forests except at the southern edge, where it abuts marshland bordering the Río Tarvo/Paragua. The grassland is subject to partial flooding from January to May, deep in the southern part (30–80 cm) and shallow in the northern part, the latter with many small elevated hummocks of dry ground with woody vegetation, on termite mounds embedded in lightly flooded grassland (hummock campo or "termite" savanna), and areas of "campo cerrado" that do not flood. In the wet season, maned wolf activity was restricted to dry and shallow-flooded parts of the habitat. In the dry season, the savanna dries completely and by late August, surface water is found only at the river or in rare waterholes and pools in forest streams. Maned wolf food resources are distributed in two phases: rodents in the grasslands, and fruits on woody plants that grow on dry ground in campo cerrado, termite hummocks, roadside berms, or "forest islands" in savanna. Both phases are critical to preserving optimal maned wolf habitat.

Maned wolves were trapped, anesthetized with Telazol®, and handled and monitored for health as described in Deem and Emmons (2005). All adults were fitted with VHF or archival GPS collars that were usually programmed to take hourly locations for 3 days of each week. Because most programmed drop-off devices failed, GPS data were recovered after recapture the next year, when collars were replaced. We recovered data from 20 radio-collar deployments on ten individuals, some collared for two to six sequential years and nearly their whole adult lives. Our results are presented in a monograph (Emmons 2012), including images and details of the study area and all methods.

Small mammals were trapped once yearly on standard plots (Emmons 2009). Plots were designed for large area coverage (17 ha) but not for density estimation, so interyear comparisons are expressed as relative numbers of captures per 1,000 trap nights. The same plots were trapped in the same calendar weeks of each year with the same trap placement and bait. Relative biomass on plots was estimated from weights of voucher specimens and captures in the study area. To follow changes in the rodent resource base of maned wolves, a savanna plot (PA) monitored from 2001 to 2007 was retrapped in 2010 and 2011. Taxon lists and trapping results from NKP plots are given in Emmons (2009) and Emmons et al. (2006).

Climate parameters of temperature, rainfall, and river levels were recorded with simple garden gauges at El Refugio. Yearly data are plotted in Emmons (2012). There are no complete precipitation records for national stations in Bolivia, as all are missing days of data for most months of each year; but I have analyzed temperature records from the weather station nearest to the study area (178 km), San Ignacio de Velasco (San Ignacio), which apparently ceased to record rainfall in 2009 (archives from www.tutiempo.net).

SUMMARY OF PREVIOUS RESEARCH

The circadian activity of maned wolves at Los Fierros was chiefly nocturnal in the dry season, but partly diurnal in months of flooding (January–March). Activity decreased at temperatures above 27°C, with travel nearly absent above 30°C. Maned wolves need to drink and in the dry season they traveled daily to water points. Adult home ranges were 40–123 km^2, with strong seasonal variations in land use. Daily travel averaged 14 km/night in the driest months and 7 km/night during flooded months, when maned wolves were active, on average, for 14.6 h and 12.0 h of the 24 h day, respectively, based on hourly GPS locations. Breeding pairs shared territories with contiguous borders that did not overlap with neighboring pairs. Borders shifted temporally, with adjacent pairs adjusting in parallel so as to avoid overlap. Young females stayed twice until adulthood on natal territories, as presumed helpers, and acquired the territory upon disappearance/death of the adult females. All helpers were female and we never identified more than one. Females were the holders of territories into which males moved to form pairs, never the reverse (four instances documented). Young males all emigrated. When 6–8 years old, maned wolves showed extreme tooth wear and dental disease was a major cause of morbidity. They showed serological evidence of exposure to pathogens of domestic dogs that can cause pup mortality (i.e., morbillivirus and parvovirus) and they had high levels of infection with potentially fatal (*Dirofilara immitis*) or debilitating (*Dioctophyme renale*) parasites (Deem and Emmons 2005; Bronson et al. 2008; Deem et al. 2012). The maned wolf diet in our study area was qualitatively and quantitatively similar to that found in other parts of the geographic range (about half each of fruit and small vertebrates; Castro 2010; Lilienfeld 2000). Calculations of calorie consumption and energy acquisition based on scat analysis and ranging behavior suggest that maned wolves are near their energetic limits in our study area. From 2002 to 2004, rodent biomass in plots on the savanna declined to 2%–7% of former levels. A major prey of maned wolves (*Cavia aperea*) became locally extinct (Emmons 2009) and other cerrado-endemic rodent taxa became extremely rare. Concomitantly, maned wolf body weight declined and home range size increased, with an associated decrease in population density; but causation has not been established. We discuss conservation issues identified from our study and from the literature, including the management of habitat by prescribed burning, with a suggested 4-year burn rotation. Finally, I speculated that the omnivorous diet has allowed the persistence of maned wolves through Late Pleistocene megafaunal extinctions, and that the tallness of maned wolves may be advantageous more for their role as frugivores than as predators.

RESOURCE VARIABILITY

Climate Change

The climate of our study area has been changing in the short-term perspective of the available weather records (about 15 years), and of the nearest local weather station, San Ignacio (Figure 16.1). The most striking change has been the reduction in May–September rainfall (Figure 16.1a). Although the annual rainfall has not shown an overall trend, there has been a slight increase in rainfall in the wet season and a decrease in the dry season (Emmons 2012). Winter (June–September) temperatures have apparently increased (Figure 16.1d), as seen also in San Ignacio weather station records (Figure 16.1c). Declining dry-season rainfall is reflected in dropping minimum river levels (Figure 16.1b). Some of these trends have persisted through several intervening ENSO (El Niño, La Niña) cycles associated with year-to-year precipitation variation. I do not think that the trends in the climate data from simple instruments at El Refugio Huanchaca are artifacts of instrumentation, but T_{min} recorded there may be elevated by sensor proximity to a building. Regardless of absolute values, the trends indicate gradual warming during this short period. The increasing dry season temperature and drought has caused a perceptible desiccation of the landscape, with a drop in the

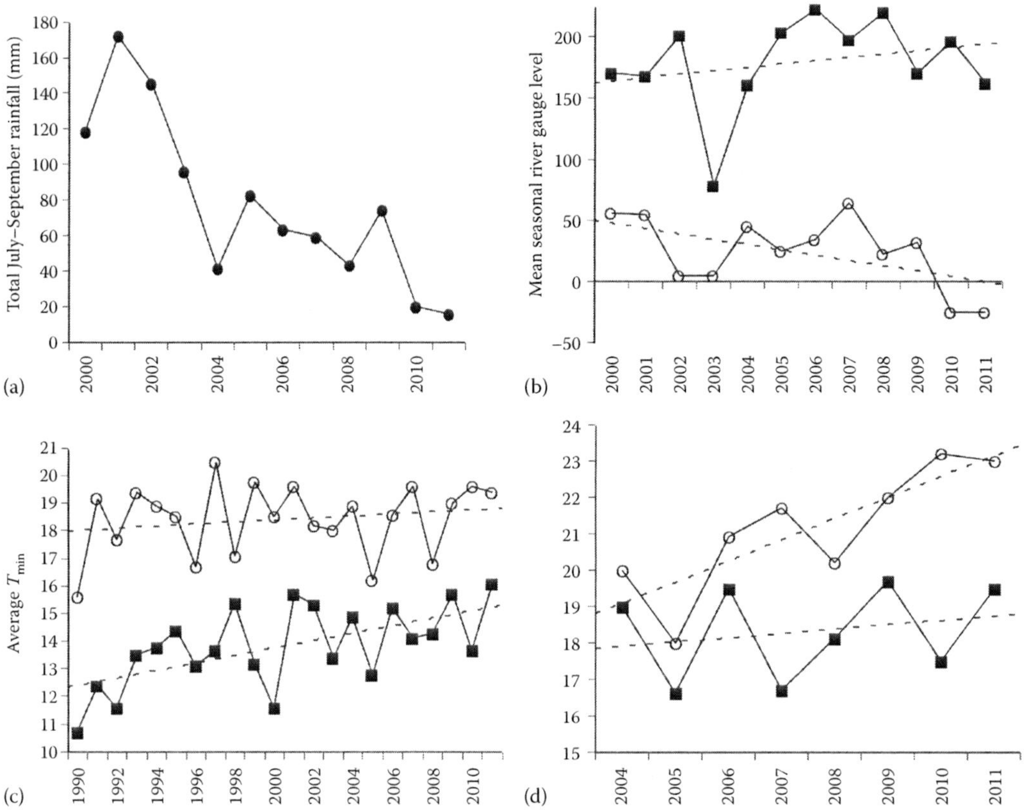

FIGURE 16.1 Changes in climate parameters during the study. (a) Mid dry-season total rainfall (July–September) during the study years, as measured with a garden gauge at El Refugio Huanchaca. (b) Highest and lowest seasonal river levels at El Refugio Huanchaca, average of January–March (filled squares), and average for October–December (open circles, adjusted to the same gauge, which creates negative values). (c) Average annual July (filled squares) and September (open circles) T_{min} (°C) for the nearest official weather station to the study area, San Ignacio de Velasco, with linear trendlines (Data from www.tutiempo.com; many days of data are missing in some years). (d) Average July and September T_{min} (°C) at El Refugio Huanchaca; symbols as in (c). All x-axis values are for years.

savanna water table indicated by the increasing depth of dug wells. A scarcity of drinking water could force an increase of maned wolf territory size so as to include water, especially as the months of least water availability are those of lactation, when females are stressed by increased metabolic needs.

Maned Wolves and Fire

Fires, both anthropic and lightning ignited, sporadically burn the NKP savannas, generally in the late dry season (September–October). Prescribed burning is not permitted in Bolivian parks, and about once or twice a decade, a hot, catastrophic fire consumes the vegetation of more than half of the Los Fierros savanna (1996, 1999, 2009). Fires burn the same areas repeatedly (northeastern half), while other zones are affected only by the largest burns (southern half). Fire in our study area is chiefly produced naturally, by lightning strikes. Electrical storms are nearly daily in October–November, when small afternoon thunderstorms produce little rain, but the vegetation is dry enough to ignite. Such fires are usually small, and because prevailing winds are northerly, they tend to burn downwind to the southern grassland edge with forest, preventing tree growth there, but almost never reaching the northern, upwind edge (Figure 16.2a: note the position of fire scar), which is experiencing invasion by shrublands and trees (Cuellar 2011). Savanna overgrown with woody vegetation is increasingly avoided by maned wolves (Figure 16.2b through d; note the absence of location points on the northwestern savanna edge). Satellite images show that the whole northern half of the savanna burned in 1996 and 1999, and the eastern half of the same area reburned in 2003 and 2006 (Figure 16.2c). In 2009, the entire savanna burned for the first time in recent decades. Burning of their habitat has both positive and negative effects on the resource base of maned wolves.

We recorded the habitat use of maned wolves before and after an almost identical part of the habitat burned, first in 2003 and again in 2006. In 2003, we could detect no change in habitat use of a female, F3, during the fortnight before, when she was first GPS-collared, and the 2.5 months after the fire. In strong contrast, after the next, late November 2006 fire, both F3 and her mate M5 avoided the entire burned area for at least 5 months (Figure 16.2c). Maned wolves normally traveled widely, covering their home ranges every few days (Emmons 2012), so the striking avoidance of the burned area (Figure 16.2c) suggests a major reduction of their effective home-range size, with temporary loss of resources. Rodents were abundant although declining (46 individuals captured

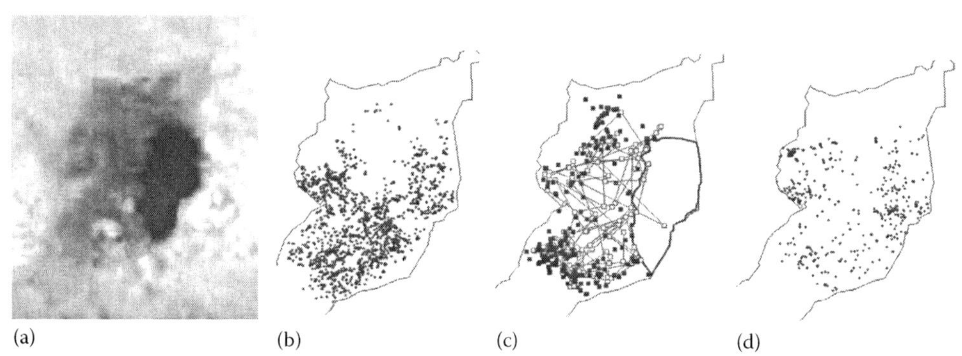

FIGURE 16.2 Maned wolf ranging before and after a fire. (a) Los Fierros savanna, fire scar on a low resolution (250 m) Modis satellite image 25 November 2006, when fire was still burning (white smoke on edge). (From NASA (GSFC), MODIS Rapid Response.) (b) Ranging of maned wolf adult female F3 prior to fire, October 2005 to July 2006 (2,900 locations). (c) Postfire ranging of maned wolf pair, F3 in January (130 locations, open symbols joined into routes) and male M5 from February to March 2007 (349 locations, closed symbols); heavy outline is extent of fire scar. (d) Maned wolf F3 ranging July–October 2008 (1,008 locations). Thin line encompassing points is the edge of Los Fierros savanna; GPS location points, black dots.

on plot PA) in 2003, when a burn did not seem to influence maned wolf range use. But prior to the 2006 fire, when maned wolves avoided the burned area, rodents had dropped 79% from levels in 2001 (10 individuals caught). I surmise that the 2006 fire temporarily pushed prey numbers below a threshold where it was worth foraging in the burned habitat.

We have no direct evidence of mortality of maned wolves by fire, as was observed by Silveira et al. (1999), but three old animals in our study area disappeared after the immense fire of 2009. This fire was survived by a single known individual, who in 2010 also used both the study savanna and a large region outside the burned area across the river (outside NKP). By February 2011, she no longer traveled outside the Los Fierros grassland during GPS samples.

Fruit Resources and Fire

Fruit resource plants that have their vegetative parts destroyed by fire (i.e., *Alibertia edulis* and *Miconia albicans*) can take 2 or more years to recover fruit production; but fruit crops after recovery from a burn can be significantly larger than those prior to the fire (Hoffmann 1998; Sanaiotti and Magnusson 1995). Species occurrences in the diet of maned wolves mirrored these effects, especially the dominant dietary fruit at Los Fierros, *A. edulis*, which peaked in the diet the second year after burns and then declined in the third and fourth years (Figure 16.3; Castro and Emmons 2012). The habitat where most scats were collected burned only partially in 2003 and 2006, and *A. edulis*, although reduced, persisted in the diet; but a year after the area burned entirely (1996), the species was nearly absent in scats sampled by Lilienfeld (2000). Plants cannot recover reproduction by fruiting if fires are too frequent, which may locally extirpate species that propagate by fruiting rather than by vegetative sprouting (Hoffmann 1998). On the other hand, fires increase cerrado soil nutrients and environmental heterogeneity (Da Silva and Batalha 2008).

Whereas some fruits are scarce the year after burns, fire stimulates immediate flowering and fruiting of a few, especially *Bromelia ignaciana* and *Duguetia furfuracea*. The giant-fruited *Solanum* species on which maned wolves rely in many regions (Rodrigues et al. 2007) have

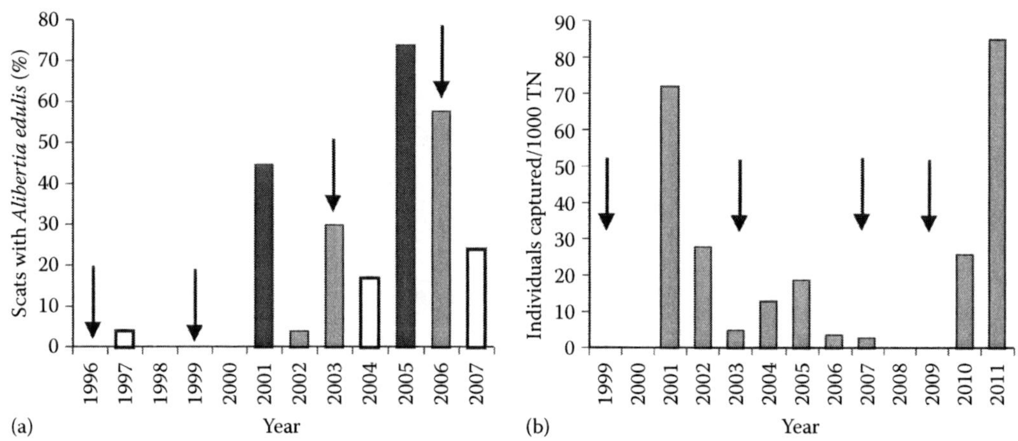

FIGURE 16.3 Temporal relationship of fires (arrows) and fruit and rodent resources. (a) Timing of fires that burned half or more of the maned wolf north territory (arrows) with the percentage of September–November scats that included *A. edulis* (N = 131 scats, 10–29 scats/sample), first year after burns (white bars), second year after burns (black bars), and third and fourth years after burns (gray bars). (Data from Castro, J. M. and Emmons, L. H., *Smithson. Contrib. Zool.*, 639, 37, 2012; data for 1997 from Lilienfeld, M. D., Unpublished Lic. Thesis, Universidad Autónoma "Gabriel René Moreno," Santa Cruz, 2000.) (b) Number of *N. lenguarum* captured on plot PA and years of fires that burned that plot. Absence of bars, no data. Burns in 2003 and 2006 occurred after plots were trapped. Burns of largest extent were in 1996, 1999, and 2009.

fire-resistant, corky bark that protects the stems from moderate fires, which destroy the current crop of leaves and fruit. When fires are too hot from strong winds and/or biomass accumulation, large *Solanum* plants can also lose their above-ground stems. Da Silva and Batalha (2008) found *S. lycocarpum* plants much more common on frequently burned sites than on infrequently burned areas.

In recent years, *M. albicans*, with fruits formerly much eaten by maned wolves in October–November, has apparently declined such that we have not found it in scats. The fruits seem to be drying on the plants without ripening (unquantified observations). This seems independent of fire events and may be linked to the increasing severity of dry-season drought (Figure 16.1a).

Rodent Prey and Fires

For rodent prey of maned wolves, several studies have shown that at least when fires are patchy and fairly small, within a matrix of unburned grassland, burning has little effect on density of common rodent species (Ghizoni et al. 2005; Vieira 1999). However, Briani et al. (2004) reported that, just as with fruits, the abundance of some rodents seems favored by certain successional stages (years) after cerrado fires. In particular, they found *Necromys lasiurus* to reach maximal numbers in the third year following burns, with declining populations in each year thereafter. Our data for the equivalent *Necromys lenguarum* are consistent with peak populations occurring two, and not three, years following fires (Figure 16.3); but our data set is too short for much inference. We recorded a steep decline in both *N. lenguarum* and *Juscelinomys huanchacae* (smaller insectivore–omnivores) beginning in 2003, 3 years postfire on two plots. The decline in *Cavia aperea* (cavies) and *Kunsia tomentosus* (larger herbivores) occurred in 2004, a year after a second fire on PA, which accounts for a higher 2003 biomass during declining rodent numbers (Figure 16.4). Our data suggest possible local depression of cavy numbers following an extensive fire that burned their whole habitat in 1999; or conversely, that the fire stimulated a delayed population increase, because numbers rose between 1 and 2 years postfire (Emmons 2009). We have

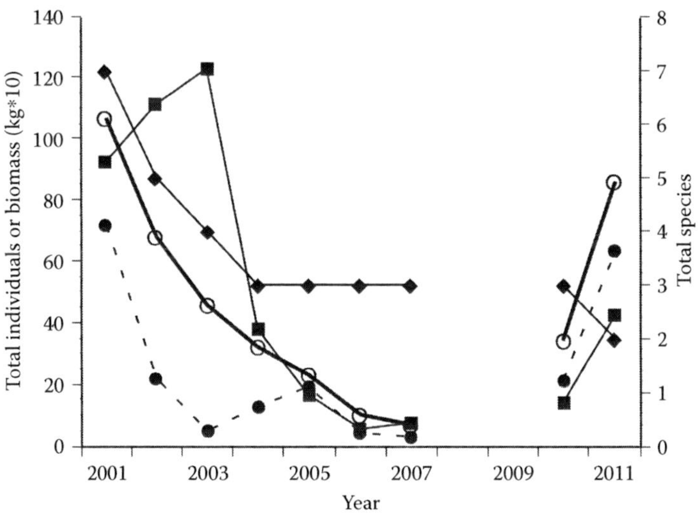

FIGURE 16.4 Long-term changes in the rodent prey community in a maned wolf territory in Los Fierros Pampa, from single yearly samples on a trapping plot. Total rodent biomass, square symbols, thin line; total individuals captured, open circles, thick line; number of *N. lenguarum* captured, dashed line, filled circles; and total species, filled diamonds, thin line (right axis). Note that the decline in species began prior to the decline in biomass, and that species richness per yearly sample did not increase in 2010–11, when increases were due to a single taxon. Compare with the timing of fires (Figure 16.3).

no prefire data other than that cavies were common for at least a decade. Cavies vanished from both burned and unburned plots after 2003, so the relationship between fire and cavy populations was not documented.

Briani et al. (2004) recorded an exceedingly slow postpeak decline in *Necromys*, on the order of over 20 years postfire to reach nearly zero; but the declines on our study plots were rapid, just 3 years to drop from 77 to 3 *Necromys* individuals captured (Figure 16.3), so numbers of mice in the two studies seem to have been influenced by different factors. A 90% decline in *N. lasiurus* over a 22-year interval, investigated by Magnusson et al. (2010), showed population density to be positively associated with fire and the linked Southern Oscillation Index, but not directly with rainfall (Magnusson et al. 2010). Our much smaller study seems generally consistent with a positive association with fire after a 2-year delay, but only for *Necromys* among maned wolf prey.

Cavies were abundant in the study area in 2003 (Emmons 2009), and on the day following the fire, they aggregated by dozens in small patches of unburned vegetation that provided little cover. We saw burned corpses and many presumably killed by foxes (*Cerdocyon thous*) and left uneaten: a fox traveling the road had dead cavies strung along behind it. By the next morning, the heads of all of the carcasses had been eaten, likely by maned wolves (which eat only the heads of birds they catch *ex situ* [C. Saffoe, National Zoological Park, pers. comm.]). Fires thus cause the death of prey both directly, by burning, and indirectly, by exposing them to intense predation pressure.

Rodent Variability

Extreme interannual population fluctuations are normal for the small mammals that comprise half of the maned wolf diet, so documenting the patterns and magnitude of excursions in the maned wolf prey base requires a multiyear effort. Changes that we recorded in the small mammal community in the Los Fierros study savanna from 2004 to 2007 included steep declines in species richness, numbers, and biomass (Emmons 2009). Species other than *Necromys*, all of which declined in habitats both burned and unburned in 2003 and 2006, and lightly flooded and unflooded in the wet season (Emmons 2009), did not seem to show any positive postfire numerical response; but other factors, such as the timing of postfire rainfall (affecting the rate of plant regrowth), could influence the magnitude of any fire-mediated effects.

We resumed small mammal trapping after a 2-year hiatus in 2010–2011. A single species, *N. lenguarum*, entirely recovered its 2001 peak numbers and was alone responsible for an increase in biomass (Figure 16.4). As this was the smallest species on the campo cerrado plot (averaging 60 g), it was of lesser value for maned wolves than the larger rodents that had not recovered to former numbers (*Kunsia tomentosus*, *Juscelinomys huanchacae*, *Cerradomy scotti*) or were apparently extinct in the study area (*Cavia aperea*). The latest sample (2011) showed the lowest species richness ever recorded on the plot (two species). We do not know if rodents have also decreased in other savannas of NKP.

Regarding the general rodent declines, I hypothesize related causes: (1) declining dry season dewfall (Emmons 2009); and (2) declining dry-season rainfall (Figure 16.1). These separately and together result in desiccation of the upper layers of soil and the biota on which both herbivorous and insectivorous micromammals depend. Although previously I did not detect increasing daily temperature maxima in the study area (Emmons 2009), another 4 years of data are indicating gradually increasing temperatures for both daily maxima and minima. This would augment desiccation of the landscape under reduced rainfall (Figure 16.1a) and also keep temperatures above the dewpoint for more of the time. The recovery of *N. lenguarum* numbers despite ever-increasing drought suggests that factors other than drought are influencing its numbers, such as its invertebrate food supply (Ghizoni et al. 2005). Why only one of five grassland-restricted rodent species has recovered its former abundance is unknown, but *N. lenguarum* is the only grassland taxon that occupies both deeply flooded, hummock savanna and never-flooded parts of the grassland, and it has many refuges from which to repopulate after large fires and flooding.

Maned wolves seem to eat nearly every small prey that they encounter, and we found all small mammals known from our study area in their scats (Motta-Junior et al. 1996; Castro and Emmons 2012). Some small mammal monitoring can therefore be done indirectly through scat analysis; but if a favored larger prey such as *Cavia* is abundant, rare or smaller species may be underrepresented. Small mammal trapping yields more quantifiable numbers and biomass, habitat distribution, reproduction, and population structure data, as well as the voucher specimens that are essential to adequate small mammal inventory.

Invasive Grasses

The flora of NKP is quite well known, with 134 species of Poaceae collected (Killeen and Schulenberg 1998; Panfil 2001; T. Killeen, pers. comm. of data set). Of these, only one species recorded by 1998 was an invasive African grass (*Melinis minutiflora*), collected only on granite outcrops and dry forest. In 2011, we found that the largest known stand of the important maned wolf food species, *Solanum gomphodes*, on a roadside berm in Los Fierros savanna, was overtopped, and many plants had apparently been killed by the African grass *Hyparrhenia rufa* (det. J. R. I. Wood). Recent forest clearing for artificial cattle pasture near NKP has led to the dispersal of invasive forage species along roads. Some newly colonized roads cut from forest near the park are now bordered by 3 m tall impenetrable walls of grass (species unidentified). *H. rufa* has taken over "vast areas" of Velasco Province of Bolivia and is spreading fast (John Wood, pers. comm.). Invasive grasses thus can suppress the growth of trees and woody shrubs upon which maned wolves depend for half of their diet (Hoffmann and Haridasan 2008), and their greater biomass potentially increases the spread rate of fires (Mistry and Berardi 2005). We do not know if *H. rufa* will affect future fruiting of *S. gomphodes* in NKP, as it did in 2011, but nonnative grasses are one of the most serious threats to entire native grassland communities, including the fruit and prey upon which maned wolves depend (Pivello et al. 1999), and in particular the roadside taxa favored by anthropic disturbance, such as *Solanum* species (Courtenay 1994).

Disease Outbreaks

> Aparentemente [*Chrysocyon*] es común en toda la zona oeste y norte de Trinidad [Beni Department, Bolivia], incluyendo las provincias Iténez, Mamoré, Ballivián y Yacuma. Pues en la mayoría de estancias visitadas en esta zona, la gente lo conoce muy bien y lo usa para curar males y hechizos. Por ejemplo el cuero lo usan para que no entre el mal cuando uno esta durmiendo ... Lo interesante es que en la zona sur de Trinidad, lo que corresponde a la provincia Marbán [*Chrysocyon*] ha desaparecido completamente. Mucha gente afirma que fue una peste que lo mato, pues se encontraban individuos moribundo o muertos en el monte, al igual que los zorros. Aunque dicen que los zorros otra vez han vuelto en cambio el Borochi desaparecido por completo. (M. Herrera, pers. comm., March 2006)*
>
> Mamerto Lopez, Oman Rivero y otras personas, cuentan que en el año 72 se podian encontrar en los caminos carreteros borochis y zorros enfermos y muertos, como que esten con una peste. Y que despues de la llenaura del 73, que hubo una gran inundación ya no quedo ninguno por la zona. Pero que hace muy poco es que nuevamente se esta volviendo a escuchar. (M. Herrera, pers. comm., April 2012)†

---

* *Translation*: [*Chrysocyon*] is apparently common in the entire area to the west and north of Trinidad [Beni Department, Bolivia] including the provinces of Iténez, Mamoré, Ballivián, and Yacuma, because in most of the estancias visited within this zone, people know the species well and use it to cure illness and sorceries. For example, the skin is used to prevent harm [or evil] from entering during sleep ... It is interesting that in the region south of Trinidad, in Marbán Province, [*Chrysocyon*] has completely disappeared. Many people state that they died from epidemic disease, because they encountered dead and dying individuals of both *Chrysocyon* and foxes [*Cerdocyon*]. Although foxes returned to the area, maned wolves completely disappeared.

† *Translation*: Mamerto Lopez, Oman Rivero, and others state that in 1972 they encountered sick and dying maned wolves and foxes, as if diseased, on the roads and trails. After the wet season of 1973, which was a year of exceptional floods, no maned wolves remained in the region. However, very recently people there have begun to hear [their calls] again.

The above reports from Bolivian conservation biologist Maurico Herrera not only note the widespread use of maned wolf parts for sorcery and folk medicine, common throughout the range in Bolivia (Muir and Emmons 2012), but also recount a local die-off in southern Beni Department in 1972, when both foxes (*C. thous*) and maned wolves ("borochi") were found moribund and dead on the roads. While the fox population recovered, maned wolves entirely disappeared from the region. This probable disease epidemic is remembered because the next year there were exceptional floods, after which no more maned wolves were seen. Recently (2012), however, maned wolves have for the first time once again been heard calling in Marbán Province.

During our decade of research in NKP, two apparent disease outbreaks caused declines of foxes that coincided with maned wolf mortality. (1) From February to July 2000, 4–6 *C. thous* (foxes) could be regularly seen circulating near the buildings around Mangabalito camp in NKP, on the Mangabalito savanna. By September 2000, only a rare individual fox was seen there, and a 1–2-year-old VHF-collared maned wolf had died there in the same interval. This probable outbreak did not extend to Los Fierros, where foxes were abundant in October 2000. (2) Foxes were the most commonly seen mammal on Los Fierros savanna: in a daily vehicle census in 2006 we noted 31 foxes, 1 jaguarundi, and 1 Geoffroy's cat during about 40 vehicle trips to check maned wolf traps. On 13 days, we saw 20 foxes on about 400 km of repeatedly driving 22 km of track, or about 1 fox/20 km driven. Other years until 2010 were similar, but the fauna was not censused comparably as we used bicycles. In September–October 2011 we saw no foxes and no felids during 360 km of checking traps by vehicle on the same 22 km of track; none during about 614 km driven over 6 weeks in the field; and none at all during many hundreds of kilometers on bicycles by five researchers. We found two dead foxes in the pampa, and a young female maned wolf about 2 years old was found freshly dead beside the river on the park boundary. A moribund fox was observed in the El Refugio camp (R. Choré, pers. comm.). The manager of the neighboring estancia (Caparú) likewise reported a major fox die-off, and vaqueros found a number of fox cadavers. The above anecdotes suggest that fatal epidemics occur abruptly among wild Canidae in the Bolivian cerrado and may cause mortality among maned wolves. In 2000, local villagers reported a possible distemper outbreak among their pets (from symptoms), but in 2011 there were no reported diseases or die-offs among local dogs at either Caparú or in the two nearest communities.

It is difficult to document epidemics among wildlife or assign them to causes, especially in the tropics with its rapid decomposition and scavenging by vultures. The only evidence of a die-off, as here, is from anecdotal reports (Fragoso 1997, Lahm et al. 2007). In each case above, foxes, which are seen daily in the maned wolf habitat, were the indicators of probable epidemics.

DISCUSSION

Because dry-season rainfall has declined in our study area through both El Niño (annual drought) and La Niña (high late wet-season rainfall) years (Figure 16.1), a long-term climate change may be in progress. Changes in climate variables of temperature and rainfall may be root determinants of productivity levels within the trophic pyramid that supports maned wolves. There are two related likely causes of reduced dry-season rainfall in NKP. First is the massive deforestation for agroindustry of the Brazilian cerrado in neighboring Rondônia and Mato Grosso. The nearest large deforested areas to Los Fierros savanna are less than 100 km away and include the entire eastern border of the park and far beyond it. Between 2001 and 2010, land under agriculture in Mato Grosso increased from 34% of the state to 43%, while forest was reduced from 36% to 32% of the land area (Marcia Macedo, pers. comm.). An estimated 25%–35% or more of Amazonian dry-season rainfall is recycled from the transpiration of forest, and this has been predicted to be lost by deforestation (Eltahir and Bras 1996; Mahli et al. 2008). Second, there is always smoke over NKP in the late dry season, and atmospheric smoke is inversely related to rainfall (Bevan et al. 2009; Davidson et al. 2012) and, I believe, to dewfall (Emmons 2009). If these phenomena are in action, cerrado deforestation in Brazil has a wide reach that may affect the biota, including maned wolf populations, in NKP and other protected areas. It is also possible that we are seeing part of a more general global warming phenomenon. Small temperature changes can also affect the fire regime, with serious consequences for both maned wolves and other taxa (Magnusson et al. 2010). Whatever the cause, for management planning, we hope for, but cannot assume, a future reversal to former conditions of wetter dry seasons. Our own climate records are far too short to establish decadal trends, but other studies appear to be showing deforestation-related drought (Bevan et al. 2009; Davidson et al. 2012).

Drought affects entire ecosystems, sometimes in unexpected ways. Munson et al. (2008) found that droughts were positively associated with lion mortality from distemper. The canid mortality from suspected disease in our study area in 2011 was associated with the driest dry season on record (Figure 16.1a and b), so stress from drought perhaps influenced disease mortality. The catastrophic fire of November 2009 was associated with the driest October–December then on record (figure 1.9 in Emmons 2012). Of all future threats to the persistence of maned wolves, extended droughts, whether caused by deforestation or climate change, may be the most serious, as they can prejudice all maned wolf food resources, promote excessive fires, and perhaps even increase susceptibility to disease, even within protected areas distant from agroindustry. Epidemic disease is likewise a serious threat to wildlife with small, isolated demes, such as maned wolves, where a few individuals represent much of the local population and gene pool (Deem et al. 2000).

I have outlined a number of bottom-up variables likely to have influenced the carrying capacity of maned wolves in our small study area, and some sporadic events that likely caused unexpected mortality (disease, large-area fires). All may be in play, but we cannot judge how they interacted during the decline in our maned wolf study population. Another maned wolf decline was recorded in 2006–2007 in Emas National Park (Goiás, Brazil). The large marked maned wolf population analyzed by Sollmann et al. (2010) showed good annual survivorship from 0.60 to 0.97 for the first 11 years of the study (1995–2005), but in the final 2 years (2006–2007), maned wolves at Emas experienced abrupt and drastic mortality, with a survivorship of only 0.28 and 0.31. Sollmann et al. (2010) did not identify a cause for low survival, but they speculated that changes in agricultural land use, from soybeans to sugarcane, may have reduced the usable habitat, as many Emas maned wolves ranged outside the park. However, the abrupt decline from one year to the next seems more suggestive of epidemic disease than of habitat change. In our study site, the decline in maned wolf numbers coincided with a rapid reduction in rodent biomass and a doubling of the maned wolf territory size, but it took place gradually, over 3 years from 2004 to 2006 (Emmons 2012). A second decline, in 2009–2010, resulted in a quadrupling of the territory and coincided with the largest fire on record, which burned the entire savanna. But it also coincided with the driest dry season on record (Figure 16.1). Although based on only a few individuals, our data indicate that scarcity in prey and fruit resources may have driven an expansion of the area needed to sustain a pair, and a catastrophic fire temporarily made the whole savanna insufficient for even a single territory. It would be interesting to see if Emas also experienced years of exceptional drought that were coincident with decreased maned wolf survivorship.

For maned wolves, fires produce a brief prey bonanza followed by temporary scarcity. Resources peak 2 or 3 years postfire, apparently followed by declining productivity. If a savanna is big enough to support many maned wolf pairs, natural fires may achieve a sustainable scenario; but in small savannas surrounded by inhospitable habitat, or large ones protected from burning, proactive habitat management with prescribed burns should prevent catastrophic megafires, conserve the extent of grasslands, and help to maintain maximal maned wolf population sizes for long-term survivorship and genetic diversity (Muir and Emmons 2012). Because the productivity of both plant and animal maned wolf food species can depend on the recent fire history of a maned wolf territory (Da Silva and Batalha 2008), diets based on scat analysis may be difficult to generalize for a particular study site when based on data from one or two annual cycles. Likewise, short surveys cannot adequately evaluate the average standing crop of food resources in a territory.

Suggestions for Action

I believe that maned wolves are usually resource limited, which determines their home-range size and consequently the number of breeding pairs in a given area and their success in raising offspring. However, this will not be true if populations are controlled below carrying capacity by exceptional mortality from road kill, hunting, disease, or predation, such as by jaguars. Conservation efforts should thus focus on maintaining ecosystems with maximal fruit and prey resources as well as on reducing specific causes of mortality. Nothing can be done about global climate change, but the local fire regime can be controlled (Myers 2006).

The importance of single ecological variables, individually and in combination, will be assessed only after long-term monitoring, on decadal scales, of the many factors that potentially influence the habitat productivity and/or the population size of maned wolves. Some suggestions follow for longitudinal research aimed at understanding the influence of environmental variables on maned wolf populations and the cerrado ecosystems in which they live.

1. Maintain good climate records within maned wolf habitat.
2. Monitor vegetation changes in maned wolf habitat with GIS analysis every 5 years.
3. Monitor small mammal populations directly by trapping and/or indirectly by multiyear analysis of scats.
4. Monitor the fruiting of key dietary fruit species.
5. Monitor fires and their effects on Items 2 and 3.
6. Study the effects of invasive alien grasses on fruiting shrubs and small mammal diversity and biomass.
7. Study the effects of invasive grasses on fire frequency and extent.
8. Census foxes to detect canid epidemics and hidden morbidity or enigmatic mortality in maned wolves.
9. Estimate the maned wolf population by capture, camera-trapping, DNA analysis of scats, or other means.
10. Fight the entrenched opposition to prescribed burning in protected grasslands.

Longitudinal environmental monitoring has been most effective when launched from universities, museums, or consortia that have hosted multiple generations of students and researchers. The fragile cerrado ecosystem needs long-term research sites in maned wolf habitat, where programs can be maintained on decadal scales.

ACKNOWLEDGMENTS

This study was a collaboration with the Museo de Historia Natural Noel Kempff Mercado (MNK), Santa Cruz, Bolivia. We thank the Dirección General de Biodiversidad (DGB) and the Servicio Nacional de Áreas Protegidas (SERNAP) of Bolivia for granting us permits to conduct our research in Noel Kempff National Park. Kathia Rivero was instrumental in obtaining permits and solving problems of logistics. Ian and Barbara Phillips made El Refugio Huanchaca a second home and kept the climate data coming. Maurico Herrera provided field notes on maned wolves in Beni Department. Essential field assistance was given by S. Angulo, A. Céspedes, J. M. Castro, S. Copetto, S. Cabrera, V. Chávez C., P. Cuellar S., F. Del Aguila, F. Huaquipa, E. Melgárez, R. Peña F., N. Rocha, and L. Sainz. Marcia Macedo kindly provided data on cerrado deforestation. Financial support was generously granted by The Douroucouli Foundation, The National Geographic Society, The Wildlife Conservation Society, The Amazon Conservation Association, W. Alton Jones Foundation, The Friends of the National Zoo, The Conservation Endowment Fund of the Association of Zoos and Aquariums, SeaWorld Busch

Gardens Conservation Fund, the Maned Wolf Species Survival Plan, The Department of Animal Health of the National Zoological Park (Smithsonian), The Maryland Zoo, and Alan Weedon.

REFERENCES

Bevan, S. L., P. R. J. North, W. M. F. Grey, S. O. Los, and S. E. Plummer. 2009. Impact of atmospheric aerosol from biomass burning on Amazon dry-season drought. *J. Geophys. Res.* 114: D09204.

Briani, D. C., A. Palma, E. M. Vieira, and R. P. B. Henriques. 2004. Post-fire succession of small mammals in the Cerrado of central Brazil. *Biodiver. Conserv.* 13(5): 1023–1037.

Bronson, E., L. H. Emmons, S. Murray, E. J. Dubovi, and S. Deem. 2008. Serosurvey of pathogens in domestic dogs on the border of Noel Kempff Mercado National Park, Bolivia. *J. Zoo Wildl. Med.* 39: 28–36.

Castro, J. M. 2010. Dieta estacional del borochi (*Chrysocyon brachyurus*) en el Parque Nacional Noel Kempff Mercado (PNK): Reserva Biológica el Refugio Huanchaca (ERH) y posibles causas de su variación temporal. Lic. Thesis, Universidad Autonoma Gabriel Rene Moreno, Santa Cruz de la Sierra.

Castro, J. M. and L. H. Emmons. 2012. Variation in diet and resources. *Smithson. Contrib. Zool.* 639: 37–55.

Courtenay, O. 1994. Conservation of the maned wolf: Fruitful relations in a changing environment. *Canid News* 2: 41–43.

Cuellar, P. 2011. Análisis multitemporal de cambios de la vegetación en la pampa "Los Fierros" a través de imágenes satelitales Landsat para la identificación del área de actividad y uso de habitat del borochi (*Chrysocyon brachyurus*) en el Parque Nacional Noel Kempff Mercado. Lic. Thesis, Universidad Autónoma "Gabriel Rene Moreno," Santa Cruz, Bolivia.

Da Silva, D. M. and M. A. Batalha. 2008. Soil-vegetation relationships in cerrados under different fire frequencies. *Plant Soil* 311: 87–96.

Davidson, E. A., A. C. de Araujo, P. Artaxo, J. K. Balch, I. F. Brown, M. M. C. Bustamante, M. T. Coe, et al. 2012. The Amazon basin in transition. *Nature* 481: 321–328.

Deem, S. L. and L. H. Emmons. 2005. Exposure of free-ranging maned wolves (*Chrysocyon brachyurus*) to infectious and parasitic disease agents in the Noel Kempff Mercado National Park, Bolivia. *J. Zoo Wildl. Med.* 36(2): 192–197.

Deem, S. L., L. H. Spelman, R. A. Yates, and R. J. Montali. 2000. Canine distemper in terrestrial carnivores: A review. *J. Zoo Wildl. Med.* 31(4): 441–451.

Deem, S. L., E. Bronson, S. Angulo Alpire, and L. H. Emmons. 2008. Monitoreo sanitario del borochi (*Chrysocyon brachyurus*) en el Parque Nacional Noel Kempff Mercado, Bolivia. *Rev. Boliv. Ecol. Conserv. Ambient.* 21: 41–50.

Deem, S. L., E. Bronson, S. Angulo, V. Acosta, S. Murray, R. G. Robbins, U. Giger, B. Rothschild, and L. H. Emmons. 2012. Morbidity and mortality. *Smithson. Contrib. Zool.* 639: 77–89.

Eltahir, E. A. B. and R. Bras. 1996. Precipitation recycling. *Rev. Geophys.* 34(3): 367–378.

Emmons, L. H. 2009. Long-term variation in small mammal abundances in forest and savanna of Bolivian Cerrado. *Biotropica* 41(4): 493–502.

Emmons, L. H. (ed.) 2012. The maned wolves of Noel Kempff Mercado National Park. *Smithson. Contrib. Zool.* 639.

Emmons, L. H., V. Chávez, N. Rocha, B. Phillips, I. Phillips, L. F. del Aguila, and M. J. Swarner. 2006. The non-flying mammals of Noel Kempff Mercado National Park (Bolivia). *Rev. Boliv. Ecol. Conserv. Ambient.* 19: 23–46.

Fragoso, J. 1997. Desapariciones locales del baquiro labiado (*Tayassu pecari*) en la Amazonía: Migración, sobre-cosecha, o epidemia. In, *Manejo de fauna silvestre en la Amazonía*, ed. T. Fang, R. E. Bodmer, R. Aquino, and M. H. Valqui, 309–312. Editorial Instituto de Ecología, La Paz.

Ghizoni, I. R. J., V. M. G. Layme, A. P. Lima, and W. E. Magnusson. 2005. Spatially explicit population dynamics in a declining population of a tropical rodent, *Bolomy lasiurus*. *J. Mammal.* 86: 677–682.

Hirota, M., M. Holmgren, E. H. Van Nes, and M. Scheffer. 2011. Global resilience of tropical forest and Savanna to critical transitions. *Science* 334: 232–235.

Hoffmann, W. A. 1998. Post-burn reproduction of woody plants in a neotropical savanna: The relative importance of sexual and vegetative reproduction. *J. App. Ecol.* 35: 422–433.

Hoffmann, W. A. and M. Haridasan. 2008. The invasive grass, *Melinis minutiflora*, inhibits tree regeneration in a neotropical savanna. *Austral Ecol.* 33:29–36.

Killeen, T. and T. Schulenberg. 1998. A biological assessment of Parque Nacional Noël Kempff Mercado, Bolivia. *RAP Working Papers* 10: 1–372.

Lahm, S. A., M. Kombila, R. Swanepoel, and R. F. W. Barnes. 2007. Morbidity and mortality of wild animals in relation to outbreaks of Ebola haemorrhagic fever in Gabon, 1994–2003. *Trans. Roy. Soc. Trop. Med. Hyg.* 101(1): 64–78.

Lilienfeld, M. D. 2000. Determinación de los habitos alimenticios del aguará guazú (*Chrysocyon brachyurus*) en dos zonas del Parque Nacional Noël Kempff Mercado. Lic. Thesis Universidad Autónoma "Gabriel René Moreno," Santa Cruz (unpublished).

Magnusson, W. E., V. M. G. Layme, and A. P. Lima. 2010. Complex effects of climate change: Population fluctuations in a tropical rodent are associated with the southern oscillation index and regional fire extent, but not directly with local rainfall. *Glob. Change Bio.* 16(9): 2401–2406.

Mahli, Y., J. T. Roberts, R. A. Betts, T. J. Killeen, W. Li, and C. A. Nobre. 2008. Climate change, deforestation, and the fate of the Amazon. *Science* 319: 169–172.

Mayle, F. E., R. P. Langstroth, R. A. Fisher, and P. Meir. 2007. Long-term forest-savannah dynamics in the Bolivian Amazon: Implications for conservation. *Philos. Trans. R. Soc. Lond.* B Biol. Sci. 362: 291–307.

Mistry, J. and A. Berardi. 2005. Assessing fire potential in a Brazilian savanna nature reserve. *Biotropica* 37: 439–451.

Motta-Junior, J. C., S. A. Talamoni, J. A. Lombardi, and K. Simokomaki. 1996. Diet of the maned wolf, *Chrysocyon brachyurus*, in central Brazil. *J. Zool.* 240: 277–284.

Muir, M. J. and L. H. Emmons. 2012. Conservation. *Smithson. Contrib. Zool.* 639: 91–115.

Munson, L., K. A. Terio, R. Kock, T. Mlengeya, M. E. Roelke, E. Dubovi, B. Summers, A. R. E. Sinclair, and C. Packer. 2008. Climate extremes promote fatal co-infections during canine distemper epidemics in African lions. *PLoS ONE* 3(6): e2545.

Myers, R. 2006. Convivir con el fuego. Mantieniendo los ecosistemas y los medíos de subsístencía mediante el manejo integral del fuego. The Nature Conservancy, Tallahassee.

Panfil, S. N. 2001. Late Holocene forest and savanna diversity and dynamics across an Amazonian ecotone. PhD Dissertation, University of Georgia, Athens.

Pivello, V. R., C. Nagako Shido, and S.-T. Meirelles. 1999. Alien grasses in Brazilian savannas: A threat to the biodiversity. *Biodiver. Conserv.* 8:1281–1294.

Queirolo, D., J. R. Moreira, L. Soler, L. H. Emmons, F. H. G. Rodrigues, A. A. Pautasso, J. L. Cartes, and V. Salvatori. 2011. Historical and current range of the Near Threatened maned wolf *Chrysocyon brachyurus* in South America. *Oryx* 45: 296–303.

Robbins, R. G. and S. L. Deem. 2002. Ticks of the genus *Amblyomma* (Acari: Ixodida: Ixodidae) from a maned wolf, *Chrysocyon brachyurus*, with the first report of *A. ovale* from this endangered canid. *Entomol. News* 113: 25–28.

Rodrigues, F. H. G., A. Hass, A. C. R. Lacerda, R. L. S. C. Grando, M. A. Bagno, A. M. R. Bezerra, and W. R. Silva. 2007. Feeding habits of the maned wolf (*Chrysocyon brachyurus*) in the Brazilian Cerrado. *Masto. Neotrop.* 14: 37–51.

Rumiz, D. and L. Sainz. 2002. Estimación del hábitat útil y la abundancia potencial del lobo de crin o borochi (*Chrysocyon brachyurus*) en Huanchaca, Santa Cruz. *Rev. Boliv. Ecol. Conserv. Ambient.* 11: 3–16.

Sanaiotti, T. and W. E. Magnusson. 1995. Effects of annual fires on the production of fleshy fruits used by birds in a Brazilian Amazonian savanna. *J. Trop. Ecol.* 11: 53–65.

Silveira, L., F. H. G. Rodrigues, and A. T. A. Jácomo. 1999. Impact of wildfires on the megafauna of Emas National Park, central Brazil. *Oryx* 33: 108–114.

Sollmann, R., M. M. Furtado, A. T. A. Jácomo, N. M. Tôrres, and L. Silveira. 2010. Maned wolf survival rate in central Brazil. *J. Zool.* 282: 207–213.

Staver, C., S. Archibald, and S. A. Levin. 2011. The global extent and determinants of savanna and forest as alternative biome states. *Science* 334: 230–232.

Vieira, E. M. 1999. Small mammal communities and fire in the Brazilian cerrado. *J. Zool.* 249: 75–81.

17 The Maned Wolf (*Chrysocyon brachyurus*) in Paraguay

José L. Cartes, Anthony J. Giordano, and M. Nathalia Mujica Cameroni

CONTENTS

Geography and Land Use 235
A Contextual Introduction 236
Historical References 237
Distribution and Habitats 238
Ecology 242
Threats and Conservation 243
The Future 245
References 246

GEOGRAPHY AND LAND USE

Paraguay is an ecologically diverse, landlocked country occupying 406,752 km^2 in south central South America (S 19° to 28°, W 54° to 63°). It is bordered to the northwest by Bolivia, South America's only other landlocked country; Argentina to the south and west; and Brazil to the north and east. Paraguay is bisected by the river bearing its name (i.e., Rio Paraguay), which divides the country into two distinct geographical regions. East of the drainage basin, most areas of the country were historically part of the inland Atlantic Forest ecoregion (Cartes and Yanosky 2003). However, an extensive cerrado–forest mosaic still exists in the north of this ecoregion, which is contiguous with the much larger Brazilian Cerrado ecoregion. Close to the Rio Parana in the southeast, pockets of grassland (pampas) and wetlands mark the northern extension of these habitats from Argentina, where they occupy a much greater area (Figure 17.1).

West of the Rio Paraguay drainage is the Gran Chaco ecoregion, which comprises >60% of the country by area. The Gran Chaco, nearly one-third of which currently occurs in Paraguay, is not a homogenous ecological region but rather a mosaic of transitioning ecotones. From the southeast to the northwest, the Chaco follows a general pattern of increasing dryness, changing from mixed open palm savanna to dense, semiarid, low-canopy scrub and thorn forest. Exceptions to this general rule exist, however, particularly in the extensive northern transition zones and close to both the Rio Paraguay and Rio Parana, where gallery or dense low-canopy forests predominate. Along the border with Brazil in Paraguay's extreme north, the dry Chaco transitions into Pantanal, which is a large flooded savanna and wetland ecoregion occurring predominantly to the east and north, beyond the border into Brazil.

More than 95% of eastern Paraguay's Atlantic Forests have disappeared over the past few decades (Macedo and Cartes 2003). These forests were originally converted for cattle-ranching but have since been maintained largely by intensive soybean farming and other forms of agriculture (Macedo and Cartes 2003). Those forests that remain, including those occurring in and adjacent to protected areas and private reserves, are under continued pressure of conversion due to increasing

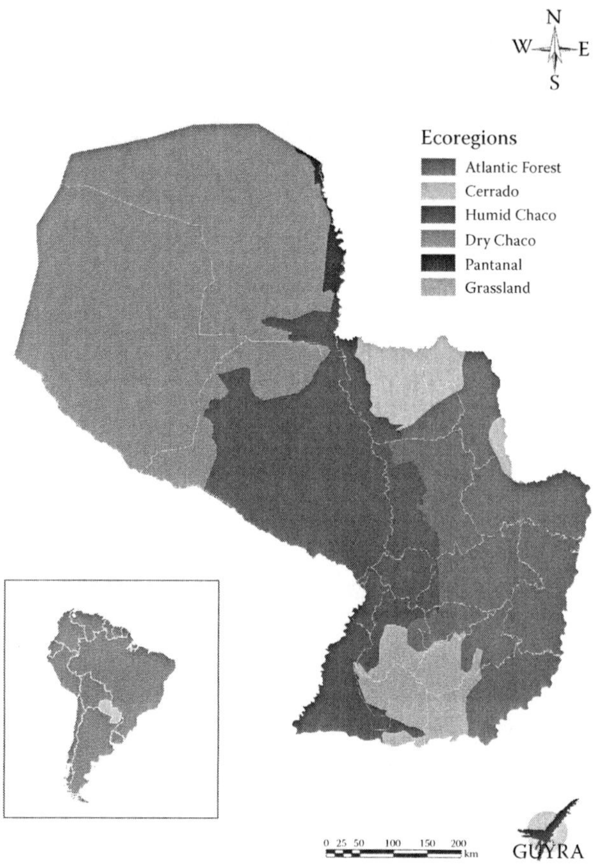

FIGURE 17.1 Map of Paraguay showing the confluence of six ecoregions. *White lines* correspond to political divisions. (From Clay, R.P., Del Castillo, H., and De Egea, J., *Áreas de Importancia para la Conservación de las Aves del Paraguay*, 1st edn, Guyra Paraguay (ed.), Guyra Paraguay/BirdLife International, pp. 31–44, 2008.)

economic pressures for more biofuel sources, as well as continued expansion of existing soybean interests. Paraguay's few remaining patches of natural cerrado in the north and grasslands to the south are facing the same threats. Conversely, only 2% of Paraguay's human population lives in the extensive Gran Chaco ecoregion. The Paraguayan Chaco is also home to more protected land by area than anywhere else in Paraguay, as well as its largest private land holdings. Despite extensive contiguous tracts of relatively natural vegetation, conversion of these areas for cattle-ranching is accelerating rapidly (Huang et al. 2009). This is particularly true for the relatively open humid Chaco savanna or low Chaco, which is frequently burned during the peak of the dry season. In the dry Chaco and Chaco–Pantanal ecotonal regions of the far north, between 500 and 1000 ha/day are currently being deforested to create more pasture (Cartes, pers. comm.; Guyra Paraguay 2012).

A CONTEXTUAL INTRODUCTION

Much of what we know about the ecology of the maned wolf stems from studies conducted in Brazil. The seminal work by Dietz (1984) was among the first, and for a long time it was the most comprehensive study of the maned wolf's life history. Among all ensuing investigations, those addressing the maned wolf's general diet and feeding ecology are perhaps the most common (Motta-Junior et al. 1996; Aragona and Setz 2001; Juarez and Marinho-Filho 2002; Bueno et al. 2002; Santos et al. 2003; Bueno and Motta-Junior 2004; Rodrigues et al. 2007). More recently, there have been detailed studies on various other aspects of the maned wolf's ecology, such as territorial behavior, resource and space use, and density (de Melo et al. 2006; Trolle et al. 2007; Coelho et al. 2008; Jacomo et al. 2009; Silveira et al. 2009; Vynne et al. 2011). In addition, the recent work of Consorte-McCrea (2011) thoroughly addresses the human dimensions of maned wolf ecology and conservation and is among the first such works to do so. In future years, as the species becomes increasingly the subject of investigation, we can expect newer, intensive efforts to reveal additional information on the maned wolf's ecology.

Outside of Brazil, however, the situation is very different. While the general distribution of the maned wolf in Bolivia, Argentina, and Paraguay is at least generally, if not specifically, known (Prevosti et al. 2004; Queirolo et al. 2011), all areas suffer from a dearth of empirical research. One exception is Noel Kempff Mercado National Park, Bolivia (Emmons 2012), where a recently completed ecological study of the maned wolf, the most in-depth outside Brazil, is likely to be seminal. Though an assessment of the maned wolf's status in Argentina and the threats it faces there was summarized by experts at a 2002 workshop (Soler et al. 2005), scientific information on the species there is still limited.

Perhaps the least is known about the status of the maned wolf in Paraguay. Compared to other maned wolf range nations, Paraguay has a younger history of traditional wildlife research. As such, there are fewer trained wildlife scientists there. Though wildlife biology is currently gaining popularity and importance in Paraguay, the country still suffers tremendously from a lack of basic information on the natural history, ecology, and conservation status of much of its fauna. In fact, through the 1980s, Paraguay's ecosystems were still relatively unexplored. Despite its relative obscurity, historical references to the maned wolf in Paraguay do exist. Furthermore, several scientific efforts to better understand the maned wolf's local ecology and wider distributional patterns are now underway in Paraguay.

Here we expand on information assembled on the maned wolf in Paraguay for the First Maned Wolf International Workshop and Population and Habitat Viability Assessment held in Serra da Canastra National Park, Minas Gerais, Brazil (de Paula et al. 2008). We start by providing an overview of historical references to the maned wolf in Paraguay. This is followed by an updated summary of what is currently known about the distribution and habitat of maned wolves in the country, a synopsis of previous and current scientific studies there, and a brief assessment of potential threats they face. Finally, we close with some suggestions for future conservation-relevant research on the maned wolf in Paraguay.

HISTORICAL REFERENCES

The first references to the maned wolf in Paraguay were made by the famous Spanish naturalist Felix de Azara during the late 1700s. Azara referred to the species by its local Guarani name by which it is still commonly known across the region, the *aguara guasu* (= "big fox"). Azara (1998) characterized the maned wolf as an "inhabitant of the flooded lands not beyond the Rio de la Plata." He was also one of the first to describe the maned wolf in Paraguay as being susceptible to common parasitic infections of the kidney, referring to "the presence of worms that feed in a mix of blood and water" encountered during postmortem kidney analyses. Azara was one of the earliest naturalists to refer to the disposition of the maned wolf, characterizing it as an "inoffensive eater of fruit, birds and rodents" that does not present a threat to livestock.

For the next century and a half, nearly nothing was written on the maned wolf in Paraguay. In 1939, A. W. Bertoni, a naturalist working in Paraguay and son of the famous Moises Bertoni, merely stated that the maned wolf was part of Paraguay's fauna. Although he said little else on the matter, the geographical focus of the Bertoni family's work was an area near Iguazu Falls amid the Atlantic Forest ecoregion. Thus, Bertoni's remarks may have been suggestive of the maned wolf occurring in this same region, and numerous recent anecdotal and confirmed observations in this region have proven this to be true.

In 1956, Carlos Gatti, a medical doctor working in Paraguay, compiled a comprehensive synopsis of local Guarani names and knowledge for the country's fauna and flora. He described the maned wolf as omnivorous and distributed widely across Paraguay's low-lying grassland areas. Interestingly, Gatti also references some indigenous medicinal use of the maned wolf tibia bone to treat venomous snake bites. He believed that the maned wolf did not readily consume beef and remarked that because of this it was difficult to maintain in captivity.

As recently as the early 1970s, live young maned wolves were still openly being caught and sold for the animal trade (Merrit 1973). During the warmer (November to January) months, they appeared in the animal markets of Asunción, apparently destined for export to Europe and the United States. Merrit (1973) also claimed that maned wolves were captured in the northern Argentinean states of Chaco and Formosa and were brought to these Asunción animal markets.

In the mid-1970s, the tagua or Chacoan peccary (*Catagonus wagneri*), the world's largest peccary, was "discovered" in the Paraguayan Chaco (Wetzel et al. 1975). Although long familiar to the indigenous tribes in the region, the tagua was believed to be a fossil form of peccary long thought extinct. While an amazing discovery in its own right, this scientific finding had implications beyond this "new" species. It spurred interest in natural science research in Paraguay, from which the National Biodiversity Inventory Project evolved. In the 1980s, this collaboration between Paraguay's then Livestock and Agriculture Ministry and the U.S. Peace Corp led to the development of the Museum of Natural History of Paraguay, which to this day holds the majority of the country's faunal distributional records (Gamarra de Fox and Martin 1996). This includes records of the maned wolf. Unfortunately, many of these records contain only generalized location data based on observations, indirect signs, or interviews, and often further details of these records are lacking. In the next section, we review these details as they pertain to the maned wolf in Paraguay, as well as incorporate more recent and detailed observations relating to the national distribution of the species.

DISTRIBUTION AND HABITATS

Although the maned wolf has previously not been the subject of formal survey efforts in Paraguay, information on the species is frequently gathered and recorded during field activities associated with other species (Cartes, pers. comm.; Guyra Paraguay 2012; Giordano, personal observations, 2009, 2011). However, formal registries and checklists of these sources of data are nonexistent. Information associated with most maned wolf records is incomplete and record details are not standardized. Furthermore, it is our opinion that few Paraguayan researchers consider an observation of the maned wolf or its sign to be academically noteworthy due to a general anecdotal acceptance that it is widely distributed in the country.

We have reviewed records as collected and maintained by the National Inventory of the Natural History Museum, Guyra Paraguay (Guyra Paraguay's Biodiversity Database BDGP), SPECIES (AJG), and collected by Neris et al. (2002). These records span the full spectrum of classifications and include observations of live animals by credible witnesses (e.g., the authors or biologists), collections of dead specimens (e.g., roadkill, opportunistic), interviews with local people, observations of signs (e.g., tracks, scats) by experienced biologists, and vocalizations. Our review supports the belief that the maned wolf is widespread in Paraguay and that it appears to occur in all of the country's major ecological regions except the extensive dry Chaco zone (Table 17.1 and Figure 17.2). Merrit's (1973, p. 53) characterization of maned wolf distribution reflected this issue well: "It appears that there are no areas of concentration of this species but rather the animal's range is broken and discontinuous." We agree that, despite ranging widely, the maned wolf's distribution in Paraguay *is* broken and discontinuous, which has perhaps compounded difficulties in describing suitable habitat for the species.

Maned wolves have been reported frequently from mixed agricultural areas in the eastern Atlantic Forest ecoregion, where actual standing forest is limited to secondary regrowth or is nonexistent.

TABLE 17.1
Recent Records of the Maned Wolf in Paraguay

Location	Department	Date	Ecoregion	Type	Source
Arroyos y Esteros County, Highway #3, km 79	Central	2004	HLC	Observed (roadkill)	BDGP
Bella Vista	Amambay	1987–1996[a]	CER	Collected	National Inventory (Museum of Natural History)
Cero Cora National Park	Amambay	2005, 2009	CER	Tracks, scats	BDGP, SPECIES
Estancia Don Luis, San Pedro	San Pedro	2002	HLC		BDGP
Estancia El Trébol, Highway #12, arrival entrance	Presidente Hayes	2004	HLC	Observed	BDGP
Estancia El Trébol, Highway #12, km 23	Presidente Hayes	2004	HLC	Observed	BDGP
Estancia Estrella	Concepcion	2007	CER	Tracks, interviews	BDGP
Estancia Golondrina	Presidente Hayes	2011	HLC	Observed, CT photos	SPECIES
Estancia Karanda	Presidente Hayes	2011	HLC	CT photos	SPECIES
Estancia La Graciela	Misiones	2004	GRA	Observed (roadkill)	BDGP
Estancia La Petrona	Presidente Hayes	2009	HLC	Tracks, scats	SPECIES
Estancia Maroma	Presidente Hayes	1994	HLC	Scats, tracks	BDGP
Estancia Pozo Azul	Presidente Hayes	1987–1996[a]	HLC	Collected	National Inventory (Museum of Natural History)
Estancia Redondo	Ñeembucu	2004	HLC	Collected, interviews	BDGP
Estancia Salazar	Presidente Hayes	2009	HLC	Tracks, scats	SPECIES
Estancia Santa Asuncion, road of entrance.	Presidente Hayes	1987–1996[a]	HLC	Collected (roadkill)	National Inventory (Museum of Natural History)
Estancia Señorita	San Pedro	2003	CER	Vocalizations	BDGP
Estancia Sta. Maria del Doce (La Piedad), riacho negro	Presidente Hayes	2005	HLC	Observed	BDGP
Highway "Ñ," km 40	Presidente Hayes	2005	HLC	Observed (roadkill)	BDGP
Highway #9 Trans-Chaco, km 60	Presidente Hayes	2005	HLC	Observed (roadkill)	BDGP
Highway #9 Trans-Chaco, km 70 close to Ea. Buena Vista	Presidente Hayes	2008	HLC	Observed (roadkill)	BDGP
Highway #9 Trans-Chaco, km 105	Presidente Hayes	1987–1996[a]	HLC	Collected (roadkill)	National Inventory (Museum of Natural History)
Highway #9 Trans-Chaco, km 175	Presidente Hayes	1987–1996[a]	HLC	Collected (roadkill)	National Inventory (Museum of Natural History)
Highway #9 Trans-Chaco, km 240 Ea. La Victoria's entrance	Presidente Hayes	2007	HLC	Observed (roadkill)	BDGP
Highway #9 Trans-Chaco, km 250 close to Pirahu	Presidente Hayes	2008	HLC	Observed (roadkill)	BDGP

(continued)

TABLE 17.1 (Continued)
Recent Records of the Maned Wolf in Paraguay

Location	Department	Date	Ecoregion	Type	Source
Highway #9 Trans-Chaco, km 286.7	Presidente Hayes	2005	HLC	Observed (roadkill)	BDGP
Isla Yacyreta Ecological Reserve	Misiones	1992, 2007	GRA	Collected, tracks, interviews	BDGP
Itaipu Dam Influence Zone	Alto Parana	1987–1996[a]	IAF	Unknown	National Inventory (Museum of Natural History)
Kai Rague Nature Reserve	Amambay	2006	IAF	Interviews	BDGP
Laguna Blanca Nature Reserve	San Pedro	Ongoing (2004–Present)	CER	Tracks, observed, vocalizations	Multiple sources: Staff of Fauna Paraguay/ Amigos de la Tierra, BDGP
Mbaracayu Forest Nature Reserve; buffer region: Mbaracayu Forest Biosphere Reserve	Canindeyu	Ongoing (1987– Present)	IAF	Observed, tracks, scats, vocalizations, CT photo	Multiple sources: Staff of Fundacion Moises Bertoni, National Inventory (Museum of Natural History), G. Zuercher, Authors, BDGP
Morombi (Ea. Golondrina) Nature Reserve	Caaguazu	1992, 2000	IAF	Observed, vocalizations	BDGP
Paso Bravo National Park	Concepcion	1999, 2008, 2009	CER	Observed, tracks, scats	BDGP, EER Paso Bravo, SPECIES
Rio Negro National Park, Pantanal Reserve/Three Giant Biological station	Alto Paraguay	2012	PAN	CT Photo	BDGP
River Tagatiyami/guasu	Concepcion	2002	CER	Observed	BDGP
San Luis Hills National Park	Concepcion	1987–1996[a]	CER	Observed	National Inventory (Museum of Natural History)
San Rafael National Park (San Clemente, Daniel)	Itapua/Caazapa	2001	IAF	Observations	BDGP
San Rafael National Park (San Pedro mi)	Itapua/Caazapa	2012	IAF	Observations	BDGP
Sombrero Nature Reserve	Cordillera	1995	HLC	Tracks	BDGP
Tapyta Nature Reserve	Caazapa	2001	IAF	Scats, vocalizations	BDGP
Tinfunke National Park	Presidente Hayes	2003	HLC	Interviews	BDGP
Toro Mocho Reserve	Presidente Hayes	2009, 2011	HLC	CT photos	SPECIES
Yabebyry Wildlife Refuge, Santa Ana Ranch	Misiones	1987–1996[a]	GRA	Collected	National Inventory (Museum of Natural History)
Yacyreta Island (actually flooded by Rio Parana)	Itapua	1987–1996[a]	GRA	Collected	National Inventory (Museum of Natural History)

IAF = Inland Atlantic Forest; CER = Cerrado; LHC = Low/Humid Chaco; GRA = Grasslands; PAN = Pantanal.

[a] Due to the lack of dates associated with individual National Inventory faunal records, we are using the broad time periods associated with inventory survey activities.

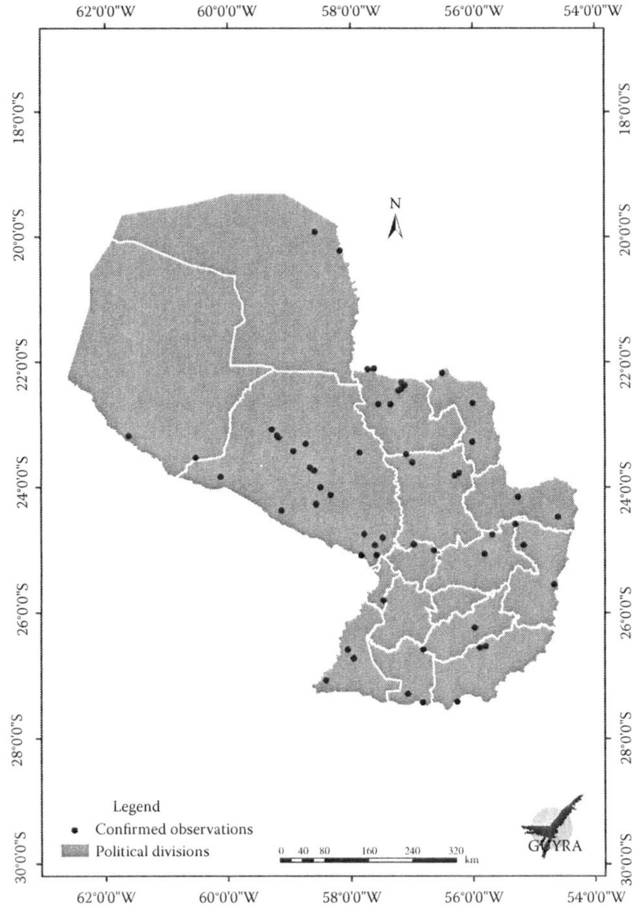

FIGURE 17.2 Map of recent records for the maned wolf in Paraguay.

Many of these areas constitute a mosaic of habitats, including the Cerrado ecoregion, particularly in the northeast part of the country. Often these mosaics transition abruptly across a small area and are so heterogeneous that broad classification of a given area's habitat can be challenging. Areas such as Mbaracayu Forest Nature Reserve (MFNR), for example, can be characterized as predominantly Atlantic forest, yet it contains small but ecologically important pockets of cerrado within. Conversely, other areas, such as Paso Bravo National Park, display the inverse pattern, where dominant cerrado vegetation is interspersed with pockets or "strips" of Atlantic forest. Maned wolves occur in both of these protected areas.

The maned wolf also occurs further south in the "pampas" grassland areas and among the wetlands of the Rio Parana drainage. In the 1990s, the Natural History Museum, in collaboration with the Japan International Cooperation Agency for Development (JICA), rescued fauna threatened by the flooding caused by the construction of the Yacyreta hydroelectric dam on the Rio Parana. Of particular geographical focus was their work in the southwestern-most Paraguayan state of Ñeembucu, a state dominated by wetland ecosystems along the border with Argentina. During this effort the maned wolf, the black howler monkeys (*Alouatta caraya*), and the marsh deer (*Blastocerus dichotomus*) were among the principle target animals for rescue and relocation as a result of flooding. This objective supposedly led to an effort by the Yacyreta Dam administrative authority in 1995 to initiate a captive breeding program for the maned wolf. Said to have occurred at the "Atinguy Zoo" near the city of Ayolas, little is known about the fate of this program or whether it still exists.

FIGURE 17.3 A maned wolf captured by a camera-trap on a trail in the Three Giant station, Pantanal Reserve.

Although several animals were allegedly transported from this program to the Asunción Zoo in the early 2000s, there are no maned wolves at the zoo at the time of writing.

In the far north of Paraguay, the maned wolf has been infrequently reported from the Pantanal ecoregion and was recently confirmed from the Reserva Pantanal Paraguayo (Cartes, pers. comm.; Guyra Paraguay 2012) (Figure 17.3). It is unclear how widely distributed the maned wolf is in this ecoregion, and reports on it are limited. One idea is that maned wolf movements in the Paraguayan Pantanal may have a seasonal component (Queirolo et al. 2011). Flooding in lower-lying Brazilian and Bolivian Pantanal zones during the wet season might drive regional populations to slightly higher and thus drier areas in Paraguay. Although information from this ecoregion of Paraguay is comparatively limited, given that maned wolves appear to occur throughout the Brazilian Pantanal, there is no reason to believe that the ecoregion is not similarly suited to the species in Paraguay.

The humid low Chaco of Paraguay, most of which occurs in the Department of President Hayes, may be the stronghold for the maned wolf in Paraguay. However, much more research is needed to determine this. Recent preliminary surveys (2011) have determined the maned wolf to be present across numerous sites in the humid Chaco, including within 30 km of Asunción (AJG, unpublished data). Despite their seemingly wide distribution across this ecoregion, local abundance may vary in accordance with the mosaic of the Chaco's ecotones, which are likely differentially suitable for the species (AJG, unpublished information). This may account for the "discontinuous" pattern that Merrit (1973) and the authors observe for the species in Paraguay. Shy of further intensive investigation, it is probable that any description of habitat preferences for the maned wolf in Paraguay will remain limited by our understanding of how the species responds to strong ecotonal heterogeneity across the landscape.

ECOLOGY

As stated earlier, the maned wolf has been the subject of limited prior scientific investigation in Paraguay. As a result, few aspects of the species' ecology are well understood. The limited work that has been completed has focused principally on its natural history in Mbaracayu Forest Nature Reserve. During one of the author's (AJG) previous visits (2005–2011) to this area, all maned wolf signs (e.g., tracks, scats) were encountered in the eastern cerrado portion of the reserve. All scats (~20–25) contained the fruit of *Solanum astrocarpum*, which is a common part of the reserve flora and is one of the most commonly reported if not dominant items in the maned wolf's diet elsewhere (Dietz 1984; Motta-Junior et al. 1996; Aragona and Setz 2001; Juarez and Marinho-Filho 2002;

Santos et al. 2003; Bueno and Motta-Junior 2004; Rodrigues et al. 2007). Similar visits by the author to both Paso Bravo and Cero Cora National Parks have resulted in casual observations of maned wolf scats containing the fruits of *S. astrocarpum.*

Zuercher et al. (unpublished data) conducted a more scientific analysis of the maned wolf's diet and activity in the MFNR. From 2005 to 2008, they opportunistically sampled the park's principal road (~40 km long) for tracks and scats. Camera-traps were also placed along parts of the road, which bisects the northern and southern parts of the reserve and extends through both the Atlantic Forest and Cerrado habitats. They concluded that both scats and tracks were detected throughout the year but were more commonly recorded during winter months (May to August). The locations of signs detected along the road appeared to vary annually. During some years (2005–2006) both tracks and scats were not infrequently encountered in the Atlantic Forest portion of the reserve far from the Cerrado, which clearly demonstrates use of these deeper forest habitats by the species. By occurrence, fruits (e.g., *Solanum lycocarpum, Butia paraguariensis, Syagrus romanzoffiana*) comprised 65% of the diet, which exhibited significant annual variation. Interestingly, although small mammals such as rats (*Clyomys* sp.) and armadillos (unspecified spp.) constituted part of the maned wolf's diet as they do elsewhere, the occurrence of medium–large mammals in scats, such as pacas (*Cuniculus paca*), tamandua anteaters (*Tamandua tetradactyla*), and red brocket deers (*Mazama americana*), was not infrequent.

A recent effort to document the diversity and location of mammals killed on the Trans-Chaco Highway (Rt. 9) in the low humid Chaco (Department President Hayes) recorded the maned wolf among the many vehicle-caused mortalities (Cartes et al. 2011). The highway is the longest and most significant paved road west of the Rio Paraguay and extends from Asunción to Fortin Infante Rivarola. The authors are aware of many additional anecdotal reports of road-killed maned wolves and on at least two occasions (2008, 2011) AJG recorded a road-killed maned wolf along the side of this highway in the same general region.

Another effort underway in the humid Chaco ecoregion seeks to map maned wolf distribution and habitat occupancy, the first such effort in the country (AJG, unpublished information). This project, involving multiple, relatively distant sites, is employing an occupancy-modeling approach relying on camera-traps to gather some of the first regional ecological information on the maned wolf in Paraguay. Although it is promising, this project has only just begun (2011).

THREATS AND CONSERVATION

It is likely that the extreme lack of information on the distribution of the maned wolf in Paraguay has influenced prior thinking regarding not only its occurrence or status in the country, but probably its overall range as well. As a result, it could be argued that the conservation potential of the maned wolf has been largely marginalized outside Brazil. Recent research seeks to address these recently completed and now underway seeks to address these important knowledge gaps for this species. In Paraguay, threats posed to the short- and long-term security of maned wolf populations are largely unknown. It is feasible that the maned wolf is susceptible to the same threats there that it is facing in Brazil and elsewhere, including habitat fragmentation and conversion, direct persecution and harvest, vehicular traffic, and diseases communicated by domestic animals (Muir and Emmons 2012). However, no formal national threat assessments for the species have been conducted. There is also a lack of consensus as to whether or not some perceived threats such as habitat conversion are as serious as previously described. Although there is little doubt that regions important to overall biodiversity conservation, such as the cerrado and Pantanal, are increasingly fragmented and converted for agriculture and cattle-ranching (Mittermeier et al. 1990; Rattner et al. 1997; Klink and Machado 2005; Carvalho et al. 2009), it is unclear what effect this is having on the maned wolf.

Vynne (2010) concluded that maned wolves can survive in industrialized cropland regions so long as low-standing crops (i.e., those most likely to harbor sufficient rodent populations) are emphasized, and at least 20% of the farm's natural cerrado vegetation is set aside as reserve. Similarly,

Queirolo et al. (2011) have proposed that the maned wolf's distribution may actually be expanding further into the Atlantic Forest ecoregion due to deforestation and conversion for livestock grazing. The notion that conversion of habitat otherwise inhospitable to the maned wolf, such as primary or thick secondary forest, could lead to the creation of new, minimally adequate habitat is not implausible. However, Muir and Emmons (2012) found no evidence of maned wolves expanding into or surviving in agricultural landscapes in Bolivia. Vynne (2010) proposed that as tall, homogenous sugarcane crops increasingly compose agricultural regions, this transformation will eliminate foraging opportunities for the maned wolf and thus negate any benefits that may otherwise have been obtained by the maintenance of other forms of agriculture.

In eastern Paraguay, soybean agriculture currently dominates the Inland Atlantic Forest ecoregion and sugarcane production is expanding. The proximity of maned wolves to "islands" of Atlantic forest that remain in Paraguay, as well as their not infrequent forays deep into forest (Zuercher et al. unpublished), may or may not be a result of recent habitat transformations. Park rangers of Fundacion Moises Bertoni charged with protecting the Mbaracayu Forest Nature Reserve claim that the maned wolf commonly uses the agricultural matrix beyond the forest and cerrado outside the reserve (Freddy Ramírez, pers. comm.). Because historical references are scarce and data nonexistent, the evidence as to how maned wolf populations use higher-canopy forest is equivocal at best. In addition, the recent discovery that maned wolves use denser, semiarid transitional Chaco forest (AJG, unpublished information) amid a landscape of mixed natural Chaco savanna and areas converted for livestock grazing presents similar important questions as to how the maned wolf uses these habitats. However, as by all accounts ecotonal habitat mosaics were historically common throughout the Gran Chaco, Atlantic Forest, and Cerrado, the maned wolf may have always used the landscape in this manner.

It is also possible that the Paraguay River watershed contains important source habitats for the maned wolf in Paraguay. Given the predominance of gallery forests in proximity to the drainage, it is not unlikely that populations along this drainage are particularly important. Gallery forest edges could host larger populations of small mammals that maned wolves feed on. Such forests may also facilitate connectivity among maned wolf populations in the Pantanal and Cerrado–Atlantic Forest ecoregions of Brazil and Paraguay with populations in Paraguay's Chaco and southern grasslands (i.e., "pampas"). It is important to add the caveat that no such connectivity has been demonstrated yet, and more research would be needed to determine if this hypothesis is true.

Another consideration in planning for the conservation of viable maned wolf populations in Paraguay is the representativeness of the country's protected area network. Unlike some countries, Paraguay has an extensive National Protected Area System (SINASIP in Spanish). Approximately 24% of the country by land area is protected as official national parks, reserves, or refuges. By area, however, the dry high Chaco ecoregion is by far the most represented in this system, yet it is the only major habitat that maned wolves do not appear to occupy in Paraguay. Few flooded savannas in the humid low Chaco ecoregion are protected. In 2011, the government designated a former military facility not far from Asunción as a new fauna refuge. Along with San Bernardino National Park, a small protected area in the Central Department also not far from Asunción, this is the extent of the formal protection the humid Chaco has thus far received. Similarly, the Paraguayan Pantanal, the country's smallest ecoregion, is represented only by a single designated protected area, Rio Negro National Park.

Of course even the conferment of protected status is still not enough to ensure the survival of key species occurring in any given area. The reasons for this vary according to the country or even region and encompass political, socioeconomic, cultural, and ecological factors. In Paraguay, only 3%–4% of the country qualifies as truly "protected," utilizing any modern definition of a protected area (i.e., nationally protected land devoid of private landownership and nonindigenous residents). This does not include biosphere reserves, the majority of which often comprise a mosaic of human-transformed land uses (e.g., Mbaracayu Forest Biosphere Reserve, Gran Chaco Biosphere Reserve). This excludes some other "protected areas" designated as national parks (e.g., Rio Negro

National Park, Tinfunque National Park), which actually comprise private land and are thus not true parks. Among the "true" parks and protected areas in Paraguay, most are unstaffed or understaffed and thus protected in name only. This includes Defensores del Chaco National Park and Medanos National Park in the dry Chaco, which are the country's two largest national parks. Overall, most of Paraguay's protected areas are severely undermanaged, a problem that leads to persistent illegal encroachment for livestock grazing, uncontrolled fires to remove woody biomass and promote new grass, poaching of wildlife, and illegal logging. Although the Secretariat of the Environment is engaged in conservation issues and staffed with many competent and willing professionals, the funding and resources they are allocated by the central government are unfortunately very limited.

THE FUTURE

Ongoing efforts to model the distribution, activity patterns, and habitat use of maned wolves across varying ecotonal gradients and land uses will hopefully shed light on how these populations respond to landscape heterogeneity. Among the specific research goals of one such project is to obtain a better understanding of how the amount and type of human activity in the humid low Chaco affect maned wolf occupancy (AJG, pers. comm.). Future studies along these lines would be most welcome elsewhere in Paraguay, including the Pantanal and Cerrado–Atlantic Forest ecoregions. Such studies are needed to properly define those types of habitat fragmentation or conversion that negatively affect the maned wolf.

Distribution modeling could also play a role in the development of a national map of likely habitat. For example, many areas contain habitat suspected to be adequate for the maned wolf, yet confirmation of the species from these regions is lacking (JLC and AJG, unpublished information). Defining and subsequently modeling suitable habitat, followed by surveys seeking to ground truth a subsample of these locations, might be extremely useful in the creation of such a map. Alternatively, if such an effort was cost- or time-prohibitive, more protected areas thought to contain suitable maned wolf habitat could be surveyed when the opportunity allows. Existing examples of such areas include the following national and privately managed protected areas: Lago Ypoa National Park, Lago Ypacarai Reserve, Ypeti Reserve, Palmar Quemado Reserve, and Tagatiya-Mi Reserve.

Research looking to better identify and resolve proximate threats to maned wolves should also be a high priority. For example, do increased use of and proximity to agricultural and grazing land lead to increased disease transmission? Large groups of feral dogs (e.g., 20–80 individuals), long abandoned by thousands of dam workers years ago, are also a known challenge to the management of Yacyreta Island Reserve in the Rio Parana, an area of confirmed maned wolf occurrence (Cartes, pers. comm.). As unattended domestic dogs (both pets and free-ranging) are also commonplace in many areas of the humid Chaco, it is possible that disease transmission as well as the effects of interspecific aggression and interference competition might serve as limiting factors to maned wolf distribution or local population persistence.

Finally, in terms of current conservation potential for the maned wolf, Paraguay has surely been overlooked. Despite being plagued with challenges to protected area management and representativeness, a lack of fundamental ecological and distributional information, and accelerating habitat and land use changes, the maned wolf appears at least to be widely distributed throughout the country and commitment to wildlife conservation issues is increasing. Furthermore, many private land holdings are currently very large, only small portions of which have been converted to pasture. This situation is not static however, and, as mentioned earlier, deforestation rates are increasing. Furthermore, landowners unaware of the maned wolf's conservation status, habitat requirements, and ecological role in natural and mixed agricultural ecosystems are unlikely to incorporate changes on their own that directly benefit maned wolf populations. We recommend that conservation outreach efforts accompany future scientific investigations or surveys of the maned wolf and should emphasize actions landowners can take to reduce basic threats. Such efforts should additionally emphasize that maned wolves pose no threat to cattle. If more landowners are aware of changes they can implement to mitigate threats to maned wolves, perhaps Paraguay's potential as a maned wolf stronghold can be realized.

REFERENCES

Aragona, M. and E. Z. F. Setz. 2001. Diet of the maned wolf, *Chrysocyon brachyurus* (Mammalia: Canidae), during wet and dry seasons at Ibitipoca State Park, Brazil. *Journal of Zoology* 254: 131–136.

de Azara, F. 1998. *Viajes por la América Meridional.* Tomo I. Elefante Blanco. Buenos Aires (ed.) (Reimpresión de la edición original de 1802).

Bertoni, A. de Winkelried. 1939. Catálogos sistemáticos de los vertebrados del Paraguay. *Revista de la Sociedad científica del Paraguay* 4(4): 1–59.

Bueno, A. de A., S. C. da S. Beletani, and J. C. Motta-Junior. 2002. Feeding ecology of the maned wolf, *Chrysocyon brachyurus* (Illiger, 1815) (Mammalia: Candiae), in the ecological station of Itirapina, Sao Paulo State, Brazil. *Biota Neotropica* 2. Accessed online at http://www.biotaneotropica.org.br/v2n2/pt/abstract?article+BN01802022002.

Bueno, A. de A. and J. C. Motta-Junior. 2004. Food habits of two syntopic canids, the maned wolf (*Chrysocyon brachuryus*) and the crab-eating fox (*Cerdocyon thous*) in southeastern Brazil. *Revista Chilena de Historia Natural* 77: 5–14.

Cartes, J. L. and A. Yanosky. 2003. Dynamics of biodiversity loss in the Paraguayan Atlantic Forest: An introduction. In Galindo-Leal, C. and I. de. G. Camera (eds), *The Atlantic Forest of South America: Biodiversity Status, Threats, and Outlook.* Island Press, Washington, DC, pp. 267–268.

Cartes, J. L., C. Morales, M. Motte, C. Vitale, and N. E. López de Kochalka. 2011. Estudios faunísticos en la Ruta 9 "Carlos A. López" (Transchaco). Conclusiones generales. *Revista de la Sociedad Científica del Paraguay* 16(2): 195–223.

Carvalho, F. M. V., P. de M. Junior, and L. G. Ferreirra. 2009. The cerrado into-pieces: Habitat fragmentation as a function of landscape use in the savannas of central Brazil. *Biological Conservation* 142: 1392–1403.

Clay, R. P., H. Del Castillo, and J. De Egea. 2008. Paraguay: Contextos eco-regionales, geográficos y socio-económicos. In Guyra Paraguay (ed.), *Áreas de importancia para la conservación de las aves del Paraguay*, 1st Edition. Guyra Paraguay/BirdLife International, pp. 31–44.

Coelho, C. M., L. F. B. de Melo, M. A. L. Sabato, E. M. Vaz Magini, A. Hirsch, and R. J. Young. 2008. Habitat use by wild maned wolves (*Chrysocyon brachyurus*) in a transition zone environment. *Journal of Mammalogy* 89: 97–104.

Consorte-McCrea, A. G. 2011. Conservation of the maned wolf (*Chrysocyon brachyurus*): Carnivore and people relationships in the southeast of Brazil. PhD Dissertation, University of Kent, Canterbury.

de Melo, L. F. B., M. A. L. Sabato, E. M. Vaz Magni, R. J. Young, and C. M. Coelho. 2006. Secret lives of maned wolves (*Chrysocyon brachyurus* Illiger 1815) as revealed by GPS tracking collars. *Journal of Zoology* 271: 27–36.

de Paula, R. C., P. Medici, and R. G. Morato (eds). 2008. Maned wolf action plan: Population and habitat viability assessment (PHVA). ICMBIO, Brasilia.

Dietz, J. M. 1984. Ecology and social organization of the maned wolf (*Chrysocyon brachyurus*). *Smithsonian Contributions to Zoology* 392: 1–51.

Emmons, L. (ed.). 2012. The maned wolves of Noel Kempff Mercado National Park. *Smithsonian Contributions to Zoology* 639:1–147.

Gamarra de Fox, I. and A. Martin. 1996. Lista de Mamíferos del Paraguay. In O. Romero Martínez (ed.), *Colecciones de flora y fauna del Museo Nacional de Historia Natural del Paraguay.* Museo Nacional de Historia Natural del Paraguay-Dirección de Parques Nacionales y Vida Silvestre, Asuncion, pp. 469–573.

Guyra Paraguay. 2012. Monthly technical reports about land use change, burning and flooding in the Great Chaco of South America. At: http://www.guyra.org.py/reportes-de-cambios-de-uso-de-la-tierra-del-gran-chaco-americano.

Huang, C., S. Kim, K. Songa, J. R. G. Townshend, P. Davis, A. Altstatt, O. Rodas, et al. 2009. Assessment of Paraguay's forest cover change using LandSAT observations. *Global and Planetary Change* 67:1–12.

Jacomo, A. T. de A., C. K. Kashivakura, C. Ferro, M. M. Furtado, S. P. Astete, N. M. Torres, R. Sollman, and L. Silveira. 2009. Home range and spatial organization of maned wolves in the Brazilian grasslands. *Journal of Mammalogy* 90: 150–157.

Juarez, K. M. and J. Marinho-Filho. 2002. Diet, habitat use, and home ranges of sympatric canids in central Brazil. *Journal of Mammalogy* 83: 925–933.

Klink, C. A. and R. B. Machado. 2005. Conservation of the Brazilian cerrado. *Conservation Biology* 19: 707–713.

Macedo, A. M. and J. L. Cartes. 2003. Socioeconomic drivers in the interior Atlantic Forest. In Galindo-Leal, C. and I. de G. Camara (eds), *The Atlantic Forest of South America: Biodiversity Status, Threats, and Outlook*. Island Press, Washington, DC, pp. 310–324.

Meritt, D. A. 1973. Some observations on the maned wolf *Chrysocyon brachyurus* in Paraguay. *Zoologica* 58: 53.

Mittermeier, R. A., I. de G. Camara, M. T. J. Padua, and J. Blanck. 1990. Conservation in the Pantanal of Brazil. *Oryx* 24: 103–112.

Motta-Junior, J. C., S. A. Talamoni, J. A. Lombardi, and K. Simokomari. 1996. Diet of the maned wolf, *Chrysocyon brachyurus*, in central Brazil. *Journal of Zoology* 240: 277–284.

Muir, M. J. and L. H. Emmons. 2012. Conservation. In Emmons, L. H. (ed.), *The Maned Wolves of Noel Kempff Mercado National Park*. Smithsonian Institution Press, Washington, DC, pp. 91–116.

Neris, N., F. Colmán, E. Ovelar, N. Sukigara, and N. Ishii. 2002. Guía de Mamíferos medianos y grandes del Paraguay. Distribución, tendencia poblacional y utilización. SEAM-Agencia de Cooperación Internacional del Japón (JICA).

Prevosti, F. J., M. Bonomo, and E. P. Tonni. 2004. La distribución de *Chrysocyon brachyurus* (Illiger, 1811) (Mammalia: Carnivora: Canidae) durante el holoceno en la Argentina: Implicancias paleoambientales. *Mastozoologia Neotropical* 11: 27–43.

Queirolo, D., J. Moreira, L. Soler, L. Emmons, F. H. Rodrigues, A. Pautasso, J. L. Cartes, and V. Salvatori. 2011. Historical and present geographic distribution of *Chrysocyon brachyurus* (Carnivora: Canidae), Fauna & Flora International. *Oryx* 45(2): 296–303.

Rattner, J. A., J. F. Ribeiro, and S. Bridgwater. 1997. The Brazilian cerrado vegetation and its threats to biodiversity. *Annals of Botany* 80: 223–230.

Rodrigues, F. H. G., A. Hass, A. C. R. Lacerda, R. L. S. C. Grando, M. A. Bagno, A. M. R. Bezerra, and W. R. Silva. 2007. Feeding habits of the maned wolf (*Chrysocyon brachyurus*) in the Brazilian cerrado. *Mastozoologia Neotropical* 14: 37–51.

Santos, E. F., E. Z. F. Setz, and N. Gobbi. 2003. Diet of the maned wolf (*Chrysocyon brachyurus*) and its role in seed dispersal on a cattle ranch in Brazil. *Journal of Zoology (London)* 260: 203–208.

Silveira, L., M. M. Furtado, N. M. Torres, R. Sollmann, G. Uhl, and A. T. de A. Jacomo. 2009. Maned wolf density in a central Brazilian grassland reserve. *Journal of Wildlife Management* 73: 68–71.

Soler, L., J. M. Carenton, D. Birochio, V. Salvatori, M. M. Orozco, M. S. Rosso, M. Carpinetto, et al. 2005. Problems and recommendations for the conservation of the maned wolf in Argentina: Results from the first workshop of *Chrysocyon brachyurus* in Argentina and surrounding countries, conservation in situ and ex situ. *Endangered Species Update* 22: 1–9.

Trolle, M., A. J. Noss, E. D. S. Lima, and J. C. Dalponte. 2007. Camera-trap studies of maned wolf density in the cerrado and the Pantanal of Brazil. *Biodiversity and Conservation* 16: 1197–1204.

Vynne, C. 2010. Landscape use by wide-ranging mammals of the cerrado. PhD Dissertation, Department of Biology, University of Washington, Seattle.

Vynne, C., J. L. Keim, R. B. Machado, J. Marinho-Filho, L. Silveira, M. J. Groom, and S. K. Wasser. 2011. Resource selection and its implications for wide-ranging mammals of the Brazilian Cerrado. *PLoS ONE* 6(12): e28939.

Wetzel, R. M., R. E. Dubos, R. L. Martin, and P. Myers. 1975. *Catagonus* an "extinct" peccary, alive in Paraguay. *Science* 189: 379–381.

Zuercher, G. L., S. Fernandez, F. R. Pinto, and M. C. Velazquez. Ecologia del Aguara Guazu (*Chrysocyon brachyurus*) en el cerrado de la Reserva Natural del Bosque Mbaracayu. Unpublished poster, Fundacion Moises Bertoni.

Part III

Perspectives for the Future

Drawing of maned wolf by Marcelo Pinto.

18 Education and Communication for the Conservation of Wild Canids

Denise Taylor

CONTENTS

Introduction 251
Past and Current Conservation and Education Initiatives 252
Education Problems and Issues 253
Environmental and Conservation Education 254
Creating and Strengthening Communities of Practice 256
Environmental Pedagogy in the Digital Age 257
Conclusion 260
References 262

INTRODUCTION

When there are conservation problems such as loss of habitat and threats to individual species, education is often seen as one of the solutions; and usually with the maxim that the more we know about a species or a particular habitat, and the more awareness we raise, the more likely we are to take action to protect and conserve it. However, implementing an education program as one of a number of conservation strategies is rarely straightforward. Almost four decades on from the Belgrade Charter (UNESCO, 1975) and Tbilisi Declaration (UNESCO, 1978), which outlined detailed principles and guidelines for environmental education, there is still very little understanding of exactly what environmental education is (Orr, 1994; Jacobson et al., 2006; Taylor, 2009). This applies across the social spectrum from grassroots projects to the formal curriculum-based education system and to policy makers; with those in the roles of educators and teachers struggling not only with some of the complex concepts of environmental education, but also with the daunting challenges of finding and implementing the most appropriate and relevant educational solutions.

The conservation issues surrounding the maned wolf (*Chrysocyon brachyurus*) are mirrored across many other species and have been discussed at length in other chapters in this book. However, in relation to education and communication it is perhaps habitat loss and the prevailing attitudes toward the maned wolf that are key to designing and developing education programs that will help to conserve future populations.

In this chapter, I will argue for a multiagency and multiple approach to environmental education that needs to be underpinned by a pedagogical framework that builds on education theory as well as taking best practice from other disciplines such as business and commerce. To begin, I will explore some of the recent studies that have been conducted on the maned wolf, the attitudes toward them and their habitats, and some of the recent education programs that have taken place, which provide valuable information that can be used for the design and development of future education initiatives, and which will build on the foundations already laid down. I will then provide a brief summary of some of the concepts and theories of education and communication, and what they have to offer education practitioners working to conserve the maned wolf and its habitats. The literature on education and learning theories is too vast for the space allowed in this section, so I will focus on just a few theories that are more relevant to environmental educators. Finally, I will discuss the rapidly changing nature of environmental education and some of the tools that are now available that make education in the twenty-first century not only more accessible as technology becomes ubiquitous, even in poorer countries, but also engaging and fun. As Orr (1994) said, perhaps it is true that we will only protect and conserve what we truly know and understand, and through knowledge and understanding we come to love.

PAST AND CURRENT CONSERVATION AND EDUCATION INITIATIVES

A Maned Wolf Species Survival Plan was first established in the early 1980s with the aim of maintaining viable self-sustaining captive populations and promoting conservation through education and field conservation initiatives. This was reviewed and discussed in a paper in 2005 that reaffirmed the goals for the conservation of the maned wolf (Songsasen and Rodden, 2010), and in the same year an international workshop was held on population and viability assessment of the maned wolf. This resulted in the publication of the Maned Wolf Management Plan in 2008 (de Paula et al., 2008). The findings from the discussions at the workshop in 2005 concurred with those of the workshop held in Argentina in 2002 (Soler et al., 2002) which were as follows: a lack of cooperation and public involvement within the species range; a lack of public interest in general; ignorance at all levels; a lack of communication between managers and management agencies; a lack of political decisiveness; a lack of economic and human resources; and a lack of education expertise and resources. Education is a key topic discussed at both workshops. From the workshop in 2002, the authors reported that the presence of "different social levels of the community from Corrientes province was enriching." The report also stated that the "maned wolf had been declared a national monument in two provinces of Argentina, and this had improved protection of the species." Other researchers agree with this approach and feel that it is one of the ways of helping to raise awareness and provide opportunities for increasing knowledge about the maned wolf (Butler, 1995; Jacobson, 1995). It has also worked successfully for other species. In the Caribbean, a conservation project used the Bahama parrot (*Amazona leucocephala bahamensis*) as a target species to foster a sense of national pride. It entailed a widespread communication campaign that incorporated an array of methods and media including 5,000 fact sheets distributed through schools, a school song produced and taught to more than 20,000 children, 26,500 bumper stickers, an eight-page newspaper supplement in a national newspaper, and 10 million letters franked with an image of the parrot (Butler, 1995). Participants at the workshop in 2005 reported that more needed to be done to develop education programs with the appropriate levels of knowledge, skills, and expertise, and furthermore that any education initiatives needed to take into account local culture, as well as the cultural differences between potential educators and learners.

Despite this encouraging start, more recent surveys suggest that there is a widespread lack of knowledge and information about the species (Consorte-McCrea and Rubin, 2009), and compared with other wild canid species, the maned wolf has also received little attention within the natural sciences discussions at conferences and symposia. There are some exceptions to this where the maned wolf is a zoo exhibit in various facilities around the world, the most popular and well known being the Smithsonian National Zoological Park which has carried out numerous studies on the maned wolf and produced education materials and programs.

Bizerril's (2004) survey investigated the attitudes of children and young adults toward one of Brazil's key biomes, the cerrado, which is a primary habitat for maned wolves. He found that there is a high level of ignorance and apathy toward its flora and fauna. The study found that teachers in the schools surveyed also displayed a lack of knowledge of and indifference toward the cerrado, perhaps compounded by the fact that the textbooks they were using in their classrooms had very little information about the cerrado and the species living there, including the maned wolf.

The maned wolf is a large mammal with a striking appearance, and while most people seemed to be able to recognize it and correctly identify it as a species, the research suggests that the majority of people were not able to accurately describe it. They were also misinformed about its diet and some people believed its body parts have magical properties, either for medicinal purposes or to bring good luck or ward off bad luck. Despite the lack of knowledge, attitudes toward the maned wolf tend to be largely positive, probably because unlike its wild canid cousin, the gray wolf (*Canis lupus*), it is not an aggressive animal, and it does not prey on either wild or domestic animals, other than the occasional chicken, and is therefore not seen as a competitive threat for food resources.

In terms of current education initiatives, efforts in the Serra da Canastra National Park in southeast Brazil have recently been focused on community education, with a project aimed at linking community communication and participation to the conservation of the maned wolf (Bizerril et al., 2011). This program adopted a multiple approach using three key tools to foster community participation including the communal production of a book about the region; the deployment of a traveling environmental cinema to local villages and farms; and the provision of training courses for the community on how to use communications tools for environmental education. This project enjoyed some success and provides other educators with a model for further developing community education programs. As the research suggests a lack of information and a sense of apathy and indifference, it is vitally important that education programs are embedded in local communities, and that these community projects have strong links with schools, policy makers, and the various stakeholders with vested interests in the grassland habitats. As well as education programs, human dimensions approaches have proven to be successful when focused on individual species within their habitats. Wolf management plans for the gray wolf have been implemented in several countries (Bath, 2009) and are discussed further in Chapter 19. It is perhaps fortunate that the maned wolf has not suffered the same maligned reputation as other wild canids such as the gray wolf, with attitudes overall tending to be positive toward this charismatic animal (Consorte-McCrea and Rubin, 2009).

EDUCATION PROBLEMS AND ISSUES

Before embarking on designing and developing education programs and communication campaigns for the conservation of the maned wolf, it is necessary to understand what the issues are. Jacobson et al. (2006) suggest that conservation issues are rarely one-sided scenarios with simple solutions. They usually involve an array of problems and multiple stakeholders. A *problem* is defined as "any situation in which something valuable is at risk," and an *issue* arises when "two or more parties or 'players' disagree about the solution to the problem."

The main threat to the maned wolf is habitat loss and habitat fragmentation. The "disagreement" about the issues involved, if it can be called this in relation to the conservation of the maned wolf, is primarily about land use for the different human activities and need for resources, which are destroying the maned wolf's habitat, and which make the conservation issues very complex. Protecting the remaining grassland and savanna habitats is of paramount importance, both for nonhuman species that rely on the resources that these habitats provide and for human communities living in and around the grasslands. This is the biggest challenge for educators, as it requires change at a political level, and perhaps on a global scale when you consider that many of the crops now grown on previous grassland habitat are for export. A global education approach was laid down in the Belgrade Charter (UNESCO, 1975) and Tbilisi Declaration (UNESCO, 1978) and in this instance also needs to have a longer-term, multiagency, multiple approach that works in partnership with other conservation strategies such as political campaigning and lobbying for social change.

The second problem, identified earlier, is the lack of knowledge and information about the maned wolf as a species. For maned wolf populations to thrive and flourish, conservation measures need to be implemented that include education and communication programs to address the issues. Having the maned wolf as an emblem species is a good start, but as there is still so little information and knowledge available, education and communication programs will need to take place on a long-term basis taking a lifelong learning approach, which has been a highly influential paradigm more recently (Field and Leicester, 2000). This would mean incorporating educational material, lessons, and activities into the formal education of the school curriculum as well as into the nonformal and informal education approaches of community-based projects. With the wealth of tools and methods now available to deliver education and communication programs, this problem ought to be relatively easy to address as long as there is the will and motivation within the various communities, calling on community champions with the passion and determination to raise awareness and educate people about the maned wolf.

Finally, an additional problem is the lack of trained professionals as educators and community project leaders, suggested in the study conducted by Bizerril et al. (2011). This is a common problem for conservation projects throughout the world working on a wide range of biodiversity conservation projects and links very strongly to the previous point on the lack of information. In gray wolf conservation, wildlife biologists often take on the additional role of educators, without any formal education training, and with education being supplementary to their main role, thus reducing the time and effort they can devote to it (Taylor, 2004, 2009). Educators, however, need not be pigeonholed to specific roles, and especially when advancements in digital and media technologies are making it easier for them to access information, acquire knowledge, and become experts in their chosen fields. However, for education to be effective it is vital that educators taking on this role have some understanding of the concepts, theories, and processes of education and learning, and how these can be used to bring about the changes desired.

ENVIRONMENTAL AND CONSERVATION EDUCATION

Environmental education is still a relatively new discipline of education, which in modern times is perhaps more concerned with the problems of environmental degradation and destruction and how education can help to find the solutions to address these problems. There have been many definitions of both environmental education and conservation education, but they all have similar underlying principles and philosophies. The aims and objectives of environmental education as stated in the Belgrade Charter (Box 18.1) continue to resonate today, even though the world has changed dramatically in the past four decades since it was first published, and it is perhaps worth revisiting some of these at this juncture, starting with its definition, and then looking at the aims and objectives:

> Environmental education, properly understood, should constitute a comprehensive lifelong education, one responsive to changes in a rapidly changing world. It should prepare the individual for life through an understanding of the major problems of the contemporary world, and the provision of skills and attributes needed to play a productive role towards improving life and protecting the environment with due regard given to ethical values. (Belgrade Charter, UNESCO, 1977)

In attempting to define environmental education as a concept, it can be said to be the process of worthwhile learning designed to affect people's awareness of and attitudes and behaviors toward natural resources, with the specific aim of developing lifelong knowledge and skills for conservation action (Jacobson, 1995; Jacobson et al., 2006). The terms "designed to affect" and "action" allude to a process of transformation to some degree or other, where a change in perspective is required in order to bring about positive change that occurs when the learner shifts from a particular way of thinking or perspective to a different one, which is achieved through the acquisition of new knowledge (Mezirow, 1991, 2000). The new knowledge is then assimilated and can be applied to help the learner find solutions to a problem or novel experience they are presented with. This process of transformative learning usually requires a shift in worldviews or perspectives, which comes about when the learner, through the education process, is equipped with the skills and experience

BOX 18.1 AIMS AND OBJECTIVES OF ENVIRONMENTAL EDUCATION FROM THE BELGRADE CHARTER, 1975

ENVIRONMENTAL EDUCATION AIMS

- To improve all ecological relationships, including the relationship of humanity with nature and people with each other.
- To develop a world population that is aware of, and concerned about, the environment and its associated problems, and which has the knowledge, skills, attitudes, motivations and commitments to work individually and collectively toward solutions of current problems and the prevention of new ones.

ENVIRONMENTAL EDUCATION OBJECTIVES

1. *Awareness*: to help individuals and social groups acquire an awareness of and sensitivity to the total environment and its allied problems.
2. *Knowledge*: to help individuals and social groups acquire basic understanding of the total environment, its associated problems and humanity's critically responsible presence and role in it.
3. *Attitude*: to help individuals and social groups acquire social values, strong feelings of concern for the environment and the motivation for actively participating in its protection and improvement.
4. *Skills*: to help individuals and social groups acquire the skills for solving environmental problems.
5. *Evaluation ability*: to help individuals and social groups evaluate environmental measures and education programmes in terms of ecological, political, economic, social, aesthetic, and educational factors.
6. *Participation*: to help individuals and social groups develop a sense of responsibility and urgency regarding environmental problems to ensure appropriate action to solve those problems.

to be able to think critically and reflexively in order that any decisions or actions taken are done so based on the knowledge, skills, and experience available at that time. The shift in perspective often arises as a result of some form of dislocation (Mezirow, 1991, 2000). This is where the learner is presented with a new situation or experience that stirs them enough to question their current value position or belief system. This incidence of dislocation may be some kind of education process itself or even something as simple as an evocative painting or poem that challenges the learner to think differently or to see things from another perspective.

Transformative learning is what environmental educators are aiming to achieve when attempting to change attitudes and behaviors. In order for education programs to be designed to have such a transformative effect, educators and project leaders need to understand the existing perspectives of their target audiences, and by this I mean understanding current attitudes, values, and belief systems. As well as understanding the cultural and sociopolitical demographics of a particular group, some conservation practitioners also draw lessons from the world of business and commerce, where it is common practice for businesses to get to know their customers and target markets: who they are, how old they are, where they live, what they do, what they like and do not like, what they believe, what they value, and what influences them. Jacobson (1995; Jacobson et al., 2006) has long understood the parallels between marketing communications and education practice, and this has helped to shape much of the research she has carried out on a wide range of conservation projects throughout the world. She recognized that conservation organizations and projects are run and managed in the same way as commercial businesses, with similar processes such as planning, implementation, and evaluation. These three key elements can be applied across any conservation education program or communication campaign, with the relevant education concepts and theories weaving through each program. For example, if we simply want to raise awareness, then a communication campaign can be planned, implemented, and evaluated on a short-term basis. This might take the form of a poster campaign or, as discussed earlier, a community video project or, in more recent times, a social media or online campaign. Alongside this, longer-term education programs can be planned, implemented, and evaluated that are likely to take the form of a curriculum-based program or a "human dimensions" approach (discussed in more detail in the next chapter) where all the stakeholders agree to a process that has the end result of reaching consensus on conservation strategies. The nature of this approach also involves transformative learning, as the process itself requires participants to have an open mind, to think critically and reflexively about the conservation problems and issues, and to challenge their own beliefs and values as well as those of others with perhaps different perspectives (Bath, 2009).

The scope and length of education and communication initiatives required are largely determined by the degree of positive or negative attitudes and also how much existing knowledge learners or participants possess. Someone with a high level of ignorance but with a positive attitude is more likely to respond positively to education programs more quickly than someone with a hostile and negative attitude (Taylor 2004, 2009). In research carried out in wolf conservation education, hostile attitudes are much harder to change and usually require an integrated mix of approaches involving multiple agencies and multiple methods. Unlike the gray wolf, however, the maned wolf does not suffer high levels of hostility and aggression toward it, although, like the gray wolf, loss of habitat is one of the primary issues affecting its future conservation. Because there is not a high level of hostility, educators need to look at ways of overcoming the levels of apathy and indifference, and at the same time building on the higher level of positive attitudes that appear to exist; turning this into a passion will help to fuel the motivation and the desire to protect the maned wolf and the habitats in which it lives.

Figure 18.1 summarizes the different approaches that could be taken depending on existing attitudes and whether these are positive or negative and to what degree.

CREATING AND STRENGTHENING COMMUNITIES OF PRACTICE

Ultimately, the success of conservation education will be measured by the biodiversity conserved worldwide (Jacobson et al., 2006), but its success relies on human communities coexisting with wild species, especially where there is a level of conflict or threat. An example is the increasing need for land for agriculture and extraction of natural resources at the expense of the species living there, which also rely on it for their survival. Bizerril et al.'s (2011) study of a community-based education program demonstrated that this approach can be effective at raising awareness of the conservation issues and empowering local people to participate in the protection and conservation of habitats and biodiversity. In essence, what this project demonstrated was a "community of practice" where groups of people share a concern or passion for something they do and learn how to do it better as they interact regularly (Wenger, c2007). Communities of practice are everywhere and most people are generally involved in a number of them: at home, at work, at school, and through leisure pursuits (Smith, 2003, 2009). Wenger (1998) argues that through regular interactions, shared enterprise learning is also taking place. For the conservation of the maned wolf, the shared concerns and passion are being demonstrated in existing communities of practice that could be extended, and new ones are being established. The community linking project described earlier highlights several communities of practice that have been effective in bringing people together to work on common goals, with learning as a key element of the process. Sharing is another fundamental characteristic of communities of practice including resources, stories, tools, and ways of addressing problems (Smith, 2003, 2009), and helps to build the strong relationships that need to exist within

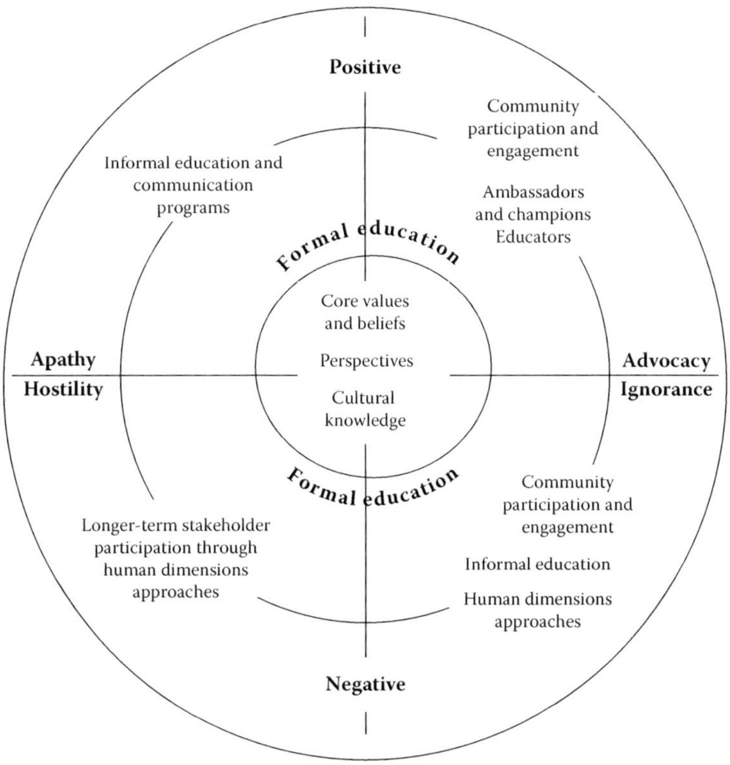

FIGURE 18.1 Attitudes toward the maned wolf and potential education approaches.

communities in order for them to flourish over time. Communities of practice provide a framework that enables links between communities in the real world and online or digital communities, such as those using social media websites, chat rooms, and forums, and even e-mail lists or listservs. Digital tools are discussed in the next section, which highlights the opportunities they present to educators.

As we can see, there are many different types of communities, and in conservation there are often attempts to link education programs with schools and communities (Table 18.1). Jacobson et al. (2006) recognized this, and although they do not refer to them as communities of practice, the principles are the same, and their model suggests a variety of approaches for community-based activities.

ENVIRONMENTAL PEDAGOGY IN THE DIGITAL AGE

We live in an increasingly networked world through advancements in information and communications technology (ICT). A number of Latin American countries are in the top ten for the use of social media platforms (Ipsos/Reuters monthly polls), and the use of mobile technology in emerging economies is also rising rapidly due to the ease of network expansion, cheaper relative costs, high demand, and the willingness of companies to invest (Banks and Burge, 2004). Across South America, there are an estimated 175,643,035 Internet users (45% penetration) and 369,723, 642 mobile cellular subscribers (92.4% penetration) (Miniwatts Marketing Group, 2012). The current generation, therefore, is growing up in a world where digital technology is ubiquitous, even in poorer countries and rural areas where there is less infrastructure. This new generation of learners has been described as the Millennials (Oblinger and Oblinger, 2004 cited in Siemens, 2008).

In the digital age, these Millennial learners have come to expect education as a participative, engaging, and active environment (Siemens, 2008), and teachers and educators need to adapt to

TABLE 18.1
Connecting Schools and Communities to Conservation

Technique	Purpose
Service learning	Applies academic learning to community needs by involving learners in planning, service, reflection, and celebration of environmental action.
Issue investigation	Involves students in defining, investigating, and analyzing an environmental issue and developing an action plan.
Project-based learning	Uses projects as an organizing framework to engage learners in addressing a real-world problem with no known solution and presenting their findings in groups.
Community-based research	Involves research conducted by and for communities with a goal of social and environmental change.
Citizen science	Uses the general public, teachers, or students in any or all steps of a scientific research project.
Mapping	Allows individuals or groups to create visual representations of resources, community, and regions. Can document either (1) knowledge and perceptions or (2) scientific data.

Source: Jacobson, S.K., McDuff, M.D., and Monroe, M.C., *Conservation Education and Outreach Techniques*, Oxford University Press, Oxford, 2006.

Note: Techniques that help conservation educators use real issues to connect schools and communities to conservation

the different approaches in learning and incorporate the new tools available (Figure 18.2). These include blogs, online learning websites, wikis, podcasts, online video, online serious games, virtual learning environments (VLEs), mobile and web applications (commonly known as apps), and social media environments. This presents real opportunities for rich blended learning experiences where experiential learning activities, perhaps programs that engage children and adult groups in "outdoor classrooms" and activities in the cerrado, are combined with a mix of activities using online games, apps, and social media, for example. Experiential learning is the process of learning through direct experience (Kolb, 1984) and is a vitally important element for any nature-based education. The

FIGURE 18.2 Children from a school in rural Bulgaria taking photographs of a poster display showing wild canids at the Large Carnivore Education Centre, Vlahi, Bulgaria. (© Chris Senior, http://www.ecoscape.org.uk.)

benefits of adopting a blended learning approach are that it involves members of different communities and programs can be designed and developed very cost effectively when infused with creativity and resourcefulness.

Table 18.2 outlines some of the additional benefits that can be achieved using available ICT technologies.

There is no doubt that education and learning in the twenty-first century will have to adapt to new approaches and opportunities, with Millennial learners already embracing participative and engaging platforms and technologies. However, with all the "shiny new toys" available it would be very easy to forget the fundamental principles of pedagogical frameworks, and educators incorporating digital technologies into their blending learning and education practice must be mindful also to incorporate the appropriate and relevant learning theories and concepts. Mayes and De Freitas (2007) suggest three sets of pedagogic priorities, which they term as perspectives: (1) the associationist perspective, where "learning occurs through the process of connecting the elementary mental or behavioral units, through sequences of activity followed by feedback" (p. 15); (2) the cognitive perspective, where "the underlying theme for learning is to model the processes of interpreting and constructing meaning" (p. 16); and (3) the situative perspective, where the "view of learning focuses on the way knowledge is distributed socially" (p. 18). The mapping of theoretical accounts

TABLE 18.2
Benefits of Using ICT for Conservation

Biodiversity Gain	Details
Practical benefits	ICT could provide simple, straightforward, practical benefits in conservation areas. By their nature these areas tend to be relatively remote and isolated, and lacking in ICT infrastructure for the needs of conservation, development, and local communities. The presence of an ICT infrastructure can save time and money by providing improved information flow and basic communications, for example.
Protects environments and wildlife	ICT could allow for more effective law enforcement by providing improvements in administrative and communications capacity, which in turn help to improve environment and wildlife protective measures.
Counteracts threats to wildlife	ICT could allow rangers, park managers, and communities to react more effectively to threats (e.g., poaching, fires, and encroaching animals).
Park management	ICT could enable a proactive, speedier, and standardized communications platform for park management, resulting in improved environmental management and planning.
Data management	ICT could provide a standard platform for recording, sharing, distributing, and analyzing data (e.g., on species populations, GIS, gene pools).
Builds better relations between parks and communities	ICT could improve liaisons between park authorities and communities living in and around parks. It could improve the speed, effectiveness, and quality of response. Practical gains for communities include assistance on health matters and communication between family members.
Improves external communications	ICT could connect conservation areas to the outside world, e.g., to headquarters of NGOs and overseas.
Supports ecotourism	ICT could encourage ecotourism in conservation areas, especially national parks, and can be used in the dissemination of location-specific information to tourists.
Supports community-based natural resource management	ICT could help small businesses in the marketing and selling of products. Improved management of conservation areas can lead to a better use of natural resources, which would have conservation, development, and standard of living gains.
Raises community awareness on local conservation issues	Community-based websites can increase knowledge and information on conservation issues (local and national), and help promote protection of the local environment.

Source: Banks, K. and Burge, R., *Mobile Phones: An Appropriate Tool For Conservation and Development?* Flora and Fauna International, Cambridge, 2004.

of learning onto pedagogical principles for the design of e-learning programs thus attempts to frame this within a familiar curriculum design model.

CONCLUSION

Educators working for the conservation of the maned wolf face many challenges. Habitat loss and degradation is perhaps the biggest challenge, and research suggests that a lack of knowledge about both the maned wolf and its habitats needs to be a key focus of conservation and education programs and campaigns.

A lack of trained education professionals is also hindering any progress in environmental education, but there have been some attempts to address this by training community leaders. Further work is needed and one approach may be to develop communities of practice where everyone plays a role in building their knowledge about the maned wolf as well as about the cerrado, its ecology, and its biodiversity.

Bringing together all the aspects of education and learning theories and concepts including transformative, experiential, and life-long learning, for example, with the multiple methods for delivery including curriculum-based education, outdoor classrooms and activities, community-based projects, and online, digital, and mobile platforms presents educators with very powerful opportunities to bring about the desired changes needed for the conservation of the maned wolf and its habitats. The challenge is taking a strategic and critically reflexive approach that will empower both educators and learners and that incorporates some of the best practices adopted in the commercial world and used by Jacobson (1995, Jacobson et al., 2006) to support conservation projects such as strong planning, implementation, and evaluation strategies throughout the process. Table 18.3 provides an example of a practical approach that I use with some success in my own commercial marketing business, which I have adapted using Jacobson's principles.

Environmental education may seem like a daunting task, but humans are natural and social learners, and all the tools needed to develop truly effective education programs and communication

TABLE 18.3
Environmental Education and Communications Planning

Organization/ project goals	These are the main organization or conservation goals that outline the overall purpose of the organization and what it aims to achieve. Many organizations outline their mission statement in this section. This section might also be used for the individual project goals, which might be more appropriate for a government department or academic institution, especially as these are likely to be different from those of a community-based organization or an NGO.
Objectives	These are the specific aims and objectives that will help to achieve the primary project goals. Make them SMART (Specific, Measurable, Achievable, Realistic, and Time-bound). Objectives may also be separated out over different education projects, as well as having overarching objectives for the organization.
Target audiences	Detailed research is vital to really understand your target audiences. Focus on demographics and the role of the participants involved in the conservation of the maned wolf. Different stakeholders require different interventions and education approaches.
Key messages	What key messages need to be conveyed that will help achieve the conservation goals? The maned wolf is not perceived negatively, but there is a lack of knowledge about it as a species and about its habitat.
	Key words: Develop a set of key words and phrases that can be used to help with communication campaigns. When used digitally and in online forums, including social media, these key words help to drive visitors to websites, online news and articles, blogs, YouTube videos, social media websites, and e-learning platforms. It is a valuable tool to help raise awareness.

TABLE 18.3 (Continued)
Environmental Education and Communications Planning

Education themes and topics	• Maned wolf as a large wild canid: Biology, ecology, and ethology • Maned wolf as an emblem: Give it a personality and unique voice • Habitat conservation: Key issues • Biodiversity: In relation to other species • Be as creative as possible and develop strong themes and campaigns around a variety of subject areas relevant to the maned wolf and its habitats.	
Implementation and delivery	**Communication** • Poster campaigns/leaflets/brochures • Public relations: feature articles • Social media campaigns • Marketing literature • Broadcast media: television and radio interviews and reports	**Education** • Community participation projects • School curricula programs • Educational materials: books, papers, films, games • Videos, photographs, graphics • Creative programs: theater, dance, performance • Link programs with zoos and other wildlife/ environmental organizations • E-learning programs
Resources required	What resources will your project require? This is an essential and fundamental part of the planning process and one that should tie in strongly with the aims and objectives. • Funding • Staffing • Training • Materials and equipment • Premises • Infrastructure • Partnerships and collaborations	
Monitoring and evaluation	There are many ways of monitoring and evaluating programs. Metrics and measurements need to be built into a project at different key points and not just after the project has been completed. • Regular review sessions/debriefing • Regular surveys pre and post projects and programs • Measure digital engagement: For education campaigns many e-learning platforms use quizzes and tests to measure the effectiveness of the learning. For digital communication campaigns use tools such as Google Alerts and social media monitoring tools, many of which are free.	
Month 1	Plan each element of your education and communications activity for each month. For example: Workshops to be delivered Talks and presentations School visits Poster campaigns Leaflets Online communication activity: blogs, etc. Media activity: press releases, radio interviews, television interviews	
Month 2	Continue the planning for each month and make sure your campaigns and education programs are cohesive. Some education programs may run over several months so make sure you plan the awareness-raising (i.e., marketing) activities around each part of the program. Create time in your scheduling and planning for evaluation. Remember to raise awareness of any successful outcome through media activity. (This is important for education and communication work about the maned wolf as lack of awareness is one of the key issues.)	
Month 3	One effective approach is to work on a rolling 3-month schedule where you will plan review and evaluation sessions. Following a critical analysis of the previous 3 months' activity, use the information to build on the successes and make any adjustments where a project did not work so well. This will help make the next 3 months' scheduling more effective. Are there areas where you need to carry out additional activities, for example?	

campaigns for the conservation of the maned wolf are already at our disposal. All we need now is enough people to care enough to champion the cause, and I hope this book will go some way to achieving this.

REFERENCES

Banks, K. and R. Burge (2004). *Mobile Phones: An Appropriate Tool for Conservation and Development?* Cambridge, Flora and Fauna International.

Bath, A. J. (2009). Working with people to achieve wolf conservation in Europe and North America. In: *A New Era for Wolves and People: Wolf Recover, Human Attitudes, and Policy*. Eds. M. Musiani, L. Boitani and P. Paquet. Calgary, AL, University of Calgary Press.

Bizerril, M. X. A. (2004). Children's perceptions of Brazilian cerrado landscapes and biodiversity. *The Journal of Environmental Education* **35**(4): 47–58.

Bizerril, M. X. A., C. C. Soares, and J. P. Santos (2011). Linking community communication to conservation of the maned wolf in central Brazil. *Environmental Education Research* **17**(6): 815–827.

Butler, P. J. (1995). Marketing the conservation message: Using parrots to promote protection and pride in the Caribbean. In: *Conserving Wildlife: International Education and Communication Approaches*. Ed. S. K. Jacobson. New York, Columbia University Press, pp. 87–102.

Consorte-McCrea, A. G. and R. Rubin (2009). Attitudes of zoo visitors and professionals towards the conservation of the maned wolf in southeast Brazil. *Canid News* **12.1** (online).

Field, J. and M. Leicester (2000). *Lifelong Learning: Education across the Lifespan*. London, RoutledgeFalmer.

Jacobson, S. K., Ed. (1995). *Conserving Wildlife: International Education and Communication Approaches*. New York, Columbia University Press.

Jacobson, S. K., M. D. McDuff, and M.C. Monroe (2006). *Conservation Education and Outreach Techniques*. Oxford, Oxford University Press.

Kolb, D. (1984). *Experiential Learning*. Englewood Cliffs, NJ, Prentice Hall.

Mayes, T. and S. de Freitas (2007). Learning and E-learning: The role of theory. In: *Rethinking Pedagogy for a Digital Age: Designing and Delivering E-Learning*. Eds. H. Beetham and R. Sharpe. London, Routledge, pp. 13–25.

Mezirow, J. (1991). *Transformative Dimensions of Adult Learning*. San Francisco, CA, Jossey-Bass.

Mezirow, J. and Associates, Eds. (2000). *Learning as Transformation: Critical Perspectives on a Theory in Progress*. San Francisco, CA, Jossey-Bass.

Miniwatts Marketing Group (2012). Internet world stats: Usage and population statistics.

Oblinger, D. and J. Oblinger (eds) (2004). *Educating the Net Generation*. Educause. e-book. http://www.educause.edu/educatingthenetgen.

Orr, D. W. (1994). *Earth in Mind: On Education, Environment, and the Human Prospect*. Washington, DC, Island Press.

de Paula, R. C., P. Medici, and R. G. Morato, Eds. (2008). *Plan de acción para la conservación del aguará guazu: Análisis de viabilidad poblacional y de hábitat.* [*Maned wolf action plan: Population and habitat viability assessment.*] Brasília, Edições IBAMA.

Siemens, G. (2008). Learning and knowing in networks: Changing roles for educators and designers. Presented to ITFORUM for Discussion.

Smith, M. K. (2003, 2009). Communities of practice, *The Encyclopedia of Informal Education*. 2012.

Soler, L., J. M. Carenton, D. Birochio, V. Salvatori, M. M. Orozco, M. Soledad Rosso, M. Carpinetto, et al. (2002). Problems and recommendations for the conservation of the maned wolf in Argentina. *Endangered Species UPDATE* **22**(1): 3–9.

Songsasen, N. and M. D. Rodden (2010). The role of the species survival plan in maned wolf *Chrysocyon brachyurus* conservation. *International Zoo Yearbook* **44**: 136–148.

Taylor, D. (2004). Conservation education and its relevance to wild canids. In: *Canids: Foxes, Wolves, Jackals and Dogs*. Eds. C. Sillero-Zubiri, M. Hoffman and D. W. Macdonald. Gland, Switzerland, International Union for Conservation of Nature, pp. 298–305.

Taylor, D. (2009). Education as a conservation strategy: Exploring perspective transformation. In: *A New Era for Wolves and People: Wolf Recovery, Human Attitudes, and Policy*. Eds. M. Musiani, L. Boitani and P. C. Paquet. Calgary, AL, University of Calgary Press, pp. 161–171.

UNESCO (1977). The Belgrade Charter. The International Workshop on Environmental Education. Belgrade, 13–22 October 1975. Available at http://unesdoc.unesco.org/images/0002/000276/027608eb.pdf.

UNESCO (1978). The Tbilisi Declaration: Intergovernmental Conference on Environmental Education: October 14–26, 1977. *Connect* III(1).

Wenger, E. (1998) Communities of practice: Learning as a social system. *Systems Thinker*, http://www.co-i-l.com/knowledge-garden/cop/lss.shtml. Accessed November 2012.

Wenger, E. (c2007) Communities of practice: A brief introduction. *Communities of Practice*. http://www.wenger.com/theory/. Accessed November 2012.

19 Introduction to the Human Dimension

A Valuable Research Tool to Achieve Wildlife Conservation Objectives and Maned Wolf Conservation

Alistair Bath

CONTENTS

Dimensions of Wildlife Resource Management ... 265
Is the Maned Wolf Considered a Wildlife Resource? ... 266
The Wildlife Management: Human Dimension Relationship ... 267
Values, Attitudes, and Perceptions: The Building Blocks of HD Understanding ... 268
Understanding Attitudes and Beliefs ... 269
Understanding the Nature of Conflict ... 271
Addressing Conflict in Wildlife Resource Management Issues ... 272
Addressing Cognitive Conflicts ... 272
Addressing Value Conflicts ... 273
Addressing Costs and Benefits Conflicts ... 273
Addressing Behavioral Conflicts ... 274
Designing and Evaluating Educational Programs and Public Awareness Campaigns ... 274
Implementing the Information and Educational Campaign ... 277
Human Dimensions Summary ... 280
References ... 280

DIMENSIONS OF WILDLIFE RESOURCE MANAGEMENT

There are many dimensions or perspectives in any natural resource management issue, and wildlife management is no different. In fact, most dimensions in any resource management issue tend to be human in nature rather than biological, although traditionally we have not learned as much about the human dimensions issues as the biophysical issues when trying to implement conservation efforts. Thus, when a wildlife manager or park superintendent asks me to do a human dimensions study, in many ways it is like asking a biophysical scientist to do some biology; both are extremely broad. Therefore, the nature of human dimensions needs to be examined, especially within the context of wildlife management. Mitchell (1989) suggests that there are seven perspectives of resource management and that these vary over space and time. Most of these perspectives are human oriented and broadly could be considered human dimensions issues. These include economic (i.e., costs and benefits and values of landscapes and wildlife, compensation issues dealing with amounts and time), social (i.e., attitudes, knowledge levels, expectations, and behavior toward fish and wildlife resources), political (i.e., the nature of political will or lack thereof), legal (i.e., policies, guidelines, national and international laws and conventions), institutional arrangements (i.e., mandates of different agencies and their interactions with each other, governance issues), and technological perspectives (i.e., complex vs. simple technologies to solve challenges in different societies). There is also the biophysical perspective (i.e., ecological understanding of habitats, wild animal populations, and prey–predator relationships). Ecological research on the maned wolf is important, but to successfully address the conservation challenges facing the maned wolf, a comprehensive approach focused on all of these human dimensions issues is needed.

Is the Maned Wolf Considered a Wildlife Resource?

From a biologist's perspective, the question above is moot; clearly the maned wolf is a valuable wildlife resource with ecosystem functions and existence values. This is, however, the perspective of only one interest group among many who are affected or are affecting the maned wolf. We tend to talk about wildlife management as if we are managing the resource, but in fact we manage the people, and we have to realize that what is a resource to some people may not be considered a resource to others. A resource is a subjective, dynamic concept that depends upon the understanding and interaction of three components: human, culture, and nature. According to Zimmerman (1933), "resources are not, they become." A resource emerges because of human wants and needs, and the available technology to make it happen. For some, the wolf may be considered "neutral stuff" in Zimmerman's words, with no value or need for them; hence the wolf does not become a wildlife resource. For others the animal may be revered as a symbol of nature and may have value and indeed be a resource. The task of a human dimensions researcher is to assess how diverse attitudes and values toward the wolf may be; does the culture want and need the maned wolf?

Humans may have the need and the ability to produce something to become a resource, however, the culture may not sanction turning the "neutral stuff" into a resource. For example, people in North America have a need and want for food and the technology to kill food (let's say a horse); the North American culture, however, does not sanction the eating of horse, thus the "neutral stuff" (horse) does not become a food resource. In this case, the cultural component of the equation prevents the neutral stuff from becoming a resource. In certain societies, wildlife may not be considered a wilderness resource; large carnivores, for example, may not be considered a tourism resource to a local village, but may be highly valued by an urban public. Large herbivores may be considered as causing damage to trees and not a wildlife resource, while in other countries they may be important in wildlife viewing as a nonconsumptive wildlife use.

The third aspect of a resource is the nature component. When developing a resource, there will be natural resistances and cultural resistances to it becoming a resource. In Canada, aboriginal people such as the Innu, Inuit, and Metis are becoming increasingly involved in natural resource management decision making; concepts of what is considered appropriate resource use, and even the nature of land ownership, can be quite different from their perspectives compared to those of the more urban white population. For issues such as the establishment of protected areas or management of wildlife, resource managers must be sensitive to these cultural values and be willing to truly listen to these interest groups when working together toward resource decisions.

The interaction of the three components of human, culture, and nature determines whether "neutral stuff" becomes a resource. A resource is very subjective and dynamic in terms of changing over time; what is a resource today may not be a resource tomorrow, likewise what is not a resource today may be a resource tomorrow. As we think about integrating the human dimension in wildlife conservation efforts, it is important to remember that the resources (e.g., wilderness, wolves, gazelles) perceived as valuable by some groups are not necessarily seen as resources at all by others.

THE WILDLIFE MANAGEMENT: HUMAN DIMENSION RELATIONSHIP

Wildlife management has been defined:

> as the science and art of changing the characteristics and interactions of habitat, wild animal populations, and people in order to achieve specific human goals by means of managing wildlife resources. In one form or another, everything done in wildlife management is done for the people. (Anderson et al. 1987)

Biophysical scientists have focused on understanding the species, and there has been considerable discussion about endangered species. Emphasis in wildlife management then shifted away from studies of the species to the need to understand the spaces where these species survived. Biophysical scientists began discussing endangered spaces (Hummel 1989). These scientists struggle, however, to adequately address the people component of the equation and with some species, particularly large carnivores, this human dimension is probably more important than the biophysical component. Social scientists (e.g., human geographers and rural sociologists) began to apply techniques to understand the people part of the wildlife management equation in the late 1960s in North America. This application of studying people and their interactions with wildlife was later termed human dimensions in wildlife management.

Today, successful wildlife management involves not only an understanding of the biology of the species and its habitat, but also an understanding of public attitudes toward and knowledge about the species, and attitudes toward possible management approaches to the species. The human dimension of wildlife resource management is particularly important to understand when designing and implementing management plans for large carnivores, which often arouse conflicting emotions among the general public. Indeed, large carnivore management is often more a sociopolitical issue than a biological one (Bath 1998). These human aspects of the wildlife resource management equation need to be understood through a scientific and objective process for successful maned wolf conservation to occur.

While wildlife management by definition has for many years realized that there is indeed a human dimension to successfully implementing species action and management plans, integrating human dimensions into daily decision making remains a challenge for many wildlife agencies. In North America, Aldo Leopold, considered the founder of wildlife management in North America, stated in 1943 that deer management was more about managing the people than managing the deer. Since those early statements, the human dimension in wildlife management has become increasingly integrated into wildlife management planning and decision making in North America and various parts of Europe, with some very positive results. Wolf restoration efforts in Yellowstone National Park included a human dimensions component, which was important in understanding the amount of support that existed for wolf restoration and the reasons why people were in favor of or against wolf restoration (Bath 1991; Bath 1989; Bath and Buchanan 1989). These latter data were useful in designing effective educational efforts and working toward conflict resolution. Wolves were eventually restored to Yellowstone National Park in 1995, with many interest groups supporting the restoration effort; only 60 years earlier poison campaigns occurred within the same national park. Human dimensions results allowed managers to successfully address the issues and concerns of many of these groups, and by working with groups, rather than against them, wolf conservation occurred. Given the limited human dimensions work done on the maned wolf, there are many aspects of human dimensions that could be employed.

Involving the public in the early development of a management plan was key to a successful wolf management plan in Croatia and more recently a wolf management plan in Bulgaria. The interest groups also worked together toward consensus on other large carnivore management plans. In this case an applied human dimensions facilitated workshop approach was taken where various interest groups chose to work together to understand each other's core values and concerns (Figure 19.1). Groups reached an acceptable compromise and produced management plans that all groups could

FIGURE 19.1 Dr. Bath uses his applied human dimension facilitated workshop approach demonstrating the visual nature of the facilitation style with cards of different colors and shapes.

support because each plan addressed the fundamental concerns of each group. The Bulgarian wolf management plan was completed in 2012 after such a human dimensions facilitated workshop approach was implemented. Such "win-win" scenarios are only possible with the implementation of a human dimensions research and applied approach. The maned wolf needs an applied human dimensions facilitated workshop approach to be implemented so that consensus can be reached by all the diverse interest groups on understanding and addressing the key issues facing the species.

VALUES, ATTITUDES, AND PERCEPTIONS: THE BUILDING BLOCKS OF HD UNDERSTANDING

Understanding attitudes is an aspect of human dimensions but understanding and being able to affect the relationship between attitudes, knowledge, and behavior is where the human dimensions field has its greatest strengths. Attitudes, perceptions, and values are all terms we use regularly but for an effective human dimensions discussion these terms must be defined. To illustrate the differences between values, attitudes, and perceptions, I will use the analogy of the use of a traditional camera, not a digital camera. One can think of values as the "camera." You are probably not going to change your camera without a fair bit of thought. Your camera may last a very long time and is of high value to you. Values are difficult to change. People tend to have a hierarchy of values where some are more important than others and will override other values. For example, protecting a rare deer may be an important value, but providing food for the family may be a higher value, thus resulting in killing the deer for food. Farmers may support maned wolf conservation, but they must also feed their families and so may fear livestock losses, whether a real or perceived threat; thus, values may be ordered in a hierarchy. For example, there is a widespread belief among the agricultural community that maned wolves attack chickens, but are not a real threat to other livestock. Maned wolves probably have taken chickens, although on rare occasions. Biological research focused on composition of scat analysis suggests that chickens are rarely eaten (see Motta-Junior,

Queirolo, and de Arruda Bueno, Chapter 6, this volume), but this does not stop rural people from believing that maned wolves are a real threat to their free-range chickens.

Attitude, in this broad context, is the "film" in the camera. Attitudes are relatively enduring beliefs about something and, like your film, can be changed, but only after some time and with some effort. Attitudes are difficult to change but can be influenced through a factual route and/or an emotional route. Changing attitudes through emotion (e.g., images of wolf pups) can change attitudes more quickly but the attitude change will not last as long as when changed through factual information, although this is harder to do. Balancing messages to create attitude change is a challenge for the environmental educator and wildlife manager. Different speeds of film used in your camera are similar to different attitudes that may exist within a hierarchy of values. And just as films change, attitudes also can change. Human dimensions researchers are interested in not only the direction of attitudes (e.g., positive or negative toward a species or an issue) but also the strength of those attitudes (e.g., strongly agreeing with a statement, moderately agreeing, slightly agreeing, or neutral toward the species or issue). Extreme attitudes toward the maned wolf may be difficult to change, but neutral attitudes lend themselves to being influenced by value persuasive messages.

Perceptions are the "individual pictures"; perceptions are temporary responses to stimuli and can be easily altered, similar to the speed at which different pictures can be taken. As mentioned earlier, it is very easy to change perceptions and even influence attitudes, particularly through emotional images; however, the attitude change does not last for very long if affected solely by emotional images. On the other hand, if attitudes can be changed through factual information, although more difficult, the attitude change will last much longer. Values, attitudes, and perceptions strive to be understood through a human dimensions approach, but of particular interest to the human dimensions researcher and practitioner are the links between values, attitudes, perceptions, and behavior. To effectively affect behavior, it is necessary to better understand the nature of attitudes; changing components of attitudes will result in attitude change and possibly behavioral change.

Understanding Attitudes and Beliefs

Fishbein and Ajzen (1975) defined attitude as "a learned predisposition to respond in a consistently favorable or unfavorable manner with respect to a given object." Attitudes have been conceptualized further into four main components that are used by human dimensions researchers when designing an initial baseline assessment study:

- *Affective*: feelings of liking or disliking the wildlife species
- *Cognitive*: beliefs about the wildlife species that may or may not be true
- *Behavioral intention*: a statement of how one would behave in a certain situation (e.g., willingness to view a wolf, shoot a Mongolian gazelle, hike in bear country, hear the bugling of a red deer, etc.)
- *Behavior*: overt or actual behavior (e.g., purchased a trip to view wolves, bought a hunting license to shoot a Mongolian antelope) and verbal behavior (i.e., what someone states he or she did). Verbal behavior may be different from overt behavior, thus requiring human dimensions practitioners to ask several questions to attest the validity and reliability of the responses to the items.

Moore (1983) further conceptualizes these aspects of attitude within a model of cognitive and affective components of behavior. Moore (1983) suggests that the components of information and experience, beliefs, attitudes, values, behavioral intention, and overt behavior exist within a hierarchal pyramid where each has strong feedback loops to each level. Beliefs are formed based upon information and experience that in turn influence attitudes. Attitudes influence values and value systems that in turn predict behavioral intention and ultimately behavior. Feedback loops between each level imply that each component is constantly being affected and can be affected by lower levels. Within human dimensions, the cognitive hierarchy is similar in several ways to that developed by Vaske and Donnelly (1999), who refined these earlier ideas and applied the hierarchy in various situations. Without understanding the relationship between these components of behavior, it is difficult to design effective ways to change attitudes and behavior and increase awareness in all sectors of society.

Public attitudes toward a wildlife species and its management are affected by a collection of beliefs that make up a belief system. Some of these beliefs are more important than others in affecting attitudes. For example, our overall attitude toward wolves is a product of our beliefs and the evaluation of that belief (whether it is good or bad), and the likelihood that the belief could occur. In forming our attitudes toward wolves, people strive to keep their affective components of attitude (liking or disliking of wolves) consistent with their cognitive component (beliefs and evaluations of those beliefs). For example, I could dislike wolves (affective); I believe they have killed people (cognitive belief), and having people killed is of course bad (evaluation of belief). The behavioral intention that could result from this is to support a wolf control program and actual behavior (fourth component of attitude) may be a history of shooting wolves. In this example, all aspects of attitude are consistent with each other, producing a negative overall attitude toward wolves.

Persuasive communication efforts strive to understand the underlying beliefs and directly affect or change the belief most strongly linked to attitude, thus causing what has been called cognitive dissonance (Fishbein and Ajzen 1975). In the above example, if information delivered by a credible messenger convinces the individual that wolves do not kill people, they may no longer have a consistent reason to dislike wolves, resulting then in a change of other components of attitude. Environmental education efforts can be most effective in these cases where the cognitive component or belief about the wildlife species is incorrect and thus can be specifically targeted by the educational effort.

Understanding belief systems and the relationship between beliefs, attitudes, and behavior can be one of the most important uses of human dimensions conservation projects. In France, a beliefs and attitudes human dimensions study was used to provide the terms of reference for a communication officer (Bath 2000). In addition to an understanding of key beliefs affecting attitudes, the human dimensions conservation project in France identified which value persuasive messages (e.g., wolves for future generations) were important in affecting overall attitudes toward wolf management options. And while something is known of public beliefs about wolves (Kellert 1985; Tucker and Pletscher 1989; Bath 1989; Bath 1991; Glikman et al. 2012), much less is known of public beliefs about other large carnivores (Jope and Shelby 1984; Braithwaite and McCool 1989), and even less is known about public beliefs about the maned wolf. In those cases where attitudes are directly being affected by beliefs, targeting such beliefs can be effective; however, attitudes toward the issue could be influenced by other factors. These factors are discussed in the next section: understanding the nature of conflict.

Unfortunately, many organizations and governments have not studied specifically the key beliefs affecting attitudes before designing educational efforts (e.g., posters, brochures, videos, etc.) and consequently the effectiveness of such efforts is unknown. More recently, in many parts of Europe human dimensions studies have been implemented that now help guide the development of new material in those cases where beliefs underlie the expressed attitudes. In many cases, knowledge of the underlying factors affecting attitudes has yet to be gained. An assessment of large carnivore educational materials (Bath and Tavares 2000) found that the available material varies considerably across species and across countries in terms of quality and quantity. Without knowing what information currently exists within the various publics, what beliefs are most directly linked to attitude, the nature of the conflict, and how different beliefs affect different target audiences, such educational materials may not be effectively working. A stronger statement may be to say that such efforts are similar to "shooting in the dark." A baseline assessment of beliefs and their relationship to attitudes could prove useful in better understanding the key issues facing the maned wolf.

UNDERSTANDING THE NATURE OF CONFLICT

Human dimensions conservation projects can work toward achieving public acceptance of wildlife species by providing a better understanding of the nature of conflict between all interest groups. Identifying where conflict exists is the first step toward conflict resolution.

There are basically four types of conflict:

- *Cognitive*: Cognitive conflict results when there is a difference in beliefs between various interest groups. These beliefs may or may not be true.
- *Value*: A value conflict occurs when there is a difference in the importance of an issue between various groups. This could result because of a hierarchy of values within a value system.
- *Costs/benefits*: A costs/benefits conflict occurs when a difference is perceived between groups of who bears the costs of implementing an issue versus who reaps the benefits of such an issue. A group may feel it is unfair that it must suffer all the costs while another group gets the benefits.
- *Behavioral*: A behavioral conflict can be a personal conflict between individuals of different agencies over issues not directly related to the issue at hand. A behavioral conflict could also exist due to a mistrust of a particular agency by another interest group based upon past history with the agency or organization.

To illustrate these types of conflict, let us examine the following example. In Brazil, the maned wolf is currently completely protected and hunting of any wildlife is against the law, but there may be interest groups that would like to see the status of the maned wolf changed. The issue is wolf management and whether wolves should be completely protected, hunted under strict seasons, or liberally killed whenever possible. Interest Group A believes there are 500 wolves in the country while Interest Group B feels there are only 100 animals. Interest Group A comes to the conclusion that wolves should be allowed to be shot whenever possible; Interest Group B believes since there are few wolves from their perspective, wolves should be completely protected. A conflict occurs because of different beliefs or factual information; this is a cognitive conflict. Environmental educational messages communicating the actual numbers of wolves to both groups delivered by a messenger who has credibility could eliminate this conflict.

Interest Group A and Interest Group B are once again disagreeing about wolf management. The two groups agree on the number of wolves that exist, each believing there are 100 wolves. Thus, there is no cognitive conflict. Group A comes to the conclusion that 100 are too many wolves and that they should be shot whenever possible; Group B feels that, since there are only 100, they should be protected. A difference in values occurs—a value conflict. Interest Group A may fear that resource extraction in the area may be limited if wolves are protected, hence the need to allow liberal hunting of them. Interest Group B may feel that while resource extraction is important, the higher value is wolf protection.

Interest Group A and Interest Group B may agree that there are 100 wolves (no cognitive conflict) and may also agree that 100 wolves are not enough and that it is important to protect them (no value conflict); however, they still might disagree about the management because of a costs/benefits conflict. Interest Group A has an ecotourism operation and would like wolves to be completely protected in their area (gain the benefits); Interest Group B would rather see them protected in their area so that they can have the ecotourism benefits. The two groups choose to disagree over whether wolves should be protected or hunted liberally not because they do not share the same information (cognitive conflict), or because they do not feel wolves are important and should be protected (value conflict), but because they disagree over who will get the benefits of wolf protection and who will not.

A similar example may involve the small farmer who believes that the number of wolves in the country is 100, and that it is important to protect them; however, the small farmer wants them to be protected in an area far away from his/her poultry. In this case, the interest group (small farmers) does not want to bear the costs of the conservation action and thus chooses to oppose the protection management option for wolves. This opposition is not the result of a lack of knowledge or a lack of a conservation value for the species, and thus environmental education messages about numbers of wolves and traditional environmental education will not work under these circumstances.

Finally, we might have Interest Group A and Interest Group B still disagreeing over the type of wolf management not because they do not share the same knowledge of wolf numbers, or the same value that 100 wolves is not enough, nor the belief that they should be protected in one area and not another. They disagree because Interest Group A does not trust Interest Group B. Several years ago, Interest Group B went ahead and protected a species without the involvement of Interest Group A and today the groups do not like each other and disagree with each other just on principle. This is a behavioral conflict. Such a conflict could also occur because a certain individual working for Interest Group A did something personally against a member of Interest Group B many years ago and Interest Group B will not forget it and thus will always disagree with anything proposed by the other individual. In these cases where a group has lost credibility and trust with another group, the first few applied human dimensions facilitated workshop sessions may need to discuss this issue of trust before being able to talk about wolf management or whatever issue is at hand.

The most common conflict in natural resource management issues is in fact behavioral conflicts, and yet traditionally we treat most conflicts as cognitive issues and design expensive educational efforts, only later wondering why they have not been effective. In addition, wildlife resource management issues would be relatively simple if in each conflict there was only one of these types of conflict happening at one time. Often the reality is that several types of conflict are happening at once. Human dimensions research can help identify the nature of the conflict and begin "peeling back the layers of the conflict," thus enabling a better understanding of the people component and the necessary direction toward a possible solution. It is important to recognize that traditionally we have assumed that all conflicts can be solved through more public awareness and educational materials. Depending upon the nature of the conflict, educational materials could be absolutely useless; in fact, unless a cognitive conflict exists, educational materials are not the way to resolve the conflict. Usually, resolving conflicts requires listening more than talking. We are born with two ears and one mouth and perhaps we should be listening at least twice as much as talking when engaging various interest groups. In addition, the first contact with groups should be a listening exercise of key issues and concerns, not a presentation of what "you" believe so as to educate them. The second session should be a presentation of what "you" heard and perhaps a beginning of connecting the interest group or community issues with those of "your" own. Only through communicating with interest groups (listening and then talking) can trust be built between groups and effective conservation occur. Human dimensions research can be used to build that trust and to understand and address the nature of the conflict between all groups.

ADDRESSING CONFLICT IN WILDLIFE RESOURCE MANAGEMENT ISSUES

Human dimensions as a research tool helps to provide insights into the nature of the conflict between various interest groups, a necessary first step toward conflict resolution. Human dimensions as an applied facilitated workshop approach helps to resolve those conflicts. Steps toward resolving each of the above-mentioned conflicts are briefly outlined in this section.

Addressing Cognitive Conflicts

1. Cognitive conflict is first identified using a quantitative human dimensions research tool that clearly illustrates that the nature of the conflict is poor knowledge levels among the various interest groups.

2. Key beliefs that are most directly linked to attitude are identified using various statistical analysis techniques including examining correlation coefficients between belief items and attitudinal items. Knowledge is determined to be a strong predictor of attitude. (Rarely is knowledge a good predictor of attitude.)
3. Beliefs may vary among the same interest group and may vary across space. For example, one educational program for hunters assuming that all hunters, wherever they live, are basically the same could be ineffective. Statistical analyses can be done to explore these differences.
4. Human dimensions as a research tool is used to identify the best modes of communication (e.g., verbal vs. written, magazines, brochures, etc.) for each interest group.
5. Focus groups can be used to determine whether designed messages are too low or too high for the target group.
6. Messages should be delivered based on assessment of the best mode of communication and also the best messenger. Human dimensions should be used to assess credibility and legitimacy of various messengers.

Basically, solving cognitive conflicts is about identifying the following: (1) *what message* should be delivered; (2) *to whom* should the message be sent, focusing on specific interest groups and specific scales; (3) *who* should deliver the message, understanding which individual or what agency has the highest credibility with that target audience; and lastly (4) *how* should the message be delivered or understanding the target medium (e.g., TV, newspaper, brochure, video, etc.).

Addressing Value Conflicts

1. Identify value conflicts using a quantitative human dimensions research tool that clearly illustrates the nature of the conflict is value based.
2. Focus on learning about key issues from each interest group individually and only later bring the groups together. Focus on principles, not positions, to understand where agreements may be found.
3. Work with all groups individually, develop a Common Ground Matrix (Bath 2000), and seek a willingness from groups to meet to share common interests and work toward solutions.
4. Use "a five times why" exercise to explore the root cause of the conflict. Basically, after each statement is given by the interest group, you ask the group to think about "why" that is an issue. By repeating this process five times, the root cause of the conflict can be determined.
5. Use human dimensions research to test for compromises that may be possible (e.g., willingness to change position under certain conditions).
6. Identify and understand the hierarchy of values (which values are more important than others) through facilitated focus group discussion with the individual interest group.

Addressing Costs and Benefits Conflicts

1. First identify costs and benefits conflict using a quantitative human dimensions research tool that clearly illustrates that the conflict is about economic issues (e.g., rate of compensation for livestock losses is too low, time taken to receive payment is too long, etc.).
2. Explore the strength of attitude toward the issue. Is the issue about differing amounts? Is the issue about who pays?
3. Negotiate first on principles (e.g., should compensation be paid), then explore the issue of positions (e.g., how much?).

Addressing Behavioral Conflicts

1. First identify behavioral conflict using a quantitative human dimensions research tool that clearly illustrates that the nature of the conflict is credibility and legitimacy of opposing interest groups, government departments, etc. Ask specific credibility questions in the questionnaire.
2. Then use focus group discussion to explore why such conflicts exist.
3. Work to gain respect and trust from the various interest groups. This will involve a lot of listening before talking. Exploring questions such as "what does your organization need to see to gain the respect or trust of the government agency again" will help.
4. Send the best messenger to listen to the interest groups. Use an independent facilitator to explore the issue.

It is easier to start slowly and head in the right direction than run into conflict and have to backtrack to fix the conflict. Human dimensions needs to be seen as a proactive approach rather than a crisis management approach to working with people. It should be considered not as a one-shot study but as a longitudinal process where working with people is integrated into the daily decision-making process. Human dimensions research must collect data that are representative of the entire resource constituency, rather than just the vocal lobby groups.

DESIGNING AND EVALUATING EDUCATIONAL PROGRAMS AND PUBLIC AWARENESS CAMPAIGNS

Human dimensions research results can be used to create better educational programs and to evaluate existing ones. Those individuals or groups responsible for communicating messages to the various publics need to know whether the designed messages are targeted too high (people cannot understand the concept), targeted too low (basic information that everyone seems to already know), or targeted at beliefs that are not that relevant in affecting attitudes (could be interesting information but it is having no effect). The best educational and awareness programs have messages (cognitive and value persuasive messages) that are targeted by interest group and by geographic area, where the messages delivered and received are the ones that are most strongly linked to attitude change and behavior.

Cognitive messages are based on factual information such as the number of wolves in the country, their population status, their breeding behavior, their weight, and their feeding habits. Depending upon the level of credibility of the messenger of these factual messages and the trust that exists between the interest groups and the messenger's agency, this factual information may or may not be believed. It is important to involve the various interest groups where possible in discussions and interpretation of these "facts." The difficulty is to overcome the reaction by certain interest groups that the cognitive information is "the agencies' facts," manipulated to suit their cause. In addition, values on those facts should be avoided if the purpose is to communicate cognitive messages. For example, there are 100 wolves in the country and there should be more. The factual information is now clouded with the value judgment that more wolves are needed.

Value persuasive messages are used purposely to address conflicts in values between interest groups. Messages that large herbivores and large carnivores should exist for future generations and that a species should have existence value (e.g., whether or not one sees a musk deer, it is important that it exists) are examples of value persuasive messages. Past human dimensions research has shown that such value persuasive messages can be as effective, if not more effective, as cognitive messages in changing attitudes.

Definitely linked with the discussion on cognitive and value persuasive messages is delivery of such messages. Attitudes change through cognitive messages and through emotion. While it is more difficult to change attitudes by factual information (cognitive route) than through emotion, the attitude change does last longer. Changing attitudes through emotion but without the supporting factual information will result in a short-term attitude change. Perhaps one of the best examples of this is in Newfoundland, Canada, where the hunting of seal pups was very effectively portrayed, particularly by animal rights groups, using images of white teary-eyed seal pups, blood-stained ice, and women in fur coats. All of these images were powerful, emotional, and effective in swaying public attitudes toward the issue; this public awareness campaign did result in a reduction of seal quotas and the ban of hunting seal pups. Factual information now exists about the large numbers of seals off the coast of Newfoundland (estimates are several million animals): the species is not endangered, and seal hunting has been a part of the Newfoundland culture and economy for poor fishermen living in the poorest province in Canada for years. While the killing of white coats is still banned, public support for hunting seals has certainly improved and the images of white coats being killed (images that are still being used today by certain interest groups even though the activity no longer occurs) no longer gain the public attention they once did. A more informed public resulted in this change in attitude toward the issue, illustrating that the impact of the emotional images produced only a short-term attitude change.

Emotional images such as wolf pups trying to howl, red deer fawns, piles of tiger bones, and dead antelope carcasses can be effective in changing attitudes through the emotional route, however, the attitude change may not last. *Death of a Legend*, by Bill Mason, produced in the early 1970s by the National Film Board of Canada, used images of wolf pups trying to howl; the film was effective in getting the wolf bounty in Ontario stopped at that time. There is still no wolf bounty system in Ontario, Canada, and today several thousand wolves exist in the province; public support for the species seems positive, if the large attendance at public wolf howls held in Algonquin Provincial Park can be used as an indicator. Factual information accompanied the emotional images and probably contributed to the longer-lasting positive attitude toward the species.

Using emotional images without the balance of factual information may negatively impact attitudes toward the issue and could particularly damage the credibility and legitimacy of the agency. For example, the portrayed slaughter of several thousand Mongolian antelope over a short period of time by night hunters could be a very powerful emotional image to communicate the demise of the population. However, when the public learns that there are hundreds of thousands of animals and that hunting does not occur constantly throughout the year, the public support that might have been gained by the initial emotional images could be lost and the public may turn more strongly against the agency portraying the images. Presenting emotional messages must be only a small part of an agency's communication effort.

Emotional images within a context can speak considerably about the issue. For example, a close-up image of a dead brown bear in Croatia with its head crushed has strong emotional appeal, but it does not communicate as strong a message as showing the dead brown bear on the railway tracks where it was killed. The latter image communicates clearly the challenges large carnivores face with ever-increasing transportation networks. If an emotional image is to be used, placing that image (positive or negative) within a larger message context is much more effective. Human dimensions research is needed in more depth regarding the maned wolf so that communication efforts can be implemented successfully.

Developing a means to assess the effect of certain messages or beliefs upon attitudes requires a quantitative and qualitative understanding of the link between specific belief items (cognitive messages) and aspects of attitude (like/dislike toward a species, support/opposition for a certain policy, and behavioral intention-poaching or protection). An inventory of messages within existing materials provides the opportunity to learn about what audiences are being communicated to and what messages are being focused on. This could be a first step toward designing effective communication material for the maned wolf. For example, in developing the Large Carnivore Initiative for Europe (LCIE) educational resource inventory (Bath and Tavares 2000), types of material (posters, pamphlets, videos, books), target audiences (shepherds, children, hunters, government officials, foresters, etc.), and geographic regions were examined. One of the results of this qualitative assessment of large carnivore material was that while a considerable amount of material exists for young children, many of the other target audiences lacked any material.

Based upon data and analysis from France (Bath 2000) and Croatia (Bath and Majic 2000; Majic and Bath 2010) regarding human dimensions in wolves and wolf management, population numbers, population status, attacks by wolves on humans, and deaths caused to humans by wolves are important pieces of information strongly linked to attitude. Unfortunately, many educational resource inventory materials do not specifically highlight these messages. Thus, by examining the nature of existing messages it is possible to make preliminary comments about the possible effectiveness of programs, realizing however that differences in factors affecting attitudes can be very local and audience specific. This highlights the importance of having site-specific human dimensions information.

Considering the huge amounts of money and time invested in educational material and public awareness campaigns, managers need to know whether such programs have had any effect. Has knowledge about the species by interest group increased (e.g., biology, feeding habits, numbers, status)? Has knowledge about the issue been altered (e.g., familiarity with the new policy, proposed project that could negatively impact habitat, proposed ecotourism project to increase jobs)? Have attitudes become more positive (e.g., stronger support or liking of the species)? Have changes occurred in behavioral intention or in actual behavior (e.g., participation in an activity like wildlife viewing, donation behavior)? Do certain materials have a stronger impact than other materials for particular interest groups (e.g., posters for teenagers, pamphlets for farmers, videos for business leaders)? And most importantly, can any changes be attributed to the implementation of an educational and public awareness program, thus demonstrating the importance of such materials. Such an evaluation of the effect of implementing a new or existing educational or awareness program requires an experimental research design.

The most effective experimental research design that can by its nature accurately represent the effect of a program is the Solomon 4-group experimental research design (Campbell and Stanley 1963). Designed initially for examining situations within a laboratory environment, the design can be adapted to a human dimensions application. This research design consists of randomly (R) allocating a population into four groups (I, II, III, IV) over three time periods (T_1, T_2, T_3).

	T_1	T_2	T_3	
R	O_1	X	O_2	I
R	O_3		O_4	II
R		X	O_5	III
R			O_6	IV

For example, a group of school children from schools that are considered similar in a particular area could be randomly assigned to the various groups. Ideally, individual children would be allocated to each group, but logistically schools could be allocated to four different groups based upon the assumption that the children within each school should not be significantly different in their attitudes, knowledge, and sociodemographic characteristics. Feeling comfortable with this assumption could be challenging. Bath and Farmer (2000) illustrated that differences in attitude and knowledge about large carnivores (bears, wolves, and otters) existed between teenagers in Northern Ireland, England, and Wales; differences were also found between the attitudes of children in London and of those in the rest of England.

Over periods of time, assessments of beliefs and attitudes are made and changes can be attributed to a specific treatment (X), in our case an educational program. There are three time periods (T_1, T_2, T_3). In time period one (T_1), a pretest is done to assess existing beliefs and attitudes toward the species and issue; this pretest is given only to Groups I and II in the same time period. The treatment (X), the educational program (pamphlet, presentation, etc.), is then administered at T_2, but only to

Groups I and III. An identical set of items measuring beliefs and attitudes are again administered at T_3 to all groups (I, II, III, and IV).

Why is it necessary to have so many groups and such a complex experimental research design? Would not just Group I and Group II suffice, with Group II acting as a control group, as it does not receive the educational program? If a test is administered once and then is repeated, whether or not any treatment has been given, knowledge scores will increase the second time solely because of the familiarity with the items. There will always be an increase in knowledge scores between O_1 and O_2 and between O_3 and O_4. To remove the pretest effect, Group III receives the treatment and the posttest, but not the pretest. The final group (IV) controls for history effects; these effects are the result of a film, book, news article, policy action, or something else that occurs between T_1 and T_3.

To attribute the effect of the educational program on the attitude or knowledge change, comparisons can be made between:

O_2 and O_1
O_4 and O_3, O_4 and O_2, O_4 and O_1
O_5 and O_4, O_5 and O_3, O_5 and O_2
O_6 and O_5, O_6 and O_4, O_6 and O_2

By examining these differences in attitude and knowledge scores using multivariate statistical procedures, one can evaluate the true effect of the education program. Managers and agencies need to evaluate their public awareness and education campaigns in this rigorous manner; such evaluations will help improve existing programs and justify further expenditures on existing successful programs.

IMPLEMENTING THE INFORMATION AND EDUCATIONAL CAMPAIGN

To successfully implement your educational and information campaign, six basic steps must be considered. Such educational assessments can be an important part of human dimensions research.

Step 1: Determine the objectives of the information campaign

- Why are you providing the information?
- Identify SMART targets: Specific, Measurable, Attainable, Realistic, Timed.

Step 2: Identify target audience(s) for the campaign

- To whom will the information be provided?
    - Do not try to reach too many audiences with the same message. Be specific, as this will aid in evaluating the effectiveness of your educational campaign in the future.
    - Segment target audiences and obtain target audience descriptors.
    - Segment audiences by geographic area (residence). In Croatia (Bath and Majic 2000), attitudes and beliefs varied not only within the same interest group but also across regions of the country, thus requiring different messages to be effective. More recently, Glikman et al. (2012) segmented their respondents by beliefs about wolves and bears.
    - Segment audiences by sociodemographic characteristics (e.g., age). Teenagers need to be reached using different methods from younger school children. Perhaps we need to try to reach children at age 1–5 and their parents, as various child psychology literature suggests that much of our later response to things is developed in this early phase of childhood.
    - Segment audiences by gender. Some activities such as hunting may be more male dominated, and on other occasions, where women are the leaders of the rural village, perhaps it would be better to target them.
    - Segment groups by lifestyle characteristics.

- Segment target audiences by skill levels, which is often done when considering outdoor recreational management issues (e.g., experienced, novice, and beginner horseback riders; frequent hikers and novice hikers).
- Understand and assess the criteria for viable target segments.
  - Is the audience directly pertinent to the goals of the educational campaign? Often we target efforts to children in the anticipation that messages will reach parents, who have the power to make decisions. Is this the most effective way to reach the parent target audience?
  - Is it possible to measure and understand the characteristics of the audience?
  - Can the target audience be reached? Is it accessible?
  - Is the target audience large enough or important enough to merit spending time and money?
  - Do we understand the target audiences well enough to know if they might respond differently to a communication strategy?

Step 3: Define response of target audience(s) to the campaign

- People respond to an information campaign in different ways. In the design of your educational campaign, it is necessary to think about the way the target audience is to respond to the designed messages.
- Attention: The purpose of the message is to raise awareness about an issue (e.g., wolves are in the area and there is a need to protect livestock).
- Interest: The purpose is to motivate the audience to learn more (e.g., go to a website, ask for a demonstration of livestock preventative measures).
- Action: The purpose is to create an actual behavior change (e.g., buying only organic foods, purchasing and using an electric fence to minimize livestock losses).

Step 4: Develop the message

What kinds of information will be provided to the target audience(s)? For locations where a variety of audiences may visit (e.g., a large carnivore center), think of messages matching three types of audiences: streakers, strollers, and studiers. There will be individuals who will walk quickly through an exhibit, and by streaking through the exhibit this group will only receive the main-level message. There will be other visitors that will stroll along, and will thus be able to pick up a secondary level of messages. Finally, there are visitors that are studiers, who are very small in number and are willing to read all the details in an exhibit or on a board or poster. Thus it is more important to design effective messages at the highest level for the streakers and at the secondary level for the strollers than to concentrate all efforts on the detailed information. There are many considerations in delivering the message. Human dimensions research has already identified specific weaknesses in knowledge that are most directly linked to attitude. In designing the message, one must follow the following guidelines:

- Use appropriate language: Avoid jargon (e.g., capacity building), technical terms (e.g., ecological integrity, viable populations), and vague terms (e.g., sustainable development, in "southern France," where does it begin and end?).
- The acronym "k.i.s.s." (keep it simple stupid) can often be applied liberally when designing the message.
- While the popular phrase "a picture is worth a 1000 words" still holds true, as a communicator you do not want people to guess at what the visual message is. Each image still requires a written message.
- Repeat the message in different ways. Teachers are often told that they should first tell students what they will talk about, secondly talk to them and deliver the information, and thirdly tell the class what was just told to them. Children and adults alike learn through

repetition. Think of the small child who can enjoy hearing the same story at least 100 times or the principles of the Suzuki method for teaching music to young children where 5-year-olds can play violin concertos just because the music has been played in the house regularly. Indeed, we learn our own language not by single words but by constantly hearing the language and the words on a daily basis.

- It is possible to deliver a lot of information in an educational campaign, but just because one can, that does not mean that one should. Clustered and complicated posters, brochures, and newsletters are not as successful as clear, concise, and simplified messages. Present only the relevant information. Use human dimensions research to identify the key beliefs directly linked to attitude. Then communicate only these important and directly relevant pieces of information.
- Consider the knowledge and experience level of the target audience. Can the target audience read maps if communicating areas closed to hiking or hunting?
- Send the best messenger to deliver the message. Credibility and trust issues are often the source of conflict and hence the message to be believed by the target audience must be delivered by a source with credibility for that audience. Consider involving the interest group in the collection of the data. Such cooperative efforts not only will increase the believability of the results, but also will build credibility and legitimacy between interest groups.
- Value persuasive messages (e.g., messages about existence value, animal rights, protection for future generations) have been found to be more important than factual messages in many cases. In terms of large carnivores, information about numbers of individuals and status of the population (whether it is increasing, decreasing, or remaining stable) and direct information about human–wildlife interaction (e.g., attacks) have been shown to be consistently important in influencing attitudes.

Step 5: Select the media strategy and implement the campaign

Educational campaigns can be short term and highly concentrated or spread out over a longer period of time and more diverse in the media used. There are various media and through focus groups or human dimensions research, there should be an understanding of what media work best for each target audience.

- Select major media category
  - Understand media habits of each target audience.
  - Consider the nature of the message and what media best suit the message being delivered. Will a poster work to deliver a visual image of habitat fragmentation better than a brochure?
  - Consider the complexity of the message and the time needed to deliver the message. Is it a message that, due to its complexity, would be better delivered in written form than in a television advertisement?
  - Costs vary greatly depending upon the type of media used.
- Select specific media vehicle
  - If a newspaper is chosen as the media outlet, consider where in the newspaper should the message be printed. Are there credibility issues depending upon where the message is placed within the paper and which paper is chosen?
- Determine media timing
  - Be aware of seasonal holidays when the public may not be focused on certain media.
  - Messages delivered earlier in the week tend to be received better than messages received later in the week or on the weekend.
  - What effect is desired? "Shock and awe" approach of an intensive period of delivering messages through a variety of media hoping for a quick impact and change, or a "sustained commitment" approach, where messages are delivered on a regular basis.

Step 6: Evaluation

Often, educational and information campaigns are delivered and never reviewed, revised, or assessed for their effectiveness. Evaluation can be organized into a goals and objectives evaluation or a content evaluation. A goals and objectives evaluation would be more a checklist approach to assess whether all specific targets are in fact reached. For example, the education campaign stated it would produce a poster and deliver educational programs to 15 schools in 4 months. A goals and objectives evaluation would check whether a poster is produced and how many presentations are given to schools.

In contrast, a content evaluation would assess whether knowledge has actually changed within the target group. This would require engaging the target group and assessing beliefs and attitudes to compare against the baseline assessment done prior to implementation of the campaign.

HUMAN DIMENSIONS SUMMARY

In the form of a summary, human dimensions research:

> focuses on the public's knowledge levels, expectations, attitudes and activities concerning fish and wildlife resources and associated habitats. There is a close tie between human dimensions and conservation education research. (Adams 1988)

Human dimensions research can address various objectives:

- Baseline assessment to begin attitudinal and belief monitoring: Has an educational effort, management policy, made a difference?
- Educational role: Targeting specific weaknesses in knowledge to affect attitudes. Working toward designing more effective educational materials.
- Partnership building: Bringing groups together around a common data set. Working toward understanding the issues of a variety of interest groups, building trust, and initiating the first steps toward working together.
- Identification of areas of support and disagreement over management options, thus assessing the feasibility of approaches that are successfully implemented.
- Identification of types of conflict (cognitive, values, costs/benefits, and behavioral conflicts): The first step toward conflict resolution.

Human dimensions research can address many questions and can provide managers with useful information for decision making. It is important for managers to clearly define how they wish to utilize the results; this will affect all aspects of the study, including formulation of the problem, design and collection of the data, analysis, and presentation. Similar to how one biophysical study cannot answer all the biophysical questions surrounding an issue, one human dimensions study cannot address all the social science questions that pertain to an issue. Such human dimensions research can, however, offer valuable insights into issues for wildlife managers. Human dimensions as a facilitated workshop approach can also resolve issues. In this capacity, human dimensions is a form of public involvement in wildlife resource management decision making.

REFERENCES

Adams, C.E. 1988. Establishing a human dimensions program. *Human Dimensions Newsletter* 7(3): 3–7.

Anderson, S.H., R.E. Beiswenger, and P.W. Purdom. 1987. *Environmental Science,* 4th edn. Prentice Hall, Upper Saddle River, NJ.

Bath, A.J. 1989. The public and wolf reintroduction in Yellowstone National Park. *Society and Natural Resources* 2: 297–306.

Bath, A.J. 1991. Public attitudes in Wyoming, Montana, and Idaho toward wolf restoration in Yellowstone National Park. *Transactions of the 56th North American Wildlife and Natural Resources Conference* 56: 91–95.

Bath, A.J. 1998. The role of human dimensions in wildlife resource research in wildlife management. *Ursus* 10: 349–355.

Bath, A.J. 2000. Human dimensions in wolf management in Savoie and Des Alpes Maritimes, France: Results targeted toward designing a more effective communication campaign and building better public awareness materials. Large Carnivore Initiative for Europe (LCIE) and Council of Europe Publication, Gland, Switzerland.

Bath, A.J. and T. Buchanan. 1989. Attitudes of interest groups in Wyoming toward wolf restoration in Yellowstone National Park. *Wildlife Society Bulletin* 17: 519–525.

Bath, A.J. and L. Farmer. 2000. Europe's carnivores: A survey of children's attitudes towards wolves, bears and otters. WWF-UK report.

Bath, A.J. and A. Majic. 2000. Human dimensions in wolf management in Croatia: Understanding attitudes and beliefs of residents in Gorski Kotar, Lika and Dalmatia toward wolves and wolf management. Large Carnivore Initiative for Europe (LCIE) Publication, Gland, Switzerland.

Bath, A.J. and D. Tavares. 2000. Toward a comprehensive large carnivore educational resources inventory: Work in progress. Large Carnivore Initiative for Europe (LCIE) Publication, Gland, Switzerland.

Braithwaite, A.M. and S.F. McCool. 1989. Social influences and backcountry visitor behaviour in occupied grizzly bear habitat. *Society and Natural Resources* 2: 273–284.

Campbell, D.T. and J.C. Stanley. 1963. *Experimental and Quasi-Experimental Designs for Research*. Rand McNally, Skokie, IL.

Fishbein, M. and I. Ajzen. 1975. *Belief, Attitude, Intention and Behavior: An Introduction to Theory and Research*. Addison-Wesley, Reading, MA.

Glikman, J.A., A.J. Bath, and J. Vaske. 2010. Segmenting normative beliefs regarding wolf management in Central Italy. *Human Dimensions of Wildlife* 15(5): 347–358.

Hummel, M. 1989. *Endangered Spaces: The Future for Canada's Wilderness*. Key Porter Books, Toronto.

Jope, K.L. and B. Shelby. 1984. Hiker behaviour and the outcome of interactions with grizzly bears. *Leisure Sciences* 6: 257–270.

Kellert, S.R. 1985. Public perceptions of predators, particularly the wolf and coyote. *Biological Conservation* 31: 167–189.

Majic, A. and A.J. Bath. 2010. Changes in attitudes toward wolves in Croatia. *Biological Conservation* 143: 255–260.

Mitchell, B. 1989. *Geography and Resource Analysis*. John Wiley, New York.

Moore, D.E. 1983. The effects of guided experiences with animals on the beliefs, attitudes, behavioral intentions and knowledge of children. Unpublished PhD dissertation. University of Wyoming, Laramie, WY.

Tucker, P. and D. Pletscher. 1989. Attitude of hunters and residents towards wolves in northwestern Montana. *Wildlife Society Bulletin* 17: 509–513.

Vaske, J.J. and M.P. Donnelly. 1999. A value-attitude-behavior model predicting wildland preservation voting intentions. *Society and Natural Resources* 12(6): 523–537.

Zimmerman, E.W. 1933. *World Resources and Industries: A Functional Appraisal of the Availability of Agricultural and Industrial Resources*. Harper & Brothers, New York.

20 The Role of Ethnoscience in the Growth of Ethnoconservation as a New Approach to Nature Conservation in the Tropics

The Case of Brazil

Antonio Carlos Diegues

CONTENTS

Introduction....283
The Conservation Approach Based on No-Take Protected Areas in the Tropics....284
Traditional Peoples' Rights and Government Policies....286
Basic Principles Guiding an Ethnoconservation Approach....287
Dialogue Needed Between Natural Sciences, Social Sciences, and Ethnoscience....289
Conclusions....290
References....290

INTRODUCTION

The objective of this chapter is to discuss the growth of a new, nonhegemonic approach to nature conservation in the tropics, herein called *ethnoconservation*. The departure point is a critique of the classic/hegemonic approach for conservation of protected areas, which is far from achieving its main ecological objectives in tropical regions, in addition to creating serious problems for indigenous and traditional peoples. According to this hegemonic approach, traditional (indigenous) peoples and traditional communities should be expelled from their territories when they are transformed into no-take protected areas or forbidden from maintaining their livelihood, which is based mainly on the use of natural resources through small-scale agriculture, fishing, and forest extractivism.

In the hegemonic approach to nature conservation, only the natural sciences, particularly biology and botany, are used in the planning and management of these no-take protected areas. It is being internationally recognized, however, that traditional peoples and communities can play a crucial role in nature conservation through their traditional knowledge of living species and habitats as well as through their socioeconomic and symbolic practices. In light of this, ethnoecology is called to play a crucial role as a bridge between the natural and social sciences. Ethnoscience, in addition to social participation, respects the social rights and livelihoods of traditional peoples and communities. It is a cornerstone of the new models and approaches to nature conservation and protected areas in the Tropics that are here defined as ethnoconservation (Diegues, 2000).

In Brazil, since the late 1980s, as a result of the struggle of rubber tappers (seringueiros) and other traditional peoples and communities to protect the tropical Amazon forest against deforestation, new sustainable use, protected areas have been established all over the country, particularly in forest and coastal/marine realms (Schwartzman, 1989). These new reserves have been requested and are supported by social movements and traditional peoples/communities, and their establishment and management are based mainly on traditional knowledge and practices as well as on social and natural sciences. They are an important field in social experimentation in *ethnoconservation*.

THE CONSERVATION APPROACH BASED ON NO-TAKE PROTECTED AREAS IN THE TROPICS

The hegemonic preservation approach to conservation based on no-take protected areas has been criticized, particularly in Southern Countries, by social movements, socioecological organizations, and a growing number of biologists and social scientists for the last 20 years (at least) (Gomez-Pompa and Kaus, 1992; Balée, 1993; Diegues, 1999; Toledo, 2001; Chapin, 2004; Sarkar, 1998; Bahuchet et al., 2000; Colchester, 1997).

The hegemonic approach to conservation is based on the following:

1. The notion of *wilderness*, which appeared mainly in the United States in the middle of the nineteenth century, in the establishment of no-take zones from which human beings, including indigenous people, should be expelled (Sarkar, 1998; Diegues, 1999).
2. The *naturalist notion* that every human being and all societies, independently of their rationality and social organization, are, per se, a threat to nature. Human beings are equivalent to threats in the language of most conservationist organizations.
3. The exclusive use of the *natural sciences* for the planning and management of these protected areas. In most cases only the natural sciences are used in spite of the presence of traditional people/communities and their traditional knowledge about forests and seas in most of the territories where these conservation areas are established. According to the hegemonic view, conservation is not a social practice but derives from the application of the natural sciences.
4. *Lack of involvement of local social groups.*

No-take protected areas, established without consultation with people living in that territory, are considered by *preservationists* to be the only way to protect biodiversity, also understood only in natural sciences terms. Police force and repression are often used to protect biodiversity and top-down approaches hinder any democratic way of managing biodiversity with indigenous people living inside those parks.

The transfer of the *wilderness model* to Third World countries resulted in the expulsion of indigenous peoples from their territories, loss of traditional knowledge, and in serious constraints on their traditional way of living, violating their land rights and compelling many of them to social marginality (Chapin, 2004).

As *wilderness* transforms territories of indigenous people into a *no place* where no social relations may exist, the only sciences considered to be useful are the natural sciences. In countries such as Brazil, the expansion of this model has been possible since the 1960s through an alliance of preservationist nongovernmental organizations (NGOs), natural scientists, and state bureaucracies, particularly during the military regime (1964–1984), when many no-take protected areas were created, mainly in the Amazon region.

A new wave of no-take occurred in Brazil from the 1980s with the arrival of the *big international NGOs* (*BINGOs*). They came with funds, influence, concepts, and strategies to implement protected areas programs. Some of them bought land to transform into no-take zones, often using local NGOs as intermediaries and creating suspicion among nationalistic political forces. It became clear that "wilderness" was not an isolated concept, but it became associated with a philosophy (deep ecology) and a natural science model—conservation biology. Initially, some of them were more involved in community-based conservation (CBC), but more recently they embarked on the so-called "global conservation" (Diegues, 2008).

In the beginning, BINGOs were interested mainly in the establishment of no-take zones protected in isolation and their science was based on elements of ecosystem theory and conservation biology (Chimere-Diaw, 2000; Sarkar, 1998; Guha, 1997; Diegues, 2010). In spite of that, many natural scientists dealing with biodiversity conservation take an ideological position far from this so-called objective approach. For many of them, biodiversity conservation became a new religion or dogma of urban atheists (Nelson, 2012). Most of the agendas proposed and conducted by the BINGOs consider conservation as a "natural" issue (protection of biodiversity and endangered species), and discussion on issues such as science and power, power and conservation is avoided (Chimere-Diaw, 2000). In fact, the discussion of conservation also has a geopolitical aspect, as in countries such as Brazil, the territory covered by protected areas is larger than many large European countries and economic activities such as food production is forbidden, particularly in no-take areas.

In the last decade, they launched the so-called "global" conservation, which was based on concepts and strategies of large scale such as hot spots, biological corridors, large-scale landscape and bioregions (Chimere-Diaw, 2008). This large-scale conservation was more in line with the concerns of global institutions such as the World Bank, some big North American foundations, and private corporations that participate in the boards of the BINGOs. The definition of the above-mentioned concepts excluded local populations, although the larger the scale, the bigger the risk of not seeing local populations that live in the forests and coastal areas of Southern Countries (Chapin, 2004; Dowie, 2006).

As an example, in February 2006, the US Agency for International Development (USAID) launched the Amazon Basin Conservation Initiative, comprising countries located in the area. The program was based on a large-scale conservation that ultimately would be implemented by North American BINGOs. The Brazilian government rejected such a plan on the argument that it was not even consulted (Diegues, 2008).

In fact, since the beginning of the 1990s, the BINGOs (particularly the World Wildlife Fund [WWF], Conservation International [CI], and The Nature Conservancy [TNC]) have organized national workshops to define biodiversity conservation priorities (hot spots). Some central concepts, such as biodiversity, were defined differently by natural and social scientists in Brazil. In the process coordinated by BINGOs for defining "hot spots" for biodiversity conservation, only natural variables are considered, in spite of the fact that in tropical forests many of these hot spots are inhabited by traditional peoples (and other traditional groups) that have used the forest for centuries (Toledo, 2001).

These historical and cultural relationships and social practices (slash and burn agriculture, traditional management, etc.) that have resulted in the "pristine" state of the forests are not taken into consideration by most conservation biologists. On rare occasions when social scientists are called on at workshops to define conservation areas, their role is to enumerate "human threats" to biodiversity (Diegues, 2008).

Social scientists were absent from the debate until the mid-1980s, when a new category of protected area—the extractive reserve—was created as a result of the fight of rubber tappers against deforestation and loss of territory and for access to natural resources. One crucial element in this process is the recognition that forest peoples have an important contribution to conservation through their traditional knowledge and management of natural resources. Since then, traditional knowledge and management have been proposed as an important contribution for the planning and implementation of the sustainable use reserves. These reserves were also supported by a political movement that was strong at that time—The National Council of Rubber Tappers—and by some ecosocially oriented local and international NGOs, which were able to influence conservation policies in the country.

From the 1980s onwards, there was a series of *conflicts* between the preservationists (government environment officers and NGOs) that controlled the state environmental bureaucracy on one side and socioenvironmentalists and social movements on the other (Diegues, 1999). This conflict became clear during the meetings that preceded the National System of Protected Areas (SNUC), which was finally approved in 2000 by the Congress. For the first time the presence of traditional people in no-take protected areas was officially recognized. It was decided that protected areas should be established through public hearings. Two categories of sustainable use protected areas were officially created: Extractive Reserves and Reserves of Sustainable Development. Since then, more than 33 forest extractive reserves and 14 marine extractive reserves have been created, occupying an area of 6 million ha and benefiting over 10,000 families. Reserves of Sustainable Development are also being established, particularly in the Amazon region. In these reserves, land is state owned and given as concessions to the users of the Associations of the Reserves. These associations and Deliberative Councils are the local instances that take decisions in association with the Chico Mendes Institute at the Ministry for Environment, which appoints the reserve director.

TRADITIONAL PEOPLES' RIGHTS AND GOVERNMENT POLICIES

Since the end of the military regime (1984) and the beginning of a new democratic period, important social movements have appeared in Brazil, among them the Landless People's Movement, Indian organizations, the National Council of Rubber Tappers, the Artisanal Fishermen's Organization, and the *quilombola* movement (descendants of African slaves). In December 2006, the federal government established the National Commission for Traditional Peoples, in which Indian and traditional communities participate. This commission comprises Indian peoples, quilombola communities, Amazonian Caboclos, artisanal fishers, Caiçaras, inter alia. In another official document, Policies for the Sustainable Development of Traditional Peoples and Communities, issued in 2007, they are defined as culturally distinct social groups, recognized by themselves as such, that have their own social organizations and livelihoods based on the use and management of natural resources.

The main features of traditional peoples are as follows:

1. Strong economic and symbolic ties with the land and the sea through continuous observation of natural cycles
2. Attachment to continual use and occupancy of a specific group territory, which allow a community to reproduce itself through ongoing traditions of communal and family land and sea tenure
3. Subsistence activities that continue to play a vital role in their economy (In these cases, production for market is also becoming important)
4. Individual/family ownership of means of production
5. Limited accumulation of capital
6. Crucial socioeconomic relations that are structured along family, domestic, and communal kinship lines
7. The use of relatively simple technology, with a limited impact on the environment
8. Positions of marginality from political power bases, which tend to be concentrated in urban centers
9. Oral traditions responsible for the production and transmission of knowledge, symbols, myths, and rituals associated with artisanal fishing, forest harvesting, and small-scale agriculture
10. Social identity is a fundamental issue

BASIC PRINCIPLES GUIDING AN ETHNOCONSERVATION APPROACH

This approach, which is in the process of elaboration in different tropical countries, is based on research, socioecological thoughts, and social practices on biodiversity–nature conservation that emphasize the need for the involvement of traditional peoples and their culture, social organizations, traditional knowledge, and management. Ethnoconservation is guided by principles such as the following.

1. Nature conservation for traditional peoples is the result of *social practices* and associated knowledge, management, perceptions, and ethics and not seen as an isolated issue from other aspects of their culture, as happens in Western industrial/urbanized societies.
2. Adequate nature conservation has to be based on the *recognition of the rights* of traditional peoples to live with dignity in their own territories. The notion of territory to which they are attached is fundamental for traditional peoples as their culture, social relations, and traditional knowledge are produced there (Nietschmann, 1987).

   The "territory" that a particular society claims as its own grants to all or to some of its members stable rights of access, control, and use for all or part of the natural resources located there (Godelier, 1984). This territory furnishes, first of all, the nature of humans as a species, but also the means of subsistence, the means of production, and the means of producing material aspects of social relations, such as kinship relations (Godelier, 1984; Cordell, 1992).

   The respect for their livelihood, besides the guarantee of staying in the territory of their ancestors, should include adequate access to health, education, and infrastructure, and respect for their decision to change aspects of their culture in contact with other cultures (Bahuchet et al., 2000).
3. *Biological diversity can only be protected when cultural diversity is respected and vice versa.* There is evidence that many forest areas put aside to be transformed into protected areas (particularly national parks) are inhabited by indigenous peoples. This overlapping is not accidental, as has been shown by many researchers (Toledo, 2001; Diegues, 1999; Alcorn, 1993; Gomez-Pompa, 2000). The high biodiversity existing in indigenous territories is the result of traditional knowledge and management practices.
4. *Traditional knowledge and management practices are fundamental for biodiversity conservation.*

   Traditional knowledge may be understood as a distinct cognitive realm. On the one hand it consists of a replicable, orally transmitted set of specialized skills and culturally shared practices and beliefs that have stood the test of time, enabling people to make a living from different environments such as forests, savannas, and coastal and marine environments, relying on artisanal techniques. On the other hand, traditional fishing knowledge exists in more encompassing symbolic and conceptual frameworks governing social relationships and spiritual connections to the various habitats of their territories (Diegues and Cordell, 2001).

   It can also be defined as a cumulative body of knowledge and beliefs handed down through generations by cultural transmission about the relationship of living beings (including humans) with one another and their environment (Gadgil et al., 1993).

   This repertory of knowledge is at the same time individual and social. On the basis of this knowledge, traditional peoples and communities take decisions on where, when, and how to plant or to fish, utilizing a multiuse strategy that maximizes the variety of goods produced and providing the family with basic requirements throughout the year (crops, fish, handicrafts, etc.).

The construction of this body of complex and detailed concepts and symbols is based on long-term empirical observation and is applied to rather small areas used by local communities and seldom can be replicated elsewhere.

As Ruddle points out (2000, p. 282),

> Resource use patterns are products not of their physical environment and its resources per se, but of their perceptions our culturally formed images of the environment and its resources. Thus, to properly understand human ecological relationships, an understanding of a society's local knowledge base, and the cognitive system that underlies it is crucial.

The social appropriation of nature does not only imply an extension of social relationships on the land and the accumulation of local environmental knowledge, but also involves the formation and symbolic expression of links with the spiritual world. Conceptions and representations of the natural world and its resources differ greatly between subsistence- and market-oriented societies. Godelier (1984) argues that these two societies have different rationales, and each displays a system of social rules consciously elaborated to best attain a set of objectives. According to this anthropologist, each economic and social system creates a specific mode of exploitation of natural resources and use of the human labor force and, consequently, utilizes specific norms of good and bad use of natural resources.

5. Biodiversity is also maintained through social values that reveal a *particular cosmovision* of traditional peoples and communities. According to this cosmovision, nature has a sacred nonmarket value and land as well as water and sea are primary sources of their livelihood. (Toledo, 2001; Berkes, 1999). In this connection, their representation and classification of the territories they occupy and their elements (forest, water, natural processes) are the core of their culture and the origin of their identity (Marques, 1995; Toledo, 2001).

According to Godelier (1984), at the heart of our material relationship with nature there is an underlying nonmaterial bond that unites the three key functions of knowledge: to simultaneously represent, to organize, and to legitimize our social relations and our relations with nature. In order to understand the process of material production, it is essential to understand symbols and myths used by traditional peoples to represent nature and its beings.

Thus, together with defining a space for economic reproduction and projecting principles of social relations, terrestrial and marine territories can also be the locus of representations and of the mythological imagination of these traditional societies. The intimate relationship of these people with their surroundings and their greater dependency on the natural world when compared with urban–industrial societies result in the cycles of nature (the arrival of schools of fish and the abundance of crops) being associated with mythical and religious explanations.

An ongoing debate surrounds the natural resource conservation function of these mythological beings. In other words, are traditional peoples aware of the ecological intentions of these cultural practices? Can such practices actually facilitate conservation or be viewed as "conservationist"?

Darrel Posey (1992) uses the emic/etic approach to discuss the issue of intentionality related to traditional practices. According to him, in some conservationists' minds traditional practices that limit overexploitation of resources can be considered to enhance or support biological conservation, in the modern, scientific sense. For Posey, this interpretation falls into the category of an etic approach that is developed by the researcher. On the other hand, in the minds of traditional peoples (the emic approach) the function of beliefs about the behavior of mythical beings may turn out to be something quite different. Fear of being punished by supernatural beings may function, for instance, to discourage capital accumulation and social differentiation in societies organized along egalitarian lines. In this connection, an emic approach to explain these practices is unlikely to be deliberately "conservationist," at least not in the sense this concept is defined in Western science.

6. *Participatory approach and empowerment.* Ethnoconservation applied to the establishment and management of protected areas requires continuous and strong participation in deciding what type of protected area is going to be established and what the role of local associations will be in the decision-making process, monitoring, and evaluation. Empowerment is primarily related to the control of the territory to be transformed into a protected area and the recognition by the authorities of the social and cultural rights of the traditional peoples and communities involved (Pimbert and Pretty, 1997).

DIALOGUE NEEDED BETWEEN NATURAL SCIENCES, SOCIAL SCIENCES, AND ETHNOSCIENCE

The recent importance given to sustainable use reserves in Brazil, in which local communities play a central role, has made it evident that conservation is a social (and epistemological) practice involving the protection and sustainable use of natural resources (Larrère and Larrère, 1997; Diegues, 2000; Toledo, 2001). It requires an organic cooperation between social sciences, natural sciences, and traditional knowledge. There are at least two questions concerning this cooperation: First, there are still few social scientists working in conservation because they think that nature conservation is a field for biologists and they do not want to be "contaminated" by geographic–biological determinism. Second, there is difficulty in finding a common ground (methodological) to work together given the different methodologies of natural and social sciences. As long as some biologists think that the central concept for nature conservation is "wilderness," from which human beings must be absent, and only "pristine environments" have to be protected, the scope for collaboration will be reduced. At the same time, as long as some social scientists consider "nature" to be only a social construction, cooperation with natural scientists is doomed to fail.

However, a few social and natural scientists have started cooperating in the planning and management of sustainable use protected areas, and this is a good starting point. Often ethnoscience (ethnoecology, ethnobiology, ethnoichthyology, etc.) is being used as a bridge between social and natural scientists.

Therefore, there is a need to improve cooperation among the social sciences in order to understand conservation as a social practice that appeared in a given historical and political context of modernity. Some disciplines have been more involved in the conservation debate, particularly history through environmental history, environmental geography, environmental sociology, and political ecology, but there is little interaction among them in the tropics. In some Southern countries, the crucial role of political ecology in understanding conservation has almost disappeared, although it was important in the 1970s in countries such as Brazil (Diegues, 2008).

Some concepts and approaches, such as the *ecosystem approach*, used by biologists were not designed to integrate social and cultural dimensions. More recently, this approach has become a kind of dogma, including in some important international conventions, for example biodiversity conventions, where it is being recommended as the only possible approach to biodiversity conservation. As far as conservation biology is concerned, even biologists (at least in Brazil) have recognized that the methodology is unable to incorporate the social and cultural dimensions of conservation. Concepts such as landscape and coevolutionary processes, as a result of interactions between society and nature, are more appropriate to deal with the relationships between society and the environment.

An important contribution to the incorporation of ethnoecology into ethnoconservation was made by Balick and Cox (1996, pp. 199, 201, 207) when they proposed that

> a new and very important branch of ethnobotany might be termed as *the incorporation of indigenous conservation models into wildlands biology* including the role of traditional healers . . . This multitiered effort has linked the mutual interests and activities of local healers, farmers, students, ethnobotanists and pharmaceutical researchers to the conservation of the main source of materials and ideas: the area's

> forests. These forests serve as both a classroom and a source of raw materials for local health practitioners. . . . Clearly, one priority for the future is to involve indigenous colleagues in ethnobotanical research as co-investigators and to train a new generation of people from a variety of cultures to initiate studies among their own people . . . We believe that indigenous peoples, if given proper information and granted status as equal partners are capable of plotting their own future

CONCLUSIONS

Ethnoconservation is a new approach to nature conservation under construction in different tropical countries, based on interdisciplinary cooperation and on experiences of traditional peoples through their knowledge and nature management practices. Ethnoecology and ethnobiology, as well as other branches of ethnoscience, are called on to play an important role as a bridge between the social and natural sciences in the buildup of ethnoconservation as a social practice. Ethnoconservation also uses the contribution of political ecology, as ultimately nature conservation is controversial, in a scenario where different social forces and classes have different and often conflictive views (nature conservation vs. agrobusiness, dam constructions, tourism, etc.). Traditional peoples are a new player in this difficult game; they are politically weak and their strength depends on different kinds of alliances both nationally and internationally. Many of them are under severe threats from large development projects as well as from ill-conceived conservation initiatives that lead to territorial expropriation and loss of identity, language, culture, and territories.

It is also becoming clear that new approaches to conservation, taking into consideration the complex relationships between societies and environments, are needed in Brazil as well as in other tropical countries based not only in protected areas. The possibility of establishing protected areas of any kind is reaching its limits in terms of land and conflicts. The simplistic and still dominant, hegemonic approach of national parks as "conservation fortresses" is not performing well in spite of the funds and political influence of BINGOs, possibly because of authoritarian practices, ideological views of wilderness, and a lack of interdisciplinarity (including ethnoecology and traditional knowledge).

New social and ecological practices are required and have to be built based on an interdisciplinary approach that takes into consideration local/traditional knowledge and social and democratic participation. In some Southern countries such as India, Brazil, Mexico, and many others, new approaches and experiences in nature conservation that take into consideration biological as well cultural diversity are being constructed. Some people call this new approach *ethnoconservation*.

REFERENCES

Alcorn, J. 1993. Indigenous peoples and conservation, *Conservation Biology* 7, 424–426.

Bahuchet, S., Grenand, F., Grenand, P., and Maret, P. 2000. *Forêts des tropiques, forêts anthropiques, sociodiveristé, biodiversité: Un guide pratique*. Brussels: APFT–ULB.

Balée, W. 1993. *Footprints of the Forest: Ka´apor Ethnobotany. The Historical Ecology of Plant Utilization by an Amazonian People*. New York: Columbia University Press.

Berkes, F. 1999. *Sacred Ecology: Traditional, Ecological, Knowledge and Resource Management*. Philadelphia, PA: Taylor & Francis.

Chapin, M. 2004. A challenge to conservationists. *World Watch Magazine* 17(6), 17–31.

Chimere-Diaw, M. 2000. Scales in conservation: Another clash of civilization. Center for International Forestry Research, Cifor. In: Diegues, A. (org.) *Etnoconservacao*. São Paulo: Hucitec.

Colchester, M. 1997. Salvaging nature: Indigenous people and protected areas. In: Ghimire, K.B. and Pimbert, M.P. (eds), *Social Change and Conservation: Environmental Politics and Impacts of National Parks and Protected Areas*. London: UNRISD Earthscan Publications.

Cordell, J. 1992. Locally managed territories in the Brazilian coast fishing. Rome: FAO.

Diegues, A. 1998. Recycled rain forest myths. *Terra Nova: Nature and Culture* 3(3).

Diegues, A. 1999. *The Modern Myth of Untamed Nature*. São Paulo: USP/Nupaub.

Diegues, A. 2000. Etnoconservaçao da natureza-enfoques alternativos. In: Diegues, A. (org.) *Etnoconservaçao: Novos srumos para aprotecao da natureza nos trópicos*. São Paulo: Hucitec.
Diegues, A. (org.) 2008. A ecologia política das grandes ONGs transnacionais conservacionistas. São Paulo: Nupaub-USP.
Diegues, A. 2010. Protecting biodiversity and artisanal fishers rights. In: Falque, M. and Lamotte, H. (eds), *Biodiversity, Property Rights, Economics and Environment*. 8th International Conference, Aix Marseille, June, 2010, Bruylant.
Diegues, A. and Cordell, J. 2001. Traditional fisheries knowledge and social appropriation of marine environment in Brazil. Paper presented at Maré Conference, Amsterdam, August/September.
Dowie, M. 2006. Conservation refugees. Available at http://www.orionmagazine.org/index.php/articles/article/161/. Orion, Great Barrington, MA.
Gadgil, M., Berkes, F., and Folke, C. 1993. Indigenous knowledge for biodiversity conservation. *Ambio* 22(2–3), 151–156.
Godelier, M. 1984. *L´* ideel et le materiel. Paris, Fayard.
Gomez-Pompa, A. and Kaus, A. 1992. Taming the wilderness myth. *Bioscience* 42(4), 271–279.
Guha, R. 1997. The authoritarian biologist and the arrogance of anti-humanism. Wildlife conservation in the Third World. *The Ecologist* 27, 14–20.
Larrère, C. and Larrère, R. 1997. Du bon usage de la nature. Pour une philosophie de l environnement. Alto/Aubier, Paris.
Marques, J.G.W. 1995. *Pescando pescadores: Ciência e etnociência em uma perspectiva ecológica*. São Paulo: Núcleo de Apoio à Pesquisa sobre Populações Humanas e Áreas Húmidas Brasileiras.
Nelson, R. 2012. Biodiversity as a religious case. In: Falque, M. and Lamotte, H. (eds), *Biodiversity: Property Rights, Economics and Environment*. XIII International Conference, Aix-Marseille, 17–19 June.
Nietschmann, B. 1989. Traditional sea territories and rights in Torres Strait. In: Cordell, J. (ed.), *A Sea of Small Boats*. Cambridge, MA: Cultural Survival.
Posey, D.A. 1992. Interpreting and applying reality of indigenous concepts: What is necessary to learn from the natives. In: Redford, K.H. and Paddoch, C. (eds), *Conservation of Neotropical Forests: Working from Traditional Resource Use*, pp. 21–34. New York: Columbia University Press.
Pimbert, M. and Pretty, J. 1997. Parks, people and professionals. Putting the participation into protected areas management. In: Ghimire, K.B. and Pimbert, M.P. (eds), *Social Change and Conservation. Environmental Politics and Impacts of National Parks and Protected Areas*. London: Earthscan.
Ruddle, K. 2000. Systems of knowledge: Dialogue, relationships and process. In Begossi, A. and Hens, L. (eds), *Environment, Development and Sustainability*, Vol. 2. Dordrecht: Kluwer.
Sarkar, S. 1998. Restoring Wilderness or reclaiming forest? *Terra Nova, Nature and Culture* 3(3), 35–51.
Schwartzman, S. 1989. Extractive reserves: The rubber tappers' strategy for sustainable use of the Amazon rainforest. In: Browder, J. (ed.), *Fragile Lands of Latin America: Strategies for Sustainable Development*, pp. 150–166. Boulder, CO: Westview Press.
Toledo, V.M. 2001. Indigenous peoples and biodiversity. In: Levin, S. (ed.), *Encyclopedia of Biodiversity*. San Diego, CA: Academic Press.

21 Policy Intervention in the Cerrado Savannas of Brazil

Changes in Land Use and Effects on Conservation

Carlos A. Klink

CONTENTS

The Importance of the Cerrado 293
Drivers of Landscape Modification 294
Agricultural Development and Changes in Land Use 295
Regional Developmental Programs 296
New Technologies 296
Minimum Price Policy 296
Expansion of Agroecosystems 297
Land Use Effects on Biodiversity and Ecosystems 297
Change of Policy Framework 298
Promoting Cerrado Sustainability in the Future 301
Restoration of Degraded Lands for Business Development 301
Applying Knowledge of Trade-Offs in Land-Use Decisions 302
References 303

THE IMPORTANCE OF THE CERRADO

At the time of writing, more than 50% of the 2,000,000 km^2 of the cerrado savanna has been transformed into pastureland, cash crops, and other types of land use. Pastureland cultivated with African grasses now covers 500,000 km^2, an area the size of Spain. Crops occupy another 100,000 km^2, 63% with soybeans alone (Klink and Machado 2005; Sparovek et al. 2010). Today the cerrado is responsible for more than half of Brazil's soybean and beef production, most of it for export. The perception of abundant land has fostered land-use change in the cerrado. Modern technology has exacerbated social inequality since it demands less labor (Mueller 2005; Mueller and Martine 1997) and there has been major environmental modification (Carvalho et al. 2009; Davidson et al. 2012; Klink and Machado 2005): landscape fragmentation, loss of biodiversity, biological invasion, soil erosion, water pollution, land degradation, heavy use of chemicals, imbalances in the carbon cycle, and regional climate modification.

The cerrado also harbors a great diversity of plants, animals, and habitats and is one of the world's hotspots for biodiversity conservation (Diniz et al. 2010; Diniz Filho et al. 2006; Myers et al. 2000; Simon et al. 2009; Werneck et al. 2012; WWF 2011). Cerrado is the Portuguese name given to central Brazil's vegetation and ecosystems. The main ecological aspect of the cerrado is its environmental diversity in soil types, geology, geomorphology, climate, and biodiversity. It is located in the Brazilian Central Plateau in the South American tectonic plate, where rocks are mostly from the Precambrian (from 4,700 to 570 million years ago), covered with laterites of the Cenozoic (approximately 65 million years ago). Most soils are clay, highly weathered, and acidic and have low organic matter, low nutrient concentration, and high concentrations of iron and aluminum; soils are deep and well drained on gentle slopes, commonly less than 3%. To become productive, soils should be fertilized and corrected with lime. While the high aluminum concentration decreases crop productivity, some native trees are able to accumulate large amounts of aluminum in their leaves (Haridasan 1982).

The cerrado has a seasonal tropical climate with an average of 1,500 mm of rain per year; 90% of the precipitation occurs during the wet season (October through March) and the extent of the dry season (usually April through September) determines the biological productivity of the cerrado and can be a major environmental constraint on rain-fed crops. Fuel accumulation (dead vegetation biomass), low humidity, and farmers' use of fire at the end of the dry season promote widespread burning in the region. The annual average temperature is 22°C in its southern part and 27°C in its northern part.

The biological wealth of the cerrado comes from its diversity of vegetation types (23 have been identified by Ribeiro and Walter 2008), consisting mostly of woodlands, tropical savannas, grasslands, forests, and dry forests. Due to its extension, habitat heterogeneity, and proximity to other tropical ecosystems, the cerrado has a rich and usually unappreciated biodiversity (Oliveira and Marquis 2002; Pacheco and Vasconcelos 2012). The number of vascular plant species exceeds that of most floras in the world: the current number of herbs, shrubs, trees, and lianas is over 12,000 species (Mendonça et al. 2008), of which 44% are endemic, that is, they are unique to the region, which makes the cerrado the richest of the world's tropical savannas. Cerrado-endemic plants represent 1.5% of the globe's total flora (Mittermeier et al. 2004).

Considering its geographical extension, the mammalian fauna is relatively poor (199 species). The avifauna is rich, with more than 800 species, 12% of which are endemic. The total numbers of fishes, reptiles, and amphibians are still uncertain. The invertebrate fauna is also poorly known. Insects are the largest group of organisms in the cerrado, with an estimated 90,000 species. According to a recent estimate, the cerrado harbors 13% of the butterflies, 35% of the bees, and 23% of the termites of the neotropics (Cavalcanti and Joly 2002).

In this chapter, I will present a brief overview of past and current human occupation and then explain the main driving forces behind agricultural expansion in and the recent transformation of the cerrado landscape and impacts on its conservation. Then I will focus on societal responses to cerrado modification and discuss issues that will arise in both the scientific and policy arenas in the near future and their implications for the sustainable use and conservation of the cerrado.

DRIVERS OF LANDSCAPE MODIFICATION

Both in Brazil and abroad, the cerrado is seen as one of the last major land frontiers of the world (Borlaug 2002; Embrapa 1995; Herrera-Estrella 1999; Paterniani and Malavolta 1999; *The Economist* 2010; Tollefson 2012; Wheatley 2010). Though human presence in central Brazil dates back more than 9,000 years, it is only in recent times that human occupation has taken place. The idea of central Brazil as a region to be conquered and transformed has been embedded in Brazilian society since colonial times. The first permanent settlements were established by the Portuguese in the early eighteenth century and were associated with gold mining (Klink et al. 2005). After the Paraguay War (1864–1870), cerrado occupation was promoted by the Brazilian authorities, who were concerned that they would be unable to defend and maintain the border because of the region's very low population density.

The first major economic boom in the cerrado came during the period 1920–1930, when coffee-growing and coffee-processing industries were booming in the state of São Paulo, which hence became the major market for cerrado cattle. Later, the Getulio Vargas government (1930–1945) actively promoted the colonization of southern Goiás, providing land, subsidies, and technical assistance, and encouraging farmers to settle on and clear fertile forested lands (Garrett et al. 2012;

Klink and Moreira 2002; Klink et al. 2005). Population growth and agricultural development had important implications for cerrado land use. Until the late 1950s, the contribution of the cerrado to Brazil's agricultural output was very low, contributing less than 10% of the national total. This state of affairs changed dramatically after the 1960s, and the cerrado became Brazil's major producer and exporter of important cash crops and beef.

The major obstacle to the occupation and development of the cerrado has been its distance from the major Brazilian urban areas along the coast and lack of a transport system. Construction of the first railway, which linked São Paulo to Mato Grosso, was initiated in 1905 but was not completed until 1947 (Klink et al. 2005). After 1946, roads replaced railways as the main link between the Brazilian regions. The construction of Brasilia and highways linking the new capital to the main Brazilian cities represented the doorway to the modern occupation of the cerrado, which began in the 1960s (Hogan et al. 2001).

New links to connect the cerrado to external markets are planned or are underway, particularly the paving of the BR-163 highway in central Amazonia to carry soybeans to the international port of Santarém in the Amazon river and the "Pacific" Highway that connects the cerrado to the Pacific Ocean through the state of Acre and Bolivia and Peru. The construction of railways is again becoming fashionable. The most noticeable is Ferro Norte, a newly built railway linking the cerrado to Brazil's largest port (Santos) to carry soybeans for external markets.

AGRICULTURAL DEVELOPMENT AND CHANGES IN LAND USE

Cerrado land use is driven by technological innovations, capital investments, and knowledge and policy interventions, applied with the objective of promoting the expansion of intensive agriculture to meet the demand for agricultural products in Brazil and abroad. The two main land uses are pastures cultivated with African grasses; cash crops, mainly soybeans, corn, cotton, and wheat; and planted forests. During the 25-year period between 1970 and 1995, the area under crops had grown 250% and planted pastures 520% (Klink et al. 2005).

Several environmental and economic conditions favored these transformations. Although the rainfall distribution within the year is uneven, the mean total annual rainfall is considered sufficient for crop production. Temperatures are mild all year and the amount of sunshine does not restrict photosynthesis. Level topography propitiates mechanization and the cerrado savannas are cheaply and more easily cleared for farming or cattle ranching than tropical rainforest.

The strong performance of the Brazilian economy during the 1960s and 1970s, associated with a national development policy aimed at integrating the "empty" spaces of the cerrado and the Amazon regions into the capitalist economy of the richer south and southeast parts of the country, created the right atmosphere for investments (Helfand and Rezende 2004; Klink et al. 2005; Mueller 2003; Rezende 2003).

Policies aimed at expanding the agriculture output of the cerrado included subsidies, developmental programs, and the use of new technologies. Subsidized credit was one of the main instruments utilized by the Brazilian government to promote agricultural development during the 1960–1980 period. The greatest incentive was provided by a policy of low-interest farm loans. Because these loans were at fixed rates of 13%–15%, they were further subsidized by ignoring rapidly rising inflation rates (averaging 40% annually during this time). Even lower rates were imposed for fertilizers, pesticides, and machinery. As a result, credit to the rural sector grew between 1969 and 1979 at a rate 188% faster than total output (99% for agriculture, 164% for livestock production) (Klink et al. 2005). After 1979, credit flows, especially long-term investment credit, slowed.

Loans were not evenly distributed among crop types. Over 75% of production loans were concentrated in six crops: soybeans, rice, coffee, wheat, maize, and sugarcane. Soybeans alone received 20% of the credit available to Brazilian farmers (Klink et al. 2005). Since loans were allocated based on area planted, it encouraged extensive and inefficient agriculture (Goedert 1990). Cerrado farmers received 70% more credit than the Amazon and northeast farmers, where the small producers of food crops are concentrated. This readily available credit increased the demand for land in the cerrado and drove up land prices (Klink and Moreira 2002).

Regional Developmental Programs

The failure of the settlement initiatives in the Amazon region (Mahar 1989) and the desire to hoist the cerrado economy led to the creation of the Program for the Development of the Cerrado (POLOCENTRO) in 1975. Its goals were to settle farmers in places with good farming potential, to improve secondary roads and electricity, and to develop agriculture research and technology. Farmers received subsidized loans and credit lines were at relatively low fixed interest rates and no monetary correction. Rising inflation and considerable grace periods effectively transformed these loans into donations (Mueller 2003).

This program had a major impact on cerrado agriculture. Medium and large farmers benefited most (Klink et al. 2005): 81% of the farms were 200 ha or larger in size and these accounted for 88% of the total funds allocated. Farms larger than 1,000 ha accounted for 39% of all projects and received more than 60% of the credit. The program's original target to give preference to foodstuffs had never been realized and instead it induced the expansion of commercial agriculture (cattle ranching and soybeans).

After 1985, many developmental programs, including the POLOCENTRO, were closed in Brazil. However, in 1979 the Brazil–Japan Cooperative Program for the Development of the Cerrado (PRODECER) was created. It selected experienced farmers from the south and southeast of Brazil to be settled in the cerrado. It was financed by loans from both the Brazilian and the Japanese governments, but contrary to the POLOCENTRO, loans were granted at real (not fixed) interest rates. PRODECER was ended in 2001.

New Technologies

The development of appropriate technologies to enable farmers to deal with the nutrient-poor, acidic soils of the cerrado has also promoted agricultural expansion. These have included application of phosphate fertilizer and lime to correct for soil nutrient deficiency and acidity, rhizobium-based nitrogen fixation, the development of crop varieties, heavy use of herbicides and pesticides, and use of modern machinery (Klink et al. 2005). The performance of the Embrapa-Cerrado Research Center has been impressive in almost every respect (Paterniani and Malavolta 1999). However, the technologies that were developed were strongly biased toward medium and large, capital-intensive farmers and cash crops, especially soybeans (Clay 2004; Kaimowitz and Smith 2001; Klink et al. 2005; Mueller 2003).

The use of modern machinery in agricultural operations decreased the use of manual labor. Between 1970 and 1985, the period of greatest agricultural expansion, employment grew 2.7% per year, whereas the farmed area expanded 5.4%, planted pastures grew 8.4%, the bovine herd increased 5.5%, and the stock of tractors increased 13.6% (Mueller 2003).

Minimum Price Policy

The price support policy in Brazil has been in effect since the 1930s, to guarantee a minimum price for agricultural products. In the 1980s, farm credit lines were restricted and were reduced or eliminated. Consequently, production costs increased substantially in the cerrado. The government then started to purchase large amounts of cerrado products, particularly soybeans, rice, and corn (Mueller 2003). This benefited farmers from remote areas in the cerrado, because they indirectly benefited from the fact that fuel had a unified price for the whole country. The net result was that it encouraged the expansion of commercial farming in areas that could not have supported profitable production without subsidies and consequently the deforestation of new land.

Expansion of Agroecosystems

Agricultural activities are not evenly distributed, with large intraregional differences. In the state of Mato Grosso do Sul, southern Mato Grosso, central and southern Goiás, the Federal District, the Triângulo Mineiro, and western Minas Gerais, where basic infrastructure is well developed and access to the more dynamic markets of the country has been established, modern farming activities are consolidated (Mueller 2003; Rezende 2003). This "modern" portion of the cerrado is responsible for most of the soybean, corn, coffee, cotton, and bean production; it produces a large share of the regional rice and cassava, as well as most of the bovine herd (Klink et al. 2005; Mueller 2003). In the remaining cerrado areas, agriculture is still developing and the road network and commercialization facilities are precarious, but as agriculture expands, further deforestation is expected. As the agricultural frontier expands into the cerrado, supply-oriented food processors follow. Grain and meat processors are built and processing capacity is increasing in the region. Consequently, the cerrado has the fastest growth of food industry in Brazil today (Azevedo et al. 2004; Vieira Filho et al. 2011).

Ranching varies from relatively modern and efficient farms to extensive operations using rudimentary methods and having low productivity. Modern techniques have been most readily adopted in the areas closest to markets where access to technical assistance, basic and support infrastructure, and relatively well-developed meat packing facilities is better (Mueller 2003). Transportation costs and processing and commercialization infrastructure have had an impact on both the size of the herd and the area planted. The total number of cattle in the cerrado is more than 48 million (Mueller 2003), representing 33% of the national herd. The increase in the number of bovines is a direct consequence of the increase in the area of planted pastures since the carrying capacity is usually low, less than one animal per hectare (Barcellos et al. 1999). Planting pastures by far the most important land use in the cerrado. To establish planted pastures, the savannas are clear-cut and burned and then seeded with grasses of African origin. Legumes are used as a source of protein (Barcellos et al. 1999; Sano et al. 1999).

LAND USE EFFECTS ON BIODIVERSITY AND ECOSYSTEMS

Transformation of cerrado landscapes continues at a fast pace. The total area and the annual rate of deforestation are greater in the cerrado than in the Amazon forest. Using MODIS satellite data, a recent survey estimates that 55% of the cerrado has been deforested, an area equivalent to half the size of Mexico (or 1,000,000 km^2). The annual rate of native vegetation clearing is greater in the cerrado than in the Brazilian Amazon forest. Between 1970 and 1975, the average annual rate of land clearing in the Cerrado was 40,000 km^2 per year, 1.8 times the estimated deforestation rate of the Amazon forest during the period 1978–1988 (Klink and Machado 2005). Current estimates (Machado et al. 2004) are that cerrado deforestation ranges between 22,000 and 30,000 km^2 per year, higher than the historical deforestation rates for the Amazonian forest (18,000 km^2 per year).

One hundred and thirty-seven animal species are threatened with extinction (Klink and Machado 2005). In Brasilia, one plant species is already extinct and 30 others are threatened due to intensive harvesting of native species of ornamental plants extracted from native vegetation (Marsicano 2001).

Extensive cultivation of African grasses is increasing their proliferation and where they persist they become detrimental to water regimes, fire cycles, and biodiversity (Miranda et al. 2002; Pivello et al. 1999). For instance, the African grass *Melinis minutiflora* (molassa or fat grass) is extensively found in disturbed areas, roadsides, abandoned pasturelands, and nature reserves (Pivello et al. 1999). This grass can produce high biomass that becomes a highly combustible fuel at the end of the dry season, initiating a grass-fire interaction capable of preventing the regrowth of natural vegetation. Given their current scale, African grasses are major agents of change in the cerrado (Klink and Machado 2005).

Agricultural expansion has led to an increase in burning, and areas still covered with natural vegetation are now burned almost every year (Klink et al. 2005). More frequent fire has a negative effect on tree and shrub establishment. Based on Landsat satellite data, it is estimated that between 20,000 and 26,000 km^2 of natural cerrado was burned during June–July of 1999. In 2000, 67% of the area that burned in Brazil was in the cerrado (Tansey et al. 2004). Burning also leads to the release of carbon dioxide and other greenhouse gases to the atmosphere (Krug et al. 2002).

Soil mechanization of large tracts of land exposes great areas of bare soil, causing erosion and compaction. Land that is cleared but is not used increases erosion (45,000 km^2 of cleared but not used land exists in Cerrado) (Mueller 2003). Soil losses under conventional soybean crop are 5 tons/ha/year and under bare fallow can be as high as 130 tons/ha annually (Goedert 1990; Rodrigues et al. 2001).

Cerrado cultivation requires a substantial use of fertilizers and lime to correct soil fertility and acidity. Erosion destroys not only the soil but also the "constructed fertility," which creates economic waste (Mueller 2003). This also causes pollution of streams and rivers. Inputs such as pesticides and herbicides are widely used, especially in soybean plantations and planted pastures to control insects and plant and animal diseases and to combat invading weeds (Mueller 2003). The short-term efficiency of these biocides has increased their use substantially, with negative impacts in terms of bioaccumulation, persistency, and toxicity to various organisms (Vieira et al. 2001).

Concern regarding water conservation for agricultural use has been increasing because irrigation is expanding. Irrigated agriculture is still limited if compared to the total planted area in the cerrado, but its expansion has been fast and disorderly. Usually riparian forests are removed (which is illegal according to Brazil's forest code), causing damage to freshwater supplies to some of the cities in the region (Mueller 2003). Modification of the hydrological cycle at the basin scale has been reported in a major tropical river (the Tocantins River). By 1998, 49% of the basin had been used as cropland and pastures, and this increased river discharge by 24% (Costa et al. 2003).

Many cerrado trees and shrubs are deep-rooted and take up water deep in the soil (Oliveira et al. 2005). These species rely on deep-stored soil water during the dry season and also during the wet season, when rainfall is below average (Oliveira et al. 2005). The ability of cerrado trees to tap water stored deep in the soil during the dry season may be critical to the maintenance of the hydrological cycle at the regional scale. Field studies, satellite observations, and modeling simulations of the effects of the conversion of natural cerrado into planted pastures (that rely on shallow water) have shown that precipitation may be reduced by up to 10%, that dry spells may become more frequent, and that mean surface air temperature may increase by 0.5°C (Hoffmann and Jackson 2000; Oliveira et al. 2005). These modifications of regional climate may have profound effects on crop production.

Considerable amounts of carbon is stored in roots and soil organic matter in the cerrado (up to 70% of cerrado biomass is underground) (Aduan 2003; Castro and Kauffmann 1998; Davidson et al. 1995, 2012; Grace et al. 2006). The size of soil carbon stocks may be as large as that of the Amazon forest on an area basis (Aduan 2003). Due to the level of degradation of cultivated pastures, soil carbon accumulation in these degraded ecosystems does not contribute to the sink of atmospheric CO_2 (Silva et al. 2004). Given the extension of cerrado vegetation already transformed into planted pastures, a significant change in the regional carbon stocks has possibly occurred (Galford et al. 2011).

CHANGE OF POLICY FRAMEWORK

In the past, land-use policies were often formulated with little attention to their implications for land degradation, social conflict, and biodiversity conservation. Despite advances in biodiversity conservation in Brazil, priority has been given to the Amazon forest. The cerrado has not received the same attention and therefore is underrepresented in the National System of Conservation Units

(SNUC): only 2.2% of the surface area is under legal conservation (Klink and Machado 2005). In addition, while Brazil's forest code establishes that 80% of the Amazon forest should be preserved as legal reserve, it establishes a cap of 35% in the cerrado.

For the implementation of the Convention on Biological Diversity (CBD), in 1994 Brazil established a general framework program (PRONABIO). In 2002, Brazil issued the main principles and guidelines for a national policy for biodiversity (PNB), and in 2003 PRONABIO's scope was reviewed to be in agreement with PNB. Thematic components were added (scientific and traditional knowledge acquisition, conservation, sustainable use, monitoring and mitigation of impacts, genetic resources, capacity building and environmental education, and institutional development) and a biogeographic approach was established to represent the Brazilian biomes.

A major achievement was the creation of the national biodiversity commission (CONABIO) that is ultimately responsible for biodiversity policies in the country. CONABIO has representatives of federal and state levels of government, nongovernment organizations (NGOs), academia, indigenous populations, and the private sector (from both agriculture and industry). Following CONABIO guidelines, technical and financial activities at the project level are carried out by PROBIO, a World Bank-GEF-Federal Government consortium.

A highly successful strategy financed by PROBIO has been the priority-setting workshops for all Brazilian biomes. The cerrado and Pantanal workshop of 1998 compiled the existing knowledge on cerrado biodiversity, conservation efforts, and socioeconomic drivers. This knowledge was integrated in order to produce a list of species and maps of endangered species, endemic species, remaining vegetation, protected areas, land use, and "zoning." A total of 87 priority areas for Cerrado conservation were selected based on biological importance (distribution of endemic, rare, threatened, or migratory species; richness of biological communities), economic and cultural values, and landscape features (such as climate, soil, geomorphology, and rivers) (PROBIO-MMA 2013).

Human pressures were considered at the local level: demography, agriculture and cattle expansion, and urban and industry expansion. Estimates of land cover by analysis of satellite imagery gave rise to conservation opportunities: regions were identified where heavy human impact had already occurred (where between 50% and 92% of the cerrado landscape had already been heavily modified) (Cavalcanti and Joly 2002; Diniz Filho et al. 2009; Silva et al. 2006).

The World Bank has recognized that little attention has been given to cerrado conservation. It proposes that ecological and economic zoning of the cerrado is a priority and should be carried out through a biome-wide project similar to the Amazon Region Protected Areas program for the Amazon (Margulis 2003). International support for such a proposal is likely. For instance, the three global-level conservation NGOs (Conservation International, The Nature Conservancy, and World Wildlife Fund [WWF]) have offices in Brasília, and have specific programs and projects for cerrado conservation.

Combining ecological and economic zoning with institutional strengthening would be a way to promote sustainable natural resource management of cerrado. Based on the existing knowledge, many researchers and policy makers have suggested that major zoning categories are possible for cerrado: areas used preferentially for crops, areas for crops and cattle raising, areas for crops and afforestation, and areas for conservation, particularly those that contain fragile ecosystems (Goedert 1990).

Civil society and local NGOs have had experience with the sustainable use and conservation of the cerrado. Since 1992, the "Rede Cerrado" network of NGOs has been promoting sustainable use of the cerrado's natural resources. The "Rede" also promotes local workshops (use of medicinal plants and ecotourism, for instance), and its best-known activity is its annual fair ("Feira dos Povos do Cerrado"), which usually takes place in Goiânia, the capital of Goiás State. In 2003, the "Rede" released a white paper (Rede Cerrado 2003) to the newly elected Lula government, which influenced the ministry of environment (MMA)'s decision to create the Cerrado Working Group, a collegiate that had the participation of NGOs, federal and state government representatives, academia, rural workers, farmers, indigenous organizations, and the rural Afro-Brazilian community ("quilombolas"). In 2004, the MMA created the Sustainable Cerrado program. Its main goal is to integrate institutional and governmental actions toward cerrado conservation and sustainable use, and its membership is similar to that of the working group.

The current trend is to coordinate cerrado policy making more with macropolicies, aiming not only at conservation but also at sustainable use and climate change mitigation. As part of this trend, we can cite Brazil's recently approved Forest Investment Program (FIP) and policies aimed at climate change mitigation.

The FIP is a targeted program to catalyze policies and measures and mobilize significantly increased funds to facilitate the reduction of deforestation and forest degradation and to promote improved sustainable management of forests, leading to emissions reductions and the protection of carbon stocks. FIP is a program of the Strategic Climate Fund (SCF), which is funded by country donors and coordinated by the multilateral banks, to finance new development approaches and scale up activities aimed at a specific climate change challenge or sector response through targeted programs (http://www.climateinvestmentfunds.org/cif).

The Brazilian investment plan for FIP (which is under final elaboration) focuses on land-use change in the cerrado. Areas to be considered by Brazil's FIP include improvement of the environmental register of farms, sustainable agriculture in already deforested cerrado (restoration of degraded lands and low-carbon agriculture), improvement of forestry and natural resources information systems for management by public and private sectors, and the creation of monitoring systems (deforestation and fire alert systems). The implementing agency of FIP is the MMA in partnership with a variety of Brazilian institutions (mainly the ministry of finance, ministry of science, ministry of agriculture, Brazilian Agricultural Research Corporation (Embrapa), and the Brazilian Forest Services), with financial and technical support from the World Bank Group and the Inter-American Development Bank. Funding from FIP should amount to US$70 million and additional funding from public and other sources is expected.

Brazil's FIP should support and leverage other Brazilian programs toward sustainable land use. By enhancing sustainable agriculture in already deforested and degraded lands, it should help reduce emissions of greenhouses gases and increase carbon sequestration by adoption of best agricultural practices, improve the supply chain, and increase farmers' access to finance. Particularly important is the support of FIP for the much larger low-carbon agriculture program (ABC Program).

ABC is a ministry of agriculture program that started in June 2010; its goal is to promote food and biofuel production that at the same time reduces emissions of greenhouse gases, through the use of modern practices and new technologies, and increases in efficiency. It has a budget of US$1.7 billion per year to farmers and cooperatives. As part of the national policy on climate change and one of Brazil's nationally appropriate mitigation actions (NAMAs), ABC has ambitious goals. By 2020, it aims to add 8 million ha of no-till agriculture, restore 15 million ha of degraded pastureland, create 4 million ha of livestock-crops-forestry systems, add 3 million ha of commercial forests, increase biological fixation of nitrogen for 5.5 million ha, and use 4.4 million m^3 of waste for energy generation. US$2 billion has been invested and 15,000 contracts have been signed (http://www.agricultura.gov.br/politica-agricola).

Another important area of support by FIP is the spread of the use of Rural Environmental Cadaster (CAR). Under the new, recently approved version of Brazil's Forest Code, every farm in Brazil will have to register its location and its compliance level with the forest code using geographical information system technologies. No reliable statistics are available, but some estimates show that there are already about 60 million ha of farmland in the CAR registry in the states of Mato Grosso and Pará alone (Ana Cristina Barros, personal communication).

CAR methodology will bring Brazilian agriculture up to a new standard toward sustainable production because it demands a landscape approach for land use in Brazil. It will allow governments (both at the federal and state levels) to monitor land use and also to develop better planning and policy. CAR will allow improved collaboration between farmers, associations, NGOs, agribusiness, financial institutions, and governments toward better compliance, forest and savanna restoration, and ultimately ecosystem conservation. Several cases in the states of Mato Grosso, Bahia, and Pará have proved that this is doable for it allows reduction of costs of compliance, establishment of scientific criteria for selection and allocation of private reserves, creation of realistic strategies for forest restoration, development of information technology for mapping and monitoring vegetation, identification of economic incentives to farmers that adopt the CAR methodology, and consolidation of crucial dialog between farmers, agribusiness, banks, NGOs, and the public sector (Klink 2009; Klink et al. 2005, 2008; Schneider 2007).

The action plan for prevention and control of deforestation and forest fires in the cerrado (PPCerrado) was launched in 2010 and is also one of Brazil's NAMAs and an instrument of the national policy on climate change. The national policy aims to reduce emissions of greenhouse gases, on a voluntary basis, by 36%–39% of Brazil's projected emissions by 2020 (http://www.mma.gov.br/clima). A 40% reduction of emissions from cerrado is expected by 2020.

The PPCerrado is coordinated by the chief of staff with the collaboration of the MMA and 17 other ministries. It combines actions on land titling, protected areas, monitoring and control of deforestation, and sustainable economic activities. It is still in its infancy but policy coordination on monitoring deforestation, and an ecological–economic zoning (to be finalized by 2013) between ministries have already been launched. The PPCerrado plan will also build upon the activities planned by the FIP program and the future developments of CAR methodology.

Most importantly, better coordination between policies improves their effectiveness, and concrete results have already been seen in both the cerrado and the Amazon (Assunção et al. 2012; Brannstrom et al. 2012; Macedo et al. 2012).

PROMOTING CERRADO SUSTAINABILITY IN THE FUTURE

Restoration of Degraded Lands for Business Development

The expansion and modernization of cerrado agriculture have generated positive socioeconomic impacts such as the contribution to Brazil's supply of agricultural products both for domestic use and for export, and the generation of income. The economic activities associated with the expansion of agriculture, the incorporation of modern technology, and the resulting gains in productivity have brought diversification of local economies, an increase of agricultural and nonagricultural incomes and municipal revenues, and improvements in welfare in a few cerrado localities, where growth and adoption of new technology have been significant (Bonelli 2001).

One of the central environmental and economic issues of our time is how to reconcile the often conflicting demands of production and conservation (Cleary et al. 2011; *The Economist* 2010). Brazil has more than enough cleared land to accommodate expansion of crops, ranching, and biofuels under a variety of economic scenarios likely to prevail over the next few years. Most of this cleared land is degraded pasture, and most of this pasture has been cleared in the cerrado (Cleary et al. 2011; Ferreira et al. 2011; ICONE 2011; Lima et al. 2012; Sano et al. 2010). If expansion is channeled into these cleared areas rather than natural forest and savanna habitats, the impacts on biodiversity and climate of agricultural expansion will be minimized (Cleary et al. 2011; Malhado et al. 2010).

According to estimates made by Embrapa and the MMA, of the cerrado's original 204 million ha (roughly 22% of Brazil's area), 98 million ha have already been deforested and are mostly under crops (21.6 million ha) and cultivated pastureland (54 million ha). At least 50% of this pastureland area is in a completely or partially degraded state. According to the World Resources Institute (WRI), more than 1 billion ha of land are degraded globally, an area bigger than Brazil, which offers an enormous restoration opportunity for both agribusiness development and conservation. The real extension, location, and even a clear definition of degraded land are nonexistent for the cerrado, which impedes a more efficient use of the land stock available for crop expansion.

Transformation of the cerrado continues at a fast pace, but today there is an overall awareness (academia, government, NGOs, and some private entities and companies) that agricultural expansion should no longer follow its historical trend. The focus should be on increasing restoration of already degraded lands (particularly degraded planted pasture), therefore limiting the expansion of agriculture into regions with high conservation value.

Mapping the magnitude and location of degraded land is a necessary first step to quantify how much underperforming lands add to the supply of agricultural land. This and other information will be used to develop a business case for the restoration of degraded land as a viable option for the expansion of cash crops (mainly soybeans), avoiding further deforestation, particularly in areas of high conservation value. It should be noted that such an effort will also benefit other private sector enterprises in the cerrado, mainly expansion of planted forests for the purpose of energy production (for the pig iron industry), and for pulp and paper production. This mapping effort would be the basis for the development of a business case for incorporating degraded land into the productive agrisystem of Brazil, therefore avoiding further deforestation of new cerrado land.

Applying Knowledge of Trade-Offs in Land-Use Decisions

Trade-offs between development and conservation have not been explicitly recognized in the cerrado, where a range of political, economic, and social interests are usually conflicting. A lack of knowledge and the existence of uncertainties about drivers of deforestation have delayed conservation and management. Overall, stakeholders and institutions suggest that cerrado conservation is a high priority given the high level of habitat and landscape modifications and degradation that have already occurred.

The current massive transformation in the cerrado from native vegetation to cattle ranching and croplands is among the most rapid in the world (Lepers et al. 2005), and the resulting agricultural production is of major importance to Brazil and the rest of the world. However, while these transformations increase production for human use, they also alter a range of other ecosystem services provided by cerrado, including carbon sequestration, maintenance of biodiversity, and climate regulation. The long-term consequences for these ecosystem functions are not well quantified.

Land use is essentially a trade-off between meeting various long- and short-term human needs through appropriation of productivity and ecosystem function (Millennium Ecosystem Assessment 2003; DeFries and Rozensweig 2010; DeFries et al. 2004a; Foley et al. 2005, 2012). Ultimately, land-use decisions should incorporate society's assessment of whether the immediate benefits of habitat transformation outweigh the longer-term losses to ecosystem services that result from such actions. The ability to make these judgments depends on a thorough understanding of the ecosystem consequences. Unfortunately, these consequences are generally not well quantified in the cerrado. To make land-use decisions that account for these consequences and that are not just driven by immediate returns, we need a scientific underpinning that quantifies and communicates the trade-offs, based on several premises (DeFries et al. 2005):

1. Land use is inherently a trade-off between meeting basic human needs and ecosystem consequences. The positive and immediate benefits (i.e., increased yields, export revenue) gained from appropriating ecosystem goods for human consumption, such as beef and grain, must be weighed against the consequences for ecosystem function and, ultimately, the impacts of changes in ecosystem function on society. The role of the conservation scientist is to provide the ecological knowledge to identify, quantify, and communicate these trade-offs, recognizing that land-use decisions reflect societal values about the appropriate trade-offs.
2. A prerequisite to assessing the trade-offs inherent in land-use decisions is quantitative analysis of both ecosystem goods and the multiple ecosystem responses, including biodiversity,

watershed protection, climate regulation, carbon sequestration, and maintenance of human health. An explicit definition of time and space scales is critical for communicating trade-offs with land-use decision makers.
3. Ecosystems are likely to respond nonlinearly to land use. These responses offer the opportunity to identify land-use alternatives that allow small losses in land-use options for large gains in ecosystem function. Nonlinear responses can also, however, bring about the opposite, negative outcome, where small gains in satisfying human needs result in large losses in ecosystem function, such as land degradation and abandonment.

Some of the ecosystem consequences are well understood: for example, the high species endemism and heterogeneity found in the cerrado has meant that land transformations have major negative consequences for species diversity. Clearing for cropland might be more detrimental than clearing for pasture if clearings are larger and fragment sizes are smaller. Other ecosystem consequences are potentially important but not well understood in the context of the cerrado, such as the feedbacks to local and regional climate through altered energy, momentum, and water fluxes with the atmosphere (Bonan et al. 2004).

The cerrado ecosystem is a complex, dynamic, and heterogeneous ecosystem with interactions between plants, animals, soils, climate, and many other components (Oliveira and Marquis 2002). It is unlikely that the ecosystem will respond linearly to land-use transformations, for example, that the number of species lost will be in direct proportion to the area transformed into cropland, or that loss of carbon from pasture conversion will be constant through time (Davidson et al. 1995). These nonlinearities in ecosystem response offer opportunities to identify "small loss–big gain" solutions in which ecological functioning can be maintained ("big gain") with minimum negative constraints for human land use ("small loss") (DeFries et al. 2004a,b). Such is the case, for example, in the well-known approach of identifying "hot spots" of biodiversity for conservation (Myers et al. 2000) or biodiversity corridors essential for migrations of large mammals. If conserved, these areas provide disproportionately high benefits to ecosystem function, while minimizing constraints on the area available for human land use.

Recent analytic work has shown that where environmental assets are a source of economically important services, environmental protection may stimulate growth instead of reducing it (DeFries and Rozensweig 2010). This analytic work is vital for future cerrado sustainability, where ecologists and decision makers are only beginning to consider what combination of policies and incentives is necessary to balance human needs and ecosystem function.

REFERENCES

Aduan, R.E. 2003. Respiração de solos e ciclagem de carbono em Cerrado nativo e pastagem no Brasil central. PhD Dissertation, Department of Ecology. University of Brasília, Brazil.

Assunção, J., Gandour, C.C. & Rocha, R. 2012. A queda do desmatamento na Amazônia Brasileira: Preços ou políticas? Sumário executivo. Climate Policy Initiative Rio de Janeiro & Núcleo de Avaliação de Políticas Climáticas, Pontifícia Universidade Católica do Rio de Janeiro.

Azevedo, P.F., Chaddad, F.R. & Farina, E.M.M. 2004. The food industry in Brazil and the United States: The effects of the FTAA on trade and investment. Intal ITD Working Paper, SITI 07.

Barcellos, A.O., Vianna Filho, A., Balbino, L.C., Oliveira, I.P. & Yokoyama, L.P. 1999. Restabelecimento da capacidade produtiva e desempenho animal em pastagens renovadas na Região do Cerrado. Embrapa Cerrados, Planaltina 4p. Comunicado Técnico, Embrapa Cerrados 22.

Bonan, G.B., DeFries, R., Coe, M. & Ojima, D. 2004. Land use and climate. In G. Gutman, J. Janetos, C.O. Justice, E.F. Moran, J. Mustard, R. Rindfuss, D.L. Skole, B.L. Turner & M.A. Cochrane (eds), *Land Change Science: Observing, Monitoring and Understanding Trajectories of Change on the Earth's Surface*, pp. 301–314. Dordrecht: Springer.

Bonelli, R. 2001. Impactos econômicos e sociais de longo prazo da expansão agropecuária no Brasil: revolução invisível e inclusão social. IPEA, Textos para Discussão 838. Rio de Janeiro.

Borlaug, N.E. 2002. Feeding a world of 10 billion people: The miracle ahead. In R. Bailey (ed.), *Global Warming and Other Eco-Myths*, pp. 29–60. Roseville, CA: Competitive Enterprise Institute, Prima Publishing.

Brannstrom, C., Rausch, L., Brown, J.C., de Andrade, R.M.T. & Miccolis, A. 2012. Compliance and market exclusion in Brazilian agriculture: Analysis and implications for "soft" governance. *Land Use Policy* 29: 357–366.

Carvalho, F.M.V., de Marco Jr, P. & Ferreira, L.G. 2009. The Cerrado into-pieces: Habitat fragmentation as a function of landscape use in the savannas of central Brazil. *Biological Conservation* 142: 1392–1403.

Castro, E.A. & Kauffmann, J.B. 1998. Ecosystem structure in the Brazilian Cerrado: A vegetation gradient of aboveground biomass, root biomass and consumption by fire. *Journal of Tropical Ecology* 14: 263–283.

Cavalcanti, R.B. & Joly, C.A. 2002. Biodiversity and conservation priorities in the Cerrado region. In P.S. Oliveira & R.J. Marquis (eds), *The Cerrados of Brazil: Ecology and Natural History of a Neotropical Savanna*, pp. 351–368. New York: Columbia University Press.

Clay, J. 2004. *World Agriculture and the Environment: A Commodity-By-Commodity Guide to Impacts and Practices*. Washington, DC: WWF and Island Press.

Cleary, D., Fry, J., Todd, M., Klink, C. & Barros, A.C. 2011. An opportunity for Brazil: Minimizing the environmental costs of biofuels expansion. The Nature Conservancy and LMC International report. Brasília, Brazil.

Costa, M.H., Botta, A. & Cardille, J. 2003. Effects of large-scale changes in land cover on the discharge of the Tocantins River, Southeastern Amazonia. *Journal of Hydrology* 283: 206–217.

Davidson, E.A., Nepstad, D.C., Klink, C.A. & Trumbore S.E. 1995. Pasture soils as carbon sink. *Nature* 376: 472–473.

Davidson, E.A., de Araújo, A.C., Artaxo, P., Balch, J.K., Brown, I.F., Bustamante, M.M.C., Coe, M.T., et al. 2012. The Amazon basin in transition. *Nature* 481: 321–328.

DeFries, R. & Rozensweig, C. 2010. Toward a whole-landscape approach for sustainable land use in the tropics. *Proceedings of the National Academy of Science* 107: 19627–19632.

DeFries, R., Foley, J. & Asner, G.P. 2004a. Land use choices: Balancing human needs and ecosystem function. *Frontiers in Ecology and the Environment* 2: 249–257.

DeFries, R., Asner, G.P. & Houghton, R.A. 2004b. Trade-offs in land-use decisions: Towards a framework for assessing multiple ecosystem responses to land use change. In R. DeFries, G.P. Asner & R.A. Houghton (eds), *Ecosystems and Land Use Change*, Vol. 153, pp. 1–12. Washington, DC: American Geophysical Union.

DeFries, R., Klink, C. & Cavalcanti, R.B. 2005. Land-use choices: Balancing human needs and ecosystem function in the Cerrado Region. In C.A. Klink, R. DeFries & R.B. Cavalcanti (eds), *Symposium Cerrado Land Use and Conservation: Assessing Trade-offs between Human and Ecological Needs*. Working Papers for the XIX Annual Meeting of the Society for Conservation Biology, University of Brasilia and Conservation International, Brasilia, Brazil, July 2005.

Diniz, I.R., Marinho Filho, J., Machado, R.B. & Cavalcanti, R.B. (organizadores) 2010. Cerrado. Conhecimento científico quantitativo como subsídio para ações de conservação. Universidade de Brasília & Thesaurus Editora, Brasília.

Diniz Filho, J.A.F., Bini, L.M., Pinto, M.P., Rangel, T.F.L.V.B., Carvalho, P. & Bastos, R.P. 2006. Anuran species richness, complementarity and conservation conflicts in Brazilian Cerrado. *Acta Oecologica* 29: 9–15.

Diniz Filho, J.A.F., Oliveira, G., Lobo, F., Ferreira, L.G., Bini, L.M. & Rangel, T.F.L.V.B. 2009. Agriculture, habitat loss and spatial patterns of human occupation in a biodiversity hotspot. *Scientia Agricola (Piracicaba, Brazil)* 66: 764–771.

Embrapa 1995. *Prospects for the Rational Use of the Brazilian Cerrado for Food Production*. Brasília: Embrapa.

Ferreira, M.E., Silva, J.R., Rocha, G.F., Antoniazi, L., Nassar, A. & Rocha, J.C.S. 2011. Caracterização das áreas desmatadas no bioma Cerrado via sensoriamento remoto: uma análise sobre a expansão de culturas agrícolas e pastagens cultivadas. In *Anais XV simpósio Brasileiro de sensoriamento remoto*—SBSR, INPE, Curitiba, PR, Brazil, pp. 6727–6733.

Foley, J., DeFries, R., Asner, G.P., Barford, C.G., Bonan, G.B., Carpenter, S.R., Chapin, F.S.I., et al. 2005. Global consequences of land use. *Science* 309: 570–574.

Foley, J.A., Ramankutty, N., Brauman, K.A., Cassidy, E.S., Gerber, J.S., Johnston, M., Mueller, N.D., et al. 2012. Solutions for a cultivated planet. *Nature* 478: 337–342.

Galford, G.L., Melillo, J.M., Kicklighter, D.W., Mustard, J.F., Cronin, T.W., Cerri, C.E.P. & Cerri, C.C. 2011. Historical carbon emissions and uptake from the agricultural frontier of the Brazilian Amazon. *Ecological Applications* 21: 750–763.

Garrett, R.D., Lambin, E.F. & Naylor, R.L. 2012. Land institutions and supply chain configurations as determinants of soybean planted area and yields in Brazil. *Land Use Policy* 31: 385–396.

Goedert, W.J. 1990. Estratégias de manejo das savannas. In G. Sarmiento (ed.), *Las sabanas Americanas: Aspectos de su biogeografia, ecologia y utilizacion*, pp. 191–218. Guanara: Acta Cientifica Venezolana.

Grace, J., San José, J., Meir, P., Miranda, H.S. & Montes, R.S. 2006. Productivity and carbon fluxes of tropical savannas. *Journal of Biogeography* 33: 387–400.

Haridasan, M. 1982. Aluminum accumulation by some Cerrado native species in central Brazil. *Plant and Soil* 65: 265–273.

Helfand, S.M. & Rezende, G.C. 2004. The impact of sector-specific and economy-wide policy reforms on the agricultural sector in Brazil: 1980–98. *Contemporary Economic Policy* 22: 194–212.

Herrera-Estrella, L. 1999. Transgenic plants for tropical regions: Some considerations about their development and their transference to the small farmer. *Proceedings of the National Academy of Science* 96: 5978–5981.

Hoffmann, W.A. & Jackson, R.B. 2000. Vegetation-climate feedbacks in the conversion of tropical savanna to grassland. *Journal of Climate* 13: 1593–1602.

Hogan, D.J., da Cunha, J.M.P. & do Carmos, R.L. 2001. Socio-environmental consequences of land cover change in Brazil's center west region. Paper presented at the 2001 Open Meeting of the Human Dimensions of Global Environmental Change Research Community, Rio de Janeiro, Brazil, October 6–8.

Instituto de Estudos do Comércio e Negociações Internacionais (ICONE) 2011. Simulating land use and agriculture expansion in Brazil: Food, energy, agro-industrial and environmental impacts. BIOEN Program, Project 2008/56156-0. Final Science Report, São Paulo, Brazil, February 2011.

Kaimowitz, D. & Smith, J. 2001. Soybean technology and the loss of natural vegetation in Brazil and Bolivia. In A. Angelsen & D. Kaimowitz (eds), *Agricultural Technologies and Tropical Deforestation*, pp. 195–212. Oxon: CABI Publishing and CIFOR.

Klink, C.A. 2009. Minimizando os custos ambientais da expansão da soja. In Fundação Mato Grosso (ed.), *Boletim Pesquisa de Soja*, no. 13, pp. 28–32. Rondonópolis, Mato Grosso.

Klink, A.C. & Moreira, A.G. 2002. Past and current human occupation, and land use. In P.S. Oliveira & R.J. Marquis (eds), *The Cerrados of Brazil*, pp. 69–88. New York: Columbia University Press.

Klink, C.A. & Machado R. 2005. Conservation of the Brazilian Cerrado. *Conservation Biology* 19: 707–713.

Klink, C.A., DeFries, R. & Cavalcanti, R.B. (eds) 2005. Symposium Cerrado land use and conservation: Assessing trade-offs between human and ecological needs. Working Papers for the XIX Annual Meeting of the Society for Conservation Biology, University of Brasilia and Conservation International, Brasilia, Brazil, July 2005, (CD Rom).

Klink, C.A., dos Santos, H.G., Campari Jr, J.S., Matsumoto, M.H., de Freitas, G.K. & Baumgarten, L. 2008. Conservação dos recursos naturais em terras privadas. O papel das reservas legais no arranjo funcional das paisagens produtivas do bioma Cerrado. In S.M. Sano, S.P. de Almeida & J.F. Ribeiro (eds), *Cerrado: Ecologia e flora*, Vol. 1, pp. 399–406. Brasília: Embrapa Cerrados.

Krug, T., Figueiredo, H., Sano, E., Almeida, C., Santos, J., Miranda, H.S., Sato, N. & Andrade, S. 2002. Emissões de gases de efeito estufa da queima de biomassa no Cerrado não-antrópico utilizando dados orbitais. Primeiro Inventário Brasileiro de Emissões Antrópicas de Gases de Efeito Estufa—Relatórios de Referência. Ministry of Science and Technology, Brasília.

Lepers, E., Lambin, E.F., Janetos, A.C., DeFries, R., Achard, F., Ramankutty, N. & Scholes, R.J. 2005. A synthesis of rapid land-cover change information for the 1981–2000 period. *BioScience* 55(2): 115–124.

Lima, R.C.A., Nassar, A., Harfuch, L., Chiodi, L., Antoniazzi, L. & Moreira, M. 2012. Agricultura de baixo impacto: Construindo a economia verde brasileira. Instituto de Estudos do Comércio e Negociações Internacionais (ICONE), São Paulo, Brazil, June 2012.

Macedo, M.N., DeFries, R.S., Morton, D.C., Stickler, C.M., Galford, G.L. & Shimabukuro, Y.E. 2012. Decoupling of deforestation and soy production in the southern Amazon during the late 2000s. *Proceedings of the National Academy of Science* 109(4): 1341–1346.

Machado, R.B., Ramos Neto, M.B., Pereira, P.G.P., Caldas, E.F., Gonçalves, D.A., Santos, N.S., Tabor, K. & Steininger, M. 2004. Estimativas de perda da área do Cerrado Brasileiro. Conservation International, Brasilia.

Mahar, D.J. 1989. Government policies and deforestation in Brazil's Amazon region. Environment Department Working Paper 7. Washington, DC: The World Bank.

Malhado, A.C.M., Pires, G.F. & Costa, M.H. 2010. Cerrado conservation is essential to protect the Amazon rainforest. *AMBIO* 39: 580–584.

Margulis, S. 2003. *Causes of Deforestation of the Brazilian Amazon*. Washington, DC: The World Bank.

Marsicano, K. 2001. Flor do Cerrado em extinção no DF. *Correio Braziliense*, 4 de fevereiro, p. 16.

Mendonça, R.C., Felfili, J.M., Walter, B.M.T., Silva Jr, M.C., Rezende, A.V., Filgueiras, T.S., Nogueira, P.E. & Fagg, C.W. 2008. Flora vascular do bioma Cerrado. Checklist com 12.356 espécies. In S.M. Sano, S.P. de Almeida & J.F. Ribeiro (eds), *Cerrado. Ecologia e flora*, Vol. 2, pp. 421–442. Brasília: Embrapa Cerrados.

Millennium Ecosystem Assessment 2003. *Ecosystems and Human Well-Being: A Framework for Assessment.* Washington, DC: Island Press.

Miranda, H.S., Bustamante, M. & Miranda, A.C. 2002. The fire factor. In P.S. Oliveira & R.J. Marquis (eds), *The Cerrados of Brazil*, pp. 51–68. New York: Columbia University Press.

Mittermeier, R.A., Robles-Gil, P., Hoffmann, M., Pilgrim, J., Brooks, T. Mittermeier, C.G., Lamoreux, J. & Fonseca, G.A.B. (eds) 2004. *Hotspots Revisited.* Mexico City: CEMEX.

Mueller, C.C. 2003. Expansion and modernization of agriculture in the Cerrado; the case of soybeans in Brazil's Center-West. University of Brasilia, Department of Economics, Brasília, Textos para Discussão n. 306, November 2003.

Mueller, C.C. 2005. Impact of the recent soybean boom on the Cerrado of Brazil's center-west region. In C.A. Klink, R. DeFries & R.B. Cavalcanti (eds), *Symposium Cerrado Land Use and Conservation: Assessing Trade-offs between Human and Ecological Needs.* Working Papers for the XIX Annual Meeting of the Society for Conservation Biology, University of Brasilia and Conservation International, Brasilia, July 2005.

Mueller, C. & Martine, G. 1997. Modernização da agropecuária, emprego agrícola e êxodo rural no Brasil: A década de 1980. *Revista de Economia Política* 17: 85–104.

Myers, N., Mittermeier, R.A., Mittermeier, C.G., Fonseca, G.A.B. & Kent, J. 2000. Biodiversity hotspots for conservation priorities. *Nature* 403: 853–858.

Oliveira, P.S. & Marquis, R.J. (eds) 2002. The Cerrados of Brazil. *Ecology and Natural History of a Neotropical Savanna.* New York: Columbia University Press.

Oliveira, R.S., Bezerra, L., Davidson, E., Pinto, F., Klink, C.A., Nepstad, D. & Moreira A.G. 2005. Deep root function in soil water dynamics in Cerrado savannas of central Brazil. *Functional Ecology* 19: 574–581.

Pacheco, R. & Vasconcelos, H.L. 2012. Habitat diversity enhances ant diversity in a naturally heterogeneous Brazilian landscape. *Biodiversity Conservation* 21: 797–809.

Paterniani, E. & Malavolta, E. 1999. La conquista del "cerrado" en el Brasil: Victoria de la investigacion científica. *Interciencia* 24: 173–176.

Pivello, V., Carvalho, V., Lopes, P., Peccinini, A. & Rosso, S. 1999. Abundance and distribution of native and alien grasses in a Cerrado (Brazilian savanna) biological reserve. *Biotropica* 31: 72–82.

PROBIO-MMA. 2013. Projeto de avaliação e ações para a conservação da biodiversidade nos biomas Cerrado e Pantanal. Available at http://sistemas.mma.gov.br/sigepro/pub/display_bibliografias.php?%20vId_Gerenciamento=949&Destaque=7. Accessed 16 July 2013.

Rede Cerrado 2003. O Cerrado e a ausência de políticas públicas. Posições da rede Cerrado de ONGs. Position paper (manuscript). Brasília.

Rezende, G.C. 2003. Ocupação agrícola, estrutura agrária e mercado de trabalho rural no Cerrado: o papel do preço da terra, dos recursos naturais e das políticas públicas. In S. Hefland & G.C. Rezende (eds), *Região e espaço no desenvolvimento agrícola Brasileiro.* Rio de Janeiro: IPEA.

Ribeiro, J.F. & Walter, B.M.T. 2008. As principais fitofisionomias do bioma Cerrado. In S.M. Sano, S.P. de Almeida & J.F. Ribeiro (eds), *Cerrado. Ecologia e flora*, Vol. 1, pp. 151–212. Brasília: Embrapa Cerrados.

Rodrigues, W., Nogueira, J. & Imbroisi, D. 2001. Avaliação econômica da agricultura sustentável: O caso dos Cerrados brasileiros. *Cadernos de Ciência e Tecnologia* 18: 103–130.

Sano, E.A., Barcellos, A.O. & Bezerra, H.S. 1999. Área e distribuição espacial de pastagens cultivadas no Cerrado brasileiro. *Boletim de Pesquisa da Embrapa Cerrados* 3: 1–21.

Sano, E.E., Rosa, R., Brito, J.L.S. & Ferreira, L.G. 2010. Land cover mapping of the tropical savanna region in Brazil. *Environmental Monitoring and Assessment* 166: 113–124.

Schneider, R. 2007. Trading in legal reserve land to meet environmental objectives. Strategic Options for The Nature Conservancy. TNC Report, Brasilia.

Silva, J., Resck, D., Corazza, E. & Vivaldi, L. 2004. Carbon storage in clayey oxisol cultivated pastures in Cerrado region, Brazil. *Agriculture, Ecosystem and Environment* 103: 357–363.

Silva, J.F, Fariñas, M.R, Felfili, J.M. & Klink, C.A. 2006. Spatial heterogeneity, land use and conservation in the Cerrado region of Brazil. *Journal of Biogeography* 33: 536–548.

Simon, M.F., Gretherc, R., Queiroz, L.P., Skemae, C., Penningtone, R.T. & Hughes, C.E. 2009. Recent assembly of the Cerrado, a neotropical plant diversity hotspot, by in situ evolution of adaptations to fire. *Proceedings of the National Academy of Science* 106: 20359–20364.

Sparovek, G., Berndes, G., Klug, I.F. & Barreto, A. 2010. Brazilian agriculture and environmental legislation: Status and future challenges. *Environmental Science and Technology* 44: 6046–6053.

Tansey, K., Grégoire, J.M., Stroppiana, D., Sousa, A., Silva, J., Pereira, J.M.C., Boschetti, L., et al. 2004. Vegetation burning in the year 2000: Global burned area estimates from SPOT vegetation data. *Journal of Geophysical Research*. 109, D14S03.

The Economist 2010. Brazil's agricultural miracle. How to feed the world: The emerging conventional wisdom about world farming is gloomy. There is an alternative. 26 August 2010.

Tollefson, J. 2012. Outside Rio, agricultural scientists work to improve production—and protect the landscape. *Nature News Blog*, 17 June 2012.

Vieira, L.M., Galdino, S. & Padovani, C.R. 2001. Utilização de pesticidas na agropecuária dos municípios da Bacia do Alto Taquari de 1988 a 1996 e risco de contaminação do Pantanal, MS, Brazil, Embrapa.

Vieira Filho, J.E.R., Gasques, J.G. & de Sousa, A.G. 2011. Agricultura e crescimentos: Cenários e projeções. Brasília, Brazil: Instituto de Pesquisa Econômica e Aplicada (IPEA), Texto para Discussão 1642.

Werneck, F.P., Nogueira, C., Colli, G.R., Sites Jr, J.W. & Costa, G.C. 2012. Climatic stability in the Brazilian Cerrado: Implications for biogeographical connections of South American savannas, species richness and conservation in a biodiversity hotspot. *Journal of Biogeography* 39: 1695–1706.

Wheatley, J. 2010. Bumper harvest as professional farming takes hold. *Financial Times*, 28 June 2010.

WWF 2011. Soya and the Cerrado: Brazil's forgotten jewel. WWF UK Report, London.

22 Maned Wolf Ecology and Conservation

The Road Ahead

Claudio Sillero-Zubiri, Adriana Consorte-McCrea, and Eliana Ferraz Santos

The preceding chapters of this thorough volume provide a tour de force of what we know about the maned wolf and its ecology. But do we know enough to be able to effectively conserve this species and help resolve, or at least mitigate, the conflicts that they face with people and human enterprise? If not, what else must be studied? The captivating patterns of maned wolf biology and ecology, wonderfully illustrated with many details in this book, and the desire to know more, enable scientists to answer many questions, and ask even more. Of course the span of this research covers many fields, and not all are immediately applicable to conservation. We may want to reframe the question and ponder whether the answers in hand include those necessary to underpin conservation action.

While the scholarly pursuit of natural history is more important than generally credited as a force for canid conservation (Macdonald and Sillero-Zubiri, 2004), conservation is a pressing matter and therefore warrants a distinction between knowledge that is interesting and that which is also useful. In *Canids: Foxes, Wolves, Jackals and Dogs – Status Survey and Conservation Action Plan*, Sillero-Zubiri et al. (2004a) provided a measure of the answer for all wild canid species. Here we will focus on the maned wolf and will review some of the salient points of preceding chapters to extract the knowledge relevant to future conservation action.

For instance, while looking at the population viability throughout the species range, de Paula and Desbiez (Chapter 2) stressed the need for populations of several hundred maned wolves to guarantee long-term viability and avoid compromising their genetic diversity. When looking specifically at the situation in Brazil, however, they suggested that the model underestimated the increasing threats to maned wolf populations from projected agricultural expansion. Furthermore, they identified the main current threats to be mortality on roads and habitat loss, which result in a reduction of carrying capacity. However, they also warned of the potential impacts resulting from domestic dogs, hunting, and persecution. The latter impacts have not yet been fully assessed, but could potentially aggravate an already complicated situation and contribute to the onset of local extinctions.

The success of conservation actions will also depend on public support, understanding how people interact with the species in question, and mechanisms to foster coexistence. Looking at the attitudes of local people toward the maned wolf and its conservation, Consorte-McCrea (Chapter 3) encountered a promising and overall positive disposition. However, when taking into account the negative attitudes of minority groups, it became apparent that there was a need for clear and accurate information to dissociate the maned wolf's image from that of the big bad wolf (*Canis lupus*). Clarifying misconceptions about the feeding ecology of maned wolves and natural history traits (e.g., size, weight, population abundance, and distribution patterns) should contribute to dispelling the fears of some people toward these animals. Concomitantly, disseminating information about the species' conservation status and about the contribution of maned wolves to the lives of local people and the environment (e.g., by propagating cerrado seeds and controlling pests), and enlisting nationalistic feelings and local pride may contribute to increasing support toward their conservation.

To all intents and purposes, the maned wolf carries the necessary attributes to be a flagship species for the conservation of the cerrado, a recognized biodiversity hotspot (Myers et al., 2000). Cooperation between mass media, local credible sources and zoos, protected areas, and environmental education could be instrumental in fostering integration between natural resources and cultural values, promoting local people's self-esteem, and contributing to feelings of ownership over the maned wolf and the cerrado. Hand in hand with efforts to enhance people's perception of these animals are practical actions that can contribute to solving local disputes, such as the adaptation and application to the maned wolf scenario of methods of prevention, mitigation, and control of people–carnivore conflicts developed elsewhere (see Sillero-Zubiri et al., 2004b).

Many zoo collections around the world include maned wolves, and zoos have been more actively involved in maned wolf conservation than in most other wild canid species conservation, with the exception perhaps of gray wolves. The mere observation of these handsome animals in captivity has inspired many nature lovers and raised public awareness of their plea and their habitat. As some of the contributors in this volume suggest, zoos have contributed in several ways to maned wolf research and conservation, and there are specific areas of endeavor where they may continue to do so. For instance, research on maned wolves in zoos has enabled a better understanding of their biology, particularly that of their reproduction and health. According to Holland (Chapter 4), there are enough maned wolves in captivity to establish a healthy self-sustainable population, although exchange between the breeding programs in different regions could be improved. Bureaucracy and differences in regulations between the countries involved pose obstacles for greater integration of these collections, especially concerning transport and veterinary requirements. On the other hand, the international studbook of the maned wolf has served as a strong organizational tool, around which a genetically and demographically healthy captive population could be managed to supplement some wild populations in the future.

Many rescued maned wolves end up in zoo collections within the species' range. Together with other incidental recipients, either captured illegally or injured in various ways, these tend to augment existing captive populations. The pioneering management plan for maned wolves in Brazil (Paula et al., 2008) proposed the use of this *ex situ* population as a source of genetic stock to augment populations in the wild. Pessutti, Chierragatto, and Vanstreels (Chapter 5) call for more unified efforts to deliver this objective. Greater coordination would also help to intensify present efforts to promote the conservation of maned wolves and the cerrado, to engage newcomers, and to promote collaboration between *in situ* and *ex situ* professionals.

Molecular approaches have provided substantial contributions to the elucidation of the maned wolf's phylogeny (e.g., Wang et al., 2004). Da Fontoura-Rodrigues and Eizirik (Chapter 6) review the relevance of molecular genetics to maned wolf conservation and conclude that, despite its imperiled status, the species retains a significant level of diversity, indicating that the size of the overall population supports genetic health. They recommend studies of connectivity and population isolation at a global scale (i.e., including all regions and biomes within the species' distribution) to investigate intraspecific structure in view of increasing human impact on the species' habitat. Molecular tools coupled with noninvasive sampling are increasingly used in the study of canid behavior and ecology in general and maned wolves in particular (see Vynne, Chapter 12).

The last few decades have seen considerable advances in the study of many aspects of the maned wolf in both captive and wild populations. Motta-Junior, Queirolo, and de Arruda Bueno (Chapter 7) review current knowledge of the feeding ecology of maned wolves and suggest that future research should focus on detailed characteristics of their feeding habits in certain less-studied regions of the species' range, in both protected and nonprotected areas. They also recommend studies leading to a better understanding of the maned wolf's digestive physiology and bioenergetics; studies correlating population density and prey/fruit abundance; and long-term studies of functional responses to seasonal and annual fluctuations, and diet selectivity. Additionally, studies of the life history and behavior of prey species, and the nutritional properties of some food items such as wolf's fruit, would improve understanding of the maned wolf's relationships with food and its sources and inform some conservation management decisions.

Songsasen et al. (Chapter 8) write about maned wolf reproduction and factors affecting reproductive success, with most data resulting from studies in captivity. Advances in our understanding of maned wolf reproductive biology must guide research initiatives aimed at improving reproduction and health of *ex situ* populations, where these are not self-sustaining, including assisted breeding techniques and study of the reproductive cycle of female maned wolves. Although little is known about the survival of maned wolves in the wild, we would expect that environmental disturbance caused by anthropogenic activities affects reproductive success negatively and thus requires detailed investigation. The authors also recommend further research into social interactions and caring for the young among maned wolves.

In carnivores, diseases play a major role in high mortality rates and the resulting reduction of populations since the impact of the disease is influenced by the high trophic level that these animals occupy (Young, 1994; Funk et al., 2001; Woodroffe et al., 2004). Populations of species living at low densities, such as the maned wolf, are particularly susceptible to an increased risk of extinction. The diseases cause not only local extinctions of carnivores, but also changes in community structure (Funk et al., 2001). In an effort to confront such issues, the understanding of the role of infectious and parasitic agents in population dynamics has improved dramatically in recent years, with a concurrent increase in information concerning maned wolf exposure to pathogens. May Jr. and Felippe (Chapter 9) call for protected area management plans to include the diagnosis of exposure to pathogens originating from wildlife, domestic dogs, and livestock. These can affect the maned wolf's morbidity and mortality and are of relevance to both biodiversity conservation and public health. To address these concerns, they suggest the implementation of buffer zones in protected areas to control pathogens from domestic stock. Further studies would also be useful to elucidate the role of the maned wolf in the epidemiological cycle of diseases such as leishmaniasis, rabies, and leptospirosis. Other research areas considered for further development include the ecoepidemiology of these pathogens, alongside maned wolf population dynamics studies, to build a better picture of their exposure to pathogens. Investigation into the diversity of major histocompatibility complex (MHC) genes, and their role in immune surveillance and host–parasite interaction in maned wolves, as well as into the Paramyxoviridae family of the RNA virus genome will also advance maned wolf conservation medicine.

The maned wolf's range is largely unprotected, with only 2.2% of the surface area under some form of legal protection (Klink and Machado, 2005). As a result of the different types of land tenure across the species' range, suitable habitat is very fragmented in many parts. The fragmentation limits the dispersal and colonization potential of the species because it creates barriers to these processes. As a consequence, species such as the maned wolf that need large areas to exercise their niche are excluded from fragments that do not provide key resources for their survival. Additionally, it can divide populations into subpopulations, which become more vulnerable to demographic and genetic fluctuations and other problems associated with small population sizes.

Rodrigues et al. (Chapter 10) argue that in this context small reserves are an essential element in the conservation of the maned wolf at landscape level and suggest that small reserves should be managed to maximize biodiversity and mitigate anthropic pressures, with special attention given to connectedness between habitat blocks. The species' ability to tolerate human impact on its habitat arguably endows the species with high dispersal potential, provided that suitable food such as wolf's fruit is available in the corridors used to move from one area to another. Since large carnivores tend to require large territories, small reserves not linked to other populations would require some level of metapopulation management, including the translocation of individuals between reserves to reduce the potential effects of isolation and increase genetic viability in isolated populations.

Practical measures to increase the survival of dispersing animals and improve long-term population viability should focus on reducing road casualties by reducing and enforcing speed limits around reserves, deploying road signs, clearing road verges to increase visibility, educating drivers, creating under- and over-crossings for wildlife on key stretches of road, and even fitting maned wolves with retroreflective collars. Further research into the species' dispersal patterns and the success of maned wolf translocations and application of "soft release" versus "hard release" techniques (IUCN, 2012) will help gain a better understanding of how small reserves can contribute to overall maned wolf conservation.

The future of maned wolves in modified landscapes is a recurring theme throughout this book. Looking at the dietary requirements of maned wolves in cattle ranches, Ferraz Santos et al.'s (Chapter 11) case study recommends further studies on the impact of diminishing dietary richness and increasing Solanaceae on survival. Vynne's study (Chapter 12) suggests that maned wolves may tolerate high levels of habitat disturbance from agriculture, as long as remnants provide enough resources for resting and denning, and they keep access to food and water. Females, however, may be more sensitive to change, since they may suffer hormone-mediated reduced reproductive success in unprotected disturbed areas. Whereas maned wolves' prospects in landscapes dominated by low-growth plantations may be cautiously optimistic, the rapid conversion to sugarcane and biofuel production, accompanied by an associated increase in anthropogenic pressures, may further isolate and threaten maned wolf populations. Vynne recommends investigating further how changes in land use affect the maned wolf populations and the role of private land in interpopulation connectivity.

De Paula et al. (Chapter 13) report on the ongoing, long-term Serra da Canastra maned wolf project. They plan to build on their achievements and aim to expand their scientific work, education campaign, conflict assessment, and management practices to other maned wolf areas. An important component of this work involves disseminating information in order to improve local people's attitudes toward maned wolves and the cerrado using environmental education and public awareness based on communication practices and forming new partnerships. This project, which has been operating for longer than any other maned wolf field initiative, has extensive experience and has developed useful materials to support maned wolf conservation among local people, subsistence farmers, media, schools, and other groups in the southeast of Brazil. It offers a significant point of reference for maned wolf conservation programs elsewhere.

The First International Maned Wolf Workshop, which involved researchers from all countries within the species range (de Paula et al., 2008), took place in Serra da Canastra National Park in 2005. Its outcomes and resulting action plan for the conservation of the maned wolf can be seen throughout in the shape of regional conservation plans for the species. However, some obstacles are yet to be overcome. Queirolo et al. (Chapter 14) suggest that maned wolf conservation in Uruguay and south Brazil must account for differences in policies and legislations between countries and among states; promote links between research institutions, government, sponsorship, local people, and NGOs; take action to address birth and mortality rates; and tackle connectedness between populations, maintenance of habitat corridors, and adoption of ecological zoning. Local environmental education projects aimed at increasing knowledge about the species and involvement of local communities to minimize conflict between people and the maned wolf have already shown positive results.

Soler (Chapter 15) presents the work of the Grupo Argentino Aguará Guazú (GAAG), engaged in the implementation of conservation actions resulting from the international workshop mentioned above. She warns that periodic changes in government and conservation policies pose significant obstacles for the implementation of long-term action plans. She recommends the establishment of conservation education campaigns and research to address priority issues, defined at national level to guarantee a coordinated use of human and economic resources. Coordinated actions should maximize maned wolf conservation efforts and benefit the future of wild populations.

Based on the premise that maned wolves' home-range sizes and number of breeding pairs are limited by resources, Emmons's (Chapter 16) study in Bolivia suggests that conservation efforts must prioritize ecosystems with maximal food supplies, as well as tackling causes of mortality. She also recommends long-term longitudinal studies to further the understanding of how environmental variables influence the cerrado ecosystem and maned wolf populations. Similarly, in

Paraguay Cartes et al. (Chapter 17) show concerns as to how the species responds to landscape variability and recommend further research on how human activity affects the species' distribution in Chaco and other ecoregions occupied by the maned wolf. They reassure us however that despite the lack of information and the challenges the maned wolves face, they seem to remain widespread in Paraguay. Cartes et al. recommend mapping suitable habitats for the maned wolf, using a combination of field surveys and models. With increasing rates of deforestation, there is a need for conservation outreach programs to involve landowners in maned wolf conservation. The impacts of domestic and feral dogs as vectors of disease and through interference competition are additional concerns for the persistence of local populations in Paraguay.

Taylor's (Chapter 18) review of a series of maned wolf studies (i.e., Bizerril et al., 2011; Consorte-McCrea and Rubin, 2009; Paula et al., 2008; Songsasen and Rodden, 2010; Rodden et al., 2004; Bizerril, 2004; Soler et al., 2002) indicates that the lack of knowledge of the maned wolf and its habitat, coupled with rapid habitat loss and degradation, is the biggest challenge, which should be targeted by outreach and education campaigns. There are, however, powerful educational tools available to empower educators and learners and help facilitate the necessary changes to tackle these challenges. Strong planning, implementation, and constant evaluation of programs are best practice in education strategies to support conservation projects. Taylor is hopeful about people's innate interest in learning and is confident that the many people who care about the maned wolf are well placed to champion their cause to others. Bath (Chapter 19) also points out the importance of the inclusion of local people in long-term conservation plans. Tying in closely with conservation education research, human dimensions research can provide maned wolf conservation managers with useful information about local people's knowledge and attitudes to inform decision making. Bath recommends the use of facilitated workshops on human dimensions to resolve important issues surrounding people–maned wolf conflict.

Interdisciplinary cooperation can often be riddled with obstacles related to the divergence of approaches and beliefs routed in different research areas. Common aims may however offer an effective bridge to bring disciplines together, affording new answers to complex problems. Diegues (Chapter 20) sees the cooperation between social and natural scientists as a positive step in the planning and management of protected areas, using ethno-science as the bridge to address concerns raised by the two fields. Diegues suggests that ethnoconservation meets the need for a new interdisciplinary approach to account for biological as well as cultural diversity, considering social and ecological practices and local and traditional knowledge.

One of the main challenges for wildlife conservation is reconciling it with the increasing demand for goods and foodstuffs and how this affects land use. The cerrado is one of the most biodiverse biomes around the globe, housing 161 species of mammals, which represent an important fraction of the diversity of neotropical mammals (Trolle et al., 2007; Mittermeier et al., 2005; Myers et al., 2000). More than 50% of the 2,000,000 km^2 of cerrado savanna has been transformed into African grass pastureland, in addition to soy beans, sugarcane, and other large-scale croplands (Klink and Machado, 2005). Klink (Chapter 21) suggests that enough land is cleared to accommodate agricultural expansion, while minimizing the need for impact on the biodiversity of natural areas such as the cerrado habitat in the future. The situation calls for an urgent halt on the fast pace of transformation of the cerrado, allied with the restoration of degraded areas to limit expansion into areas of high conservation value. Cerrado conservation is a priority due to its high species endemism and heterogeneity, and high degree of degradation, which is one of the fastest in the world (Klink and Machado, 2005; Mittermeier et al., 2005; Myers et al., 2000). Trade-offs between land use to meet immediate human needs and the maintenance of long-term ecosystem function (including carbon sequestration, biodiversity benefits, and climate regulation) need to be better understood in the cerrado.

Most areas where the maned wolf continues to live today have seen the original vegetation replaced by monoculture with soy and sugarcane. While this shows the ability of maned wolves to adapt to man-made changes, it also serves as an important reminder of the importance of these disturbed areas for the conservation of local wildlife, and maned wolves can play a flagship role. The agricultural matrix that engulfs native vegetation remnants is used by maned wolves. Therefore, the incorporation of conservation practices in agricultural production processes would offer some promise in maintaining or increasing the use of these areas by wildlife. Conservation biologists should work closely with agriculture and forestry experts to gain a better understanding of the use of these production systems by wildlife, and the habitat can be improved through more sustainable agricultural practices.

The reviews and case studies in this book illustrate the depth of knowledge about maned wolves. Likewise, they bring to the fore many commonalities with other wild canid species. It is the similarity between the biology of different canids and the types of conservation problems they face that make it unlikely that some unexpected feature of their basic biology will be revealed that crucially changes the prospects for their conservation. On the other hand, for species after species it is clear that their conservation hits a brick wall—a barrier that might be overcome if we knew more. The nature of the barrier seems familiar—in the case of conflict over predation, it is necessary to find methods of preventing canids eating certain prey or to get a better grasp of the human dimensions of whether it is actually perceived as being the problem. In the case of disease, the challenge is to prevent contraction or communication of infection. The priorities would therefore seem to be for ingenious nonlethal methods of control and for the development of vaccination strategies. Mitigation measures that benefit both maned wolves and people, such as using chicken coops to shelter chickens at night and providing health care for dogs and domestic animals, may contribute to reducing maned wolf mortality and improving people's attitudes. Both examples are conservation problems that are likely to require carefully controlled field studies. Planning these and cracking other conservation problems will undoubtedly be aided by research into diverse aspects of maned wolf biology, including spatial requirements and population processes.

While the value of fundamental research is something that we would not contest, in the context of limited conservation funding some argue that it may be an indulgence to pursue basic research on species for which there is no more than an unspecified suspicion of threat. While there may be a strong conservation case for such research, making that case comes with a strong caveat. First, one of the greatest contributions biological research may make to conservation is through its influence on the priorities of the public. For instance, discoveries of the fine intricacies of animal behavior, as presented through films, photography, and books, have captivated and informed an ever wider public (at least amid a predominantly urban, educated sector of the population), radically changing their values, opinions, and, ultimately, their politics. We must come to terms with the fact that conservation action has to be paid for one way or another, and willingness to pay will be directly influenced by the public's understanding and appreciation of the value of biodiversity. Therefore, engaging the public in the wonders of maned wolf biology and the species' role in the ecosystem must surely remain a priority for all of us working to protect the species.

Different arguments may have more force with rural populations in the developing world, where the best hope for conservation may lie in convincing people of the value of having maned wolves and other wildlife as part of the process that enables the ecosystem on which their livelihoods depend to function. Furthermore, few people make a greater concrete impact on the ground for the protection of a species than a highly motivated field worker. Conservation biologists have a role as ambassadors for the species they work with, and this role should not be ignored. Additionally, there is a tendency for ranting in conservation biology (Macdonald and Sillero-Zubiri, 2004). All too often we hear that studying some aspect of a species' biology is "crucial" for its conservation, without any credible argument being made as to why this is so. Conservation biology will be ruined by such feeble arguments. It would be far better to admit that while a carefully thought-out study of a little known creature may have only tangential bearing on its conservation, it may produce results that help secure the commitment to conservation of a generation of decision makers and generate necessary financial support. Since it is only the very best science that is likely to captivate the public, dwelling on the private lives of maned wolves to continue to answer intriguing questions about their natural history and behavioral ecology may not provide immediate applied solutions, but be crucial to conservation.

REFERENCES

Bizerril, M.X.A. 2004. Children's perceptions of Brazilian cerrado landscapes and biodiversity. *The Journal of Environmental Education*, 35(4): 47–58.

Bizerril, M.X.A., Soares, C.C., and Santos, J.P. 2011. Linking community communication to conservation of the maned wolf in central Brazil, *Environmental Education Research*, 17(6): 815–827.

Consorte-McCrea, A.G. and Rubin R. 2009. Attitudes of zoo visitors and professionals towards the conservation of the maned wolf in southeast Brazil, *Canid News* 12.1 (online).

Funk, S.M., Fiorello, C.V., Cleaveland, S., and Gomper, M.E. 2001. The role of disease in carnivore ecology and conservation. In J.L. Glittleman, S.M. Funk, D. Macdonald, and R.K. Wayne (eds), *Carnivore Conservation*, Vol. 5, pp. 443–466, Cambridge University Press, London.

IUCN SSC 2012. IUCN guidelines for reintroductions and other conservation translocations. Adopted by SSC Steering Committee at Meeting SC 4 6, 5 September 2012. IUCN-RSG, ISSG. http://www.issg.org/pdf/publications/Translocation-Guidelines-2012.pdf.

Klink, C.A. and Machado, R.B. 2005. Conservation of the Brazilian cerrado, *Conservation Biology*, 19(3): 707–713.

Macdonald, D.W. and Sillero-Zubiri, C. 2004. *The Biology and Conservation of Wild Canids,* Oxford University Press, Oxford.

Mittermeier, R.A., da Fonseca, G.A.B., Rylands, A.B., and Brandon, K. 2005. A brief history of biodiversity conservation in Brazil, *Conservation Biology*, 19(3): 601–607.

Myers, N., Mittermeier, R.A., Mittermeier, C.G., da Fonseca, G.A.B., and Kent, J. 2000. Biodiversity hotspots for conservation priorities, *Nature,* 403: 853–858.

Paula, R.C., Medici, P., and Morato, R.G. (eds). 2008. *Plano de ação para a conservação do lobo-guará: Análise de viabilidade populacional e de habitat*. Centro Nacional de Pesquisas para Conservação dos Predadores Naturais-CENAP/ICMBio, Brasília.

Rodden, M., Rodriques, F., and Bestelmeyer, S. 2004. Maned wolf: *Chrysocyon brachyurus*. In: C. Sillero-Zubiri, M. Hoffman, and D.W. Macdonald (eds), *Canids: Foxes, Wolves, Jackals and Dogs*, International Union for Conservation of Nature, Gland, Switzerland.

Sillero-Zubiri, C., Hoffmann, M., and Macdonald, D.W. (eds), 2004a. *Canids: Foxes, Wolves, Jackals and Dogs—Status Survey and Conservation Action Plan*, 2nd edn, IUCN, Gland, Switzerland.

Sillero-Zubiri, C., Reynolds, J., and Novaro, A. 2004b. Management and control of canids near people. In D.W. Macdonald and C. Sillero-Zubiri (eds), *The Biology and Conservation of Wild Canids*, pp. 107–122, Oxford University Press, Oxford.

Soler, L., Carenton, J. M., Birochio, D., Salvatori, V., Orozco, M.M., Soledad Rosso, M., Carpinetto, M., et al. 2002. Problems and recommendations for the conservation of the maned wolf in Argentina. *Endangered Species UPDATE* 22(1): 3–9.

Songsasen, N. and Rodden, M.D. 2010. The role of the species survival plan in maned wolf *Chrysocyon brachyurus* conservation. *International Zoo Yearbook*, 44: 136–148.

Trolle, M., Noss, A.J., de Lima, E.S., and Dalponte, J.C. 2007. Camera-trap studies of maned wolf density in the Cerrado and the Pantanal of Brazil. In D. Hawksworth and A. Bull (eds), *Vertebrate Conservation and Biodiversity,* pp. 371–378, Springer, Dordrecht.

Wang, X., Tedford, R.H., van Valkenburgh, B., and Wayne, R.K. 2004. Ancestry: Evolutionary history, molecular systematics, and evolutionary ecology of Canidae. In D.W. Macdonald and C. Sillero-Zubiri (eds), *The Biology and Conservation of Wild Canids,* pp. 39–54, Oxford University Press, Oxford.

Woodroffe, R., Cleaveland, S., Courtenay, D.O., Laurenson, M.K., and Artois, M. (2004). Infectious disease in the management and conservation of wild canids. In D.W. Macdonald and C. Sillero-Zubiri (eds), *The Biology and Conservation of Wild Canids,* pp. 123–142, Oxford University Press, Oxford.

Young, T.P. 1994. Natural die-offs of large mammals: Implications for conservation. *Conservation Biology* 8(2): 410–418.

Index

A

African Game Service Skeerport, South Africa, 59
African wild dogs, 10–11
Agricultural expansion
  Brazilian legislation, 165
  croplands, 166
  glucocorticoid concentration, 167, 171
  physiological markers, 166–167
  progestin concentrations, 171–172
  thyroid hormone levels, 167, 171
Agricultural practices, 313–314
Águas Emendadas Ecological Station (ESECAE) maned wolves, 132–133; *see also* Brazilian cerrado maned wolves
Altina Wildlife Experience, 58
American Zoo and Aquarium Association (AZA), 53
Animal Records Keeping System (ARKS) taxon report, 54
Apple of Peru (*Nicandra physaloides*), 155
Argentina maned wolf, *see* Grupo Argentino Aguara Guazu
  conservation projects
    central region, 209–211
    central–northeastern region, 211–212
    northeastern and northwestern regions, 207–209
  threats
    ancestral beliefs, 204–205
    conflicts, 205
    dogs, 204
    floods and droughts, 206–207
    illegal trade, 206
    negative perception, 205–206
    parasites, 204–205
    road kills, 206
Artificial insemination (AI), 105
Asian wild dog, 11
Assisted reproductive technologies, 105
Association of Zoos and Aquariums (AZA), 63

B

Bell peppers (*Capsicum annuum*), 155
Black-backed jackal (*Canis mesomelas*), 7
Bolivian maned wolf habitat; *see also* Los Fierros savanna maned wolf habitat
  biogeographical regions, 221
  fire and flooding, 221
Brazilian cerrado maned wolves
  conservation management, 146–148
  small reserves
    large home ranges, 142–143
    small populations, 143
  study methods, 133–134
  threats
    domestic animals, 144–145
    road kills, 139, 144
  vortex simulation model
    age distribution, 141
    diet, 138–139
    habitat loss estimates, 31–32
    harvest, 27
    home ranges, 136–137
    initial population size, 25
    mean population size, 140–142
    mortality rates, 136
    population and dispersal estimates, 27–28, 138–139
    population characteristics, 134–135
    population viability, 145
    road kill, 30–31
    road mortality, decrease of, 140, 142
    supplementation scenarios, 136
Brazilian Institute of Environment and Renewable Natural Resources (IBAMA), 64
Brazilian Management Plan for the Maned Wolf, 64
Brazilian Studbook for the Maned Wolf (BSMW), 53
Brazilian Zoological Society (SZB), 63
Brazil–Japan Cooperative Program for the Development of the Cerrado (PRODECER), 296
Breeding pairs, 99
Bush dog (*Speothos*), 3

C

Canid Workgroup, 64
Canids
  adaptability, 6
  body size and shape, 6
  canid–human conflict, 8–9
  consensus tree, 4
  conservation action, 12
  diversity, 5–6
  and humans, 7–8
  origins, 3–5
  phylogeny, 78–80
  rabies, 9
  red list assessments, 10
  smell and hearing, 6
  spatial and social organization, 6–7
  status and conservation, 9–12
  tourism, 9
  versatility, 8
Canids: Foxes, Wolves, Jackals and Dogs – Status Survey and Conservation Action Plan, 309
Canine parvovirus (CPV), 181
Captive maned wolves, *see Ex situ* maned wolf populations
Cerrado savannas, 313
  biological wealth, 294
  degraded lands, restoration of, 301–302
  landscape modification, 294–295
    agricultural development, 295–297
    biodiversity and ecosystems, 297–298

policies, change of, 298–301
trade-offs, 302–303
seasonal tropical climate, 294
Chemical restraint, 110
hematology and serum biochemistry, 111–112
monitoring, 111
Coffee (*Coffea arabica*, Rubiaceae), 155
Committee for the Conservation and Management of Brazilian Canids, 64
Communities of practice, 256–257
Computer modeling, 15–16, 132
CONABIO, 299
Conservation Assessment and Management Plan (CAMP) workshop, 63
Conservation Breeding Specialist Group (CBSG), 63
Conservation biology, 314–315
Conservation programs
endangered species, 177–178
Paraguay maned wolf, 244–245
Convention on Biological Diversity (CBD), 299
Convention on International Trade in Endangered Species of Wild Fauna and Flora (CITES), 39, 193
Crab-eating fox (*Cerdocyon thous*), 3
Critically Endangered canids, 9–11
Critically Endangered Darwin's fox, 9–10

D

Dioctophyme renale, 119–120, 153
Distemper, 116

E

Ectoparasites, 118–119
Education programs
communities of practice, 256–257
habitat loss and fragmentation, 253
knowledge and information, lack of, 253–254
Serra da Canastra National Park, Brazil, 253
trained professionals, lack of, 254
Emas National Park study, Brazil
distribution across landscape, 167–168, 170–171
glucocorticoid concentration, 171
hormone extraction and processing, 168
progestin concentrations, 171–172
sampling, 167
statistical analysis, 168–170
thyroid hormone levels, 171
Enclosure characteristics, 69–70
Environmental education
aims and objectives, 254–255
information and communications technology, 257–260
Millennial learners, 257–259
scope and length, 256
transformative learning, 255–256
Environmental enrichment, 70
Environmental variation (EV), 18
Ethiopian wolves, 9
Ethnoconservation
natural and social sciences, 289–290
no-take protected areas, 284–286
principles
biodiversity, 288–289
nature conservation, 287
participatory approach and empowerment, 289
traditional knowledge and management practices, 287–288
traditional peoples, main features of, 286
European Association of Zoos and Aquaria (EAZA), 53, 63
European Endangered Species Program (EEP), 53–54, 56
Evolutionarily significant units (ESUs), 80
Evolutionary biology, 77
Ex situ maned wolf population
Africa, 59
Asia, 58–59
Brazil
captive population, 65, 68
environmental education strategies, 68–69
husbandry protocols, 69–72
scientific research studies, 68
studbook-recorded captive population, 64–66
vortex population viability analysis, 66–67
European region, 57
management, problem of, 54
North America, 57–58
reproduction, 104–106
zoos, 59–60
Extinction, 19

F

Falklands wolf (*Dusicyon australis*), 5, 8
Feeding ecology
diet, frequency of occurrence, 89–91
food intake and bioenergetic constraints, 92, 95
functional responses, 92–93
habitat, 94
vs. other carnivores, 94
prey and fruit
habitat, 94
selection, 93–95
size distribution, 91–92
seasonality, 92
seed dispersal, 94
telemetry studies, 95
Feeding habits study, Matias Barbosa farm
animal food items
frequency of occurrence of, 157
rarefaction curves, 158
seasonal variation, 157
vs. Cerrado and Upland Meadow studies, 160
germinated seeds, 158
Goodman's test, 155
mammal food items, 155
scats, 154
seed germination trials, 155
First International Workshop for Maned Wolf Conservation, 17, 312
Forest Investment Program (FIP), 300
Frankfurt Zoo, 57

G

Genetic structure and diversity, 19
conservation efforts, 82
ecological studies, 83–84
fast-evolving segments, 83

genetic surveys, 82
microsatellite loci, 81
molecular diversity indices, 80–81
Geographic range, 153
GPS collar technology, 222
Gray wolf (*Canis lupus*), 5
Grupo Argentino Aguará Guazú (GAAG), 312
aims and achievements, 212–213
legislative statements, 204
Maned Wolf: Face and Cross of an Endangered Species, 213–214
National Action Plan workshops, 213
priority issues, 214, 215
Guava (*Psidium guajava*), 155

H

Habitat loss estimates, Brazilian maned wolf populations, 31–32
Hand rearing, 71–72
Hoary foxes (*Pseudalopex vetulus*), 8
Human dimensions conservation projects, 274–280
attitudes, perceptions and values, 268–269
behavioral conflicts, 274
beliefs and attitudes, 269–270
cognitive conflicts, 272–273
conflict types, 271–272
costs and benefits conflicts, 273
educational and awareness programs
cognitive messages, 274
effective communication material designing, 275–277
emotional images, 275
implementation, 277–280
value persuasive messages, 274
value conflicts, 273
Human–canine relationship, 7
Husbandry protocols
cubs, hand rearing of, 71–72
enclosure characteristics, 69–70
environmental enrichment, 70
nutritional management, 70–71
reproduction management, 71
wild-caught cubs, 72

I

In situ maned wolf populations, reproduction, 103–105
Inbreeding, 18
Indian fox (*V. bengalensis*), 7
Indian nightshade (*Solanum aculeatissimum*), 155
Infectious diseases
Dioctophyme renale, 119–120
distemper, 116
ectoparasites, 118–119
leishmaniasis, 115
leptospirosis, 114–115
neosporidiosis and toxoplasmosis, 119
parvoviruses, 117
rabies, 117–118
Interdisciplinary cooperation, 313
Intergovernmental Panel on Climate Change (IPCC), 195
International Species Information System (ISIS), 54
International Studbook (ISB), 53, 57
International Union for Conservation of Nature (IUCN) Red Data Book, 39, 64
Island foxes (*Urocyon littoralis*), 5, 10
IUCN/SSC Canid Specialist Group, 9

J

Jackal-headed god, Anubis, 7
Japan International Cooperation Agency for Development (JICA), 241
Jararaca (*Bothrops jararaca*), 160

L

Lago Ypacarai Reserve, 245
Lago Ypoa National Park, 245
Land use, Cerrado savannas, 294–295, 311
agricultural development
environmental and economic conditions, 295
expansion of, 297
modern machinery, use of, 296
price support policy, 296
production loans, 295–296
regional developmental programs, 296
biodiversity and eco systems, 297–298
policies, change of, 298–301
trade-offs, 302–303
Large Carnivore Initiative for Europe (LCIE) educational resource inventory, 275
Leishmaniasis, 115
Leptospirosis, 114–115
Local people's attitudes, *see* Sao Paulo state field study
Long-term population viability, 311
Los Fierros savanna maned wolf habitat
climate change, 224–225, 230
drought, 231
fires
fruit resource plants, 226–227
rodent prey, 227–228, 231
resource variability
climate change, 224–225
disease outbreaks, 229–230
fires, 225–228
invasive grasses, 229
rodent variability, 228–229
study method, 222–223
suggestions, longitudinal environmental monitoring, 232
Low-carbon agriculture program (ABC Program), 300

M

Major histocompatibility complex (MHC), 121
Management units (MUs), 80
Maned Wolf Action Plan, xiii, 43, 195
Maned Wolf Conservation Project, 178
Maned Wolf Program of the Sao Paulo Zoological Society, 68
Maned Wolf Species Survival Plan, 252
Maned wolf–human relations
anthropogenic causes, 43–44
bio/education professional and local people, 48
chicken pens and livestock, 42–43
conservation awareness, 44–45

environmental education programs, 45
habitat impoverishment, 42
human-related threats, 43–44
hunting, 44
local people support, 46–47
misconceptions clarification, 47
negative attitudes clarification, 48
pharmaceutical value, 38–39
prevention, mitigation and control, 47
rural people and urban administrations, 45–46
Maned wolves (*Chrysocyon brachyurus*), 3, 5, 15, 64, 87, 131, 146, 153, 154, 159, 193–198, 203–204, 207, 212, 214, 215, 235–245, 251
Mating maned wolves, 60
Maximum age of reproduction, 18
Maximum number of progeny per year, 18
Mbaracayu Forest Nature Reserve (MFNR), 241–244
Mean rate of stochastic population growth, 19
Mean time to population extinction (MTE), 19
Microsatellite loci, 81
Minimum convex polygon (MCP) method, 134
Minimum viable population, 16
Modified landscapes, 294–295, 312
agricultural development, 295–297
biodiversity and eco systems, 297–298
policies, change of, 298–301
trade-offs, 302–303
Molecular diversity indices, 80–81
Molecular ecology, 77–78
Museum of Natural History of Paraguay, 238

N

National Biodiversity Inventory Project, Paraguay, 238
National Inventory of the Natural History Museum, Guyra Paraguay, 238
National List of Threatened Species of Brazilian Fauna, 38
National Maned Wolf Day, 69
National Museum of Natural History in Montevideo, Uruguay, 194
National Protected Area System (SINASIP), 193, 244
National Research Center for Carnivore Conservation (CENAP), 181
National System of Conservation Units (SNUC), 299
Nationally appropriate mitigation actions (NAMAs), 300
Near Threatened species (NT), 193
Neosporidiosis, 119
Neuwied's lancehead (*Bothrops neuwiedii*), 160
Noel Kempff Mercado National Park (NKP), 103, 221, 237
Noninvasive hormone monitoring, 105
North American Maned Wolf Species Survival Plan (SSP), 53–54, 64
North American zoos, 57–58
Nutritional management, 70–71

P

Painted Dog Conservation Project (PDC), 147
Palmar Quemado Reserve, 245
Papaya (*Carica papaya*, Caricaceae), 155
Paraguay maned wolf
challenges, protected area management, 245
conservation, 244–245
distribution records and habitats, 238–242
ecology, 242–243
ecoregions, 235–236
historical references, 237–238
life history, 236–237
threats, 243–244
Parvoviruses, 117
Philadelphia Zoo, 57
Phylogeny, 310
South American canids, 78–80
Physical restraint
foot snares, 109–110
metal box traps, 109
wooden box traps, 109
Plastic Expression Workshops, 198
Policies for the Sustainable Development of Traditional Peoples and Communities, 286
Population and Habitat Viability Assessment (PHVA) workshop, 63, 64
Population management program (PMx), 54
Population viability analysis (PVA), 19
baseline model
parameters, 17–19
sensitivity analysis, 19
vortex simulation results, 19
Potato tree (*Solanum erianthum*), 155
PPCerrado plan, 301
Prague Zoo, Czech Republic, 57–58
Pretoria National Zoological Gardens, South Africa, 59
Prey and fruit
frequency of occurrence, 89–91
habitat, 94
selection, 93–95
size distribution, 91–92
PROBIO, 299
Program for the Development of the Cerrado (POLOCENTRO), 296
PRONABIO, 299
Protected area management plans, 311
Public support, 309–310

R

Rabies, 9, 117–118
Raccoon dog (*Nyctereutes procyonoides*), 7
Radio telemetry, 134
Red Riding Hood and the Maned Wolf (play), 69
Red wolf (*Canis rufus*), 10
Region-wide culling, 8
Reproduction
assisted reproductive technologies, 105
breeding season, 100
ex situ populations, 104–106
influencing factors
ex situ populations, 104–105
in situ populations, 103–104
scent marking and roar barks, 100
in situ populations, 103–105
Reproduction management, 71
Reproductive endocrinology, 311
ejaculates, 102
noninvasive fecal endocrine monitoring
progestagen metabolites, 101–102
testosterone metabolite concentration, 101

seasonality and age, 103
seminal parameters and testicular volume, 102–103
Reserva Pantanal Paraguayo, 242
RNA virus genomes, 121
Road kills, 18
Argentina maned wolf, 206
Brazilian maned wolf populations, 30–31
Roar barks, 100
Rural Environmental Cadaster (CAR), 300–301

S

São Paulo state field study
maned wolf–human relations
anthropogenic causes, 43–44
bio/education professional and local people, 48
chicken pens and livestock, 42–43
conservation awareness, 44–45
environmental education programs, 45
habitat impoverishment, 42
human-related threats, 43–44
hunting, 44
local people support, 46–47
misconceptions clarification, 47
negative attitudes clarification, 48
pharmaceutical value, 38–39
prevention, mitigation and control, 47
rural people and urban administrations, 45–46
Portuguese settlers, 38
questionnaires, 37
respondents' beliefs, 41
statistical associations, 40
target groups, 36
wildlife protection, 38
Scent marking, 100
Seed Bank of the Integral Technical Assistance Coordination of Campinas (CATI), 154
Seed dispersal, 154
Serra da Canastra maned wolf project, 312
aims, 179
capture events, 192
conservation actions, 180–181
education, 180
history phases
gathering, informing, expanding, 187–189
healthy system, 186–187
research, education and conservation, 182–186
tourism and altered behavior, 182
local people and wildlife relationship, 178
maximum number of progeny per year, 18
project remarks, 190
research, 179–180
research, education and conservation
capturing and monitoring, 183
environmental education, 184–185
feeding ecology and habitat use, 184
health, reproduction and epidemiology, 184
human–maned wolf conflicts, 185–186
site ranges, 178
trapping success, 192
Shanghai Zoological Gardens, China, 58
Short-eared dog (*Atelocynus microtis*), 3
Singapore Zoological Gardens, Singapore, 58
Single Population Analysis and Records Keeping System (SPARKS) program, 54
Small reserves, *see* Brazilian cerrado maned wolves
Smithsonian National Zoological Park, Washington, 57
Solanaceae fruits (*Solanum lycocarpum*), 153
South American canids, *see* Canids
South American foxes (*Pseudalopex*), 3, 5
South American hypercarnivorous canids, 5
Structure studies, 80
Studbooks, 54

T

Tagatiya-Mi Reserve, 245
Third Maned Wolf Workshop, Brazil, 68
Tiletamine–zolazepan, 110
Toxoplasmosis, 119

U

Ueno Zoological Gardens, Tokyo, 58
Uruguay maned wolf
Action Plan for Maned Wolf Conservation, 198–199
historic presence, 194
molecular ecology techniques, 197
occurrence and distribution, 195
potential distribution scenarios, 195–197
predictive models, geographic distribution, 196
threats and conservation
environmental education, 198
hunting, 197
road hit, 198

V

Vortex simulation model
Brazilian cerrado maned wolves
age distribution, 141
diet, 138–139
home ranges, 136–137
mean population size, 140–142
mortality rates, 136
population and production estimates, 138–139
population characteristics, 134–135
population viability, 145
road mortality, decrease of, 140, 142
supplementation scenarios, 136
Brazilian maned wolf populations
habitat loss estimates, 31–32
harvest, 27
initial population size, 25
population and dispersal estimates, 27–28
road kill, 30–31
deterministic output, 19
harvesting *vs.* habitat loss, 23–25
initial population size
vs. habitat loss, 22–23
vs. harvesting, 22
parameters, 17–19
sensitivity analysis, demographic rates, 19–20
stochastic growth rate, 19
threats and modeling conditions, 22

W

Wild canids, 6–7
Wild maned wolves, *see In situ* maned wolf populations
Wild-caught cubs husbandry, 72
Wildlife resource management
  biophysical scientists, 267–268
  human, culture and nature, 266
  human dimension facilitated workshop approach
    attitudes, perceptions and values, 268–269
    behavioral conflicts, 274
    beliefs and attitudes, 269–270
    cognitive conflicts, 272–273
    conflict types, 271–272
    costs and benefits conflicts, 273
    educational and awareness programs, 274–280
    objectives, 280
    value conflicts, 273
Winter cherry (*Solanum pseudocapsicum*), 155
Wolf-like hyaenids, 5
World Association of Zoos and Aquariums (WAZA), 63
World Resources Institute (WRI), 301

Y

Yacyreta Island Reserve, Rio Parana, 244–245
Ypeti Reserve, 245

Z

Zoo Negara, Malaysia, 58
Zoos, 59–60, 310